CEREBRAL CORTEX

Volume 10
Primary Visual Cortex
in Primates

CEREBRAL CORTEX
Series Editors: Alan Peters and Edward G. Jones

Advisory Committee
Antonio R. Damasio, *Iowa City, Iowa*
Irving T. Diamond, *Durham, North Carolina*
J. C. Eccles, *Contra, Switzerland*
Patricia Goldman-Rakic, *New Haven, Connecticut*
Jon H. Kaas, *Nashville, Tennessee*
John H. Morrison, *New York, New York*
Adam M. Sillito, *London, England*
Wolf Singer, *Frankfurt, Germany*
R. D. Terry, *La Jolla, California*
P. Ulinski, *Chicago, Illinois*

Series Editors: Alan Peters and Edward G. Jones

CEREBRAL CORTEX

Volume 10
Primary Visual Cortex in Primates

Edited by

ALAN PETERS
Boston University School of Medicine
Boston, Massachusetts

and

KATHLEEN S. ROCKLAND
The University of Iowa
Iowa City, Iowa

Plenum Press • New York and London

Library of Congress Cataloging in Publication Data

(Revised for vol. 10)
Cerebral cortex.
   Vol. 2, 5–8 edited by Edward G. Jones and Alan Peters.
   Vol. 10 edited by Alan Peters and Kathleen S. Rockland.
   Includes bibliographies and indexes.
   Contents: v. 1. Cellular components of the cerebral cortex—v. 2. Functional properties
of cortical cells—[etc.]—v. 10. Primary visual cortex in primates.
   1. Cerebral cortex. I. Peters, Alan, 1929–   . II. Jones, Edward G., 1939–   . [DNLM:
1. Cerebral Cortex—anatomy and histology. 2. Cerebral Cortex—physiology. WL 307
C4136]
QP383.C45   1984   612′.825   84-1982

ISBN 0-306-44605-7

© 1994 Plenum Press, New York
A Division of Plenum Publishing Corporation
233 Spring Street, New York, N.Y. 10013

Printed in the United States of America

Contributors

D. L. Benson   Department of Anatomy and Neurobiology, University of California, Irvine, California 92717. *Present address:* Department of Neuroscience, University of Virginia School of Medicine, Charlottesville, Virginia 22908

Jean Bullier   Cerveau et Vision, INSERM Unité 371, 69500 Bron/Lyon, France

Vivien A. Casagrande   Department of Cell Biology, Vanderbilt University, Nashville, Tennessee 37232-2175

J. DeFelipe   Department of Anatomy and Neurobiology, University of California, Irvine, California 92717. *Present address:* Instituto Cajal, 28002 Madrid, Spain

Ron D. Frostig   Department of Psychobiology, University of California, Irvine, California 92717

Pascal Girard   Cerveau et Vision, INSERM Unité 371, 69500 Bron/Lyon, France

S. H. C. Hendry   Department of Anatomy and Neurobiology, University of California, Irvine, California 92717. *Present address:* Krieger Mind/Brain Institute, Johns Hopkins University, Baltimore, Maryland 21218

E. G. Jones   Department of Anatomy and Neurobiology, University of California, Irvine, California 92717

Jon H. Kaas   Department of Psychology, Vanderbilt University, Nashville, Tennessee 37232-2175

Rodrigo O. Kuljis   Department of Neurology, The University of Iowa, Iowa City, Iowa 52242-1053

Jonathan B. Levitt   Department of Visual Science, Institute of Ophthalmology, University of London, London EC1V 9EL, England

Jennifer S. Lund   Department of Visual Science, Institute of Ophthalmology, University of London, London EC1V 9EL, England

John W. McClurkin   Laboratory of Sensorimotor Research, National Eye Institute, National Institutes of Health, Bethesda, Maryland 20892

Lance M. Optican   Laboratory of Sensorimotor Research, National Eye Institute, National Institutes of Health, Bethesda, Maryland 20892

Guy A. Orban   Laboratorium voor Neuro- en Psychofysiologie, K.U. Leuven, Medical School, B-3000 Leuven, Belgium

Alan Peters   Department of Anatomy and Neurobiology, Boston University School of Medicine, Boston, Massachusetts 02118

Matthew Rizzo   Division of Behavioral Neurology and Cognitive Neuroscience, Department of Neurology, The University of Iowa College of Medicine, Iowa City, Iowa 52242

Kathleen S. Rockland   Department of Neurology, The University of Iowa College of Medicine, Iowa City, Iowa 52242-1053

Paul-Antoine Salin   Cerveau et Vision, INSERM Unité 371, 69500 Bron/Lyon, France

Eric L. Schwartz   Department of Cognitive and Neural Systems, Department of Electrical and Computer Systems, College of Engineering, Boston University, and Department of Anatomy and Neurobiology, Boston University School of Medicine, Boston, Massachusetts 02215

Margaret T. T. Wong-Riley   Department of Cellular Biology and Anatomy, Medical College of Wisconsin, Milwaukee, Wisconsin 53226

Takashi Yoshioka Krieger Mind/Brain Institute, Johns Hopkins University, Baltimore, Maryland 21218

Jennifer A. Zarbock Laboratory of Sensorimotor Research, National Eye Institute, National Institutes of Health, Bethesda, Maryland 20892

Preface

Volume 10 is a direct continuation and extension of Volume 3 in this series, *Visual Cortex*. Given the impressive proliferation of papers on visual cortex over the intervening eight years, Volume 10 has specifically targeted visual cortex in primates and, even so, it has not been possible to survey all of the major or relevant developments in this area. Some research areas are experiencing rapid change and can best be treated more comprehensively in a subsequent volume; for example, elaboration of color vision; patterns and subdivisions of functional columns.

One major goal of this volume has been to provide an overview of the intrinsic structural and functional aspects of area 17 itself. Considerable progress has been made since 1985 in unraveling the modular and laminar organization of area 17; and this aspect is directly addressed in the chapters by Peters, Lund *et al.*, Wong-Riley, and Casagrande and Kaas. A recurring leitmotif here is the evidence for precise and exquisite order in the interlaminar and tangential connectivity of elements. At the same time, however, as detailed by Lund *et al.* and Casagrande and Kaas, the very richness of the connectivity implies a multiplicity of processing routes. This reinforces evidence that parallel pathways may not be strictly segregated. Further connectional complexity is contributed by the various sets of inhibitory neurons, as reviewed by Lund *et al.* and Jones *et al.*

Another goal, relevant to the functional properties of area 17, is what this area contributes to the visual cortical system as a whole. As already presented by Van Essen in Volume 3 (and Casagrande and Kaas in the current volume), this system is a network of multiple interconnected areas. Area 17 has often been viewed as a pivotal "switch point," relaying segregated geniculocortical streams through specific subdivisions of extrastriate cortex. Inactivation studies, as described by Bullier *et al.*, raise some queries, supported by anatomical results, as to the paramount importance of area 17 as a funnel for visual processes. From a different point of view, the integrative, as opposed to separatist role of area 17 is highlighted by the abundance of nongeniculate afferents to this area. These were treated in 1985 by Tigges and Tigges and, in the current volume, in the chapter by Rockland, which emphasizes feedback cortical connections.

The computational functions of area 17 have been viewed from various perspectives. In one modeling approach, represented here by Schwartz, cortical topography and columnarity are analyzed as supraneuronal patterns approximated by a conformal map. Further on in this volume, the functional properties of area 17 are treated at the neuronal level by Orban and, with emphasis on temporal encoding, by McClurkin *et al.* Optical imaging technologies, surveyed by Frostig, have been developing rapidly over the last decade, and now promise to provide a long-sought-after bridge between recordings obtained by single-unit physiology and those reflecting the combined activity of populations of neurons. Advances in PET and MRI have likewise been rapid, and have the potential of probing cortical organization in the human with a resolution approaching that of experimental primate work. Two chapters, by Kuljis and Rizzo, review different aspects of human visual cortex in the normal and pathological condition. These chapters provide background for comparisons between the monkey and human visual system, and also provide a basis for the interpretation of imaging results.

We would like to thank the contributors to this volume for their enthusiasm and hard work; and the staff of Plenum Press for doing their customary excellent job of producing this volume.

Alan Peters
Kathleen S. Rockland

Boston and Iowa City

Contents

Chapter 1

The Organization of the Primary Visual Cortex in the Macaque

Alan Peters

Chapter 2

Substrates for Interlaminar Connections in Area V1 of Macaque Monkey Cerebral Cortex

Jennifer S. Lund, Takashi Yoshioka, and Jonathan B. Levitt

Chapter 3

GABA Neurons and Their Role in Activity-Dependent Plasticity of Adult Primate Visual Cortex

E. G. Jones, S. H. C. Hendry, J. DeFelipe, and D. L. Benson

Chapter 4

**Primate Visual Cortex: Dynamic Metabolic Organization and Plasticity
Revealed by Cytochrome Oxidase**

Margaret T. T. Wong-Riley

Chapter 5

**The Afferent, Intrinsic, and Efferent Connections of Primary Visual
Cortex in Primates**

Vivien A. Casagrande and Jon H. Kaas

Chapter 6

The Organization of Feedback Connections from Area V2 (18) to V1 (17)

Kathleen S. Rockland

Chapter 7

The Role of Area 17 in the Transfer of Information to Extrastriate Visual Cortex

Jean Bullier, Pascal Girard, and Paul-Antoine Salin

Chapter 8

What Does *in Vivo* Optical Imaging Tell Us about the Primary Visual Cortex in Primates?

Ron D. Frostig

Chapter 9

Computational Studies of the Spatial Architecture of Primate Visual Cortex: Columns, Maps, and Protomaps

Eric L. Schwartz

Chapter 10

Motion Processing in Monkey Striate Cortex

Guy A. Orban

Chapter 11

Temporal Codes for Colors, Patterns, and Memories

John W. McClurkin, Jennifer A. Zarbock, and Lance M. Optican

Chapter 12

The Human Primary Visual Cortex

Rodrigo O. Kuljis

Chapter 13

The Role of Striate Cortex: Evidence from Human Lesion Studies

Matthew Rizzo

The Classification of Primates

The structure and connections of the primary visual cortex have been examined in a number of different primate species, and for those readers who are not familiar with their common and scientific names and the relations between those species, we hope that the following table will be helpful. The table is very general and makes no attempt to give specific examples from all of the families of Primates, but it does include the common and scientific names of most of the primates mentioned in this volume.

Order Primates

(12 families, 52 genera, 181 species)

Suborder Prosimii

(Lower primates; 6 families)

Cheirogaleidae: dwarf and mouse lemurs
Lemuridae: lemurs
Indriidae: indri and sifakas
Daubentonidae: aye-aye
Lorisidae: bush babies or galagos; pottos and lorises
Tarsiidae: tarsiers

Suborder Anthropoidea

(Higher primates; 6 families)

Cebidae: capuchin-like monkeys; sometimes called New World or catarrhine monkeys.
  Examples:
  Night or owl monkey: *Aotes trivirgatus*
  Squirrel monkey: *Saimiri sciureus*

Capuchin monkeys: *Cebus*
Spider monkeys: *Ateles*

Callitrichidae: marmosets and tamrins

Cercopithecidae: Old World or platyrrhine monkeys
Examples:
Japanese macaque: *Macaca fuscata*
Rhesus monkey: *Macaca mulatta*
Pig-tailed monkey: *Macaca nemestrina*
Crab-eating macaque: long-tailed macaque; cynomolgus monkey:
   Macaca fascicularis
Baboons: *Papio*
Guenons: *Cacopitheus*
Mangabeys: *Cercocebus*
Talapoins: *Miopithecus*

Pongidae: great apes
Examples:
Gorilla: *Gorilla gorilla*
Chimpanzee: *Pan troglodytes*
Orangutan: *Pongo pygmaeus*

Hylobatidae: lesser apes
Example:
Gibbons

Hominidae: one species
Man: *Homo sapiens*

Scandentia
(1 family: Tupaiidae)

Tree shrews: There is some controversy about the classification of tree shrews; the current consensus is that they are not specifically related to either the primates or the insectivores, but represent a separate lineage.

The Organization of the Primary Visual Cortex in the Macaque

ALAN PETERS

The two species of macaque most commonly used in studies of the visual cortex are *Macaca mulatta* (rhesus monkey) and *M. fascicularis* (crab-eating or long-tailed macaque). The brains of these two species are very similar in appearance, although the brain of *M. mulatta* is generally somewhat larger than that of *M. fascicularis*.

1. Location of Primary Visual Cortex

In the macaque, the primary visual cortex, also variously known as area 17 (Brodmann, 1905, 1909), striate cortex, or V1, is located in the occipital pole of the cerebral hemisphere. As shown in Brodmann's (1909) map and in the atlas of von Bonin and Bailey (1947), the primary visual cortex occupies most of the domed operculum on the lateral surface of the hemisphere, where it is bounded rostrally by the lunate sulcus and inferiorly by the inferior occipital sulcus (Fig. 1). Crossing part of the opercular surface and included entirely within the primary visual cortex there is the external, or lateral calcarine sulcus. This is variable in extent and depth among individual brains, and generally appears to be deeper in the brains of *M. fascicularis* than in the brains of *M. mulatta*. Medially in

ALAN PETERS • Department of Anatomy and Neurobiology, Boston University School of Medicine, Boston, Massachusetts 02118.

Cerebral Cortex, Volume 10, edited by Alan Peters and Kathleen S. Rockland. Plenum Press, New York, 1994.

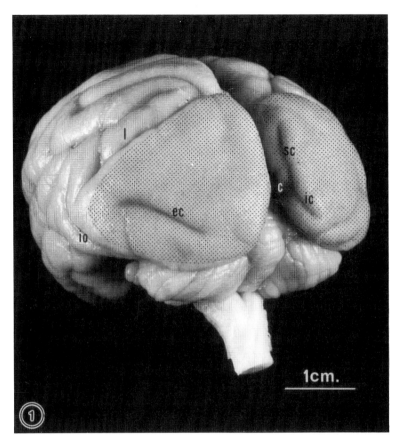

Figure 1. The brain of *Macaca fascicularis*. The location of the primary visual cortex is shown by stippling on the surface of the occipital pole. The lunate sulcus (1) is rostral to the primary visual cortex and the inferior occipital sulcus (io) is below it. The external calcarine sulcus (ec) lies within the territory of the primary visual cortex, on the opercular surface. Medially, the superior (sc) and inferior (ic) rami of the calcarine fissure (c) are evident. Much of the primary visual cortex lies within the calcarine fissure, as shown in Fig. 2.

the occipital pole there is a well-defined calcarine fissure, and at its caudal end this divides into superior and inferior rami. Much of the primary visual cortex is concealed within the depths of the fissure, and rostral to its superior and inferior rami most of the medial surface of the occipital pole is occupied by area 18, as can be seen when Nissl-stained sections are examined (see Fig. 2).

2. Lamination of Primary Visual Cortex

In the primates, the visual cortex stands out in Nissl-stained preparations because of its remarkably distinct and unique lamination (Figs. 2 and 3), which is emphasized because of the close packing of the small neurons contained within this cortical area. The primary visual cortex contains about twice the number of neurons per unit volume than any other neocortical area (see Peters, 1987), and it also looks different from other cortices when slices of unstained hemispheres are examined because of the presence of a prominent white band of myelinated axons, the outer stripe of Baillarger, or the line of Gennari. The band is located just less than halfway through the depth of the cortex and, of course, it is the presence of this stripe that leads to the cortex being called "striate cortex."

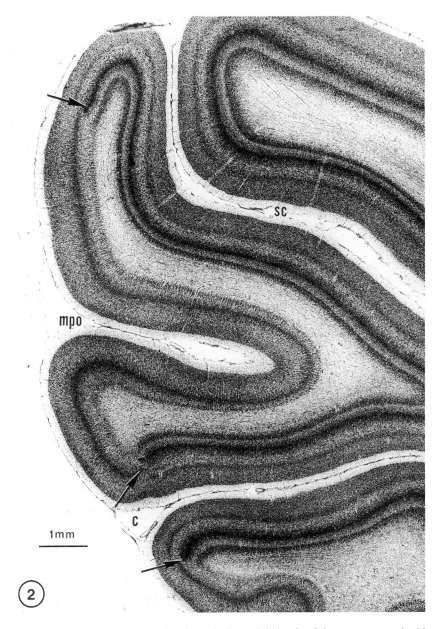

Figure 2. A Nissl-stained coronal section through the occipital pole of the macaque cerebral hemisphere. The medial surface of the hemisphere, on the left, is occupied by area 18 (V2), and at this level it is indented by the medial parieto-occipital sulcus (mpo). Below, area 17 starts (arrows) at the lips of the calcarine fissure (c), where the dark layer IV of area 18 splits to form the complex layer IV of primary visual cortex. Another transition between areas 18 and 17 is also apparent just below the superior ramus of the calcarine fissure (sc).

Billings-Gagliardi *et al.* (1974) and Valverde (1985), among others, have reviewed the various schemes for numbering the layers of the primate primary visual cortex, and the article by Billings-Gagliardi *et al.* (1974) is especially useful since it contains photographs showing the direct comparison between the various numbering schemes for the layers. But it is important to iterate that there are two schemes for numbering that have been commonly used in recent publications dealing with macaque cortex. The difference between these schemes

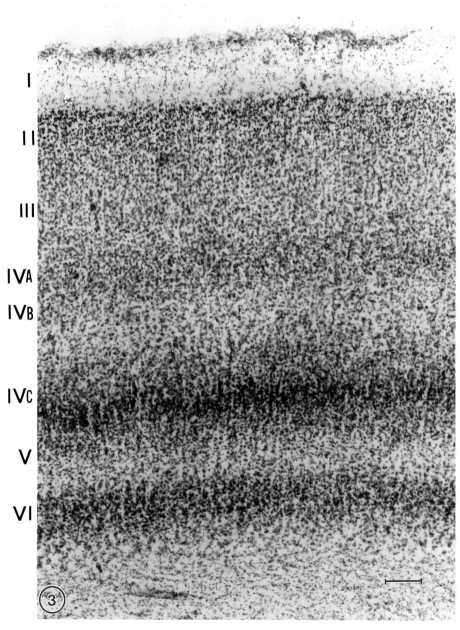

Figure 3. Nissl-stained section of macaque primary visual cortex to show the lamination. The numbering scheme of the layers is the one suggested by Brodmann (1905). Bar = 100 μm.

hinges on what constitutes layer IV. Here it should be mentioned that some authors prefer to use Roman numerals to designate the layers, while others prefer to use Arabic numbers, and some authors use lowercase letters to designate sublayers, while others prefer to use capital letters, but these differences are much less material than the decision of what to include in layer IV.

The numbering scheme that seems to have been most commonly used in recent years is basically the one proposed by Brodmann (1905). This is illustrated in Fig. 3. Brodmann divided layer IV into three parts. The upper part is layer IVA, which is a dark band of small rounded cells. It occurs just below the deepest of the pyramidal cells in layer III, but it is not always possible to precisely define the upper boundary of layer IVA, because the neurons within it are not evenly distributed. Below layer IVA is layer IVB, which is a relatively cell-poor layer. Layer IVB contains pyramidal cells, and within it are the large outer Meynert cells. Its position coincides with that of the line of Gennari, so that the layer is easy to define in myelin-stained preparations.

Below layer IVB is layer IVC, which contains closely packed stellate cells that tend to be arranged in vertical strings. But layer IVC is not a homogeneous layer because it consists of an upper portion in which the neurons are more dispersed than in the lower portion. This led Polyak (1957) to suggest that layer IVC consists of two sublayers, and he designated them layer 4Ca and layer 4Cb. Justification for subdividing layer IVC into two sublaminae was demonstrated in the lesion experiments of Hubel and Wiesel (1972), in which they examined the termination of the geniculate input to striate cortex. They found that lesions in the most dorsal layers of the dorsal geniculate nucleus, the parvocellular or small-celled layers, produced degeneration of axon terminals within layer IVA and within the lower part of layer IVC. In contrast, lesions of the ventral, magnocellular or large-celled layers of the geniculate nucleus, led to terminal degeneration in the upper part of layer IVC. Hubel and Wiesel (1972) encountered no terminal degeneration in layer IVB after having made lesions in the geniculate nucleus. Neither did Lund (1973) encounter degeneration of thalamic afferents in layer IVB, and following Polyak (1957), Lund (1973) strongly emphasized that layer IVC should be divided into two sublaminae, which she designated layers IVCa and IVCb, layer IVCb being the deepest cell-rich zone, and layer IVCa being the more superficial part less densely populated with neurons. Lund's (1973) justification for this separation of layer IVC into two sublaminae also rested on the basis that the projections of the spiny stellate cells in the two halves of layer IVC are different. The spiny stellate cells in the upper part of layer IVC have axons that do not leave layer IV, while those in the lower part project heavily to layer III. Most authors have now accepted the fact that layer IVC should be subdivided, especially on the basis of the projections of the geniculate axons to the various sublayers.

Below layer IVC is another rather cell-sparse layer. This is layer V, which is mainly composed of medium-sized pyramidal cells. And finally there is layer VI. This is made up of two sublaminae. The upper one, layer VIA, contains densely packed pyramidal cells, and this layer of neurons, together with the densely packed neurons of layer IVC, give rise to the two heavily stained bands of cells, which are so prominent when Nissl-stained preparations are examined at low magnification (see Fig. 2). The lower part of layer VI, usually designated as layer VIB, is at the border between the cortex and the white matter and it contains sparsely distributed neurons surrounded by myelinated axons. These neurons mostly have rounded or horizontally elongate cell bodies. Finally, it should be mentioned that in lower layer V and within layer VI there is a sheet of large and widely separated neurons. These are the largest neurons within this cortex and they are referred to as the deep or inner, solitary cells of Meynert.

The other common scheme for numbering the cortical layers is the one devised by von Bonin (1942). He basically recognizes the same layers as the ones described above, but ascribes two parts to layer III. The upper part coincides with layer III of the Brodmann numbering scheme, while von Bonin's lower layer III, layer IIIB, coincides with the layer IVA of Brodmann.

Where layer III stops and layer IV starts in the macaque visual cortex has been a constant source of debate and interest. The problem revolves around how to deal with the layer that coincides with the location of the line of Gennari, in which the axonal plexus is formed by the axons of pyramidal and spiny stellate cells contained within the layer, as well as by the axons of pyramidal cells in the layers above and the ascending axons of some spiny stellate cells in layer IVC (see Lund, 1973). Unlike parts of layer IV in other species, this lamina receives no direct geniculate input, and its neurons behave more like those of layer III in other species, because many of them are projection neurons. This led Colonnier and Sas (1979), for example, to suggest that in area 17 of the macaque layers III and IV subdivide and interdigitate, so that the large border pyramids (the outer cells of Maynert), which are normally present in lower layer III, have become interposed between geniculate receiving layers, as seems to be the case when the junction between areas 17 and 18 is examined at low magnification in Nissl-stained preparations (see Fig. 2). An extension of the same problem is what to do with the thin cell-dense layer that corresponds to layer IVA in Brodmann's scheme. This layer is present in macaques and in squirrel monkeys (Fitzpatrick *et al.*, 1983) and in both it receives input from the parvocellular layers of the lateral geniculate nucleus. But there is no evidence of a similar projection in the visual cortices of either galago or the owl monkey (*Aotus*) (Diamond *et al.*, 1985; Florence and Casagrande, 1987). This has led workers dealing with these other monkeys to suggest that the only layer that should be designated as layer IV is the one which receives the strong geniculate input. This layer corresponds to layer IVC in the Brodmann numbering scheme for macaques, and in all cases the upper portion of the layer receives input from the magnocellular layers of the lateral geniculate nucleus, and the more cell-dense layer receives input from the parvocellular layers of the geniculate nucleus. This scheme, in which the layer containing the line of Gennari is designated as layer IIIC, is based on the description by Hässler (1967). It has not found favor among those working on macaque primary visual cortex, but it is very important when reading articles on the primary visual cortices of primates to be aware of the numbering schemes being adopted by various authors.

3. Quantitative Data

3.1. The Area Occupied by Striate Cortex

As pointed out above, much of the striate cortex is embedded within the folds of the calcarine fissure, so that the only substantial portion of striate cortex revealed on the outer surface of the hemisphere is that of the smooth-surfaced operculum (Fig. 1). The fact that most of area 17 is hidden makes it very difficult to determine how much of the cortical surface is devoted to primary visual cortex, and this may be one reason why the calculated values for its surface area differ so widely, although another reason is that some investigators have examined the brains of *M. mulatta,* while others have examined the somewhat smaller brains of *M. fascicularis.*

The first estimate of the surface area occupied by macaque primary visual cortex appears to have been made by Clark (1942), who calculated that it occupies 1445 mm.[2] This is similar to the area of 1320 mm[2] determined by Daniel and Whitteridge (1961). Tootell *et al.* (1982) estimated that it occupies 1525 mm[2], while in a recent study of the brains of 31 *M. fascicularis* Van Essen *et al.* (1984) obtained values ranging between 690 and 1560 mm[2], with a mean value of 1200 mm.[2] At the low end of the scale is the estimate by O'Kusky and Colonnier (1982) who obtained a mean value of 841 mm[2] for five macaque brains that they examined. It is of interest that of the five brains, three were from *M. mulatta* and two from *M. fascicularis,* but they found no differences among them in the area occupied by primary visual cortex.

3.2. The Number of Neurons and Synapses in Primary Visual Cortex

The striate cortex is about 1.5 to 1.6 mm thick (O'Kusky and Colonnier, 1982; Vincent *et al.,* 1989; Peters and Sethares, 1991a), and according to stereological estimations it contains about 120,000 neurons/mm.[3] This translates into 180,000 to 190,000 neurons beneath 1 mm[2] of cortical surface, and according to the values of O'Kusky and Colonnier (1982) it means that the number of neurons within the striate cortex of one hemisphere is about 160×10^6.

O'Kusky and Colonnier (1982) have also determined the number of synapses in striate cortex and they estimate that the average neuron receives about 2300 synapses, which is significantly fewer than the ratio between synapses and neurons in the primary visual cortices of other animals, such as the cat and rat, in which the values are about 5800 and 12,500, respectively (see Peters, 1987). Presumably the low value in the macaque is largely related to the fact that the concentration of neurons is greater than in the other species that have been examined (see Peters, 1987).

It is of interest to note that in the macaque primary visual cortex, as in other cortical areas in this species, there is an initial overproduction of synapses. A study by Rakic *et al.* (1986) shows that during fetal development and continuing into the fourth postnatal month, there is a continual increase in the number of synapses as determined on the basis of the number of profiles per 100 mm[2] of thin sections taken through the cortex. This is followed by a decrease in the frequency of synaptic profiles, with adult numbers being attained by about 3 years of age. Since the proportion of neuropil does not change appreciably during the period when the number of synapses is decreasing in the rhesus monkey, Rakic *et al.* (1986) conclude that there is a selective survival of some synapses and the elimination of others. They suggest that this elimination may be necessary to attain full functional maturity of the cerebral cortex.

In cerebral cortex the neurons are frequently separated into two subpopulations, pyramidal and nonpyramidal cells (see Peters and Jones, 1984). It is easy to recognize a pyramidal cell, because it has an apical dendrite and the dendrites are spiny, but the problem is what to do with the neurons that remain, for some of them have spiny dendrites and others have smoother dendrites. In primates, most of the remaining neurons with spiny dendrites are in layer IV, and they are usually termed spiny stellate cells, but it is becoming increasingly evident that these neurons should be grouped together with the pyramidal cells, because except for the absence of an apical dendrite their features are similar to those of the pyramidal cells. They have spiny dendrites, their axons usually emerge from that pole of the cell that is facing the white matter, and they are excitatory neurons with axons that form asymmetric synapses, and like pyramidal cells their cell bodies form only symmetric synapses (see Lund, 1984; Saint-Marie and Peters, 1985). The neurons that remain, after the spiny neurons have been accounted for, have either smooth or sparsely spinous dendrites and for the most part these nonpyramidal cells are inhibitory neurons that use γ-aminobutyric acid (GABA) as their neurotransmitter.

With the availability of antibodies to GABA, it has therefore been possible to determine what proportion of the neurons in primate cerebral cortex are inhibitory in function. As part of a study of the GABAergic neuronal populations in several areas of the cerebral cortex of *M. fascicularis,* Hendry *et al.* (1987) examined area 17 and found that in strips of sections passing through the entire depth of the cortex, area 17 contains about 50% more GABAergic neurons than other cortical areas, including the adjacent area 18. But when the proportion of GABAergic neurons is determined, it turns out that in area 17 about 20% of the total population of neurons are GABAergic, as compared to about 25% of the total population of neurons in other cortical areas. If the distribution of GABAergic neurons through the primary visual cortex is examined on the basis of the proportion of these neurons contained in the various layers, Hendry *et al.* (1987) find that practically all of the neurons in layer I are labeled. Throughout the other layers there in little variation in the proportion of GABAergic neurons, except in layer IVA, in which inhibitory neurons account for 35% of the total population, and in the other two parts of layer IV, layers IVB and IVC, in which only 15 to 20% of the neurons are GABAergic.

In many respects the results obtained by Fitzpatrick *et al.* (1987) differ from those cited above. Fitzpatrick *et al.* (1987) estimate that GABAergic neurons make up at least 15% of the neurons in striate cortex, but they do not find that there is a higher proportion of labeled neurons in layer IVA as compared with other cortical layers. Instead they conclude that the highest proportion of labeled neurons is in layer 2–3, in which some 20% of the neurons are GABAergic, as compared with layers 5 and 6, in which the proportion of labeled neurons is only 12%. Fitzpatrick *et al.* (1987) have also examined the sizes of the GABA-antibody labeled neurons and find that there are obvious differences in the sizes of the neurons in various layers. For example, layer IVCa and layer VI contain populations of large GABAergic neurons, some of which have horizontally spread dendritic trees, while the GABAergic neurons within the more superficial layers of the cortex are generally smaller than those in other layers.

Such observations are of interest, because at the present time, we have little information about how different types of nonpyramidal cells are distributed throughout the cortex. And even more fundamentally, we have no clear perception about what kinds of nonpyramidal cells are present in primary visual cortex. It is generally assumed that the nonpyramidal cell types present at least include basket cells, double bouquet cells, bipolar cells, chandelier cells, and neuro-gliaform cells, but how to recognize individual neurons of each of these types among the total population of the cells labeled by GABA antibodies, for exam-ple, is an unsolved problem. In terms of morphology, the identification of indi-vidual types of neurons depends on knowing the characteristics of their cell bodies, dendritic trees, and axonal distribution, and how those features differ from the ones displayed by other kinds of neurons. Unfortunately, all of these parts of neurons are never displayed in continuity with each other in prepara-tions such as GABA antibody-labeled sections. Normally, all that is seen is the cell body with a few pieces of dendrites connected to it, and that is usually insufficient to make a positive identification. In an attempt to better determine the features displayed by the cell bodies of various types of nonpyramidal cells, Werner *et al.* (1989) examined the cell bodies of 298 neurons that had been impregnated by the Golgi–Kopsch technique in areas 17 and 18 of *M. mulatta.* After the neurons had been examined and identified as to type, they were deimpregnated so that the structural features of the cell bodies could be exam-ined. Their drawings indicate that taken as groups, the cell bodies of double bouquet cells, bipolar cells, chandelier cells, and neurogliaform cells are of rela-tively uniform sizes and they have centrally placed nuclei. In contrast, the cell bodies of basket cells seem to come in a variety of sizes, with the largest display-ing eccentrically placed nuclei. Whether the characteristics given by Werner *et al.* (1989) can be used to make positive identifications of nonpyramidal cells on the basis of sections through their cell bodies as they appear in either Nissl-stained preparations, semithin plastic sections, or antibody-labeled preparations, re-mains to be seen. The only alternative approach seems to be through labeling of subpopulations of nonpyramidal cells with antibodies which are specific for them. But, unfortunately, no such antibodies appear so far to have been gener-ated. Perhaps the antibody that comes nearest to qualifying is the one to calbin-din, because in layer II–III this antibody seems to specifically label double bouquet cells (e.g., Hendry *et al.,* 1989; De Felipe *et al.,* 1990) in the nonpyrami-dal cell population (see Jones *et al.,* this volume).

4. General Organization of Pyramidal Cells

Although Nissl-stained preparations show that the cell bodies of neurons in cerebral cortex are arranged in horizontal layers, a more interesting question from the point of view of function is whether the neurons are organized in vertically oriented units. It would be expected that there is a vertical neuronal organization, because the early studies of Hubel and Wiesel (e.g., 1968, 1972, 1977) clearly showed that the cortex contains vertically oriented functional col-umns in which all of the neurons respond to the same visual stimulus. These columns are of various types. One type is the eye preference or ocular domi-nance column system in which bands of neurons preferentially respond to stim-uli received by one eye rather than the other. In the monkey these columns are about 300 to 400 mm wide (e.g., LeVay *et al.,* 1975, 1985; Tootell *et al.,* 1988a; Wiesel *et al.,* 1974). Another type is the orientation column, in which the neurons respond best to a bar of light of a given orientation (Hubel and Wiesel, 1974, 1977). Such columns are more irregular in shape than the eye preference columns, and a 180° change in orientation occurs over a distance of about every 570 mm across the cortical surface (Hubel *et al.*, 1978). Other columns contain neurons that respond best to specific colors. The color columns extend throughout the entire depth of the cortex and they appear to be shaped like slabs that are 100 to 250 mm wide (Michael, 1981), although in addition to these slabs there are color responsive neurons that are contained in the cytochrome oxidase blobs in layer II/III. As far as can be seen at the present time, these various systems of columns bear little or no geometric relationship to each other (see Blasdel, 1992), but it is clear from studies such as the one carried out by Michael (1985) that when a single electrode penetration is made through the depth of the visual cortex, the neurons encountered all respond to a stimulus received by the same eye, a stimulus of the same orientation and of the same color. This means that the three columnar systems described above overlap each other, and that individual neurons encountered in single electrode penetrations are included within each columnar system. Of course, this has to be the case, since the different systems of columns occupy the same cortical space. This suggests that there should be sets of vertically oriented neuronal units that must be each as small as the smallest functional column, and that these units must be recruited in different combinations, some through the thalamic input to the cortex, and perhaps others by the intrinsic connections between the neurons, to generate the different sets of functional columns.

To investigate whether such units of vertical oriented neurons exist requires a preparation to be generated that will allow all of the neurons in the cortex to be visualized simultaneously, and in sufficient detail to allow their cell bodies and some of their processes to be seen. It became possible to generate such preparations when antibodies to microtubules and their associated proteins became available. The particular antibodies that seems to be most suited to such studies are those against microtubule-associated protein 2 (MAP2), and the one that we have used is MAP2:5F9 (Kosik *et al.*, 1984). This antibody binds to the microtubules within cell bodies and dendrites of neurons (de Camilli *et al.*, 1984; Bernhardt and Matus, 1984), so that these parts of neurons can be visualized when the sites of antibody binding are revealed by using the enzymatic reaction with horseradish peroxidase (HRP) to oxidize diaminobenzidine (DAB). An example of a MAP2 antibody-labeled vertical section through area 17 of macaque cerebral cortex is shown in Fig. 4. In such vertical sections there is dark labeling of layers V and VI and another dark band of labeling that coincides with layer IVB. Above that there is a pale reticulated zone where layer IVA is located and another darkening in the staining pattern corresponding to layers II and III, while layer I is extremely dense, because of the presence of the apical dendritic tufts of the pyramidal cells.

4.1. Pyramidal Cell Modules

In such preparations it becomes apparent that the apical dendrites of pyramidal cells are organized into distinct vertically oriented groups (Figs. 4 and 5, arrows), which we have referred to as dendritic clusters (Peters and Walsh, 1972; Peters and Sethares, 1991a). These dendritic clusters are initiated as the apical dendrites of the layer V pyramidal cells come together and ascend into layer IV (Fig. 5). As a consequence of this clustering, the apical dendrites of the layer V

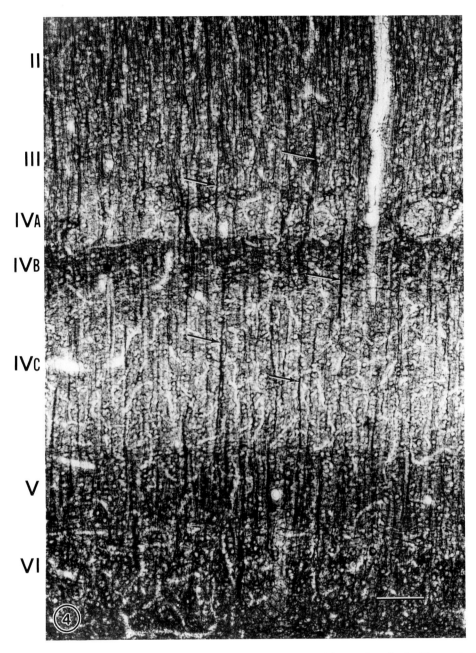

Figure 4. Vertical section through primary visual cortex reacted with MAP2 antibody. The cortical layers are indicated on the left. Note the dark staining of layer IVB, and of layers V and VI. Layer IVA has a reticulated appearance with thick bundles of dendrites passing through a pale neuropil. The clusters of apical dendrites (arrows) extend vertically from layer V into layer II/III. Bar = 100 μm.

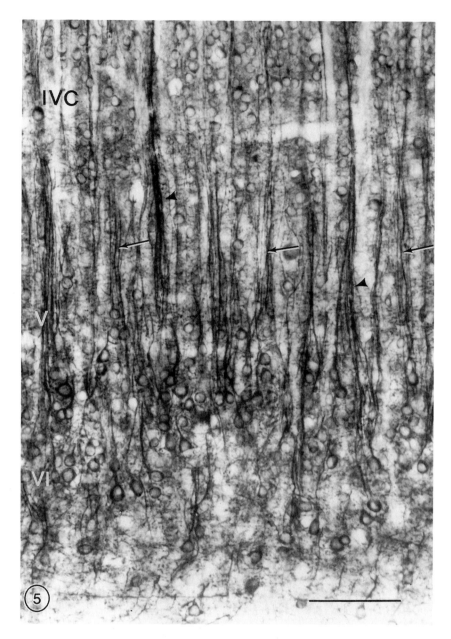

Figure 5. Vertical section through the deep layers of primary visual cortex stained with MAP2 antibody. The pyramidal cells of layer VI have darkly staining cell bodies, which frequently incline at an angle so that their apical dendrites can join fascicles of apical dendrites (arrowheads) that ascend through layer V and enter layer IVC, where they terminate. the apical dendrites of the layer V pyramidal cells are thinner, and they form clusters (arrows), which ascend into layer I. Bar = 100 μm.

pyramidal cells are quite distinct in MAP2-stained tangential sections through area 17 taken at the level of layer IVC (Fig. 6), and counts show that there are some 1270 clusters of layer V apical dendrites per mm² in the tangential plane. Thus, the clusters of apical dendrites have a center-to-center spacing of about 31 μm. It is suggested that these clusters represent the axes of modules of pyramidal cells, and that such modules are the fundamental units of neurons of which the cortex is composed. Given that the clusters of apical dendrites form the axes of the pyramidal cell modules, then the modules themselves must have diameters of 31 μm.

As the clusters of apical dendrites pass through layer IVC and into the upper strata of layer IV they are still intact, but they are more difficult to define because of the intense MAP2 staining of neuronal elements in layers IVB and IVA (see Fig. 4). These layer IV components will be referred to later, but the clusters pass through them and reach the bottom of layer III, where additional apical dendrites derived from the layer III pyramidal cells are added to them. As a result the clusters become thicker (Fig. 7) and more prominent as they pass through layer II/III and finally reach just below layer I, where they break up as the apical dendrites form their apical tufts and splay out. This branching of the apical dendrites results in a high concentration of dendrites in layer I, with the result that in MAP2-stained preparations layer I is very dark.

Although the layer V and layer II/III pyramidal cell apical dendrites come together to form the clusters, not all of the pyramidal cells in area 17 send their apical dendrites into these clusters. For example, the apical dendrites of the pyramidal cells in layer VIA come together to form fascicles that are independent of the clusters initiated by the layer V pyramidal cells. The apical dendrites of the layer VIA pyramidal cells are thicker than most of those of the layer V pyramids, and in horizontal sections taken through the lower part of layer V it can be seen that their fascicles are irregular (Fig. 8). In some fascicles the layer VIA apical dendrites are in clumps, while in others they are in sheets, so that the distribution of apical dendrites lacks the regularity displayed by the clusters, and it seems likely that the fasciculation of the apical dendrites of the layer VIA pyramidal cells is brought about because the apical dendrites have to wend their way between the groups of cell bodies belonging to the layer V pyramidal cells above them. Having passed between the cell bodies of the layer V pyramidal cells, the fascicles of apical dendrites belonging to the layer VIA pyramidal cells ascend in parallel with the thinner dendrites in the clusters (Fig. 5), but whereas the apical dendrites in the clusters pass through layer IV, most of the apical dendrites belonging to the layer VIA pyramids form their apical tufts in layer IVC. Consequently, within layer IV the fascicles of layer VIA apical dendrites become thinner and gradually disappear.

It should also be pointed out that the neurons in layer IV do not fit neatly inside the pyramidal cell modules. Most of the neurons in layer IVC, at least, are spiny stellate cells, which have all of the basic features of pyramidal cells although they lack apical dendrites (e.g., see Saint-Marie and Peters, 1985). In Nissl-stained and in MAP2-reacted sections, these small neurons often appear to be arranged in vertical rows with spaces between them, and this arrangement seems to be imposed on them by both the ascending clusters of apical dendrites of the layer V pyramidal cells and the bundles of myelinated axons that pass prominently though the lower layers of the cortex. These two sets of neuronal processes essentially form vertical strands between which the cell bodies of the

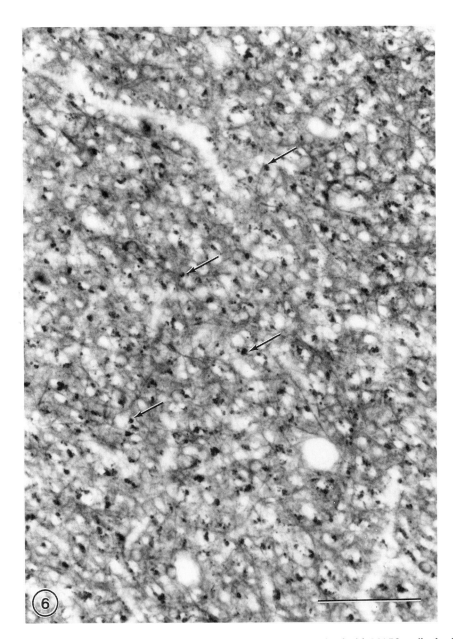

Figure 6. Tangential section though layer IVC. In this preparation stained with MAP2 antibody, the cross sections of the clusters of apical dendrites of the layer V pyramidal cells appear as dark dots (arrows). Bar = 100 μm.

layer IVC neurons are fitted and the vertical processes occupy the unstained clear spaces evident between the cell bodies in Nissl-stained preparations (see Fig. 3).

The origins of the bundles of myelinated axons that pass through the lower layers of the cortex have not yet been established, and it is not clear whether they are afferent or efferent axons, or a combination of both. But in counts made from horizontal semithick plastic sections, Peters and Sethares (1991b) find that the number of bundles of myelinated axons per unit area of section is similar to the number of apical dendritic clusters. It is possible, therefore, that each bundle of myelinated axons is generated from the pyramidal cells that are contained within one pyramidal cell module. However, this has yet to be proven.

Since it is known that there are 1.8 to 1.9 × 10^5 neurons beneath each 1 mm^2 of cortical surface, as noted in Section 3.2, and that the pyramidal cell modules have an average diameter of 31 μm, it can be calculated that there are some 142 neurons associated with each pyramidal cell module. A diagrammatic representation of such a module is shown in Fig. 9, which also gives the number of neurons associated with a module in each layer of the cortex.

Of these 142 neurons associated with a module, not all are pyramidal cells. Using the data generated by Hendry *et al.* (1987) for the proportion of GABAergic neurons in monkey striate cortex, it can be concluded that about 30 of the neurons, including most of the ones in layers I and VIB, are inhibitory in function. One point that would be worth examining is whether the inhibitory nonpyramidal cells are also disposed in a nonrandom fashion, and how they

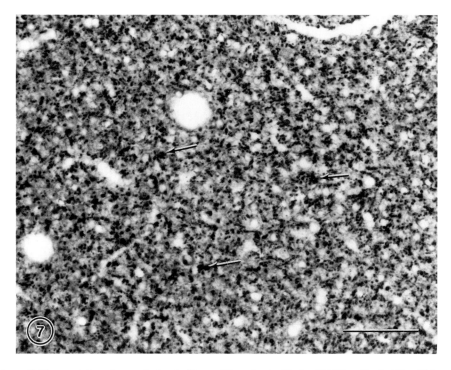

Figure 7. Tangential section at the level of layer III, stained with the MAP2 antibody. The thick and dark clusters containing the apical dendrites of the layer V and layer III pyramidal cells (arrows) pervade the neuropil at this level. Bar = 100 μm.

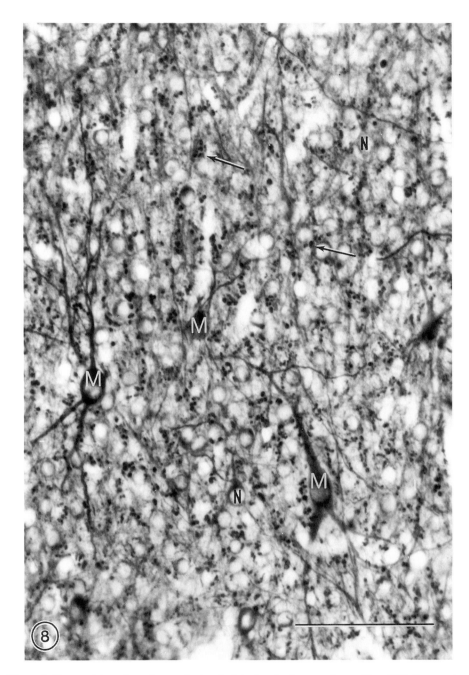

Figure 8. Tangential section taken at the level of the border between layers V and VI. Preparation stained with the MAP2 antibody. The cross-sectioned apical dendrites of the layer VIA pyramidal cells are in fascicles (arrows) between the cell bodies of the neurons (N). At this level, some of the deep Meynert cells (M) are apparent. Bar = 100 μm.

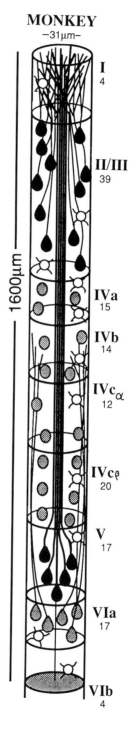

MONKEY

−31μm−

1600μm

I
4

II/III
39

IVa
15

IVb
14

IVc$_\alpha$
12

IVc$_\beta$
20

V
17

VIa
17

VIb
4

TOTAL   142

Figure 9. A diagrammatic representation of a module of pyramidal cells. The module is 31 μm in diameter. It extends through the entire depth of the cortex and contains 142 neurons. The number of neurons present in each layer is shown in Arabic numerals on the right. The neurons whose apical dendrites come together to form the clusters are blackened, while spiny neurons that do not have dendrites contributing to the clusters are stippled. The neurons with open cell bodies represent the nonpyramidal cells.

relate to the pyramidal cell modules. If all areas of the neocortex contain pyramidal cell modules similar to those in striate cortex, it might be expected that there are fundamental spatial relationships between the pyramidal cell modules and the inhibitory neurons to generate neuronal circuits that are basic to all cortical areas. One indication of this possibility is that all of the cortical areas so far examined contain similar proportions of GABAergic neurons (e.g., Hendry *et al.*, 1987), and despite an increase in the total number of neurons per unit volume in macaque striate cortex as compared with other cortical areas, essentially the ratio between GABAergic and other neurons is retained.

It should be pointed out at this stage that a module of neurons would not only receive information about the color, orientation, and the eye receiving an image, but since the visual field is mapped on the cortex, each module would be expected to respond better than its neighbors to an image in a specific part of the visual field. Also, a module would probably respond better than its neighbors to an image with a certain orientation, because as shown by Hubel and Wiesel (1974, 1977), there is a 10° shift in image orientation for every 25 to 30 mm that a recording electrode travels across the visual cortex of the macaque cortex, a distance that coincides with the sizes of the pyramidal cell modules.

4.2. Layers IVA and IVB

In MAP2-stained preparations there is a very dense band of labeling at the level of layer IVB (Fig. 4), and this is largely produced by a meshwork of closely packed horizontally aligned dendrites. Many of these dendrites belong to the numerous medium-sized spiny stellate cells that occur at the layer IVA/IVB border. In addition, there are thick dendrites that emanate from some large spiny neurons within layer IVB. Some of these are large spiny stellate cells and others are large pyramidal neurons, which tend to occur in groups. Together they are referred to as the outer solitary cells of Meynert (e.g., Valverde, 1985), and their axons project to area V5, or MT, of extrastriate cortex, a cortical area that is concerned with the analysis of motion (see Shipp and Zeki, 1989), as well as to area V2.

In MAP2-stained vertical sections examined at low magnification, it is evident that while the lower darkly stained border of layer IVB is relatively smooth, the upper border has a scalloped appearance, and from the peaks of the scalloping bundles of apical dendrites arise and pass into layer III (Fig. 4). Closer examination shows that these bundles of apical dendrites arise from groups of pyramidal cells (Fig. 10). The cell bodies of the deepest neurons in the groups are located at the border between layers IVA and IVB, while the more superficial cells are contained within layer IVA. In reality, the cell bodies of the neurons in these groups are arranged to form conical stacks (Fig. 10, *C*), and their apical dendrites emerge from the top of each stack to funnel together into a bundle (Fig. 10, arrows) that reaches into layer III.

Because of the arrangement of the cell bodies of the neurons in these groups they have been referred to as pyramidal cell cones (Peters and Sethares, 1991a,b), and in horizontal sections of MAP2-stained material taken at the level of layer IVA, these cones are seen as islands of cells separated from each other by a pale neuropil (Figs. 11 and 12). When the sections pass through the lower parts of the cones, groups of pyramidal cell bodies are the principal constituents, but

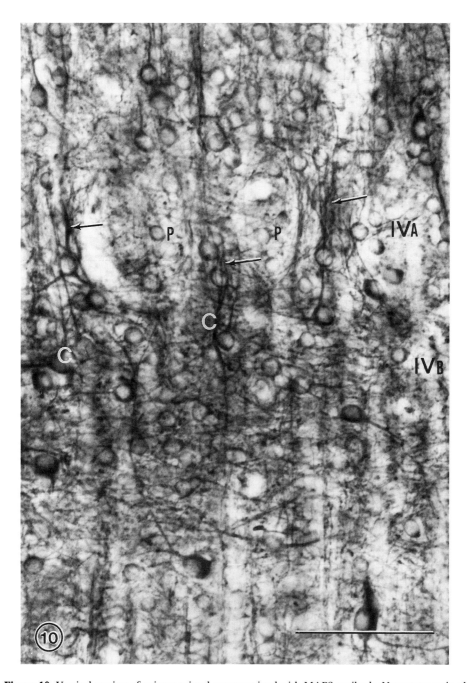

Figure 10. Vertical section of primary visual cortex stained with MAP2 antibody. Numerous stained dendrites are present in layer IVB. In the upper stratum of the layer, and extending into layer IVA, there are conelike stacks of pyramidal cell bodies (C), whose apical dendrites come together to form thick bundles (arrows). Between these pyramidal cells and their dendrites are pale zones (P), which coincide with the location of the cytochrome oxidase-positive lattice. Bar = 100 μm.

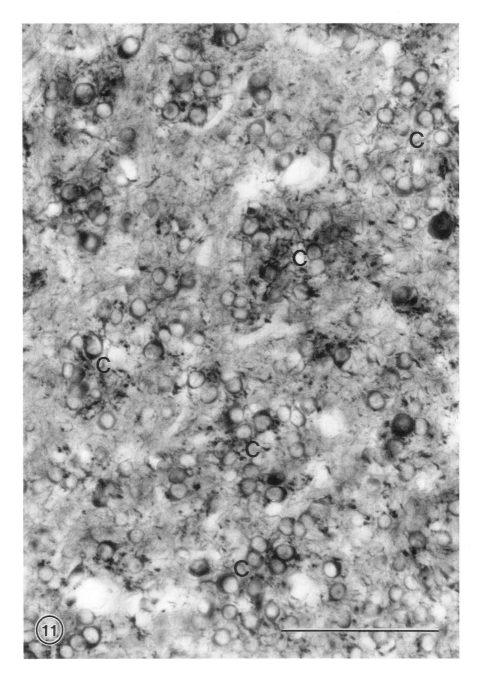

Figure 11. Horizontal section taken at the level of layer IVA. The MAP2 antibody reveals groups, or cones (C), of dark-staining pyramidal cells, surrounded by a pale-staining neuropil. Bar = 100 μm.

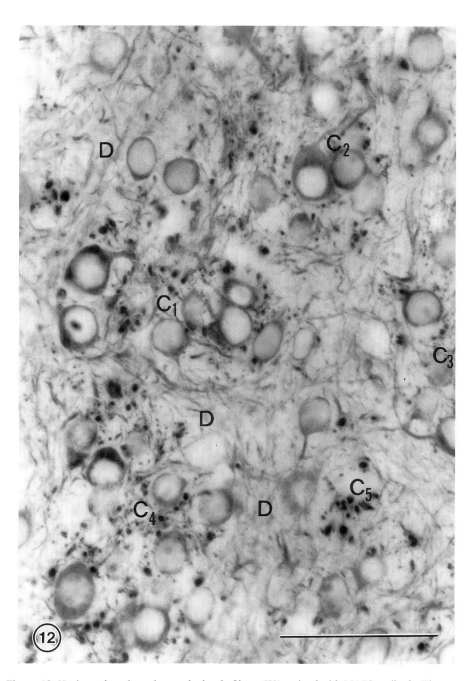

Figure 12. Horizontal section taken at the level of layer IVA stained with MAP2 antibody. The cones of pyramidal cell bodies (C_1–C) are surrounded by a neuropil in which pale-staining dendrites (D) are evident. One cone (C_5) is sectioned close to its apex so that the bundle of apical dendrites extending from the pyramidal cell bodies is clearly defined. Bar = 50 μm.

in sections taken at a slightly higher level, some of the pyramidal cells can be seen to give off apical dendrites that are inclined toward the centers of the cones. The apical dendrites become more frequent toward the tops of the cones, and the cell bodies become fewer in number, until eventually at the border with layer III, thick bundles of apical dendrites are the only evidence of the cones of pyramidal cells situated below. But quite soon these bundles of dendrites blend in with the others that are ascending through the cortex. The ultimate destination of these bundles of apical dendrites has not been determined, but it is presumed that they reach layer I.

Since the cones of layer IVA pyramidal cells are very obvious in tangential sections of MAP2-labeled material, it is evident that they are relatively evenly spaced with their centers about 0.1 mm apart (see Fig. 11). The average diameter of the cones is about 60 μm and counts made in a number of different rhesus monkeys show that there are about 120 pyramidal cell cones beneath 1 mm^2 of cortical surface (Peters and Sethares, 1991a,b).

Use of the MAP2 antibody provided the first indication of the existence of these cones of layer IVA pyramidal cells, but knowing that they exist it is then possible to recognize them in thick, Nissl-stained horizontal sections (Peters and Sethares, 1991b).

Another interesting feature of layer IVA that has been known for some time is the presence of a honeycomb pattern, or lattice, of dark cytochrome oxidase staining, which is evident when tangential sections through layer IVA are examined (see Horton, 1984; Fitzpatrick *et al.*, 1985; Wong-Riley, this volume). The dark honeycomb lattice surrounds spaces, and when cytochrome oxidase-reacted sections through layer IVA are examined using interference optics (Fig. 13), it is apparent that the spaces in the lattice are occupied by groups of neuronal cell bodies (Fig. 13, N). These are the neurons in the cones of layer IVA pyamidal cells (Peters and Sethares, 1991b).

4.3. A Summary of the Pyramidal Cell Groupings

From the above it is evident that there are three basic groupings of pyramidal cells in macaque primary visual cortex. First, there are the groups of layer VIA pyramidal cells whose apical dendrites come together to form irregular fascicles that pass between the cell bodies of layer V pyramids. Most of the apical dendrites in these fascicles form their terminal tufts in layer IVC. Second, there are sets of layer V pyramidal cells whose apical dendrites generate the clusters. These apical dendrites ascend through layer IV as discrete entities and enter layer II/III in which the apical dendrites of pyramidal cells in that layer are added to the clusters. The clusters have a center-to-center spacing of 31 μm, and it is suggested that they represent the axes of repeating modules of pyramidal cells, which are the fundamental neuronal units of the cortex. It should be pointed out that similar clusters are present in other cortices that have been examined, and that they have the same basic organization as those in primate visual cortex (e.g., see Peters and Walsh, 1972; Feldman and Peters, 1974; Fleischhauer, 1974; Schmolke, 1989). Lastly, there are the cones of pyramidal cells of layer IVA, which have their centers about 0.1 mm apart, so that there is a ratio of about 10:1 between modules and cones.

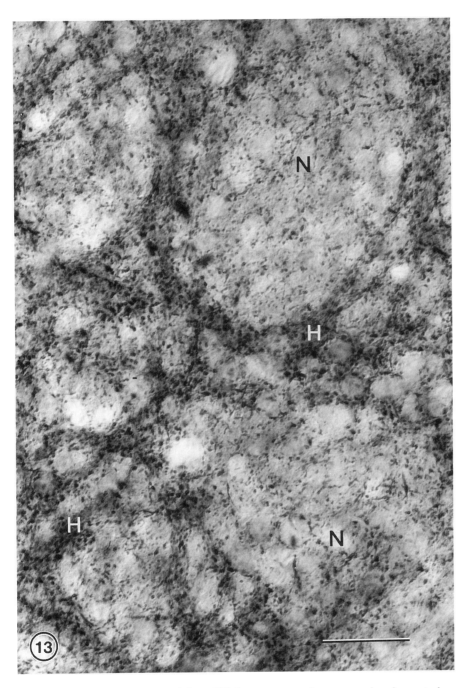

Figure 13. Horizontal section through layer IVA from a preparation processed to show cytochrome oxidase activity. In this photograph, which was taken using interference optics, the cytochrome oxidase activity appears as a honeycomb pattern of staining (H) that surrounds the pyramidal neurons (N) within the cones. Bar = 50 μm.

4.4. Numbers of Pyramidal Cell Modules in Striate Cortex

As pointed out in a previous section, there have been various estimates for the area of the cortical surface occupied by striate cortex, but if the mean area of 1200 mm^2 estimated by Van Essen *et al.* (1984) is taken as a working figure, then the striate cortex of one hemisphere would contain about 1.5×10^6 of the 31-μm-wide neuronal modules that are centered around the layer V and layer II/III pyramidal cells. As pointed out previously (Peters and Sethares, 1991b), this number is very similar to the estimates of 1.2 to 1.8×10^6 that have been given for the number of nerve fibers in the optic nerve, of which 90% project to the dorsal lateral geniculate body (Perry and Cowey, 1984). And interestingly, it has been estimated that there are 1.1 to 1.8×10^6 neurons in the lateral geniculate body (e.g., Williams and Rakic, 1988). Thus, the numbers of nerve fibers in the optic nerve, of neurons in the dLGN, and of pyramidal cell modules in the striate cortex are similar. This may have some functional significance, or it may just be coincidence. But it certainly does not mean that the action potential generated by one ganglion cell in the retina is projected via a neuron in the dLGN to one pyramidal cell module in striate cortex.

5. Possible Significance of the Pyramidal Cell Cones in Layer IVA

As pointed out above, the pyramidal cell cones in layer IVA lie within the interstices of the thin honeycomb lattice of high cytochrome oxidase activity in that layer, and the position of the cytochrome oxidase lattice coincides with a number of other features in layer IVA. This coincidence has several functional implications. For example, by filling geniculate axons with markers, Blasdel and Lund (1983) have shown that the arbors of the terminals of geniculate axons from the parvocellular layers of the dLGN to layer IVA surround spaces or lacunae, and these are of a similar size to the pyramidal cell cones. Consequently, it may be inferred that geniculate axons terminate within the neuropil that surrounds the cones. This input from the parvocellular layers of the geniculate carries information that is concerned with color, and a number of investigators have indeed shown that neurons in layer IVA respond to color (e.g., Michael, 1985; Vautin and Dow, 1985; Lennie *et al.*, 1990). As an aside it should be mentioned that the other portion of layer IV that receives input from the parvocellular layers of the cortex is layer IVCb. Like the honeycomb lattice in layer IVA, this layer also gives a strong cytochrome oxidase reaction, as do the "blobs" that are situated in layer III, and similarly contain color-responsive neurons (e.g., Livingstone and Hubel, 1984; Tootell *et al.*, 1988b; Ts'o and Gilbert, 1988; see also Wong-Riley, this volume).

Hendrickson *et al.* (1981), and Fitzpatrick *et al.* (1987) have also shown that the position of the cytochrome oxidase lattice in layer IVA coincides with the disposition of axon terminals that label with antibodies to glutamic acid decarboxylase (GAD), an antibody that is used to indicate which axon terminals are GABAergic. A similar pattern of latticelike staining is also produced when the locations of GABA$_A$ receptors are revealed by immunocytochemistry (Hendry *et al.*, 1990), and when horizontal sections through layer IVA are examined after labeling with antibodies to parvalbumin (van Brederode *et al.,* 1990; Peters and Sethares, 1991b), which is a calcium-binding protein that usually labels subsets of GABAergic neurons (e.g., Celio, 1986). van Brederode *et al.* (1990) suggest that these parvalbumin-labeled axon terminals in layer IVA are, in fact, derived from the geniculate nucleus, because the labeling with parvalbumin antibodies is greatly reduced in intensity after enucleation, and because there are thick bundles of parvalbumin-positive axons within the neuropil underlying the striate cortex. These data suggest that the geniculate input to layer IVA is through axons that terminate in the neuropil between the cones of pyramidal cells, and that, unusually, some of this input may be inhibitory.

A possible role for layer IVA in color processing is further indicated by Schein and de Monasterio (1987), who have proposed that a basic unit of organization of the color opponent system in striate cortex is what they refer to as a P-cell module. This is defined as a module of cortex containing a layer II/III cytochrome oxidase "blob" or "puff" (see Figs. 14 and 15). These blobs are relatively evenly spaced. They have been found in the primary visual cortices of all primates so far examined, and in macaques there is a single row of them along the long axis of each ocular dominance column (Horton and Hubel, 1981;

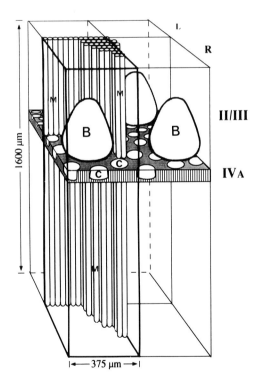

Figure 14. A three-dimensional representation of "blob-centered" modules. The modules extend through the depth of the cortex and are centered around the cytochrome oxidase blobs (B) in layer II/III. These blobs are aligned in rows along the 400-μm-wide ocular dominance columns and they have a center-to-center spacing of about 375 μm. Such a piece of cortex would also contain the pyramidal cell cones of layer IVA (C), and the pyramidal cell modules (M). For further explanation see text. After Peters and Sethares (1991a).

Hendrickson *et al.*, 1981; see Wong-Riley, this volume). Anatomically the neurons in the blobs appear to receive a direct input from the dLGN, because Livingstone and Hubel (1982) showed that when horseradish peroxidase or tritiated proline is injected into the dLGN, there are puffs of periodic labeling in layer II/III of striate cortex. This input appears to arise from the intercalated layers, which are layers of small neurons interposed between the magno- and parvocellular layers of the dLGN; thus, when Livingstone and Hubel (1982) confined their injections to the parvo- and magnocellular layers, there was no periodic labeling in striate cortex. Recently, the involvement of the small neurons in the dLGN in the innervation of the blobs has been more conclusively demonstrated by Lachica and Casagrande (1992) in the prosimian primate *Galago*, in which the small-celled component of the dLGN is grouped into two distinct layers, making them more accessible for injecting tracers. After making injections of WGA-HRP into the small-cell layers of the dLGN, Lachica and Casagrande (1992) obtained very nice labeling of axons, which branch into terminals fields that are confined to the zones of the blobs in layer III. In addition to this direct input from the dLGN, however, there is evidence that another input to the blobs is derived from neurons in layer IVCB (Michael, 1988), in which axons from the parvocellular layers of the geniculate terminate.

The cells in the blobs are color-selective, and as shown by Livingstone and Hubel (1984), they have receptive fields of three main types, namely broad-band center-surround, red-green double-opponent, or yellow-blue double-opponent. Livingstone and Hubel (1984) find that the cells within the blobs do not show orientation selectivity, whereas those between the blobs are very orientation selective. This suggests that there is a spatial separation in layer II/III between cells responding to color and cells that respond to spatial information.

A blob-centered module as defined by Schein and de Monasterio (1987) is shown diagrammatically in Figs. 14 and 15. Essentially it corresponds to a piece

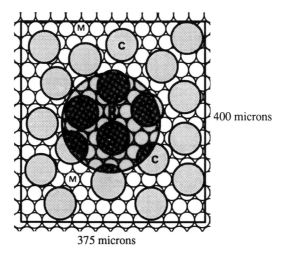

400 microns

375 microns

Figure 15. Diagrammatic representation of a "blob-centered" module, as seen from the surface of the cortex. In addition to the cytochrome oxidase blob (B), such a module would contain about 20 pyramidal cell cones (C) and 190 pyramidal cell modules (M). For further explanation see text. Area of "blob" centered module: 0.4×0.375 mm^2 (0.15 mm^2); cytochrome oxidase cell modules: diameter = 31 μm.

of cortex below 400 × 375 μm of cortical surface, because the blobs lie along the axes of the ocular dominance columns, which are about 400 μm wide, and the center-to-center spacing between adjacent blobs is 375 μm. Schein and de Monasterio (1987) estimate that there is an average of 162 P (parvocellular) neurons in the dLGN associated with such a module, and they further suggest that in addition to the neurons in the blobs themselves, the blob-centered modules contain submodules that are also responsible for color perception.

Schein and de Monasterio (1987) define these submodules as being the spaces in the cytochrome oxidase staining pattern in layer IVA, spaces which we have shown to be occupied by the pyramidal cell cones (Peters and Sethares, 1991a,b). Although Schein and de Monasterio (1987) suggest that there are 10 honeycomb spaces associated with a blob-centered module, our data on the frequency of the pyramidal cell modules show that there are about 20 of them in the space occupied by a blob-centered module, since there are 120 cones beneath 1 mm² of cortical surface. This same cortical space would be occupied by about 190 pyramidal cell modules, so that there are 9 to 10 pyramidal cell modules per cone (see Fig. 15). Since, as mentioned previously, there is a ratio of 162 P-cell afferents for each blob-centered module, this may mean that there are 9 to 10 afferents per cone. Schein and de Monasterio (1987) suggest that such a number of P-cell afferents corresponds to a minimal complete set of P-like ganglion cells in the retina, because a minimal complete set of P-like afferents would have to contain at least one blue-on-center ganglion cell, and such ganglion cells represent 6 to 10% of all ganglion cells in the retina (de Monasterio and Gouras, 1975; de Monasterio, 1978, 1979). Thus, a complete set of ganglion cells would number about 10 to 17, which is similar to the number of P-cell afferents associated with a layer IVA pyramidal cell cone. One possible conclusion is that a pyramidal cell cone represents a set of cortical neurons that receives color information from a minimal number of ganglion cells in the retina.

Obviously this is conjecture and it is unlikely that cones of layer IVA pyramidal cells receive their input from only a single set of ganglion cells, because as shown by Blasdel and Lund (1983), the geniculate input to layer IVA is provided by axons that branch to surround several lacunae or spaces, which it can now be seen to correspond to the locations of the pyramidal cell cones.

In any event, the accumulated evidence does suggest that the pyramidal cell cones in layer IVA receive input from the color-opponent pathway of the visual system. And it raises the question of what is the importance of this input to layer IVA in macaques, a layer which seems to be absent in most primates. It would be worthwhile examining other primates with the MAP2 antibody to determine if there are any corresponding groups of neurons in those primates which have not yet been revealed.

6. The Meynert Cells

There are two sets of Meynert cells in the macaque striate cortex, the outer Meynert cells in layer IVB and the inner Meynert cells in layers V and VI. The inner Meynert cells are the most obvious ones (see Fig. 8), and in human visual cortex, Meynert (1872) described these large neurons as being present in the infragranular layers. Ramón y Cajal (1899) also encountered these cells in human visual cortex and described them as having broad bases and prominent apical dendrites that reach to layer I. When he saw similar cells in rhesus monkey visual cortex, Clark (1942) agreed that they basically fitted this description but pointed out that their location is somewhat different in the human and monkey striate cortices, because while they are at the interface between layers V and VI in the human, in the monkey they tend to be more in layer VI.

One question that has arisen in recent years is what particular neurons were being described by Meynert (1872), because as pointed out by Tigges *et al.* (1981), it turns out that there are really two different kinds of large neurons in layers V and VI. One is a large pyramidal cell with a rather wide cell body and horizontally oriented basal dendrites, that tends to occur at the border between layers V and VI, while the other is a large pyramidal cell that occurs in layer VI. It has a more slender cell body and basal dendrites that extend downwards toward the white matter. Tigges *et al.* (1981) suggest that only the cells with wide cell bodies and horizontally spreading basal dendrites should be described as Meynert cells and Valverde (1985) agrees with this interpretation. But it is not clear that the two types of neurons are sufficiently different to warrant making an issue about distinguishing between them, because as far as is known, they have similar connections. As a consequence, in recent years the tendency has been to collectively refer to all large pyramidal cells in layers V and VI as Meynert cells (see Chan-Palay *et al.*, 1974; Lund, 1973; Winfield *et al.*, 1981).

The fine structure of Meynert cells in the monkey has been examined by Chan-Palay *et al.* (1974), and in addition to giving a complete cytological description of these cells, they also analyzed the distribution of spines along their dendrites. It may be mentioned that we have just completed a study of the effects of aging on the deep Meynert cells in rhesus monkey visual cortex (Peters and Sethares, 1993), and find that these large neurons are largely unaffected by age. There is no indication that any of them are lost and surprisingly they accumulate hardly any lipofuscin, even in 30-year-old monkeys.

In Nissl-stained preparations, the deep or inner Meynert cells are usually seen to be widely separated from each other, although occasionally they occur in small clumps. This suggested to Winfield *et al.* (1981) that these neurons are disposed randomly, and they found that on average there are about 25 inner Meynert cells beneath 1 mm^2 of cortical surface in *M. fascicularis.* Fries (1986) and Payne and Peters (1989) encountered similar numbers of neurons in *M. fascicularis* and in *M. mulatta* but they found that these large neurons are not randomly distributed. Instead, they are arranged to form a horizontal sheet in which there are gaps. When horizontal sections are stained using the cytochrome oxidase reaction, the cell bodies of the Meynert cells are clearly shown, as well as the cytochrome oxidase blobs in layer II/III, so that it can then be seen that the gaps in the Meynert cell distribution lie beneath the blobs (see also Shipp and Zeki, 1989). Payne and Peters (1989) suggest that this is another example of the separation between the color and movement streams of information in the visual system, because as pointed out above, the cytochrome oxidase blobs contain neurons that are color-responsive, while the interblob regions contain neurons that respond to orientation and movement. It is in the latter compartment that the apical dendrites of the inner Meynert cells are located, and this fits in with the fact that these large cells project to the superficial layers of the superior colliculus (Fries and Distel, 1983) in which the neurons are strongly binocular and sensitive to movement. The Meynert cells also project through their bifurcating axons to area V5 (Lund and Boothe, 1975; Fries *et al.,* 1985; Shipp and Zeki, 1989), in which the neurons are sensitive to movement and orientation of a stimulus.

The large spiny neurons that form the population of outer Meynert cells in layer IVB of macaque visual cortex are of two kinds. Most of them are stellate cells, but a few of them are pyramidal in shape (Lund, 1973; Valverde, 1985), an observation that has been confirmed by Shipp and Zeki (1989) who labeled many of these cells, in addition to the large neurons in layer VI, after making injections of horseradish peroxidase into area V5 or MT. Interestingly, these latter authors point out that in the owl monkey the frequency of occurrence of spiny cell types is the reverse of that in the macaque, because in the owl monkey most of the large neurons in layer IVB are pyramids. In their study, Shipp and Zeki (1989) examined the distribution of the large labeled neurons in layer IVB and the cytochrome oxidase blobs in layer II/III, but they were unable to discover any systematic relationship between the two, even though, as pointed out above, there is a relationship between the distribution of the layer VI Meynert cells and the blobs.

Interestingly, the distribution of the large spiny neurons, or outer Meynert cells, in layer IVB seems not to be related to the blobs, but to the pyramidal cell cones in layer IVA. Thus, when the relative positions of these two entities are plotted in MAP2-stained preparations, it is found that the large layer IVB cells tend to occur in groups that are located beneath the edges of the cones and beneath the spaces between the cones (Peters and Sethares, 1991a). This again suggests a separation of neurons that are responsive to color and form perception from neurons that are concerned with movement and direction sensitivity, and it is another example of the fact that the neurons in primary visual cortex are not distributed randomly, but are arranged to form regularly repeating patterns, in which neurons with different functions tend to be segregated from each other.

Another example of the separation between neurons having different functions is revealed by an antibody generated by McKay and Hockfield (1982). This antibody, which they term Cat-301, recognizes an antigen on the surfaces of some neurons, and when the antibody is applied to area 17 of macaque cortex it labels patches of neurons in layers IVB and VI (Hockfield *et al.,* 1983). The patches are 100 to 300 μm wide and are separated by spaces of 150 to 500 μm. Radially, the patches of labeled cells in layers IVB and VI line up with each other. Furthermore, the patches of labeled neurons lie along the axes of the orientation columns in area 17, and they are in register with the cytochrome oxidase patches in layer II/III. The functional significance of the Cat-301-labeled patches of neurons in layers IVB and VI is not yet known, but it is of interest that the Cat-301 antibody also labels neurons in the magnocellular layers of the lateral geniculate body, while as pointed out earlier, the cytochrome oxidase blobs in layer II/III, which are above them, receive some of their input indirectly from the parvocellular layers of the dLGN.

7. Summary

Much of the information presented in this account is shown diagrammatically in Fig. 16, and it may be summarized as follows. It is suggested that the basic unit of neuronal organization in the visual cortex is the pyramidal cell module, which extends through the depth of the cortex. Similar pyramidal cell modules are also present in the visual cortices of other mammals, and in each case the modules are centered around the clustered apical dendrites of pyramidal cells in layers V and II/III. Also, as in other visual cortices, the apical dendrites of the pyramidal cells in layer VI of macaque area 17 do not enter the clusters but ascend through the cortex in an independent system of fascicles (f), which seem to be formed so that the apical dendrites of the layer VI pyramidal cells can pass between the cell bodies of the layer V neurons. Most of the apical dendrites of the layer VI pyramidal cells form their terminal tufts in layer IV, so

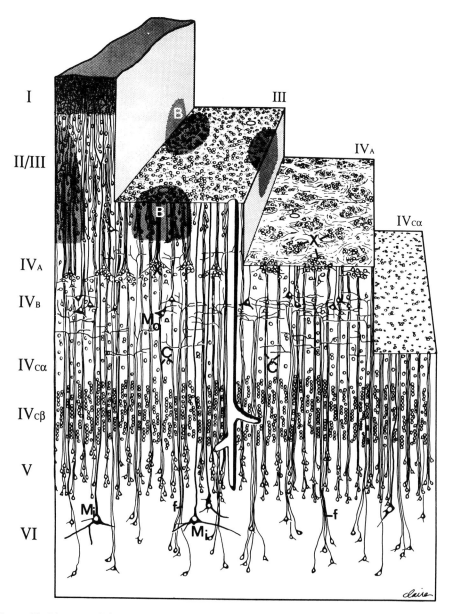

Figure 16. Diagram of the arrangement of neurons in striate cortex of the macaque. A further explanation is given in Section 7. From Peters and Sethares (1991a).

that the fascicles cannot be traced above layer IV, whereas the apical dendrites in the more regularly spaced clusters ascend to layer I to form their terminal tufts.

Given that the pyramidal cell modules are the fundamental neuronal units in the visual cortex, it is suggested that different combinations of them are activated by various inputs to produce the systems of functional columns, such as the ones concerned with eye preference, orientation, and color. Since these functional columns occupy the same cortical space and overlap each other, it follows that an individual pyramidal cell module is part of a column in each of these three functional systems.

In addition to the functional columns, a variety of neuronal units are superimposed on, and encompass groups of the pyramidal cell modules. In layer II/III for example, there is the system of cytochrome oxidase blobs (B), and below them in layer IVA there are other groups of pyramidal cells whose cell bodies aggregate into conelike groupings (X) (see also Figs. 14 and 15). These pyramidal cell cones occupy the spaces in the cytochrome oxidase-positive lattice that is evident in layer IVA, and as with the neurons in the cytochrome oxidase blobs in layer II/III, it is likely that the neurons in the cones are color-responsive.

The inner and outer Meynert cells form two other neuronal systems and their distribution appears to be related to the neurons in the cones and in the blobs. Thus, the inner Meynert cells (M_i) are distributed such that their cell bodies lie beneath the spaces between the blobs in layer II/III, while the outer Meynert cells (M_o) in layer IVB lie beneath the spaces between the pyramidal cell cones (X) in layer IVA, and beneath their edges. It is suggested that through these arrangements neurons responding to color and neurons responding to motion are geometrically separated from each other.

What is becoming apparent, then, is that the neurons in primate visual cortex are not merely arranged horizontally into layers, as the Nissl stains indicate; there are also vertically oriented modules of neurons extending through the depth of the cortex, and within some layers there are highly organized aggregates of neurons with different functions.

From what is presently known, it would seem that the organization of the neurons into functional units is much more complex in primate visual cortex than in any other cortical area. However, at the same time, it must be admitted that no other cortical area, with the possible exception of the barrel field in the rodent cortex, has been subject to such an extensive scrutiny as area 17 in the primate.

ACKNOWLEDGMENTS. The author thanks Claire Sethares and Karen Josephson for their skilled assistance in the preparation of the materials and photographs for this chapter, and Dr. Ken Kosik for the generous gift of the MAP2 antibody used for much of the study. Our analysis of the macaque visual cortex was supported by a Javits Award (NS 07016) from the National Institutes of Neurological Disorders and Stroke of the National Institutes of Health.

8. References

Bernhardt, R., and Matus, A., 1984, Light and electron microscopic studies of the distribution of microtubule-associated protein 2 in rat brain. A difference between dendritic and axonal cytoskeletons, *J. Comp. Neurol.* **226:**207–222.

Billings-Gagliardi, S., Chan-Palay, V., and Palay, S. L., 1974, A review of lamination in area 17 of the visual cortex of *Macaca mulatta*, *J. Neurocytol.* **3:**619–629.

Blasdel, G. G., 1992, Differential imaging of ocular dominance and orientation selectivity in monkey striate cortex, *J. Neurosci.* **12:**3115–3138.

Blasdel, G. G., and Lund, J. S., 1983, Termination of afferent axons in macaque striate cortex, *J. Neurosci.* **3:**1389–1413.

Brodmann, K., 1905, Beitrage zur histologischen Lokalisation der Grosshirnrinde. IIIte Mittelung: Die Rinderfelder der niederen Affen, *J. Psychol. Neurol.* **4:**177–226.

Brodmann, K., 1909, *Vergleichende Lokalisationlehre der Grosshirnrinde in ihren Prinzipien dargestellt auf Grund des Zellenbaues*, J. A. Barth, Leipzig.

Celio, M. R., 1986, Parvalbumin in most gamma-aminobutyric acid containing neurons in the rat cerebral cortex, *Science* **231:**995–997.

Chan-Palay, V., Palay, S. L., and Billings-Gagliardi, S., 1974, Meynert cells in the primate visual cortex, *J. Neurocytol.* **3:**631–658.

Clark, W. E. LeGros, 1942, The cells of Meynert in the visual cortex of the monkey, *J. Anat.* **76:**369–376.

Colonnier, M., and Sas, E., 1979, An anterograde degeneration study of the tangential spread of axons in cortical areas 17 and 18 of the squirrel monkey (*Saimiri sciureus*), *J. Comp. Neurol.* **179:**245–262.

Daniel, P. M., and Whitteridge, D., 1961, The representation of the visual field on the cerebral cortex in monkeys, *J. Physiol. (London)* **159:**203–221.

de Camilli, P., Miller, P. E., Navone, F., Theurkauf, W. E., and Vallee, R. B., 1984, Distribution of microtubule-associated protein 2 in the nervous system of the rat studied by immunofluorescence, *Neuroscience* **11:**819–846.

De Felipe, J., Hendry, S. H. C., Hashikawa, T., Molinari, M., and Jones, E. G., 1990, A microcolumnar structure of monkey cerebral cortex revealed by immunocytochemical studies of double bouquet cell axons, *Neuroscience* **37:**655–673.

de Monasterio, F. M., 1978, Properties of concentrically-organized X and Y ganglion cells of the retina of macaques, *J. Neurophysiol.* **41:**1394–1417.

de Monasterio, F. M. 1979 Asymmetry of on and off-pathways of blue-sensitive cones of the retina of macaques, *Brain Res.* **19:**441–449.

de Monasterio, F. M., and Gouras, P., 1975, Functional properties of ganglion cells of the rhesus monkey retina, *J. Physiol. (London)* **251:**167–196.

Diamond, I. T., Conley, M., Itoh, K., and Fitzpatrick, D., 1985, Laminar organization of geniculocortical projections in *Galago senegalensis* and *Aotus trivirgatus*, *J. Comp. Neurol.* **242:**584–610.

Feldman, M. L., and Peters, A., 1974, A study of barrels and pyramidal dendritic clusters in the cerebral cortex, *Brain Res.* **77:**55–76.

Fitzpatrick, D., Itoh, K., and Diamond, I. T., 1983, The laminar organization of the lateral geniculate body and the striate cortex in the squirrel monkey (*Saimiri sciureus*), *J. Neurosci.* **3:**673–702.

Fitzpatrick, D., Lund, J. S., and Blasdel, G. G., 1985, Intrinsic connections of macaque striate cortex. Afferent and efferent connections of lamina 4C, *J. Neurosci.* **5:**3329–3349.

Fitzpatrick, D., Lund, J. S., Schmechel, D. E., and Towles, A. C., 1987, Distribution of GABAergic neurons and axon terminals in the macaque striate cortex, *J. Comp. Neurol.* **264:**73–91.

Fleischhauer, K., 1974, On different patterns of dendritic bundling in the cerebral cortex of the cat, *Z. Anat. Entwicklungsgesch.* **143:**115–126.

Florence, S. L., and Casagrande, V. A., 1987, Organization of individual efferent axons in layer IV of striate cortex in a primate, *J. Neurosci.* **7:**3850–3868.

Fries, W., 1986, Distribution of Meynert cells in primate striate cortex, *Naturwissenschaften* **73:**557–558.

Fries, W., and Distel, H., 1983, Large layer VI neurons of monkey striate cortex (Meynert cells) project to the superior colliculus, *Proc. R. Soc. London Ser B.* **219:**53–59.

Fries, W., Keizer, K., and Kuypers, H. G. J. M., 1985, Large layer VI cells in macaque striate cortex (Meynert cells) project to both superior colliculus and prestriate visual area V5, *Exp. Brain Res.* **58:**613–616.

Hässler, R., 1967, Comparative anatomy of central visual systems in day- and night-active primates, in: *Evolution of the Forebrain* (R. Hässler and H. Stephen, eds.), Plenum Press, New York, pp. 419–434.

Hendrickson, A. E., Hunt, S. P., and Wu, J.-Y., 1981, Immunocytochemical localization of glutamic acid decarboxylase in monkey striate cortex, *Nature* **292:**605–607.

Hendry, S. H. C., Schwark, H. D., Jones, E. G., and Yan, J., 1987, Numbers and proportions of GABA-immunoreactive neurons in different areas of monkey cerebral cortex, *J. Neurosci.* **1:**1503–1519.

Hendry, S. H. C., Jones, E. G., Emson, P. C., Lawson, D. E. M., Heizmann, C. W., and Streit, P., 1989, Two classes of cortical GABA neurons defined by differential calcium binding protein immunoreactivity, *Exp. Brain Res.* **76:**467–472.

Hendry, S. H. C., Fuchs, J., De Blas, A. L., and Jones, E. G., 1990, Distribution and plasticity of immunochemically localized GABA$_A$ receptors in adult monkey visual cortex, *J. Neurosci.* **10:**2438–2450.

Hockfield, S., McKay, R. D., Hendry, S. H. C., and Jones, E. G., 1983, A surface antigen that identifies ocular dominance columns in the visual cortex and laminar features of the lateral geniculate nucleus, *Cold Spring Harbor Symp. Quant. Biol.* **48:**877–889.

Horton, J. C., 1984, Cytochrome oxidase patches: A new cytoarchitectural feature of monkey visual cortex, *Philos. Trans. R. Soc. London Ser. B.* **304:**199–253.

Horton, J. C., and Hubel, D. H., 1981, Regular patchy distribution of cytochrome oxidase staining in primary visual cortex of macaque monkey, *Nature* **292:**762–764.

Hubel, D. H., and Wiesel, T. N., 1968, Receptive fields and functional architecture of monkey striate cortex, *J. Physiol. (London)* **195:**215–243.

Hubel, D. H., and Wiesel, T. N., 1972, Laminar and columnar distribution of geniculo-cortical fibers in the macaque monkey, *J. Comp. Neurol.* **146:**421–450.

Hubel, D. H., and Wiesel, T. N., 1974, Sequence regularity and geometry of orientation columns in the monkey striate cortex, *J. Comp. Neurol.* **158:**267–294.

Hubel, D. H., and Wiesel, T. N., 1977, Functional architecture of macaque monkey visual cortex, *Proc. R. Soc. London Ser. B* **198:**1–59.

Hubel, D. H., Wiesel, T. N., and Stryker, M. P., 1978, Anatomical demonstration of orientation columns in macaque monkey, *J. Comp. Neurol.* **177:**361–380.

Kosik, K. C., Duffy, L. K., Dowling, M. M., Abraham, C., McClusky, A., and Selkoe, D. J., 1984, Monoclonal antibody to microtubule-associated protein 2 (MAP2) labels Alzheimer neurofibrillary tangles, *Proc. Natl. Acad. Sci. USA* **81:**7941–7945.

Lachica, E. A., and Casagrande, V. A., 1992, Direct W-like geniculate projections to the cytochrome oxidase (CO) blobs in primate visual cortex: Axon morphology, *J. Comp. Neurol.* **319:**141–158.

Lennie, P., Kranskopf, J., and Sclar, G., 1990, Chromatic mechanisms in striate cortex of macaque, *J. Neurosci.* **10:**649–669.

LeVay, S., Hubel, D. H., and Wiesel, T. N., 1975, The pattern of ocular dominance columns in macaque visual cortex revealed by a reduced silver stain, *J. Comp. Neurol.* **159:**559–576.

LeVay, S., Connolly, M., Houde, J., and Van Essen, D. C., 1985, The complete pattern of ocular dominance stripes in the striate cortex and visual field of the macaque monkey, *J. Neurosci.* **5:**486–501.

Livingstone, M. S., and Hubel, D. H., 1982, Thalamic input to cytochrome oxidase-rich regions in monkey visual cortex, *Proc. Natl. Acad. Sci. USA* **79:**6098–6101.

Livingstone, M. S., and Hubel, D. H., 1984, Anatomy and physiology of a color system in the primate visual cortex, *J. Neurosci.* **4:**309–356.

Lund, J. S., 1973, Organization of neurons in the visual cortex, area 17, of the monkey (*Macaca mulatta*), *J. Comp. Neurol.* **147:**455–496.

Lund, J. S., 1984, Spiny stellate cells, in: *Cerebral Cortex,* Volume 1 (A. Peters and E. G. Jones, eds.), Plenum Press, New York, pp. 255–308.

Lund, J. S., and Boothe, R. G., 1975, Interlaminar connections and pyramidal neuron organisation in the visual cortex, area 17, of the macaque monkey, *J. Comp. Neurol.* **159:**305–334.

McKay, R., and Hockfield, S., 1982, Monoclonal antibodies distinguish antigenetically discrete neuronal types in the vertebrate CNS, *Proc. Natl. Acad. Sci. USA* **79:**6747–6751.

Meynert, J., 1872, in: *Sticker's Handbuch der Gewebelehre,* Volume II.

Michael, C. R., 1981, Columnar organization of color cells in monkey's striate cortex, *J. Neurophysiol.* **46:**587–604.

Michael, C. R., 1985, Serial processing of color in the monkey's striate cortex, in: *Models of the Visual Cortex* (D. Rose and V. G. Dobson, eds.), Wiley, New York, pp. 301–309.

Michael, C. R., 1988, Retinal afferent arborization patterns, dendritic field orientations, and the segregation of function in the lateral geniculate nucleus of the monkey, *Proc. Natl. Acad. Sci. USA* **85:**4914–4918.

O'Kusky, J., and Colonnier, M., 1982, A laminar analysis of the number of neurons, glia and synapses in the visual cortex (area 17) of adult macaque monkeys, *J. Comp. Neurol.* **210:**278–290.

Payne, B. R., and Peters, A., 1989, Cytochrome oxidase patches and Meynert cells in monkey visual cortex, *Neuroscience* **28:**353–363.

Perry, V. H., and Cowey, A., 1984, Retinal ganglion cells that project to the superior colliculus and pretectum in the macaque monkey, *Neuroscience* **12:**1125–1127.

Peters, A., 1987, Number of neurons and synapses in primary visual cortex, in: *Cerebral Cortex,* Volume 6 (E. G. Jones and A. Peters, eds.), Plenum Press, New York, pp. 267–294.

Peters, A., and Jones, E. G., 1984, Classification of cortical neurons, in: *Cerebral Cortex,* Volume 1 (A. Peters and E. G. Jones, eds.), Plenum Press, New York, pp. 107–122.

Peters, A., and Sethares, C., 1991a, Organization of pyramidal neurons in area 17 of monkey visual cortex, *J. Comp. Neurol.* **306:**1–23.

Peters, A., and Sethares, C., 1991b, Layer IVA of rhesus monkey primary visual cortex, *Cereb. Cortex* **1:**445–462.

Peters, A., and Sethares, C., 1993, Aging and the Meynert cells in rhesus monkey primary visual cortex, *Anat. Rec.* **236:**721–729.

Peters, A., and Walsh, M. T., 1972, A study of the organization of apical dendrites in the somatic sensory cortex of the rat, *J. Comp. Neurol.* **144:**253–268.

Polyak, S., 1957, *The Vertebrate Visual System,* University of Chicago Press, Chicago.

Rakic, P., Bourgeois, J.-P., Eckenhoff, M. F., Zecevic, N., and Goldman-Rakic, P. S., 1986, Concurrent overproduction of synapses in diverse regions of the primate cerebral cortex, *Science* **232:**232–235.

Ramón y Cajal, S., 1899, Estudios sobre la corteza cerebral humana I: Corteza visual, *Rev. Trim. Micrograf.* **4:**1–63.

Saint-Marie, R. L., and Peters, A., 1985, The morphology and synaptic connections of spiny stellate neurons in monkey visual cortex (area 17): A Golgi–electron microscopic study, *J. Comp. Neurol.* **233:**213–235.

Schein, S. J., and de Monasterio, F. M., 1987, Mapping of retinal and geniculate neurons onto striate cortex of macaque, *J. Neurosci.* **1:**996–1009.

Schmolke, C., 1989, The ontogeny of dendritic bundles in rabbit visual cortex, *Anat. Embryol.* **180:**371–381.

Shipp, S., and Zeki, S., 1989, The organization of connections between areas V5 and V1 in macaque monkey visual cortex, *Eur. J. Neurosci.* **1:**309–332.

Tigges, M., Tigges, J., and Sporborg, C. D., 1981, Will the real Meynert cell please stand up? *Soc. Neurosci. Abstr.* **7:**831.

Tootell, R. B. H., Silverman, M. S., Switkes, E., and De Valois, R. L., 1982, Deoxyglucose analysis of retinotopic organization in primary striate cortex, *Science* **218:**902–904.

Tootell, R. B. H., Hamilton, S. L., Silverman, M. S., and Switkes, E., 1988a, Functional anatomy of macaque striate cortex. I. Ocular dominance, binocular interactions, and baseline conditions, *J. Neurosci.* **8:**1500–1530.

Tootell, R. B. H., Silverman, M. S., Hamilton, S. L., De Valois, R. L., and Switkes, E., 1988b, Functional anatomy of macaque striate cortex. III. Color, *J. Neurosci.* **8:**1569–1593.

Ts'o, D. Y., and Gilbert, C. D., 1988, The organization of chromatic and spatial interactions in the primate striate cortex, *J. Neurosci.* **8:**1712–1727.

Valverde, F., 1985, The organizing principles of the primary visual cortex in the monkey, in: *Cerebral Cortex,* Volume 3 (A. Peters and E. G. Jones, eds.), Plenum Press, New York, pp. 207–257.

Van Brederode, J. F. M., Mulligan, K. A., and Hendrickson, A. E., 1990, Calcium-binding proteins as markers for subpopulations of GABAergic neurons in monkey striate cortex, *J. Comp. Neurol.* **298:**1–22.

Van Essen, D. C., Newsome, W. T., and Maunsell, J. H. R., 1984, The visual field representation in striate cortex of the macaque monkey: Asymmetries, anisotropies and individual variability, *Vision Res.* **24:**429–448.

Vautin, R. G., and Dow, B. M., 1985, Color cell groups in foveal striate cortex of the behaving macaque, *J. Neurophysiol.* **54:**273–292.

Vincent, S. L., Peters, A., and Tigges, J., 1989, Effects of aging on the neurons within area 17 of rhesus monkey cerebral cortex, *Anat. Rec.* **223:**329–341.

von Bonin, G., 1942, The striate area of primates, *J. Comp. Neurol.* **77:**405–429.

von Bonin, G., and Bailey, P., 1947, *The Neocortex of Macaca Mulatta,* University of Illinois Press, Urbana.

Werner, L., Winkelmann, E., Koglin, A., Neser, J., and Rodewohl, H., 1989, A Golgi deimpregnation study of neurons in the rhesus monkey visual cortex (areas 17 and 18), *Anat. Embryol.* **180:**583–597.

Wiesel, T. N., Hubel, D. H., and Lam, D. U. K., 1974, Autoradiographic demonstration of outer dominance columns in monkey striate cortex by means of transneuronal transport. *Brain Res.* **74:**273–279.

Williams, R. W., and Rakic, P., 1988, Elimination of neurons from the rhesus monkey's lateral geniculate nucleus during development, *J. Comp. Neurol.* **272:**424–486.

Winfield, D. A., Rivera-Dominguez, M., and Powell, T. P. S., 1981, The number and distribution of Meynert cells in area 17 of the macaque monkey, *Proc. R. Soc. London Ser. B* **213:**27–40.

2

Substrates for Interlaminar Connections in Area V1 of Macaque Monkey Cerebral Cortex

JENNIFER S. LUND, TAKASHI YOSHIOKA, and JONATHAN B. LEVITT

1. Introduction

On the basis of our earlier studies of macaque visual cortical area V1 (Lund, 1973; Blasdel *et al.*, 1985; Fitzpatrick *et al.*, 1985) using Golgi impregnations and small intralaminar injections of horseradish peroxidase (HRP), a schema (reproduced here in Fig. 1) was outlined for spiny stellate neuron relays out of the thalamic recipient divisions of layer 4C (Lund, 1990). This diagram illustrates the finding that although the thalamic axons from magnocellular and parvocellular divisions of the lateral geniculate nucleus (LGN) terminate in the α and β divisions respectively of layer 4C (Hubel and Wiesel, 1972; Blasdel and Lund, 1983), the relays out of layer 4C seemed to fall into three sets. The lowermost set seemed to project in a very narrowly focused fashion to layer 4A and lower layer 3B, a set in the middle depth of layer 4C seemed to project to

JENNIFER S. LUND and JONATHAN B. LEVITT • Department of Visual Science, Institute of Ophthalmology, University of London, London EC1V 9EL, England. TAKASHI YOSHIOKA • Krieger Mind/Brain Institute, Johns Hopkins University, Baltimore, Maryland 21218.
Cerebral Cortex, Volume 10, edited by Alan Peters and Kathleen S. Rockland. Plenum Press, New York, 1994.

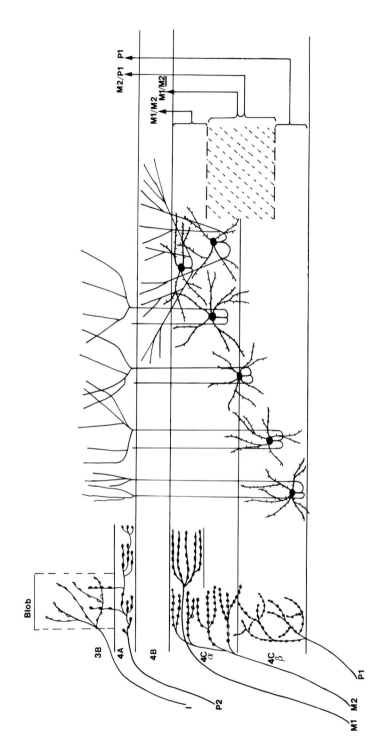

Figure 1. Diagram summarizing current information concerning the laminar distribution of thalamic inputs (to the left of diagram) from the lateral geniculate nucleus (LGN) to cortical area V1 in the macaque, and their further relays by thalamic recipient spiny stellate neurons of layer 4C. Thalamic axons P1 (to layer 4Cβ) and P2 (to layer 4A) appear to arise from different thalamic neuron populations, both situated in the LGN parvocellular laminae (Fitzpatrick *et al.*, 1983; Blasdel and Lund, 1983). Thalamic axons M1 (to upper 4Cα) and M2 (terminals throughout 4Cα) arise from the magnocellular laminae of the LGN; axon population I, relaying to the upper layer blobs, comes from the intercalated layers of the LGN (Hendrickson *et al.*, 1978; Fitzpatrick *et al.*, 1983). The output from layer 4C is suggested to be a gradient, created by dendritic sampling of both parvocellular and magnocellular inputs by the postsynaptic spiny stellate neurons; 4C neurons are suggested to shift their projection target from layers 4A–3B to layer 4B as their primary LGN input shifts from the parvocellular layers to the magnocellular layers. (See Lund, 1990, for further discussion.) Diagram modified from Lund (1990) with permission.

both layers 3B and 4B with more spreading axon arbors, and an upper set seemed to project principally to layer 4B. It was suggested that this distribution of projections could derive from a gradient of sampling of the magnocellular and parvocellular afferents by continuously overlapped dendritic fields of spiny stellate neurons through the depth of layer 4C; the three efferent sets of cells in layer 4C were seen as partially overlapping in depth.

We have followed these projections in greater detail in new studies in our laboratory (Yoshioka *et al.*, 1994); in these studies and in this chapter we consider how the relays from layer 4C relate to regions receiving separate thalamic inputs—layer 4A and the cytochrome oxidase-rich "blob" zones of layers 2 and 3 (see Fig. 1). Layer 4A receives a relay (P2) from the parvocellular layers that is a separate population of axons from that (P1) innervating layer 4Cβ (Blasdel and Lund, 1983; Fitzpatrick *et al.*, 1983). The blob zones are believed to be innervated by axons of cells of the intercalated layers (between and ventral to the parvocellular and magnocellular LGN laminae: labeled "I" in Fig. 1— Hendrickson *et al.*, 1978; Livingstone and Hubel, 1982; Fitzpatrick *et al.*, 1983). We have used the tracer substance biocytin, an orthograde tracer more sensitive than HRP that is also transported retrogradely (Horikawa and Armstrong, 1988; King *et al.*, 1989; Lachica *et al.*, 1991), to add to (and to some extent modify) the observations of Lachica *et al.* (1992) on channeling of projections from layer 4C to particular compartments in the more superficial layers. In addition, we briefly describe the local interlaminar connections—both excitatory and inhibitory—of neurons in the other laminae of area V1. The importance of understanding these relays lies in the information they may yield as to how different qualities of the visual image are analyzed in parallel or serial fashion, and how different functions are parcellated laterally across the cortex to different subsets of efferent projection neurons.

2. Relays of Spine-Bearing Neurons

2.1. Projections of Spiny Stellate Neurons of Layers 4C and 4A

2.1.1. Layer 4C Projections to Layer 4A, and Layer 4A Relays to Layer 3B Blob Zones

We have traced the outflow from 4C spiny stellate neurons in biocytin transport studies. Figures 2 and 3 illustrate representative injection sites on which we base our analyses of interlaminar connectivity. Figure 2 shows an injection into layer 4Cβ, in both bright (A)- and dark (B)-field illumination. The prominence and laminar specificity of interlaminar connections are apparent, as are the distinct laminar patterns of cell and fiber labeling. Figure 3 also shows a dense columnar focus of projections from an injection site in layer 3B. In addition, this section illustrates the prominence of horizontal connectivity in the superficial layers. Biocytin labeling was carried out in conjunction with staining for cytochrome oxidase so that relays could be examined in relation to blob or interblob territories in the superficial layers. Recent results confirm our earlier suggestions from HRP studies (Fitzpatrick *et al.*, 1985; Blasdel *et al.*, 1985; Lund, 1990)

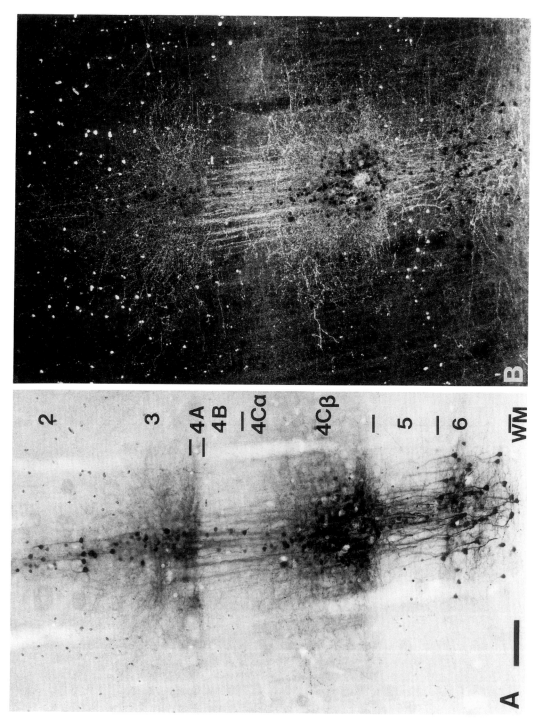

Figure 2. Photomicrograph of a biocytin injection site in layer 4Cβ, seen in both bright-field (A) and dark-field (B) illumination. This section was also reacted for cytochrome oxidase, and the injection site was underneath a blob/interblob border zone. Labeled fibers can be seen emanating from the injection site in a dense columnar focus. Orthogradely labeled terminals are observed within layer 4Cβ, as well as in layers 3, 4A, and 6, while retrogradely labeled cells are found primarily in layers 2–3 and 6. Bar = 100 μm.

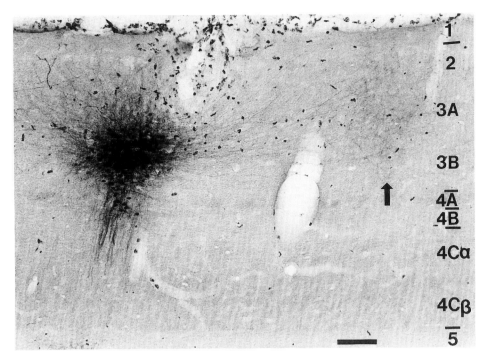

Figure 3. Photomicrograph of a biocytin injection site in a layer 3 blob. This section illustrates the prominence of horizontal connectivity in the supragranular layers. The arrow points to a cluster of terminals, which were also located in a blob zone (again determined by cytochrome oxidase staining of the same or adjacent sections). Bar = 100 μm.

that layer 4C seems to be made up of a number of neuron populations; they appear to be distributed in overlapping fashion through the depth of layer 4. One neuron set, which extends from the base of 4Cβ to mid-4Cα (see Fig. 4A), projects principally to layer 4A; information is then relayed by layer 4A spiny stellate neurons (summarized in Fig. 5) to cytochrome oxidase-rich blob zones in layers 3B–1. We note that some blobs may not receive this layer 4A input. Some layer 3B injections in blobs fail to retrogradely label neurons in any division of layer 4, even though layer 5 cells are well labeled (see Fig. 6B). The projection from the lower half of 4Cβ to layer 4A seems very narrowly focused, and rises vertically to layer 4A from under both blob and interblob territories. However, the projection from lower layer 4Cα provides axons to layer 4A that turn and run laterally within the narrow cytochrome oxidase-rich "honeycomb" of layer 4A, establishing small clumps of terminals at intervals along the layer. Small tracer injections within 4A also label laterally spreading fibers within the layer, with fibers turning up into blob zones, or sometimes making a more direct trajectory to the blobs as oblique rising fibers from 4A (Fig. 5). The degree of lateral axon spread in layer 4A does not seem to relate to the cells of origin being under blob or interblob territories. We believe that most projections from this population of spiny stellate neurons of 4Cβ plus lower 4Cα must relay to the

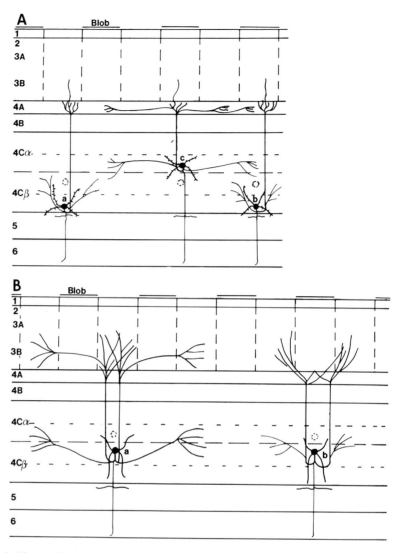

Figure 4. Diagram illustrating our conclusions concerning orthograde projections from 4Cβ and lower 4Cα spiny stellate neurons. These conclusions are based on our ongoing studies making small iontophoretic injections of biocytin in individual layers of area V1, and tracing the resulting orthograde and retrograde label (see Figs. 2 and 3 for representative injection sites). A fuller account of this work will be published soon (Yoshioka *et al.*, 1994). The biocytin-reacted sections (or adjacent unstained tissue sections) were also reacted for cytochrome oxidase (Lachica *et al.*, 1991) to reveal the position of transported biocytin relative to the blob and interblob territories in the superficial layers. Dotted circles above or below each neuron indicate the range of depths at which the type of neuron drawn can also occur.

(A) A population of neurons in 4Cβ and 4Cα projects predominantly to layer 4A. Spreading projections to 4A (neuron c) arise from cells in mid-4C, while more narrowly focused projections to 4A arise from neurons under either interblob (cell a) or blob (cell b) territories in the superficial layers.

(B) A population of spiny stellate neurons of mid-4C projects to interblob zones of layer 3B. If these cells lie under a blob (cell b), their projections spread to the sides of the blob in 3B, avoiding its territory. The neurons under interblobs (cell a), as well as those beneath blobs, make lateral connections within mid-4C as well as making laterally spreading connections at the level of layer 3B. The terminal zones in layers 4C and 3B are vertically aligned with one another. From Yoshioka *et al.* (1994) with permission.

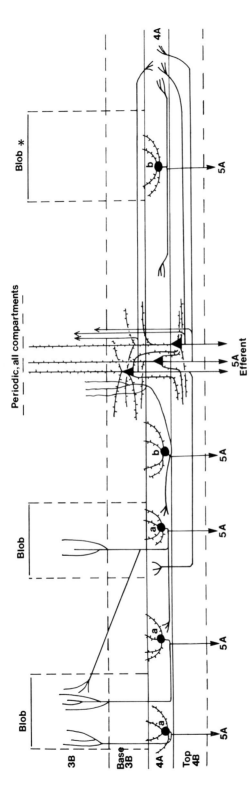

Figure 5. Diagram of projection pathways of neurons associated with layer 4A from the same biocytin-labeled material as described in Fig. 4. Spiny stellate neurons of layer 4A (cells a, b) send out long laterally spreading axon collaterals that can terminate in layer 4A, or turn up into the blob zones of 3B; some blob zones (asterisk) may not receive these layer 4A inputs (cell b). The layer 4A spiny stellate neurons may also contribute to the superficial layers at positions where clusters of pyramidal neurons bracketing layer 4A (illustrated at the center of the figure) project to the superficial layers and send out long lateral axon collaterals terminating within layer 4A. These pyramidal neurons surrounding layer 4A label intermittently after layer 3A–3B injections placed in either blob or interblob zones; they may also provide efferent projections to area V2 cytochrome oxidase-rich "thick" (Cat-301 positive) stripes since this group of layer 4A neurons are retrogradely labeled following injections of biocytin in these V2 territories (Levitt *et al.*, 1994). From Yoshioka *et al.* (1994) with permission.

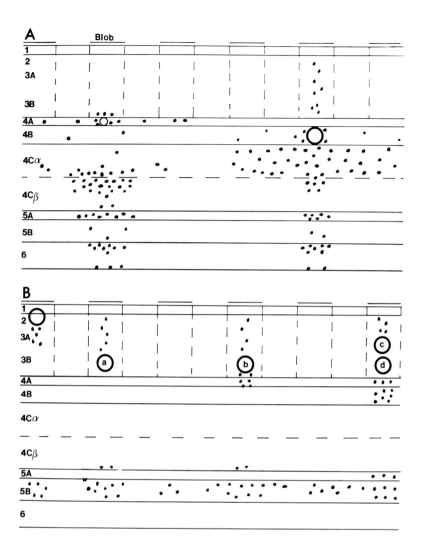

Figure 6. This figure illustrates the patterns of retrograde cell label (black dots) typical of biocytin injections (open rings) into layers 2–3, 4A, and 4B, in or under blob zones.

(A) Injections in layer 4A immediately under a blob zone give retrogradely labeled cells laterally spread in layer 4A and at the base of layer 3B immediately adjacent to the injection. Most retro-gradely labeled cells occur in mid-layer 4C with a few scattered in layers 4B and 4Cα, as well as in lower layer 4Cβ immediately beneath the injection. Labeled cells are also laterally spread in layer 5A, very few are labeled in 5B, and a cluster of cells are well labeled in upper layer 6. Sometimes a few cells appear at the layer 6–white matter boundary. Injections in layer 4B under a blob label a column of cells above the injection in layers 2–3, a laterally spread scatter of cells in layer 4B, and wide-spread, well-labeled neurons throughout the depth of 4Cα. A patch of cells can label in upper layer 4Cβ immediately under the injection site; cells are also labeled in layer 5A, occasionally in 5B, and upper layer 6 cells are well labeled. We do not believe all the retrogradely labeled cells necessarily have terminal axon fields in the injection site; on the basis of orthograde biocytin label we suspect that cells in upper 6 and 5A may be projecting to layer 4A and the injection has caught their axon arborizations just before they enter layer 4A.

(B) Injections in layers 1 and 2 in blob zones rarely give retrograde label in any layer but 5B. Injections in layer 3B in blob zones can give several patterns of retrograde label: one pattern is shown for injection (a). Here, retrogradely labeled neurons occur in 5B and in 2/3 immediately above the injection, with occasional cells at the base of layer 4Cβ. A second pattern (b) adds labeled cells in layer 4A and at the top of layer 4B to those seen in (a). A third pattern, seen after injections at variable blobs indirectly via layer 4A neurons, since injections of tracer into the layer 3B blob zones in many cases label layer 4A neurons, but cells in 4C are very rarely retrogradely labeled. Very occasionally, a few cells in lower layer 4Cβ are labeled under 3B blob injection sites, as if these cells can make a weak direct projection to the blobs only if they lie directly underneath them (Fig. 6B, injections a and b).

2.1.2. Layer 4C Projections to Layer 3B Interblob Zones

A second population of 4C neurons (Fig. 4B), lying principally in midlayer 4C, projects directly to layer 3B to terminate in interblob zones, with spreading terminal axon arbors. If these midlayer 4C cells lie directly under a blob, collaterals from the rising axon trunks split around the blob and rise up its sides. Inputs to the upper layer interblobs seem to arise primarily from these cells of midlayer 4C. However, following injections into layer 3B interblobs, we also find a few labeled cells lying directly under the injection site in upper 4Cα and lower 4Cβ (see Fig. 8B). The axon projections of the midlayer 4C neurons include lateral projections within midlayer 4C that establish punctate terminal zones offset from the cells of origin. These terminal zones in 4C are in vertical registration with axon collaterals in 3B from the same cells (Fig. 4B). The lateral projections of these axons within layer 3B are to neighboring interblob territories, bypassing intervening blobs in layer 3B.

2.1.3. Layer 4C Projections to Layer 4B

A third population of spiny neurons in layer 4C, largely restricted to layer 4Cα, projects upon layer 4B (Fig. 7A), often in periodic tufts, with sparser rising projections to layer 3B. The sparse outputs from layer 4Cα to layer 3B are principally to interblob regions (Fig. 7A, cell b), and so may arise from the mid-4C neurons described above since there is overlap of the populations in lower 4Cα, and our injections, even in upper 4Cα, may involve this overlapped set. Very light outputs to blobs are also occasionally visible following injections into layer 4Cα. While this is consistent with our retrograde results (following injections into layer 3B blobs, we find very occasional retrograde label of cells in layer 4C), there is also the possibility that our injections in upper 4Cα involve descending dendrites of layer 4B neurons, which project heavily upon the blobs. This contrasts with our injections into interblob zones in layer 3B, where we find many well-labeled cells in mid-4C as well as a few in lower 4Cβ (Fig. 8B). The projections of layer 4Cα spiny neurons are heaviest within layer 4Cα itself, with some degree of terminal periodicity in layer 4Cα (Fig. 7A). This periodicity within 4Cα appears to line up with either blob or interblob zones in the superficial layers, but this issue is still not fully resolved. Injections of biocytin in layer 4B can give widespread retrograde cell labeling in 4Cα; this labeling in some cases appears as cell clusters, in other cases as uniformly distributed cells (Fig. 6A, 8A).

Figure 6. (*Continued*) depths in layer 3 (c, d), adds cells through the depth of 4B to pattern (b). It is possible that these different patterns are simply caused by inefficient biocytin labeling; however, in all cases layer 5B cells are well labeled, which makes this possibility seem less likely. From Yoshioka *et al.* (1994) with permission.

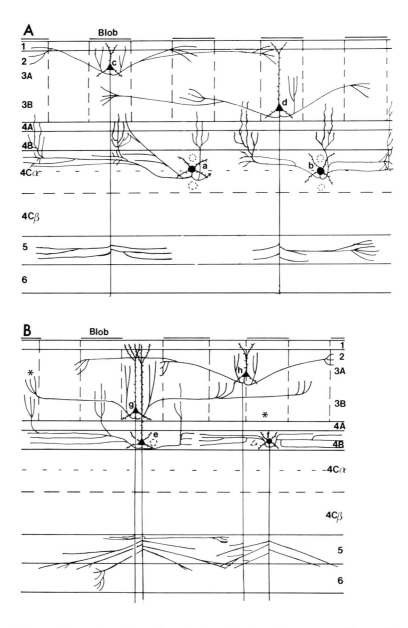

Figure 7. Diagram illustrating orthograde projections of spiny stellate neurons of layers 4Cα and 4B, and of pyramidal neurons in layer 4B and layers 2–3B. Dotted circles above and below each neuron indicate the range of depths at which the type of neuron drawn can also occur.

(A) 4Cα spiny stellate neurons project strongly laterally within layer 4Cα itself with periodic terminal fields. We are still unsure of the alignment between these terminal fields and blob or interblob territories in the superficial layers; this issue is still under investigation. Periodic terminal fields are also established in layer 4B, with weaker collaterals extending into layer 3B; 4Cα cells underneath interblobs (cell b) have stronger projections to layer 3B than do cells underneath blobs (cell a). Pyramidal neurons in the blob zones of the superficial layers 2–3 project mainly to other blob territories (cells c, d); their descending axons give off collaterals in layer 5B, some of which establish periodic terminal fields in layer 5 underlying the superficial blob zones.

(B) Spiny stellate neurons (f) and pyramidal neurons (e) of layer 4B make long laterally spreading connections within layer 4B with some evidence of periodic terminal zones. The layer 4B pyramidal neurons (e) make projections into blob zones in the superficial layers, but not all blobs

2.2. Projections of Spiny Neurons of Layer 4B

Orthograde projections from layer 4B (Fig. 7B) spread densely and for long distances within layer 4B itself; weaker rising projections spread to many but not all layer 3B blob zones. Some injections into layer 3B blobs give retrogradely labeled neurons in layer 4B as well as in layer 4A; other such injections either retrogradely label neurons in layer 4A alone, or give no retrograde label in either layer 4A or 4B. While Lachica *et al.* (1992) recently reported that there are *no* projections from layer 4B to the layer 3 interblobs (see their Fig. 2), we suspect that projections from layer 4B to the layer 3B interblob regions can occur, and arise from a special group of pyramidal neurons that lie high in layer 4B, and that are associated with layers 4A and 3B which are described in the next section.

2.3. Projections of Pyramidal Neurons Associated with Layer 4A

A group of pyramidal neurons, whose somata and proximal dendrites are closely associated with layer 4A, are intermittently labeled after injections into either blob or interblob compartments of layer 3B (see Fig. 5). Their somata can lie immediately above the narrow cytochrome oxidase-rich layer 4A in the base of 3B, or at the very top of 4B, as well as within 4A itself; their dendrites penetrate through and wrap along the upper and lower boundaries of the 4A zone. They send long horizontal axon collaterals along the borders of 4A that turn and terminate in small foci within 4A. These pyramids also provide rising axon collaterals to the superficial layers, and descending axon trunks that provide collaterals to layer 5. Layer 4A pyramids also provide extrinsic projections, and we have found that they can label retrogradely as a spatially distinct cell population from neurons deeper in layer 4B following biocytin injections into the cytochrome oxidase "thick" stripes of V2; however, this labeling is generally spatially coincident with retrogradely labeled neurons in layer 4B (Levitt *et al.*, 1994).

2.4. Projections of Pyramidal Neurons of Layers 2–3

Pyramidal cells within layers 2–3 (Fig. 7) project laterally for long distances with periodic terminations, predominantly in regions similar to those occupied by the cells of origin, i.e., blob to blob, or interblob to interblob, as observed earlier by Livingstone and Hubel (1984b). We find, however, in an extensive study of tangentially sectioned material (Yoshioka *et al.*, 1992, 1994) that these connections are also made regularly, but to a lesser degree, with the opposite "unlike" compartment; in the case of projections originating from blobs, which

---

Figure 7. (*Continued*) (right asterisk) are targets of layer 4B projections (see variable patterns of retrograde cell label in 4B after 3B blob injections illustrated in Fig. 6). Pyramidal neurons in interblob regions of layers 2–3 (cell g) project primarily to other interblob regions, but also (left asterisk) occasionally innervate blobs. Pyramidal neurons at blob edges (h) project mainly to other blob edges, but do also project into blobs or interblob zones (Yoshioka *et al.*, 1992, 1994). From Yoshioka *et al.* (1994) with permission.

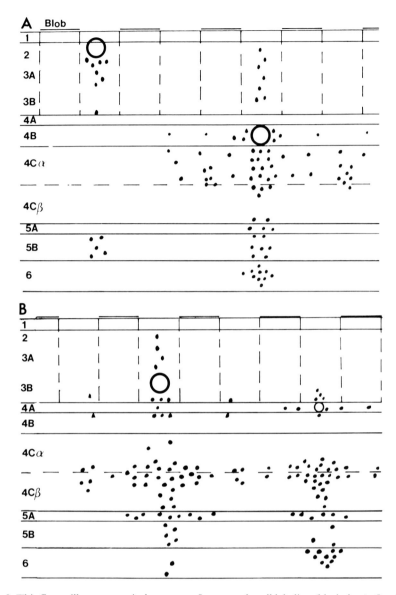

Figure 8. This figure illustrates typical patterns of retrograde cell labeling (black dots) after biocytin injections (open rings) in or under interblob territories.

(A) Injections in layers 3A and 2 do not produce retrograde label in any layers but 5B (and occasionally 5A). Injections in layer 4B label cells in all layers, but only occasionally in layers 4A and 4Cβ. Within 4B and 4Cα, labeled cells can be spread laterally.

(B) Injections in 3B always label a broad spread of cells in midlayer 4C, sometimes also in lower layer 4Cβ and upper 4Cα directly under the injection. Cells are labeled in layer 4A and upper 4B intermittently. A column of cells always labels directly above the injection; we believe this may be caused by interruption of their descending axon trunks. Cells in upper layer 4Cα and lower layer 4Cβ are frequently, but not always, labeled. A broad spread of cells is labeled in 5A, fewer and more narrowly focused cells are labeled in 5B, and rare cells are labeled in 6. After injections in layer 4A, labeled cells are spread laterally in layer 4A and on its immediate borders; a broad spread of labeled cells is found in mid-4C and in 5A, and a few narrowly focused cells label in 4Cβ and 5B. Upper layer 6 has a cluster of well-labeled cells. From Yoshioka *et al.* (1994) with permission.

obey their origins most strictly, approximately 70% of projections are to other blobs, the rest are to interblobs or to blob/interblob borders. During the course of our studies of interlaminar projections, we used the pattern of projections lateral to injection sites to cytochrome oxidase-rich and -poor zones in the superficial layers—as well as matching of the injection site within the same and adjacent sections stained for both cytochrome oxidase and biocytin to help determine the locus of the injection sites.

2.5. Relationship between Relays of Different Thalamic Inputs and Intrinsic Relays to Different Compartments in Area V1

Our conclusion, based on these studies of the interlaminar projections from divisions of layer 4 to the superficial layers (and summarized in Fig. 11), is that layer 4C cannot be viewed as having clearly parcellated territories with different projections to particular compartments in the overlying layers. Cells in 4Cβ, or cells in 4Cα that have dendrites entering layer 4Cβ and therefore access to afferents from the LGN parvocellular laminae, can provide two projection pathways: one to 4A (where their relays join another LGN axon population from the parvocellular laminae—Fitzpatrick *et al.*, 1985; Blasdel *et al.*, 1985), and a second passing directly to the interblob regions of layer 3B. We do not know if these two pathways are served by separate or identical cell populations; however, we suspect that they may be, at least in part, separate populations with perhaps some unevenness in distribution since injections into layer 4C can produce labeled axon projections favoring one or the other pathway (see summary diagram in Fig. 11). A third pathway projects heavily to layer 4B. It arises from cells in 4Cα which partially overlap (in lower 4Cα) the cells furnishing the projections to 4A and the 3B interblobs. The finding that layer 4Cα injections also label a weak projection to the layer 3B blob and interblob zones could reflect a genuine weak projection, or it could reflect uptake of label by the spatially overlapped dendritic processes of cell populations in lower 4Cα and 4B (whose dendrites enter 4Cα), cells which make strong contributions to the layer 3B interblob and blob zones, respectively. These possibilities cannot be distinguished without single cell filling. Signals from the LGN magnocellular laminae may therefore uniquely influence layer 4B via layer 4Cα, but also enter into the interblob territories of layers 2 and 3 via the projections of cells in mid-4C, which also receive parvocellular relays.

The layer 3B blob zones have direct LGN inputs from the intercalated layers, and probably also receive input from collaterals of parvocellular axons to layer 4A (Fitzpatrick *et al.*, 1983; Blasdel and Lund, 1983). They receive LGN magnocellular layer-driven relays, predominantly via layer 4B, and additional parvocellular relays via layer 4A. On the basis of the patterns of retrograde label we observe after blob injections, it seems likely that there are several sets of blobs influenced in different ways by these relays. One set of blobs seems primarily driven by the thalamus, a second set has an additional layer 4A input, and another set receives both layer 4A and layer 4B input in addition to thalamic relays. The interblob regions seem clearly dominated by mid-4C input, apparently sharing both magno- and parvocellular inputs, with horizontal links established by the same spiny stellate neuron axons both in mid-4C and in layer 3B. Weak inputs to the interblob regions from upper 4Cα and lower 4Cβ are not always visible by retrograde label, and may be periodic within the interblob territory.

In relation to the report of Lachica *et al.* (1992), we would emphasize from the findings of our own studies (Yoshioka *et al.*, 1992, 1994) that the blobs appear to receive their chief magnocellular input via layer 4B (rather than from 4C as described by Lachica *et al.*), and that their chief parvocellular input seems to derive from 4A rather than directly from 4C. We would also emphasize that we presume the mid-4C cells to receive both magnocellular and parvocellular inputs, and therefore relays from mid-4C cells to the interblob regions should reflect this combined input (rather than being a purely parvocellular input to interblobs as stressed by Lachica *et al.*)—see summary diagram of Fig. 11.

2.6. Intrinsic Relays of Pyramidal Neurons in Layers 5 and 6

2.6.1. Relays of Layer 6 Pyramidal Neurons

The relationship of pyramidal neurons of layers 5 and 6 to the overlying layers has previously been explored using Golgi and HRP tracing techniques (Lund, 1973; Lund and Boothe, 1975; Fitzpatrick *et al.*, 1985; Blasdel *et al.*, 1985). These earlier studies, together with our recent biocytin tracing experiments (Yoshioka *et al.*, 1992, 1994) and ongoing studies by Fitzpatrick (personal communication), have found layer 6 to have three subpopulations of pyramidal neurons, distributed with partial overlap through the depth of layer 6, that each project to different regions of layers 4C and 4A. Pyramidal neurons of deeper 6 project to upper layer 4Cα, neurons in the middle depth of layer 6 project to midlayer 4C, and neurons in the upper half of layer 6 project to layer 4Cβ and to layer 4A. This parcellation is seen most clearly by orthograde transport of biocytin label after small injections into layer 6 (Fitzpatrick, personal communication), but it is also evident in the changing positions of retrogradely labeled neurons in layer 6 following injections into layer 4A and different regions of layer 4C. The lateral spread of layer 6 pyramidal neuron axons in layer 4A and upper 4Cα is extensive. We suggested in our earlier HRP studies of area V1 (Lund *et al.*, 1975), on the basis of retrograde cell label in layer 6 after making injections of HRP into different divisions of the LGN, that upper layer 6 projected to the LGN parvocellular layers and lower layer 6 projected to the LGN magnocellular division; this has been confirmed with better labeling (using WGA–HRP) by Fitzpatrick in his ongoing work. Layer 6 therefore follows layer 4C in having a gradient of projections through its depth relating in inverted fashion to the intrinsic efferent outflow of 4C; underlying this gradient, layer 6 also shows the same segregated pattern of efferent projections to different divisions of the LGN and the LGN presents to layer 4C.

2.6.2. Relays of Layer 5 Pyramidal Neurons

Our earlier studies (Lund, 1973; Lund and Boothe, 1975; Fitzpatrick *et al.*, 1985; Blasdel *et al.*, 1985) have provided considerable evidence for layer 5 being divided into a superficial zone, 5A, and a broader, deeper division 5B. Layer 5A receives projections from spiny neurons in all divisions of 4C and 4A. The 5A pyramidal neurons project to layer 3B (but we suspect on the basis of retrograde label that the 5A projection to blob zones may be weak). Layer 5B projects chiefly to the most superficial layers 3A–1, and it receives projections from pyramidal neurons in both blob and interblob zones. Axon collaterals emerge from the descending trunks of the superficial pyramids and spread widely in layer 5. Within layer 5, they often make patchy terminal clusters directly under terminal clusters established in the superficial layers by collaterals from the same pyramidal cells. The descending axons of layer 4B spiny neurons also give off collaterals in layer 5B; these collaterals often descend obliquely in layer 5B, eventually entering layer 6 and traveling laterally at the layer 5–6 border.

3. Relays of Nonspiny, Local Circuit Neurons

Our studies on interneuron (largely GABAergic) circuitry within area V1 are still ongoing but we have so far described such circuitry for layers 3B–6 (Lund, 1987; Lund *et al.*, 1988; Lund and Yoshioka, 1991). Recent summaries of this work are shown in Figs. 9 and 10. In our interneuron studies, in many cases we have not found it easy to match these V1 neurons with descriptively named interneuron types described by other workers (see for instance the review of Fairén *et al.*, 1984). Some cells, such as "chandelier" cells (see Peters, 1984), are readily recognizable between species—or in different regions of a single species. In other cases, e.g., "basket" neurons (see Jones and Hendry, 1984), some features of the axon seem comparable between species or regions, but the spatial extent of single axons may differ, and the presence of readily recognized pericellular "basketlike" arrays of terminals is rare. For these reasons, we have given each local circuit neuron variety in V1 a number to identify it and its laminar location—rather than a name—but the text of our published descriptions points out similarities to "named" varieties if we find them.

Figure 9A summarizes our findings concerning interneuron relays projecting to layers 4A and 3B from divisions of layer 4C, and layer 6 cells (both pyramidal and nonpyramidal) with projections associated with these relays. It is clear that layer 4C local circuit neuron projections, presumably largely inhibitory in action, mirror the projections that we have just described for the excitatory spine-bearing cells (see the local circuit neurons to the right of Fig. 9A). Layer 5A local circuit neurons are also intimately involved with the thalamic recipient layers in that their axons innervate individual strata within layer 4C, and layers 4A and 6 (see Figs. 9A and 10A). It would be of great interest to know how the local circuit neuron projections to layers 2 and 3 relate to blob and interblob zones, but unfortunately our biocytin tracer rarely labels interneurons (and cytochrome oxidase immunocytochemistry interferes with good Golgi impregnation). It would seem likely, since spine-bearing neurons of 4C show few direct projections to the blobs, that most local circuit neuron relays out of layer 4Cβ and lower 4Cα would follow the projections of the spiny stellate neurons to layer 4A and the interblob regions of layer 3B (Fig. 9A: see projections of neurons β2 and α2 to layer 4A; and neurons β4,5 and α4 to layer 3B—these are perhaps restricted to interblob regions).

Figure 10A summarizes our findings in regard to spine-bearing neuron and local circuit neuron projections from upper 4Cα, both of which target princi-

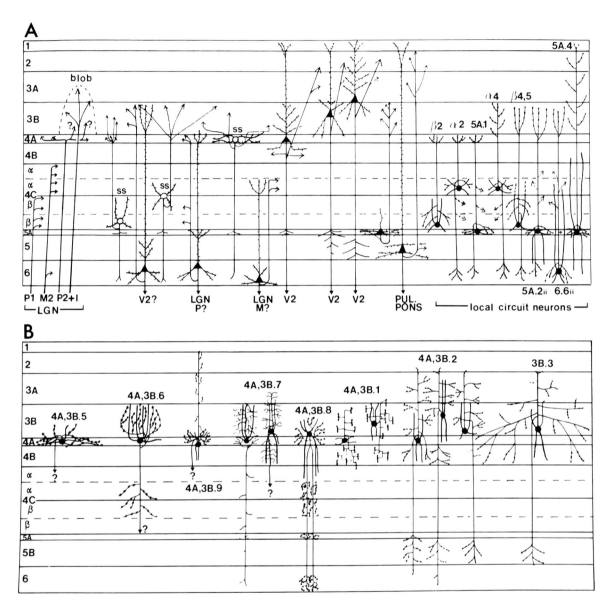

Figure 9. (A) Diagram summarizing the patterns of afferent projections from intrinsic and extrinsic sources to layers 3B, 4A, and 4C. Thalamic afferents are illustrated to the left. Spine-bearing neurons, both pyramidal and spiny stellate (SS), occupy the center and left-hand side of the diagram. Two main streams pass from middle and lower 4C, one to terminal arbors in 4A, the other to layer 3B. The local circuit neurons (shown to the right) have axon projections that follow one or the other of these two streams. Efferent axon destinations are indicated at the base of the diagram.

(B) Summary of the projection patterns of local circuit neurons found in layers 4A and 3B. Reprinted with permission from Lund and Yoshioka (1991).

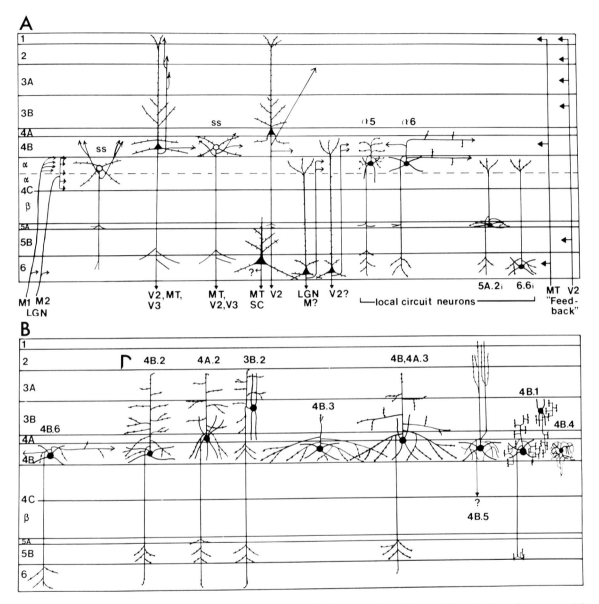

Figure 10. (A) Diagram summarizing the patterns of afferent projections from intrinsic and extrinsic sources to layer 4B. At the left are indicated thalamic inputs to 4Cα, the principal source of intrinsic afferents to layer 4B. Spine-bearing neurons, both pyramidal and spiny stellate (SS), occupy the left half of the diagram; the axon destinations of efferent projection neurons are indicated below the figure. Local circuit neurons are shown to the right, and laminar distributions of corticocortical projections from areas V2 and MT (V5) are shown at the right-hand margin.

(B) Diagram summarizing patterns of local circuit neuron projections associated with layer 4B, both from neurons with soma and dendrites within layer 4B and from neurons in layer 3B. Reprinted with permission from Lund and Yoshioka (1991).

pally layer 4B (see neurons α5 and α6 of Fig. 10A). Upper layer 4Cα receives input from layer 5A and layer 6 local circuit neurons (Fig. 10A neurons 5A.2i and 6.6i), as well as from pyramidal neurons of lower layer 6.

Figures 9B and 10B summarize our findings (Lund and Yoshioka, 1991) in regard to local circuit neurons of layers 4B, 4A, and 3B—a crucial region of interface between all relays leaving layer 4C and the thalamic inputs to layer 4A and blob zones of layer 3. At the moment, we can only guess at the possible locations of these neuron classes in relation to blob and interblob zones, and given the complexities of these layers, much more remains to be learned. However, it is of considerable interest that only two cell classes in the superficial layers—both local circuit neuron varieties—project back specifically into layer 4C (see cells 4A,3B.6 and 4A,3B.8 of Fig. 9B). They both target the mid-4C region by terminal collaterals emerging off their descending trunks and at least one (4A,3B.8) also shows prominent terminals in layers 5A and 6. It is the projection from this mid-4C region to the superficial layers that is most likely to combine properties of LGN magnocellular and parvocellular afferents, and most likely to target the interblob zones. The activity of these local circuit "feedback" relays to mid-4C could control the activity of the interblob regions by modifying the activity of the mid-4C neuropil.

There are clearly local circuit neurons that interconnect layers 4B and 3B, and which also provide relays to layer 5B (see neurons 4A,3B.2 and 3B.3 in Fig. 9B and neurons 4B.2, 4B.3, 4B,4A.3, and 4B.1 in Fig. 10B); however, it will require some considerable technical advances to allow us to place these inhibitory relays correctly in the overall circuitry, particularly in relation to the blob and interblob zones. The substrates exist for a variety of inhibitory relays between layers 4B and 3B. The spreading "basket"-type axon arbors of cells 3B.3, 4B.3, and 4B.4A3 (Figs. 9B, 10B) are of particular interest in that they reach laterally roughly three times the diameter of single blob (or interblob) regions. This would suggest that they have a role in providing spread of inhibition between "unlike" zones (e.g., from blob to surrounding interblob region, or from an interblob region to surrounding blobs). We have recently discussed possible roles of such inhibition (see Lund *et al.*, 1993) in relation to the constraints that might operate during development of tangential patchy connections of pyramidal neu-

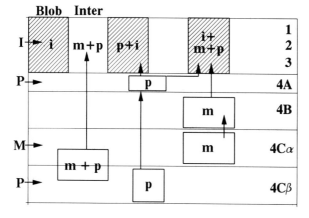

Figure 11. Diagram summarizing major routes of input from different divisions of the dorsal lateral geniculate nucleus (LGN) of the thalamus to cytochrome oxidase (CO)-rich blob regions and CO-poor interblobs in superficial layers of area V1. Capital letters indicate thalamic inputs; small letters indicate further relays of these signals-within cortex: I = intercalated LGN layers; P = parvocellular LGN layers; M = magnocellular LGN layers.

rons in the superficial layers. We suggest that the spread of such inhibition might force connections of pyramidal neurons colocalized and coactivated with basket neurons to occur outside a zone of local inhibition, thus making the pyramidal neuron connections "step" by a fixed distance to find coactivated neurons on which to terminate. A uniform presence of these spreading "basket" neuron axons would also force the "stepping" connections into discrete patches since each coactive patch of pyramidal neurons would have a similar sized zone of inhibition around it from its own local basket cell axons. These basket cells are most obvious in Golgi preparations in layers 4B and 3B, and their axons and dendrites seem to freely cross the 4A region; the basket axon arbor dimensions are such that single axons should enter into both blob and interblob territories— as well as crossing boundaries of any parcellation of territory which may exist in layer 4B (Shipp and Zeki, 1989; Lund and Yoshioka, 1991).

While we have yet to study the local circuit neurons of layers 3A–1 in any detail, this stratum is clearly different in character from layer 3B. It is not reached by most of the direct excitatory relays from 4C and 4A (Fitzpatrick *et al.*, 1983, 1985), and is likely to represent a region of unique properties. However, it is apparent that even the most superficial pyramidal neurons of layer 2 continue to make clearly focused sidestepping projections, mainly to nearby "matching" compartments, in the same manner as the deeper pyramids of layer 3B; the boundaries of the blob and interblob territories seem therefore to extend to the pia, even if thalamic relays to the blobs do not extend into these most superficial layers.

4. Functional Correlates of V1 Internal Circuitry

It is clear that many different types of visual information are processed in striate cortex. Prior work in area V1 has described neurons with responses selective for such stimulus attributes as orientation, size, color, direction of motion, or binocular disparity. Neurons with distinct functional properties seem segregated both tangentially in different columns (Hubel and Wiesel, 1977), and vertically in different laminae, even within layer 4 (Dow, 1974; Livingstone and Hubel, 1984a; Hawken and Parker, 1984; Blasdel and Fitzpatrick, 1984; Dow and Vautin, 1987; Hawken *et al.*, 1988). However, what remains ambiguous is the precise degree to which neurons having different response selectivities are segregated in the cortex, and the extent to which functional differences among different sets of neurons might reflect the dominance of signals from the magnocellular and parvocellular divisions of the LGN. The circuits we have described bear directly on these questions, as they serve as the substrates for different functions in area V1. In this section, we briefly discuss some of the possible functional consequences of this circuitry.

1. *Layers 2/3:* One feature of the superficial layers of striate cortex that has guided much research is the inhomogeneous staining pattern of cytochrome oxidase. Earlier anatomical and physiological studies, showing that the blobs and interblobs project to the thin cytochrome oxidase-rich and pale cytochrome oxidase-poor stripes of V2 (which then project to V4), and that color- and orientation selectivity are segregated in these zones, led to the suggestion that the upper layers of V1 are part of a parvocellular-dominated pathway through

visual cortex (Livingstone and Hubel, 1984a, 1988). More recently, Hubel and Livingstone (1990) have shown that many cells in layers 2/3 are more sensitive to contrast than are cells in layer 4Cβ, the parvocellular input zone; while this could indicate magnocellular input to the upper layers, the increase in sensitivity might rather be a consequence of neural convergence among the layer 2/3 cells receiving and relaying parvocellular input. However, earlier in this chapter we have described circuits by which magnocellular signals can reach the upper layers, either via mid-4C relays to interblob zones, or via 4Cα to 4B to the blobs. Recent physiological studies have demonstrated that this convergence of magnocellular and parvocellular signals can be detected in V1; Nealey *et al.* (1991) have shown that reversible block of activity in either magnocellular or parvocellular LGN divisions can lead to a reduction or block of neuronal activity in both blob and interblob zones in the upper layers of V1. Tootell *et al.* (1988b) also observed 2-DG labeling of the upper layer blobs following visual stimulation with low-contrast (~ 8%) targets, which should preferentially activate magnocellular neurons. In retrospect, we should perhaps not be surprised to find such convergence of magnocellular and parvocellular signals in the superficial layers. Over 10 years ago, Malpeli *et al.* (1981), using reversible block of activity in particular LGN laminae, found that single V1 cells may receive inputs from *both* LGN divisions; the location of the cells receiving both types of input was not determined, however. These results have clear implications for studies trying to trace the flow of visual information from the LGN, through V1, and out to the various extrastriate visual areas.

Our observations on the interlaminar relays from layer 4 also lead us to conclude that the superficial layer blobs may be a heterogeneous population. We have found that some blobs appear to receive no inputs from any division of layer 4, and are therefore presumably dominated by their thalamic input from the LGN intercalated layers, while other blobs may receive an additional, indirect LGN drive via V1 layers 4A or 4B. The notion that there may be different types of blobs also finds support from recent physiological work. Both 2-DG (Tootell *et al.*, 1988a) and single-unit recording studies (Dow and Vautin, 1987; Ts'o and Gilbert, 1988; Yoshioka, 1989) suggest that blobs may preferentially process certain colors, and that individual blobs may have different color opponency. Furthermore, not all cells in the blobs are color-selective (Livingstone and Hubel, 1984a; Ts'o and Gilbert, 1988), nor do all blobs show reduced orientation selectivity (Livingstone and Hubel, 1984a). Clearly, blobs analyze other kinds of visual information in addition to color, and they are not necessarily a homogeneous group. This inhomogeneity may also be reflected by substructure in their V2 target (cytochrome oxidase-rich thin stripes) and in the further projections of this V2 region to V4 (Van Essen *et al.*, 1990; Rockland, 1992).

While we have focused here on the interlaminar specificity of intrinsic connections, we have also traced the specificity of clustered horizontal connections in the upper layers. In agreement with Livingstone and Hubel (1984a,b), we find preferential connections between blob zones, as well as between interblob zones. In cat primary visual cortex, the superficial layer lateral projections appear to favor connections between regions of similar orientation preference (Gilbert and Wiesel, 1989), but Matsubara *et al.* (1987) stress *orthogonal* orientation connections in cat area 18. Ts'o *et al.* (1986) found correlated firing between neurons in cat V1 only when their ocular dominance and preferred orientation matched. In macaque V1, results from combined optical recording and biocytin transport studies indicate that while connections are made primarily between regions of similar orientation preference (within 45° of one another) or ocular dominance, connections clearly exist between regions of dissimilar function as well (Blasdel *et al.*, 1992; Yoshioka *et al.*, 1992; Malach *et al.*, 1992).

2. *Layer 4:* Our anatomical studies have demonstrated a rich diversity of connection patterns through the depth of layer 4; here we highlight some evidence that receptive field properties have a corresponding sublaminar specificity.

i. Layer 4A: Previous studies (Hubel and Wiesel, 1972; Hendrickson *et al.*, 1978; Fitzpatrick *et al.*, 1983) indicated that layer 4A receives direct LGN parvocellular input; we have discussed earlier in this chapter that 4A also receives relays from lower and mid-4C, and contains pyramidal neurons whose dendrites enter both 4B and 3B. Consistent with these avenues for receiving inputs from both LGN divisions, layer 4A has been reported to contain neurons whose responses are selective for either the color or direction of motion of visual stimuli (Michael, 1985; Hawken *et al.*, 1988; Lennie *et al.*, 1990). We note a projection from layer 4A to the cytochrome oxidase thick stripes of V2 (Levitt *et al.*, 1994), described earlier as part of a magnocellular-dominated pathway to area MT and posterior parietal cortex (Livingstone and Hubel, 1987, 1988). We also note that layer 4A cells exhibit both high and low contrast sensitivities (Hawken and Parker, 1984; Blasdel and Fitzpatrick, 1984), again suggesting the possible convergence of magnocellular and parvocellular signals.

ii. Layer 4B: This division of layer 4 seems clearly dominated by magnocellular signals; we have confirmed that its predominant input is from layer 4Cα (Lund and Boothe, 1975). This layer contains many neurons that are direction-selective and highly sensitive to stimulus contrast, while color selectivity is rare (Dow, 1974; Livingstone and Hubel, 1984a; Blasdel and Fitzpatrick, 1984; Hawken *et al.*, 1988). We also observed clustered lateral connectivity within layer 4B. Similar patchy patterns relating to both feedforward and feedback connections with areas V2 and MT have been observed in layer 4B, though these patterns do not correlate with cytochrome oxidase staining in the superficial layers (Livingstone and Hubel, 1987; Shipp and Zeki, 1989). Layer 4B contains both direction-selective and nondirectional populations, and it is known that the MT-projecting cells are themselves direction-selective (Movshon and Newsome, 1984). It is therefore tempting to speculate that these patchy connections within 4B interconnect groups of neurons with similar directional properties. In contrast to the superficial layers 2/3 where both magnocellular and parvocellular influences are mingled, there seems to be little parvocellular input to 4B. Thus, the magnocellular pathway through layer 4B to the thick stripes of V2 and beyond to MT may remain relatively distinct. This agrees well with the report of Maunsell *et al.* (1990) that parvocellular signals in area MT, though present, are rather weak (perhaps channeled via layer 4A relays to V2 thick stripes and thence on to MT).

iii. Layer 4C: Layer 4C can be subdivided into α and β subdivisions on the basis of LGN inputs. Patterns of intrinsic circuitry show, however, that layer 4C may also be considered tripartite, with mid-4C as a distinct source of relays out of the layer in addition to, and overlapping relays from, 4Cα and 4Cβ. As the position of spiny stellate cells in 4C shifts from the bottom of 4Cβ to the top of 4Cα (and therefore also their relative sampling of magnocellular versus parvocellular inputs), the projection zones of those cells also shift continuously from layer 3B to layer 4B. Some functional characteristics also appear to vary continuously through layer 4C. For example, although contrast sensitivity, on average, is higher in layer 4Cα than in 4Cβ, there is evidence that contrast sensitivity actually changes smoothly through the depth of layer 4C (Blasdel and Fitzpatrick, 1984; Hawken and Parker, 1984).

3. *Layers 5 and 6:* We have demonstrated that the connectivity patterns of cells in layers 1–3 depend on whether the cells lie in blobs or interblobs. Faint cytochrome oxidase-rich zones can also be seen in deeper layers 5/6 in register with the upper layer blobs (Horton and Hubel, 1981); we observed that clustered axon terminals in layer 5 were in register with those in the upper layers, and we earlier observed periodicity of lateral connections within layer 6 (Blasdel *et al.,* 1985). It would therefore be interesting to know if any response properties in the infragranular layers depend on whether cells lie in blob or interblob "columns." Cells in layer 6 are frequently direction-selective (Livingstone and Hubel, 1984a; Hawken *et al.,* 1988), and reports indicate that there are color-selective cells as well (Michael, 1985; Lennie *et al.,* 1990). We have shown that layer 6 can be subdivided into parvocellular- and magnocellular-related sublaminae on the basis of connections with the LGN and layer 4C; perhaps layer 6 neurons with particular selectivities lie in different portions of the layer. However, no physiological studies have clearly demonstrated subdivisions within either layer 5 or 6.

Overall, the neuropil of area V1 in the macaque monkey presents an anatomical architecture of exquisite precision and specialization. A strict channeling of thalamic inputs is avoided; the intrinsic relays by thalamic recipient neurons suggest that they provide substrates for graded interactions between properties provided by different LGN channels. The anatomy of local circuit neuron projections suggests that the balance of activity in the gradient could be altered by their inhibitory activity to emphasize the properties of single thalamic inputs even while access to other inputs is retained. Curiously, what might be regarded as the two extremes—the inputs to lowermost 4Cβ and to uppermost 4Cα—seem eventually to converge via 4B and 4A relays to the blob zones, while a more immediate merger of the parvocellular and magnocellular LGN input seems to occur in mid-4C for relay to the interblob regions.

ACKNOWLEDGMENTS. We thank Suzanne Holbach and Tom Harper for excellent technical assistance; Marianne Davis for manuscript preparation; Edward Lachica, Vivien Casagrande, and David Fitzpatrick for sharing their knowledge in areas of this study and for advice on the biocytin technique. We are most grateful to David Lewis for his laboratory support during our move to London. This work was supported by National Eye Institute grants EY05282, EY010021-01, EY08098 (Eye & Ear Institute of Pittsburgh), MRCG9203679N, an ARVO/ALCON postdoctoral research fellowship (T.Y.), and postdoctoral NRSA award EY06275 (J.B.L.).

5. References

Blasdel, G. G., and Fitzpatrick, D., 1984, Physiological organization of layer 4 in macaque striate cortex, *J. Neurosci.* **4:**880–895.
Blasdel, G. G., and Lund, J. S., 1983, Termination of afferent axons in macaque striate cortex, *J. Neurosci.* **3:**1389–1413.

Blasdel, G. G., Lund, J. S., and Fitzpatrick, D., 1985, Intrinsic connections of macaque striate cortex: Axonal projections of cells outside lamina 4C, *J. Neurosci.* **5:**3350–3369.

Blasdel, G. G., Yoshioka, T., Levitt, J. B., and Lund, J. S., 1992, Correlation between patterns of lateral connectivity and patterns of orientation preference in monkey striate cortex, *Soc. Neurosci. Abstr.* **18:**389.

Dow, B. M., 1974, Functional classes of cells and their laminar distribution in monkey visual cortex, *J. Neurophysiol.* **37:**927–946.

Dow, B. M., and Vautin, R. G., 1987, Horizontal segregation of color information in the middle layers of foveal striate cortex, *J. Neurophysiol.* **57:**712–739.

Fairén, A., DeFelipe, J., and Regidor, J., 1984, Nonpyramidal neurons: General account, in: *Cerebral Cortex*, Volume 1 (A. Peters and E. G. Jones, eds.), Plenum Press, New York, pp. 201–253.

Fitzpatrick, D., Itoh, K., and Diamond, I. T., 1983, The laminar organization of the lateral geniculate body and the striate cortex in the squirrel monkey (Saimiri sciureus), *J. Neurosci.* **3:**673–702.

Fitzpatrick, D., Lund, J. S., and Blasdel, G. G., 1985, Intrinsic connections of macaque striate cortex: Afferent and efferent connections of lamina 4C, *J. Neurosci.* **5:**3329–3349.

Gilbert, C. D., and Wiesel, T. N., 1989, Columnar specificity of intrinsic horizontal and corticocortical connections in cat visual cortex, *J. Neurosci.* **9:**2432–2442.

Hawken, M. J., and Parker, A. J., 1984, Contrast sensitivity and orientation selectivity in lamina IV of the striate cortex of Old World monkeys, *Exp. Brain Res.* **54:**367–372.

Hawken, M. J., Parker, A. J., and Lund, J. S., 1988, Laminar organization and contrast sensitivity of direction-selective cells in the striate cortex of the Old World monkey, *J. Neurosci.* **8:**3541–3548.

Hendrickson, A. E., Wilson, J. R., and Ogren, M. P., 1978, The neuroanatomical organization of pathways between the dorsal lateral geniculate nucleus and visual cortex in Old World and New World primates, *J. Comp. Neurol.* **182:**123–136.

Horikawa, K., and Armstrong, W. E., 1988, A versatile means of intracellular labeling: Injection of biocytin and its detection with avidin conjugates, *Neurosci. Methods* **25:**1–11.

Horton, J. C., and Hubel, D. H., 1981, Regular patchy distribution of cytochrome oxidase staining in primary visual cortex of macaque monkey, *Nature* **292:**762–764.

Hubel, D. H., and Livingstone, M. S., 1987, Segregation of form, color, and stereopsis in primate area 18, *J. Neurosci.* **7:**3378–3415.

Hubel, D. H., and Wiesel, T. N., 1972, Laminar and columnar distribution of geniculo-cortical fibers in the macaque monkey, *J. Comp. Neurol.* **146:**421–450.

Hubel, D. H., and Wiesel, T. N., 1977, Functional architecture of macaque monkey visual cortex, *Proc. R. Soc. London Ser. B* **198:**1–59.

Jones, E. G., and Hendry, S. H. C., 1984, Basket cells, in: *Cerebral Cortex*, Volume 1 (A. Peters and E. G. Jones, eds.). Plenum Press, New York, pp. 309–336.

King, M. A., Lewis, P. M., Hunter, B. E., and Walker, O. W., 1989, Biocytin: A versatile anterograde neuroanatomical tract-tracing alternative, *Brain Res.* **497:**361–367.

Lachica, E. A., Mavity-Hudson, E. A., and Casagrande, V. A., 1991, Morphological details of primate axons and dendrites revealed by extracellular injection of biocytin: An economic and reliable alternative to PHA-L, *Brain Res.* **564:**1–11.

Lachica, E. A., Beck, P. D., and Casagrande, V. A., 1992, Parallel channels in macaque monkey striate cortex: Anatomically defined columns in layer III, *Proc. Natl. Acad. Sci. USA* **89:**3566–3570.

Lennie, P., Krauskopf, J., and Sclar, G., 1990, Chromatic mechanisms in striate cortex of macaque, *J. Neurosci.* **10:**649–669.

Levitt, J. B., Yoshioka, T., and Lund, J. S., 1994, Intrinsic cortical connections in macaque visual area V2: evidence for interaction between different functional streams. *J. Comp. Neurol.*, in press.

Livingstone, M. S., and Hubel, D. H., 1982, Thalamic inputs to cytochrome oxidase-rich regions in monkey visual cortex, *Proc. Natl. Acad. Sci. USA* **79:**6098–6101.

Livingstone, M. S., and Hubel, D. H., 1984a, Anatomy and physiology of a color system in the primate visual cortex, *J. Neurosci.* **4:**309–356.

Livingstone, M. S., and Hubel, D. H., 1984b, Specificity of intrinsic connections in primate primary visual cortex, *J. Neurosci.* **4:**2830–2835.

Livingstone, M. S., and Hubel, D. H., 1987, Connections between layer 4B of area 17 and the thick cytochrome oxidase stripes of area 18 in the squirrel monkey, *J. Neurosci.* **7:**3371–3377.

Livingstone, M. S., and Hubel, D. H., 1988, Segregation of form, color, movement, and depth: Anatomy, physiology, and perception, *Science* **240:**740–749.

Lund, J. S., 1973, Organization of neurons in the visual cortex, area 17, of the monkey (Macaca mulatta), *J. Comp. Neurol.* **147:**455–496.

Lund, J. S., 1987, Local circuit neurons of macaque monkey striate cortex: I. Neurons of laminae 4C and 5A, *J. Comp. Neurol.* **257:**60–92.

Lund, J. S., 1990, Excitatory and inhibitory circuitry and laminar mapping strategies in the primary visual cortex of the monkey, in: *Signal and Sense: Local and Global Order in Perceptual Maps* (G. M. Edelman, W. E. Gall, and W. M. Cowan, eds.), Wiley, New York, pp. 51–66.

Lund, J. S., and Boothe, R., 1975, Interlaminar connections and pyramidal neuron organization in the visual cortex, area 17, of the macaque monkey, *J. Comp. Neurol.* **159:**305–334.

Lund, J. S., and Yoshioka, T., 1991, Local circuit neurons of macaque monkey striate cortex: III. Neurons of laminae 4B, 4A, and 3B, *J. Comp. Neurol.* **311:**234–258.

Lund, J. S., Lund, R. D., Hendrickson, A. H., Bunt, A. H., and Fuchs, A. F., 1975, The origin of efferent pathways from the primary visual cortex, area 17, of the macaque monkey as shown by retrograde transport of horseradish peroxidase, *J. Comp. Neurol.* **164:**287–304.

Lund, J. S., Hawken, M. J., and Parker, A. J., 1988, Local circuit neurons of macaque monkey striate cortex: II. Neurons of laminae 5B and 6, *J. Comp. Neurol.* **276:**1–29.

Lund, J. S., Yoshioka, T., and Levitt, J. B., 1993, Comparison of intrinsic connectivity in different areas of macaque monkey cerebral cortex, *Cereb. Cortex* **3:**148–162.

Malach, R., Amir, Y., Bartfeld, E., and Grinvald, A., 1992, Biocytin injections, guided by optical imaging, reveal relationships between functional architecture and intrinsic connections in monkey visual cortex, *Soc. Neurosci. Abstr.* **18:**389.

Malpeli, J. G., Schiller, P. H., and Colby, C. L., 1981, Response properties of single cells in monkey striate cortex during reversible inactivation of individual lateral geniculate laminae, *J. Neurophysiol.* **46:**1102–1119.

Matsubara, J. A., Cynader, M. S., and Swindale, N. V., 1987, Anatomical properties and physiological correlates of the intrinsic connections in cat area 18, *J. Neurosci.* **7:**1428–1446.

Maunsell, J. H. R., Nealey, T. A., and DePriest, D. D., 1990, Magnocellular and parvocellular contributions to responses in the middle temporal visual area (MT) of the macaque monkey, *J. Neurosci.* **10:**3323–3334.

Michael, C. R., 1985, Laminar segregation of color cells in the monkey's striate cortex, *Vision Res.* **25:**415–423.

Movshon, J. A., and Newsome, W. T., 1984, Functional characteristics of striate cortical neurons projecting to MT in the macaque, *Soc. Neurosci. Abstr.* **10:**933.

Nealey, T. A., Ferrera, V. P., and Maunsell, J. H. R., 1991, Magnocellular and parvocellular contributions to the ventral extrastriate cortical processing stream, *Soc. Neurosci. Abstr.* **17:**525.

Peters, A., 1984, Chandelier cells, in: *Cerebral Cortex,* Volume 1 (A. Peters and E. G. Jones, eds.), Plenum Press, New York, pp. 361–380.

Rockland, K. S., 1992, Configuration, in serial reconstruction, of individual axons projecting from area V2 to V4 in the macaque monkey, *Cereb. Cortex* **2:**353–374.

Shipp, S., and Zeki, S., 1989, The organization of connections between areas V5 and V1 in macaque monkey visual cortex, *Eur. J. Neurosci.* **1:**309–332.

Tootell, R. B. H., Silverman, M. S., Hamilton, S. L., De Valois, R. L., and Switkes, E., 1988a, Functional anatomy of macaque striate cortex. III. Color, *J. Neurosci.* **8:**1569–1593.

Tootell, R. B. H., Hamilton, S. L., and Switkes, E., 1988b, Functional anatomy of macaque striate cortex. IV. Contrast and magno-parvo streams, *J. Neurosci.* **8:**1594–1609.

Ts'o, D. Y., and Gilbert, C. D., 1988, The organization of chromatic and spatial interactions in the primate striate cortex, *J. Neurosci.* **8:**1712–1727.

Ts'o, D. Y., Gilbert, C. D., and Wiesel, T. N., 1986, Relationships between horizontal connections and functional architecture in cat striate cortex as revealed by cross-correlation analysis, *J. Neurosci.* **6:**1160–1170.

Van Essen, D. C., Felleman, D. J., DeYoe, E. A., Olavarria, J., and Knierim, J., 1990, Modular and hierarchical organization of extrastriate visual cortex in the macaque monkey, *Cold Spring Harbor Symp. Quant. Biol.* **55:**679–696.

Yoshioka, T., 1989, Color and luminance organization in visual cortical areas V1, V2, and V4 of the macaque monkey, *Ph.D. thesis,* SUNY at Buffalo.

Yoshioka, T., Blasdel, G. G., Levitt, J. B., and Lund, J. S. 1992. Patterns of lateral connections in macaque visual area V1 revealed by biocytin histochemistry and functional imaging. *Soc. Neurosci. Abstr.* **18:** 299.

Yoshioka, T., Levitt, J. B., and Lund, J. S., 1994, Independence and merger of thalamocortical channels within macaque monkey primary visual cortex: anatomy of interlaminar projections, *Vis. Neurosci.,* in press.

GABA Neurons and Their Role in Activity-Dependent Plasticity of Adult Primate Visual Cortex

E. G. JONES, S. H. C. HENDRY, J. DeFELIPE, and D. L. BENSON

1. Introduction

Since its identification more than 30 years ago (see Roberts, 1975), as a naturally occurring brain substance that inhibited neural activity, γ-aminobutyric acid (GABA) has become recognized as one of the major neurotransmitter agents of the mammalian central nervous system. The ubiquitous distribution of GABA neurons and GABA receptors and the demonstrable role of GABA in shaping, through inhibition, the stimulus–response properties of neurons at many levels of the central nervous system attest to its profound functional importance. In most of the major organized systems of the neuraxis, GABA can probably be regarded as occupying a status equal in importance to that of the excitatory amino acid transmitters.

E. G. JONES, S. H. C. HENDRY, J. DeFELIPE, and D. L. BENSON • Department of Anatomy and Neurobiology, University of California, Irvine, California 92717. *Present address of S.H.C.H.:* Krieger Mind/Brain Institute, Johns Hopkins University, Baltimore, Maryland 21218. *Present address of J.D.:* Instituto Cajal, 28002 Madrid, Spain. *Present address of D.L.B.:* Department of Neuroscience, University of Virginia School of Medicine, Charlottesville, Virginia 22908.

Cerebral Cortex, Volume 10, edited by Alan Peters and Kathleen S. Rockland. Plenum Press, New York, 1994.

The cerebral cortex is no exception to this general rule and both normal adult cortical function and the series of developmental and maturational stages that lead up to it are critically dependent on GABA- and excitatory amino acid-mediated mechanisms. In the primary visual cortex, GABA, acting via bicuculline-sensitive $GABA_A$ receptors, mediates visually evoked inhibitory effects on single cortical neurons (Sillito, 1974, 1975a,b); in some neurons, it is involved in reducing the efficacy of inputs from one eye, thus ensuring a monocularly driven response; in others it is involved in shaping directional selectivity in the responses to moving stimuli (reviewed in Sillito, 1984). Its involvement in the mechanisms that shape the selectivity of responses of visual cortical neurons to stimuli of a preferred orientation is still debated (Ferster, 1986, 1987, 1988). While all of this work has been carried out on nonprimates, there is reason to believe that analogous mechanisms operate in the primate visual cortex. Furthermore, it is now clear that even in adult primates, the regulation of GABA transmitter and receptor levels is a finely tuned process dependent on activity in the optic nerve (Hendry and Jones, 1986, 1988) and potentially having importance for understanding the activity-dependent plasticity of visual cortical organization and function (Jones, 1990). In this chapter, therefore, we will review the large body of information that has been built up on GABAergic neurons and their receptors in area 17 of the primate, along with their role in activity-dependent plasticity and we will attempt to highlight their overall importance in visual cortical function.

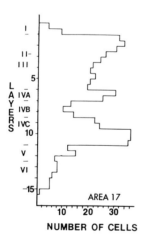

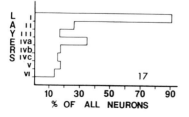

Figure 1. (Upper) Number of GABA-immunoreactive cells in different layers of monkey area 17, as counted in 500-μm-wide columns extending through the full thickness of area 17. (Lower) GABA-immunoreactive neurons as a percentage of the total neuronal population, determined by counting all neurons in 50-μm-wide columns through the thickness of area 17. From Hendry *et al.* (1987).

2.1. Number and Density of GABA Cells

GABA neurons are found throughout the full thickness of area 17 of Old World monkeys, extending from just beneath the pial surface to some distance into the white matter (Figs. 1 and 2). These cells make up a sizable proportion of the total neuronal population of area 17: one out of every five neurons displays immunoreactive staining for GABA (Hendry *et al.*, 1987) (Table I). A smaller proportion (15%) is revealed by immunoreactivity for the GABA-synthesizing enzyme, glutamate decarboxylase (GAD). This probably reflects the greater difficulty in localizing GAD by immunocytochemistry. There are at least two molecular forms of GAD, with molecular masses of 65 and 67 kDa (Kaufman *et al.*, 1989; Bu *et al.*, 1992). Both are presumably found in all cortical GABA neurons but this has not been confirmed. *In situ* hybridization histochemistry specifically localizing the mRNA coding for the 67-kDa form reveals a population similar in

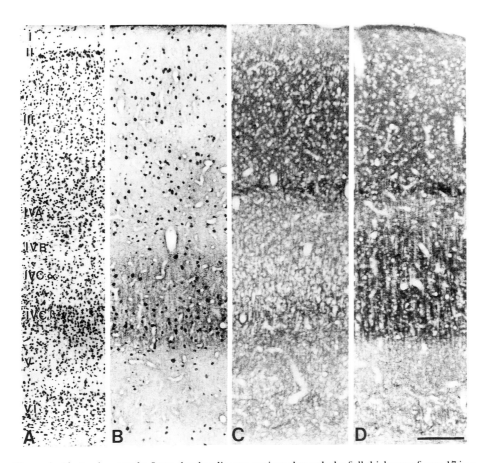

Figure 2. Photomicrographs from closely adjacent sections through the full thickness of area 17 in a macaque monkey, stained by the Nissl stain (A), immunocytochemically for GABA (B) or for β2/β3 subunits of the GABA$_A$ receptor (C) and histochemically for cytochrome oxidase (D). Bar = 250 μm. Hendry *et al.* (1987, 1990).

size to that shown by immunoreactivity for GABA (Benson *et al.*, 1991a). Quantitative studies (Table I) show that in a vertical column extending through the full thickness of area 17 and measuring 1 mm² at the cortical surface, there are approximately 35,000 GABA neurons (23,000 per mm³). This number of GABA cells is 50% greater than in a comparable column in any other area of monkey cerebral cortex, but because of the greatly increased density of all neurons in area 17 (Rockel *et al.*, 1980; O'Kusky and Colonnier, 1982), the proportion is 5% lower than in other areas (20 versus 25%).

2.2. Laminar Distribution of GABA Cells

GABA neurons in monkey area 17 are unevenly distributed across layers (Figs. 1 and 2). As seen in immunocytochemical preparations (Hendry *et al.*, 1987), GABA-immunoreactive somata are most numerous in layers II–III, IVA, and IVC where they form from 15 to 40% of the total neuronal population of those layers. They form 95% of the neurons in the cell-sparse layer I (Fig. 1). Neurons in the white matter subjacent to layer VI are also GABA-

Table I. Number and Density of GABA Cells[a]

A. Mean percentage (±SD) of the total GABA-immunoreactive neuronal population in 50-μm-wide columns through the thickness of areas 4, 3b, 1–2, 5, 7, 18, and 17[b]

Area	CM 181	CM 183	CM 184	CM 187	CM 189
4	24.2 ± 2.1	25.6 ± 1.2	24.9 ± 1.4	25.1 ± 1.2	24.9 ± 2.0
3b	24.6 ± 1.8	20.3 ± 2.8	21.1 ± 3.0	20.7 ± 2.7	24.4 ± 1.3
1–2	24.6 ± 1.9	24.4 ± 2.1	24.8 ± 1.2	25.1 ± 1.9	23.7 ± 2.2
5	24.8 ± 2.0	25.2 ± 1.4	25.4 ± 2.1	24.9 ± 1.4	23.9 ± 1.7
7	24.2 ± 2.2	24.6 ± 1.9	25.4 ± 1.3	24.9 ± 2.5	24.8 ± 2.0
18	25.2 ± 1.3	25.4 ± 1.9	24.4 ± 2.4	23.7 ± 2.4	25.3 ± 1.9
17	19.4 ± 1.5	19.1 ± 1.7	20.2 ± 1.2	19.7 ± 1.5	18.9 ± 0.9

B. Mean number (±SD) of GABA-immunoreactive neurons calculated to be present in a volume of 1 mm³ or to underlie surface 1 mm² or 50 × 50-μm squares[c]

Area	Count/mm³		Count/mm²		Count/50 × 50 μm	
	CM 181	CM 187	CM 181	CM 187	CM 181	CM 187
4	10,400 ± 190	11,200 ± 210	22,800 ± 840	24,100 ± 990	57.0 ± 2.1	60.3 ± 2.5
3b	17,200 ± 310	14,900 ± 270	23,600 ± 1100	19,900 ± 970	59.0 ± 3.8	49.8 ± 4.7
1–2	16,100 ± 210	15,800 ± 230	24,200 ± 1200	23,800 ± 1080	60.5 ± 3.3	59.2 ± 3.8
5	15,200 ± 240	14,600 ± 230	23,500 ± 1050	23,100 ± 1160	58.8 ± 1.9	57.8 ± 2.3
17	23,160 ± 390	23,700 ± 400	34,500 ± 1700	36,300 ± 1660	86.3 ± 7.0	90.8 ± 7.7

[a]Data taken from Hendry *et al.* (1987) showing: (A) the percentages of GABA-immunoreactive neurons relative to the total population of cortical neurons found in arbitrary 50-μm-wide colums extending through the full thickness of different cortical areas, including area 17, in five macaque monkeys; (B) calculations, derived by application of stereological formulas to the data from two monkeys in A, showing the density of GABA neurons and the numbers occupying arbitrary columns measuring 1 × 1 mm or 50 × 50 μm at the cortical surface and extending through the full thickness. Note that despite an unusually high density of GABA cells relative to other areas, area 17 has a lower overall percentage of such cells.
[b]Counts were made of GABA-immunoreactive and nonimmunoreactive neurons with nuclei present in 1-μm-thick plastic sections. The percentage of GABA-positive neurons was calculated from these values.
[c]Calculations were made from measurements in 1-μm-thick plastic sections through the thickness of areas 4, 3b, 1–2, 5, and 17.

immunoreactive but these make up a small fraction of the total population of neurons in the white matter. A laminar pattern very similar to that in GABA-immunoreactive preparations is seen when neurons are stained immunocytochemically for GAD (Fitzpatrick *et al.*, 1987) or labeled by *in situ* hybridization histochemistry for 67-kDa GAD mRNA (Benson *et al.*, 1991a) (see below). Both GAD immunocytochemistry and *in situ* hybridization histochemistry, especially the latter, reveal sublaminar aggregations of GABA cells based on different densities and sizes of labeled somata in layers IVC and VI.

2.3. Laminar Distribution of GABA Terminals and GABA Receptors

Small punctate GABA- or GAD-immunoreactive profiles, which electron microscopic studies have shown to be axon terminals, are also most densely packed in layers II–III, IVA, and IVC. They are moderately dense in layers I and VI and sparse in layers IVB and V (Fitzpatrick *et al.*, 1987; Hendrickson *et al.*, 1981; Hendry *et al.*, 1987). GABA$_A$ receptors, localized either by radioligand binding or by immunocytochemistry, display a very similar laminar distribution (Fig. 2) (Hendry *et al.*, 1990; Shaw and Cynader, 1986; Rakic *et al.*, 1988). Immunocytochemical studies also demonstrate that members of each of the major subunit families (α, β, and γ) of the GABA$_A$ receptor have similar laminar distributions (Huntsman *et al.*, 1991). This pattern differs markedly from that seen when binding of [^{3}H]baclofen is used to localize GABA$_B$ receptors: they are diffusely distributed through all layers and display a particularly low density in layer IV (Shaw *et al.*, 1991).

2.4. Relationship to Geniculocortical Axon Terminations

A consistent pattern that emerges from studies of the laminar distributions of GABA cell somata, terminals, and GABA$_A$ receptors is that all of them are densest in layers in which geniculocortical axons terminate. Neurons in the dorsal lateral geniculate nucleus (LGN) of the monkey thalamus send their axons principally to layers IVA and IVC with other contributions to layers I, II–III, and VI (Blasdel and Lund, 1983; Hendrickson *et al.*, 1978; Hendrickson, 1982; Hubel and Wiesel, 1972). These axons arise from functionally and morphologically distinct groups of neurons in the LGN, with the two magnocellular layers (1 and 2) projecting to the superficial half of layer IVC (layer IVCα) and the four parvocellular layers (3–6) projecting to the deeper half of layer IVC (layer IVCβ) and layer IVA. Geniculocortical terminations in layer VI arise from collaterals of axons that innervate layer IVC while the terminations in layers II and III may arise both from collaterals of axons ending in layer IVA (Blasdel and Lund, 1983) and from a selective input originating from somata in the S and intercalated layers of the LGN (Fitzpatrick *et al.*, 1983). The distributions of GABA neurons and their terminals follow very closely the laminar distribution of the geniculocortical axons and, as discussed below, where the axons are unevenly distributed within a layer, GABA neurons and terminals tend to adopt a comparable uneven distribution. Monkey area 17 is the clearest example of a general trend within the mammalian cerebral cortex in which GABA neurons and their axonal terminations are most prominent in layers and sublaminar compartments that receive thalamocortical inputs (Hendry *et al.*, 1987).

Geniculocortical axons arising from neurons in different layers of the LGN terminate in stripelike regions within layers IVA and IVC in area 17 of Old World, and certain species of New World monkey (Hubel and Wiesel, 1972; Hendrickson *et al.*, 1978). These stripelike terminations form the basis of the well-known ocular dominance columns. Within layers IVA and IVC of adult monkeys, most individual neurons receive excitatory drive from only the left or right eye (Hubel and Wiesel, 1968, 1977). This can be accounted for by the restricted distribution of geniculocortical axons in these layers and also by the restricted dendritic branching of layer IV cells: neurons near the borders of ocular dominance columns have highly skewed dendritic fields that tend to remain within a single ocular dominance stripe (Katz *et al.*, 1989).

GABA cell somata in normal monkeys display no differences in number or distribution across ocular dominance borders and, thus, no sign of the ocular dominance organization is evident in the normal patterns of GABA or GAD immunostaining or in GAD *in situ* hybridization histochemistry (Hendrickson *et al.*, 1981, 1983; Hendry *et al.*, 1987; Fitzpatrick *et al.*, 1987; Benson *et al.*, 1991a).

2.5. Laminar Distributions of Chemically Defined GABA Subpopulations

2.5.1. Calcium-Binding Proteins

GABA neurons in area 17 of monkeys are chemically heterogeneous and the chemically defined subpopulations are unevenly distributed. Two major, largely nonoverlapping populations, immunoreactive for different calcium-binding proteins are found in area 17 (Figs. 3 and 4). GABA cells immunoreactive for the 28-kDa, vitamin D-dependent calcium-binding protein, calbindin, are mainly distributed in layers II and VI while GABA cells immunoreactive for parvalbumin are concentrated in layers III, IVA, and IVC (Hendry *et al.*, 1989; Blümke *et al.*, 1990; Van Brederode *et al.*, 1990). A few GABA neurons in which calbindin and parvalbumin immunoreactivity coexists are found in layers V and VI (Hendry *et al.*, 1989) and a small population of non-GABA, calbindin-positive, pyramidal neurons is found in layer II (DeFelipe and Jones, 1992).

2.5.2. Neuropeptides

Immunoreactivity for most neuropeptides in the monkey cerebral cortex is found in GABA neurons (Fig. 5) (Hendry *et al.*, 1984a; Jones and Hendry, 1986; Jones *et al.*, 1988). Each subpopulation defined by colocalization of a particular peptide or peptides is commonly present in only one or two layers. Small subpopulations of GABA cells that contain somatostatin (SRIF)-, neuropeptide Y (NPY)-, or cholecystokinin (CCK)-like immunoreactivity are present mainly in layers II and VI and in the subcortical white matter and tend to be absent from layer IVC (Hendry *et al.*, 1984a,c; Campbell *et al.*, 1987; Kuljis and Rakic, 1989a,b; Tigges *et al.*, 1989). SRIF- and NPY-like immunoreactivity often coexist and occur in GABA neurons with relatively large somata (Hendry *et al.*, 1984c), but CCK-like immunoreactivity is not colocalized with these two. A very small number of cortical neurons display immunoreactivity not only for SRIF and NPY but also for two members of the tachykinin neuropeptide family (substance P and neurokinin A; Jones *et al.*, 1988). Unlike the majority of SRIF/NPY cells, or

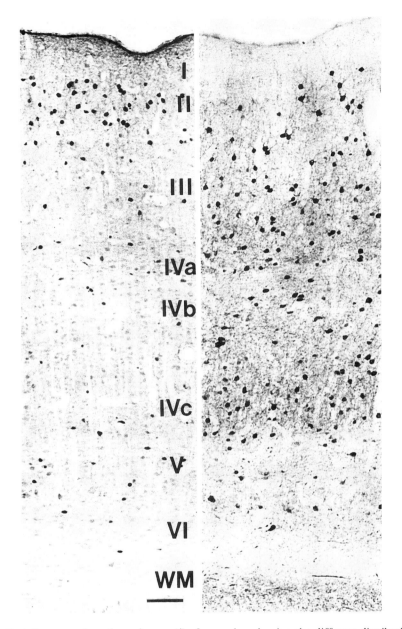

Figure 3. Adjacent sections through area 17 of a monkey showing the different distributions of calbindin- (left) and parvalbumin-immunoreactive (right) cells. Calbindin cells tend to be concentrated in superficial and deep layers and parvalbumin cells in middle layers. Bar = 100 μm. From Hendry *et al.* (1989).

tachykinin cells, the cells containing the three peptides are usually not GABA-immunoreactive (Jones *et al.*, 1988).

The largest GABA/neuropeptide subpopulation is made up of small somata that display tachykininlike immunoreactivity (Jones *et al.*, 1988). These are distinguished from the large GABA/tachykinin-positive somata by size, by absence of colocalized SRIF- and NPY-like immunoreactivity, and by the laminar distri-

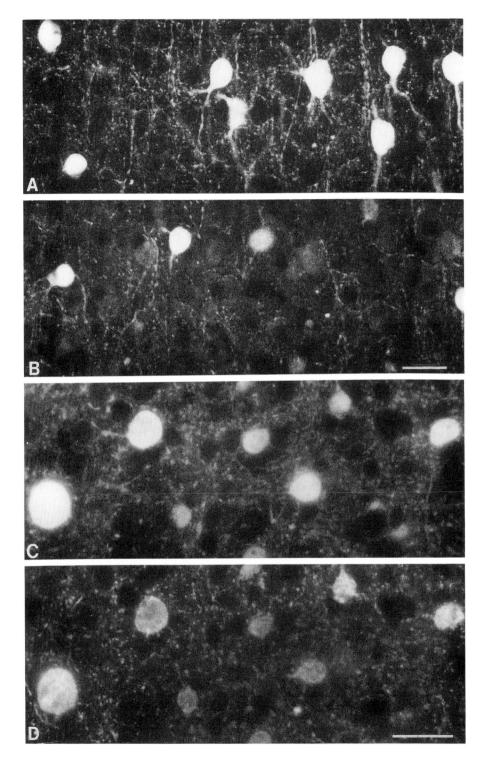

Figure 4. (A, B; C, D; E, F) Pairs of fluorescence photomicrographs, each pair from the same microscopic field of area 17 and showing: (A, B) lack of colocalization of parvalbumin (A) and

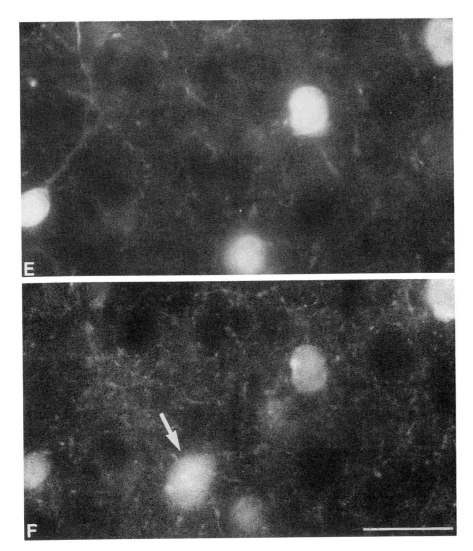

Figure 4. *(Continued)* calbindin immunoreactivity (B); (C, D) colocalization of immunoreactivity for parvalbumin (C) and GABA (D); (E, F) colocalization of immunoreactivity for calbindin (E) and GABA (F). Bars = 50 μm. From Hendry *et al.* (1989).

bution of their somata (Fig. 5A). Unlike other GABA/neuropeptide cells, most of the small GABA/tachykinin cells are found in layer IVC (Hendry *et al.*, 1988a; Jones *et al.*, 1988). These GABA/tachykinin neurons are particularly numerous and make up half the total GABA population in layer IVC (Hendry *et al.*, 1988a; Jones *et al.*, 1988).

2.5.3. Cell-Surface Molecules

Some GABA neurons in area 17 of monkeys express cell-surface glycoproteins and can thus be recognized by the use of specific monoclonal antibodies or by binding of plant lectins. The monoclonal antibodies Cat-301 and VC1.1

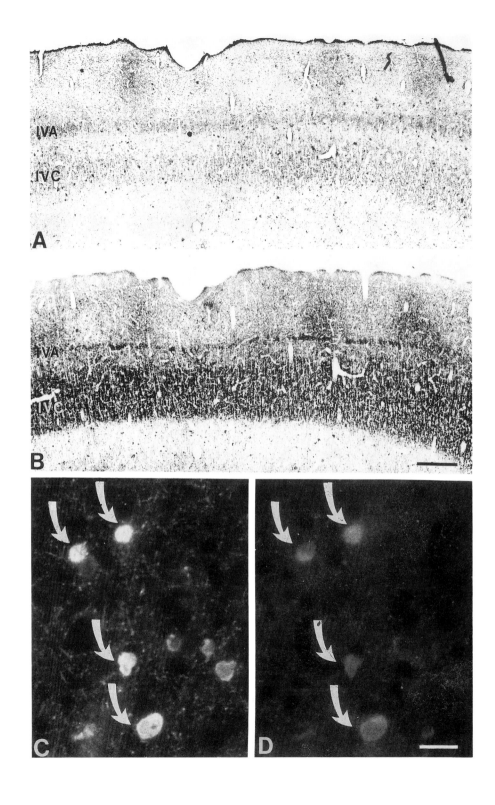

recognize cell-surface antigens that are predominantly or entirely present on GABA neurons (Naegle and Barnstable, 1989; Hendry *et al.*, 1988b), while GABA neurons that express complex surface carbohydrates with *N*-acetylgalactosamine termini can be identified by binding of the lectin *Vicia villosa* (VVA; Nakagawa *et al.*, 1986). GABA subpopulations immunostained by Cat-301 (Fig. 6) or labeled with VVA are very dense in the middle layers of area 17: layers IVB and superficial IVCα for Cat-301 (Hockfield *et al.*, 1984; Hendry *et al.*, 1984b; DeYoe *et al.*, 1990) and layers IIIB–IVCα for VVA (Mulligan *et al.*, 1989). Cat-301/GABA cells are also numerous within layers II–III and VI (Hockfield *et al.*, 1984; Hendry *et al.*, 1984b; DeYoe *et al.*, 1990). Each of these layers is a site of geniculocortical terminations. However, both Cat-301 and VVA recognize few cells in layers IVA and IVCβ (Hockfield *et al.*, 1984b; Hendry *et al.*, 1984; Mulligan *et al.*, 1989; DeYoe *et al.*, 1990), which also receive a dense geniculocortical innervation (Hubel and Wiesel, 1968, 1972; Hendrickson *et al.*, 1978). The suggestion has been made for Cat-301 in particular that the laminar patterns of cell staining may be related to the functional differences displayed by cells in different layers of area 17 (DeYoe *et al.*, 1990).

2.6. Sublaminar Organization of GABA Cells and GABA_A Receptors

GABA somata and terminals and GABA_A receptors are distributed in distinct sublaminae or preferentially occupy distinct compartments within layers IVC, IVA, and II–III. The sublaminar distribution is apparent at a relatively coarse level, since neurons immunoreactive for GABA and GAD are more densely distributed in layer IVCβ than in layer IVCα (Hendrickson *et al.*, 1981; Hendry *et al.*, 1987; Fitzpatrick *et al.*, 1987). Within layer IVCα, a narrow, superficial band of cells stands out because it contains GAD-positive neurons that are usually large (somal area greater than 120 μm², Fitzpatrick *et al.*, 1987). The immunocytochemical results suggest that layer IVC may be subdivided into bands that do not correspond to the usual division into layers IVCα and IVCβ. The most marked subdivision of layer IVC is seen when GABA somata are labeled by *in situ* hybridization histochemistry for GAD mRNA (Fig. 7). In that case, layer IVC is divided into a series of five alternating, densely and lightly labeled bands (Benson *et al.*, 1991a). A wide superficial band of dense hybridization and a narrow deep band of light hybridization are present in layer IVCα. Two dense bands occupy superficial and deep aspects of layer IVCβ, separated by a third, light band. Alternating superficial, and deep bands of light and dense hybridization are also seen in layer VI. It is not clear why the pattern of GAD mRNA localization differs so markedly from the pattern of immunocytochemical localization of GAD or GABA, although it is possible that the greater sensitivity of *in situ* hybridization histochemistry and its ability to detect differences in the relative amounts of GAD mRNA may permit it to demonstrate hetero-

Figure 5. (A, B) Adjacent sections through area 17 of a macaque monkey stained immunocytochemically for tachykinins (A) and histochemically for cytochrome oxidase (CO) (B). Tachykinin-immunoreactive cells and processes are concentrated in the CO patches of layers II–III and in the CO-rich layers IVA and IVC. Bar = 200 μm. From Hendry *et al.* (1989a). (C, D) Fluorescence photomicrographs of the same microscopic field in layer IVC showing colocalization of GABA (C) and tachykinin (D) immunoreactivity in four somata (arrows). Bar = 10 μm. From Jones *et al.* (1988).

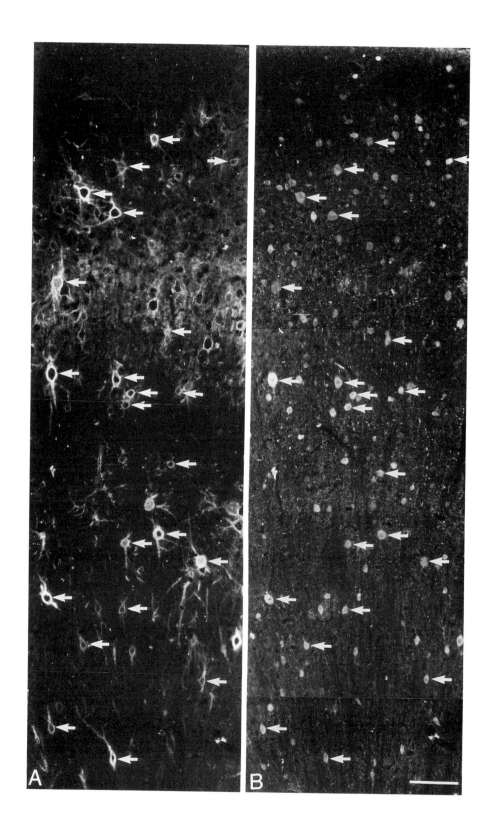

geneities that are not revealed by other methods. The significance of the many GABAergic sublaminae in layer IVC may be related to the precisely stratified pattern of connections formed by neurons in this layer with those in supragranular and infragranular layers (Fitzpatrick *et al.*, 1985; Blasdel *et al.*, 1985). Different types of GABA neurons may contribute to the different patterns of interlaminar projections (Lund, 1990; and see Section 4.3).

GABA$_A$ receptors and the subunit proteins that make up these receptors are also unevenly distributed within layer IVC. Earlier radioligand binding studies suggested that either layer IVCα (Rakic *et al.*, 1988) or IVCβ (Shaw and Cynader, 1986) contained the highest density of GABA$_A$ receptor sites, but a later combined examination of radioligand binding and immunocytochemical staining for the β2/β3 subunits of the receptor indicated that the receptors are densest in layer IVCβ (Hendry *et al.*, 1990) (Fig. 1). Immunostaining for other subunit proteins (α1 and γ2) of the GABA$_A$ receptor is intense in the two halves of layer IVC, with α1 immunostaining being somewhat denser in layer IVCβ (Fig. 8) (Huntsman *et al.*, 1991).

Layers II–III and layer IVA contain two markedly different compartments that stain differentially for cytochrome oxidase (CO) (Horton and Hubel, 1981; Wong-Riley and Carroll, 1984a,b; Wong-Riley and Norton, 1988). In layer IVA, intense CO staining takes the form of a delicate lattice or honeycomb in which less well-stained lacunae are embedded (Hendrickson, 1982; Lund, 1987; Wong-Riley and Norton, 1988) (Fig. 9). The laths of the lattice are the sites of geniculocortical terminations in layer IVA (Wong-Riley and Norton, 1988; Lund, 1987; Hendrickson, 1982) and a region in which a high density of large GAD-immunoreactive terminals (Fitzpatrick *et al.*, 1987) and intense immunoreactivity for the β2/β3 subunits of the GABA$_A$ receptor are found (Hendry *et al.*, 1990) (Fig. 9). The lacunae, on the other hand, receive no geniculocortical terminations and are deficient both in large GAD-positive terminals and in GABA$_A$ receptor immunoreactivity. The somata of GABA neurons, however, appear to be evenly distributed within layer IVA (Fitzpatrick *et al.*, 1987). These data indicate that based on their position, two different kinds of GABA neurons are present in layer IVA: some occupy one compartment (the latticework), where their somata are contacted by large GABA-containing terminals, covered with GABA receptors and likely to be densely innervated by geniculocortical axons. Other GABA neurons in layer IVA occupy a different compartment (the lacunae) where large GABA terminals, GABA receptors, and geniculocortical terminals are either absent or present in very low densities. There are no indications whether different morphological or chemical classes of GABA cells are specifically located in the lattice or the lacunae.

A similar division of GABA neurons, based on the locations of their somata, is found in layers II and III. CO histochemistry reveals in these layers periodic patches of intense staining, variously referred to as patches, puffs, periodicities, or blobs, that line up in rows (Fig. 10) (Horton and Hubel, 1981; Humphrey and Hendrickson, 1983; Horton, 1984; Livingstone and Hubel, 1984; Wong-Riley and Carroll, 1984b). The patches contain neurons whose functional properties differ from those displayed by cells in the surrounding interpatch regions (Toot-

Figure 6. Fluorescence photomicrographs from the same microscopic field through the thickness of area 17, illustrating immunocytochemical staining by the Cat-301 antibody (A) and for GABA (B). Arrows indicate double-labeled cells. Bar = 100 μm. From Hendry *et al.* (1988b).

ell *et al.,* 1988a–c; Humphrey and Hendrickson, 1983; Livingstone and Hubel, 1984; Ts'o and Gilbert, 1988). They receive terminations of geniculocortical axons (Livingstone and Hubel, 1982; Hendrickson, 1982; Itaya *et al.,* 1984) and contain large GAD-positive terminals (Fitzpatrick *et al.,* 1987), both of which are missing from the interpatch regions. The patches also show more intense immunoreactivity for the β2/β3 subunits of the GABA$_A$ receptor than do the interpatch regions (Hendry *et al.,* 1990). GABA somata are equally concentrated in the patches and in the interpatch regions. Although the distribution of GABA- and GAD-immunoreactive somata in layers II and III is not uniform, the non-uniformities do not coincide with CO-stained patches or CO-weak interpatch regions (Fig. 10) (Fitzpatrick *et al.,* 1987; Hendry *et al.,* 1987) and the high density of GABA terminals and GABA$_A$ receptors in the CO patches is not associated with specific concentrations of GABA cells.

In layers II and III, GABA cells in the patches are chemically distinct from GABA cells in the interpatch regions. GABA neurons that express a cell-surface

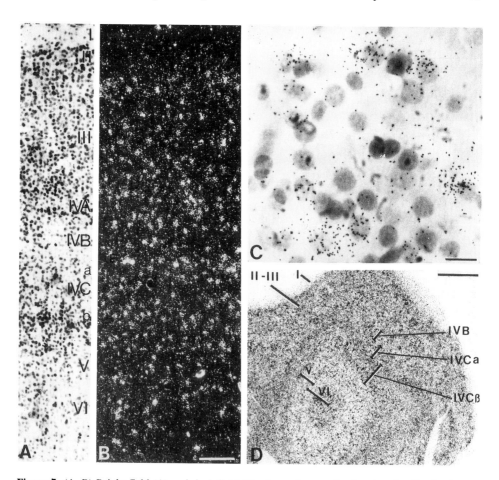

Figure 7. (A, B) Bright-field (A) and dark-field (B) photomicrographs showing the distribution of GAD-expressing cells, labeled by *in situ* hybridization histochemistry (B) in relation to the cell layers (A) of area 17 of a monkey. Bar = 100 μm. (C) Higher-power bright-field photomicrograph from B, showing radioactive label over Nissl-stained cell nuclei. Bar = 15 μm. (D) GAD *in situ* hybridization autoradiograph, showing pattern of sublaminar organization of GABA cells in area 17. Bar = 2 μm. From Benson *et al.* (1991a).

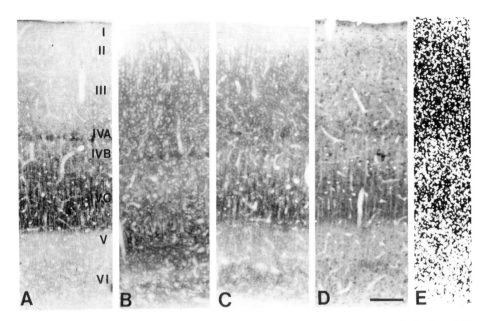

Figure 8. Matched sections stained for cytochrome oxidase (A), immunocytochemically for α1 (B), β2/β3 (C), and γ2 (D) subunits of the GABA$_A$ receptor, and an autoradiograph showing [^{3}H]muscimol binding (E) in area 17 of macaque monkeys. Bar = 200 μm. S. H. C. Hendry, M. M. Huntsman, A. L. deBlas, H. Möhler, and E. G. Jones (unpublished).

proteoglycan recognized by the monoclonal antibody Cat-301 are found preferentially within the patches (Hendry *et al.*, 1984b, 1988b; Hockfield *et al.*, 1984). A similar preference for the patches is shown by GABA somata that are parvalbumin-immunoreactive. By contrast, GABA somata in which immunoreactivity for calbindin (Celio *et al.*, 1986) or NPY coexists (Kuljis and Rakic, 1989b) are distributed principally around the patches. For the two GABA subpopulations that contain one or other of the calcium-binding proteins, the selective distribution is evident only in relation to patches in layer III; in layer II, GABA/calbindin neurons are uniformly distributed (Van Brederode *et al.*, 1990) and GABA/parvalbumin neurons are virtually absent. Layer II–III patches are also intensely immunostained for tachykininlike immunoreactivity, but as in the case of GABA and GAD, it is the staining of terminals and not of cell bodies which produces the selective pattern (Hendry *et al.*, 1988a). All of these immunostaining patterns are evident in normal adult monkeys. The distribution patterns are changed in adult monkeys following monocular deprivation (Section 6).

Periodic patches of relatively intense CO staining are also found in layers IVB, V, and VI (Horton and Hubel, 1981; Horton, 1984; Livingstone and Hubel 1984; Wong-Riley and Carroll, 1984a). In these layers the density of GABA neurons is low and the distribution is uniform (Fitzpatrick *et al.*, 1987; Hendry *et al.*, 1987). However, certain subpopulations of GABA neurons are selectively distributed around or in the deep layer patches. The somata of NPY-immunoreactive GABA neurons, located principally in layers V and VI and in the subjacent white matter (Hendry *et al.*, 1984c; Kuljis and Rakic, 1989b; Tigges *et al.*, 1989), are preferentially distributed outside the patches. Other groups of neurons in layers IVβ, IVCα, and VI that are immunostained by Cat-301 selectively occupy the patches (Fig. 6) (Hendry *et al.*, 1984b; Hockfield *et al.*, 1984; DeYoe *et al.*, 1990), and while these groups include some pyramidal cells, they are made up principally of GABA neurons (Hendry *et al.*, 1988b). GABA/Cat-301 cells thus form something resembling a radial unit running through the full thickness of area 17, interrupted only by relatively light and uniform Cat-301 staining in layers IVA and IVCβ.

From the laminar positions of Cat-301-positive neurons in area 17, from the intense Cat-301 immunostaining of the magnocellular layers of the monkey LGN (Hockfield *et al.*, 1984; Hendry *et al.*, 1984b; Hendry *et al.*, 1988b), and from the presence and position of Cat-301-immunostained bands in area 18 (Hendry *et al.*, 1988b; DeYoe *et al.*, 1990), it has been argued that the antibody preferentially recognizes neurons in the monkey central visual system that are responsive to broadband visual stimuli (Hendry *et al.*, 1984b, 1988b; Hockfield *et al.*, 1984) and concerned with the detection and analysis of moving objects (DeYoe *et al.*, 1990). Support for the suggestion that Cat-301 cells are part of a magnocellular

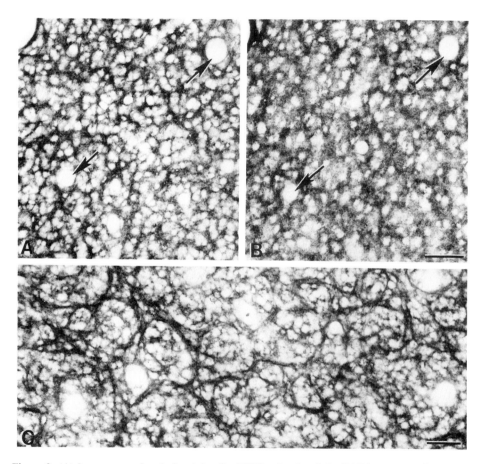

Figure 9. (A) Immunocytochemical staining for β2/β3 subunits of the GABA$_A$ receptor in tangential sections through layer IVCβ matched to an adjacent cytochrome oxidase-stained section (B), showing the fine, latticelike distribution of receptor immunostaining in that layer. Arrows indicate same blood vessels. Bar = 50 μm. (C) β2/β3 immunostaining of the GABA$_A$ receptor lattice in layer IVA. Bars = 50 μm. Hendry *et al.* (1990).

or M channel has come from studies of the cat visual system (Sur *et al.*, 1988). Consistent with this suggestion, the GABA/Cat-301 neurons of the CO-stained patches in superficial layers of monkey area 17 could be considered part of the broadband population that has been recorded in these regions (Livingstone and Hubel, 1984). However, significant numbers of Cat-301-immunostained neurons are also present in the parvocellular layers of the monkey LGN and in layer IVCβ of area 17, both of which are considered to be part of a parvocellular or P channel, whose neurons are involved in the analysis of form and color, and the densest Cat-301 immunostaining occurs in layer VI which is not generally assigned to either channel. These data indicate that while preferentially distributed in layers or compartments thought to be part of an M channel, the Cat-301 cells of area 17 and the LGN are not exclusive to these layers or compartments.

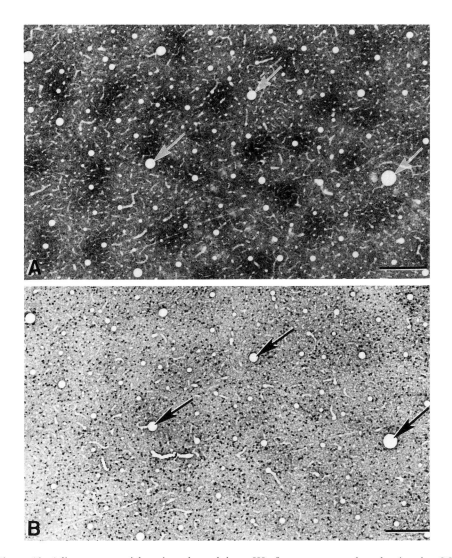

Figure 10. Adjacent tangential sections through layer III of a macaque monkey, showing that CO-stained patches (A) are not associated with focal concentrations of GABA-immunoreactive somata (B). Arrows indicate same blood vessels. Bars = 100 μm. Hendry *et al.* (1987).

2.7. GABA Neurons and Receptors in Human Visual Cortex

Ocular dominance columns and periodic patches of CO activity have been identified in human visual cortex (Horton and Hedley-White, 1984; Horton *et al.*, 1990). Golgi studies of the human visual area have also identified morphological classes of neurons that are characterized by GABA immunoreactivity in other primates. These include chandelier cells (Marin-Padilla, 1987) and double bouquet cells (DeFelipe and Jones, 1988). In general, the size, shape, and distribution of these cells are similar in human and monkey visual cortex, although human chandelier cells are smaller and their axons appear to contact fewer pyramidal cells than in monkeys (Marin-Padilla, 1987).

Direct identification of GABA neurons in area 17 and in other areas of the human brain has been impeded by the fact that GABA and GAD are unusually sensitive to postmortem autolysis. Biopsied samples have been used to study GAD-immunoreactive neurons in other regions of the human visual system (Zinner-Feyerabend and Braak, 1991) and in other areas of the human cortex (Schiffmann *et al.*, 1988; Kisvárday *et al.*, 1990), but no reports of GABA or GAD immunostaining are available for human area 17. Studies that have investigated the morphology and distribution of immunocytochemically stained calbindin

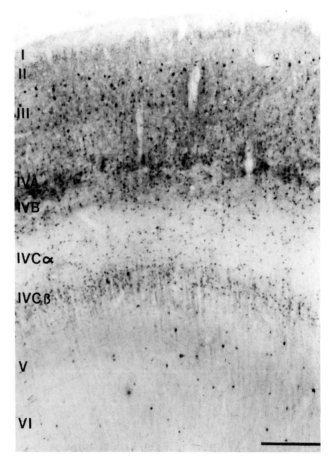

Figure 11. Calbindin-immunoreactive cells in area 17 of a human brain. Bar = 500 μm. S. H. C. Hendry and E. G. Jones (unpublished).

and parvalbumin cells, however, can be taken as a useful guide to the characteristics of GABA neurons in human visual cortex. In neurologically normal adult humans, parvalbumin neurons occupy all layers of area 17, with somata and terminal-like puncta being most numerous in layer IVC (Blümke *et al.*, 1990), as in macaque visual cortex (Blümke *et al.*, 1990; Van Brederode *et al.*, 1990; Hendrickson *et al.*, 1991). The proportion of parvalbumin-immunoreactive somata is reported to be smaller and their size larger in humans than in macaques (Blümke *et al.*, 1990). Calbindin-immunoreactive cells and puncta in humans (Fig. 11) are concentrated in regions surrounding the CO-rich periodicities of layer III (Hendry and Cander, 1993), as in monkeys (Celio *et al.*, 1986).

The distribution of GABA$_A$ receptors as determined by ligand binding studies in human area 17 also resembles that seen in macaque area 17. The laminar distribution of receptors in the human cortex revealed by binding of [^{3}H]flunitrazepam appears identical to that in the macaque, with the highest densities in layers IVCβ and II–III (Zezula *et al.*, 1988). However, immunocytochemical localization of specific GABA$_A$ receptor subunits shows a relatively low level of receptor immunostaining in layer IVA of the human (Fig. 12)

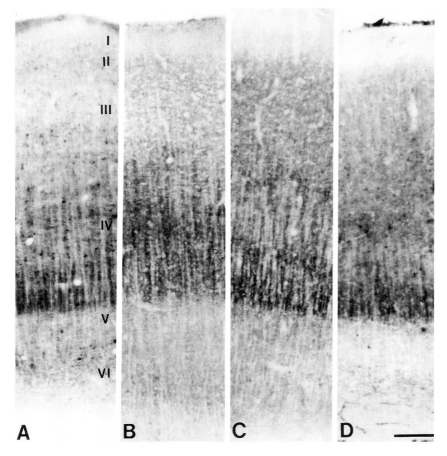

Figure 12. Portion of a CO-stained section (A) though area 17 of a human brain, matched to sections showing the distribution of immunoreactivity for α1 (B), β2/β3 (C), and γ2 (D) subunits of sections showing the distribution of immunoreactivity for α1 (B), β2/β3 (C), and γ2 (D) subunits of the GABA$_A$ receptor. Bar = 250 μm. S. H. C. Hendry, A. L. deBlas, H. Möhler, and E. G. Jones (unpublished).

in comparison with the monkey (Hendry *et al.*, 1990). The latticework of immunostaining for β2/β3 subunits that is a prominent feature of layer IVA in the monkey is absent from layer IVA of the human. This correlates with an absence of intense CO staining in layer IVA of the human (Horton and Hedley-White, 1984).

3. Contributions of GABA Neurons to Circuitry of Primate Area 17

3.1. Morphological Characterization of Aspiny Nonpyramidal Neurons in Primate Cerebral Cortex

Morphologically, the primary visual cortex is typical of all areas of the primate neocortex in possessing two fundamental categories of cell: pyramidal and nonpyramidal. The GABA cells without exception belong to the nonpyramidal class and although there are some variations in detail from area to area, a basic description of the nonpyramidal cells can be made for all areas. All of the forms mentioned in this section have at some time been described in primate area 17 although the most comprehensive descriptions have not always come from studies of primates or of area 17.

The nonpyramidal cells of the cerebral cortex form the short-axon cells or interneurons and comprise two classes: *spiny nonpyramidal cells,* which represent the typical neurons of the middle cortical layers (especially layer IV), and *aspiny nonpyramidal cells* or neurons with smooth or sparsely spiny dendrites. The latter constitute the majority of the cortical interneurons and it is to this class that the GABA cells belong (Ribak, 1978; Houser *et al.*, 1983). Although occasional aspiny nonpyramidal cells have proven to be GABA-negative in the primate cortex (DeFelipe and Jones, 1992), a general principle seems to have emerged that all the morphologically well-characterized forms of aspiny cortical cell are GABAergic and in this section we shall review much of the data leading up to this conclusion.

Types of Aspiny Nonpyramidal Neurons

Aspiny nonpyramidal neurons are morphologically heterogeneous, and eight or more different types can be recognized, primarily on the basis of their patterns of axonal arborization (e.g., Lund, 1973; Jones, 1975; Fairén *et al.*, 1984). For the purposes of description, they can be divided into three principal groups (Fig. 13): (1) neurons that form very local connections, (2) neurons with long horizontal axonal collaterals, and (3) neurons that form vertical connections.

Group 1. This group includes the majority of aspiny nonpyramidal cells and is made up of neurons whose axons are distributed in the vicinity of the parent cell body. The best known types are the neurogliaform cells and chandelier cells but other forms also exist. *Neurogliaform cells* have very local axonal ramifications, and were first described by Ramón y Cajal (1899a), who also called them arachniform or spidery cells (Fairén *et al.*, 1984; Jones, 1984; see also DeFelipe and Jones, 1988). It seems to us that the majority of the "clutch cells" of Kisvárday *et al.* (1986, 1990) probably belong to the neurogliaform group although others have superficial resemblances to small basket cells. The cells are characterized by a small cell body that emits five or more short primary dendrites that divide only once or twice and recurve to give rise to a small, dense, and spherical dendritic field. The axon is thin and divides profusely, forming a very dense axonal plexus that is frequently coincident with the dendritic field. These cells are common in monkey and human visual cortex (Ramón y Cajal, 1899a; Kisvárday *et al.*, 1986, 1990). *Chandelier cells* are distinguished by their axons which terminate in vertical rows of boutons resembling candlesticks in the visual and other cortical areas. The axonal fields have variable shapes. They usually form a plexus close to the parent soma but occasionally a displaced axonal field is seen (Lund *et al.*, 1979; Fairén and Valverde, 1980; Freund *et al.*, 1983). The dendritic fields are variable in shape but commonly bitufted (Jones, 1975; Fairén and Valverde, 1980; Peters *et al.*, 1982; Somogyi *et al.*, 1982).

Cells with axonal arcades (Jones, 1975) or cells with *sparsely spinous dendrites forming local axonal connections* (Peters and Saint-Marie, 1984) have less stereotyped dendritic fields and axonal aborizations. These are the generic form of the short-axon cell of Cajal and are present in large numbers in the primate visual cortex (Lund, 1973). They probably form the majority of the small-cell population in layer I but are found in other layers as well.

Group 2. Two types of neurons have long horizontal axons: large or classic basket cells and Cajal–Retzius neurons. *Large basket cells* represent one of the most distinctive aspiny nonpyramidal cells of the primate sensory-motor cortex, but a similar, smaller type is also found in area 17 (Marin-Padilla, 1969; Jones, 1975; for reviews see Jones and Hendry, 1984, and Fairén *et al.*, 1984). The cells in the sensory-motor areas are characterized by very large somata, long dendrites that tend to run vertically, and a series of lengthy, horizontally oriented, myelinated axon collaterals. At intervals the collaterals give rise to short terminal branches contributing to the typical basket formations on pyramidal cell somata and proximal dendrites. These basketlike terminations, first discovered by Ramón y Cajal (1899a) in the human visual and motor cortices, are also found in the visual areas of other primates (Somogyi *et al.*, 1983b; DeFelipe *et al.*, 1986b; Kisvárday *et al.*, 1987). Classic basket cells tend to give rise to their densest axonal plexus in the vicinity of the cell body, but a secondary plexus is sometimes seen below or above the parent cell body (e.g., Somogyi *et al.*, 1983b; DeFelipe *et al.*, 1986b; Kisvárday *et al.*, 1987). *Cajal–Retzius neurons,* or special cells of layer I, were first described in 1890 by Ramón y Cajal in the cortex of neonatal small mammals, and confirmed by Retzius in 1891 in the fetal human cortex (see DeFelipe and Jones, 1988). The general belief is that Cajal–Retzius neurons are a transient embryonic or neonatal form, but this is not accepted by all authors (Marin-Padilla and Marin-Padilla, 1984). They are characterized by very long, horizontally oriented processes arising from a moderately large cell body. The processes are difficult to identify as dendrites or axons and give off numerous short, vertically oriented branches ending underneath the pia mater in knotty thickenings. According to Ramón y Cajal, as the cerebrum develops, the distinction between dendritic and axonal processes becomes clearer.

Group 3. This group is made up of neurons with long descending and/or ascending axons. The *Martinotti cell* is the name given by Ramón y Cajal (1899a) to small cells that possess ascending axons. In the modern literature there are only a few references to these cells and they have not been well characterized by morphological studies using modern methods. Cells situated in deep cortical layers and labeled retrogradely after small injections of [³H]-GABA in superficial layers, are thought to be a type of Martinotti cell (Figs. 13 and 14). They probably belong to an aspiny type observed in a few Golgi studies and which forms a dense focal axonal arborization above the parent cell body (Somogyi *et al.*, 1983a; DeFelipe and Jones, 1985). The *double bouquet cell* is the name given by Ramón y Cajal to a type of cell with a bitufted dendritic arborization and particularly characterized by an axon that forms tightly intertwined bundles of long vertical collaterals. According to Ramón y Cajal (1899a), the axon: "*resolves itself rapidly into an infinity of vertical threads, so abundant that they engender truly small*

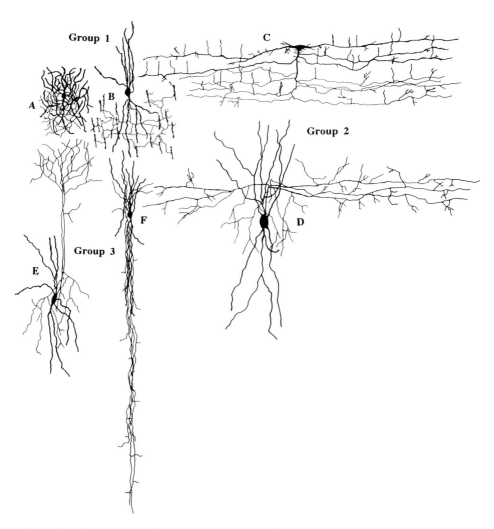

Figure 13. Principal types of aspiny nonpyramidal cells in the primate cerebral cortex. Group 1 is made up of neurons that form very local connections (A, neurogliaform or arachniform cell; B, chandelier cell). Group 2 consists of neurons with long horizontal axon collaterals (C, Cajal–Retzius neuron; D, large or classic basket cell). Group 3 is made up of neurons that form vertical connections (E, Martinotti cell; F, double bouquet cell).

bundles, comparable to locks of hair, and so long that they extend through almost the whole thickness of the gray matter." Recently, it has been shown that double bouquet cell axons form a widespread and regular microcolumnar structure in the monkey cerebral cortex: in tangential sections passing through layer III (Fig. 15), there are 7–15 double bouquet cell axon bundles/10,000 μm^2 (DeFelipe *et al.*, 1990). It is probably these cells whose somata are labeled by retrograde transport of [^{3}H]-GABA from injections in deeper layers of the visual cortex (Somogyi *et al.*, 1981a,b, 1983a; DeFelipe and Jones, 1985).

A further less well-defined cell type that can probably be included in this group are the *elongated fusiform, bipolar or bitufted cells* whose dendrites can extend vertically across several or virtually all layers of the cortex. The axon which is loosely ramifying, tends also to follow the elongated, vertical orientation of the dendrites. Cells of this type tend to be selectively stained by neuropeptide immunoreactivity.

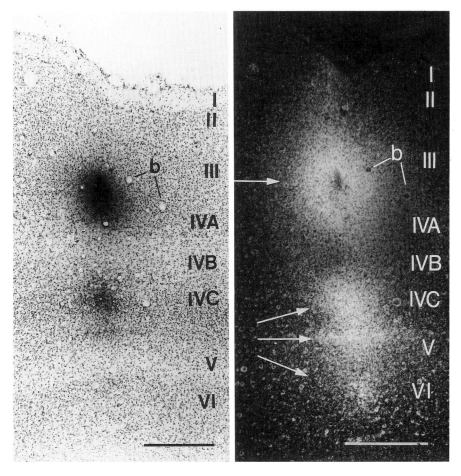

Figure 14. Bright-field (left) and dark-field (right) photomicrographs through area 17 of a macaque monkey, showing specific retrograde labeling of GABA cells in layers IVC, V, and VI (lower three arrows) following injection of [^{3}H]-GABA in layer III (upper arrow). b, blood vessel profiles used for localization purposes. Bars = 500 μm. From DeFelipe and Jones (1985).

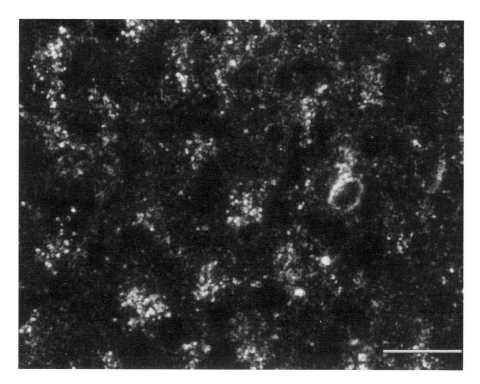

Figure 15. Tangential section through layer III of area 17 of a monkey showing regularly spaced bundles of double bouquet cell axons stained immunocytochemically for calbindin. Bar = 20 μm. DeFelipe *et al.* (1990).

3.2. GABAergic Nature of Aspiny Nonpyramidal Neurons

3.2.1. General

The GABAergic nature of the aspiny population of cortical neurons was established primarily on the basis of electron microscopic immunocytochemical studies (Ribak, 1978; Hendry *et al.*, 1983a), with strong support from studies of uptake and transport of [³H]-GABA (Hendry and Jones, 1981; Somogyi *et al.*, 1981a,b, 1983a; DeFelipe and Jones, 1985), and from correlative morphological investigations (Freund *et al.*, 1983; Peters *et al.*, 1982). The number of direct demonstrations of the GABAergic character of individual morphological types of aspiny nonpyramidal cell, especially in primate cortex, is, however, rather small (Table II). This is understandable on account of the limited extent of dendritic and axonal staining achieved in immunocytochemical preparations and the technical demands of co-staining injected or Golgi-impregnated neurons for immunoreactive GABA or GAD. Neurogliaform cells, chandelier cells, double bouquet cells, and large basket cells have been positively identified as GABAergic, although not all the evidence has come from the visual cortex. Layer I cells (other than Cajal–Retzius cells), the generic form of aspiny cell, and the long, stringy bipolar or bitufted cells can be regarded as GABAergic from indirect evidence or from correlative studies in other species.

Table II. Positive Chemical Characteristics of Identified Aspiny Nonpyramidal Cells in the Primate Cerebral Cortex

Cell type	Species	Cortical region	Identified substance[a]	References
Neurogliaform	Human	Superior temporal gyrus	GABA	Kisvárday et al. (1990)
	Monkey	Visual cortex (area 17)	GABA	Kisvárday et al. (1986)
Chandelier	Monkey	Primary sensory-motor, association regions of the prefrontal and occipital cortices[b]	PV	DeFelipe et al. (1989a), Lewis and Lund (1990), DeFelipe and Jones (1991), Hendrickson et al. (1991), Williams et al. (1992)
	Monkey	Association regions of the prefrontal, temporal, parietal, and occipital cortices	CRF[c]	Lewis and Lund (1990)
Double bouquet	Human	Temporal cortex, visual and auditory cortices	CaBP	Ferrer et al. (1992), Hayes and Lewis (1992)
	Monkey	Parietal, temporal, and visual areas	CaBP	DeFelipe et al. (1989b, 1990), Hendry et al. (1989)
	Monkey	Superior and inferior temporal gyrus, visual cortex (area 18)	SRIF	de Lima and Morrison (1989)
	Monkey	Primary auditory cortex and surrounding areas, visual cortex (area 18)	TK	DeFelipe et al. (1990)
	Monkey	Somatic sensory cortex	GABA	DeFelipe and Jones (1992)
Large basket	Human	Temporal cortex	GABA	Kisvárday et al. (1990), Schiffmann et al. (1988)
	Monkey	Visual cortex (area 17)	GABA	Fitzpatrick et al. (1987), Mulligan et al. (1989)
	Monkey	Primary sensory-motor cortex, visual cortex (area 17)	GAD	Houser et al. (1983), DeFelipe et al. (1986b), Fitzpatrick et al. (1987)
	Human	Visual cortex (area 17)	PV	Blümke et al. (1990)
	Monkey	Primary sensory-motor cortex and surrounding areas, temporal cortex and visual cortex	PV	Hendry et al. (1989), DeFelipe et al. (1989a), Blümke et al. (1990), Hendrickson et al. (1991), Akil and Lewis (1992)
	Monkey	Visual cortex (area 17)	VVA	Mulligan et al. (1989)
	Monkey	Occipital, parietal, temporal, and frontal areas	Cat-301	Hendry et al. (1988b)
Cajal–Retzius	Monkey	Visual, prefrontal, sensory-motor, and temporal cortices	CaBP	Huntley and Jones (1990), Hendrickson et al. (1991)
	Monkey	Visual cortex	PV	Huntley and Jones (1990)

[a]Abbreviations: CaBP, calbindin; Cat-301, monoclonal antibody Cat-301; CRF, corticotropin-releasing factor; GABA, γ-aminobutyric acid; GAD, glutamic acid decarboxylase; PV, parvalbumin; SRIF, somatostatin; TK, tachykinin; VVA, lectin *Vicia villosa* agglutinin.

3.2.2. Neurogliaform Cells

Neurogliaform (or "clutch") cells have been shown to be GABAergic by combining Golgi staining and postembedding GABA immunocytochemistry in both human and monkey visual cortex (Kisvárday *et al.*, 1986, 1990). These authors demonstrated that nine Golgi-impregnated neurogliaform cells, located in layers III–VI of the human temporal cortex, and nine in layer IVC of the monkey visual cortex, were GABA-positive (Fig. 16). It is not known whether they contain any of the several neuropeptides and calcium-binding proteins commonly colocalized with GABA in cerebral cortical neurons.

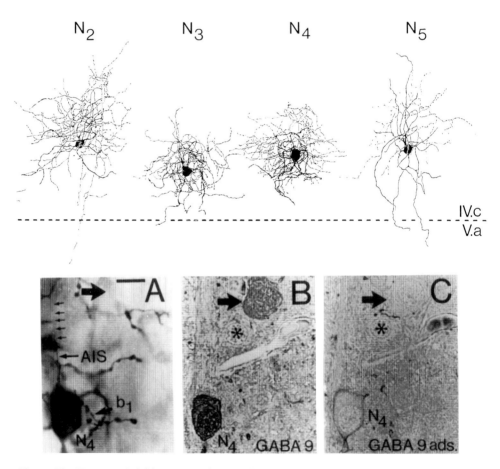

Figure 16. (Top row) Golgi-impregnated neurogliaform (or clutch) cells in layer IVC of monkey striate cortex. All neurons were also immunoreactive for GABA. From Kisvárday *et al.* (1986). (Bottom row) GABA immunoreactivity in the Golgi-impregnated neurogliaform cell N_4. (A) Photomicrograph of cell N_4. (B, C) 0.5-μm-thick consecutive sections of N_4, B stained for GABA by postembedding immunocytochemistry, C stained after the GABA antibody was preadsorbed with GABA coupled to polyacrylamide beads. Arrow indicates a second GABA cell. Bar = 10 μm. From Kisvárdy *et al.* (1986).

3.2.3. Chandelier Cells

Chandelier cell axons end in terminals that form symmetric synapses, which are characteristic of GABAergic axon terminals (Somogyi, 1977, 1979; Fairén and Valverde, 1980; Somogyi *et al.*, 1982; Peters *et al.*, 1982; Freund *et al.*, 1983; DeFelipe *et al.*, 1985). At the site of termination on pyramidal axon initial segments the majority of terminals are immunoreactive for GAD or GABA and form vertically oriented rows identical to those of chandelier cell terminals identified in Golgi material (Peters *et al.*, 1982; Freund *et al.*, 1983; DeFelipe *et al.*, 1985). Combined Golgi impregnation and immunocytochemistry for GAD or GABA in the visual cortex of the cat revealed that boutons derived from a chandelier cell axon (Freund *et al.*, 1983) and chandelier cell somata themselves (Somogyi *et al.*, 1985) were immunoreactive for GAD and GABA, respectively.

Immunocytochemical studies have failed to how the presence of vasoactive intestinal polypeptide (VIP), CCK, SRIF, or tachykinin immunoreactivity (Peters *et al.*, 1982; Hendry *et al.*, 1983b; Freund *et al.*, 1985; Jones *et al.*, 1988; de Lima and Morrison, 1989), or calbindin immunoreactivity (Hendry *et al.*, 1989; De-Felipe *et al.*, 1989b) in chandelier cell axons. However, in the monkey visual and other areas, immunoreactivity for the neuropeptide, corticotropin-releasing factor (CRF) (Lewis *et al.*, 1989; Lewis and Lund, 1990), as well as parvalbumin (DeFelipe *et al.*, 1989a; Lewis and Lund, 1990; Hendrickson *et al.*, 1991) is found in chandelier cell axons. However, only a subpopulation of chandelier cell axons are immunoreactive for parvalbumin or CRF, depending on the cortical area and primate species (reviewed in DeFelipe and Fariñas, 1992).

3.2.4. Double Bouquet Cells

Early indirect evidence suggested that these cells are GABAergic (Somogyi and Cowey, 1981, 1984; DeFelipe *et al.*, 1989b, 1990) (Fig. 17). In a recent study involving the colocalization of GABA and calbindin immunoreactivity, DeFelipe and Jones (1992) found that double bouquet cell axons identified by calbindin immunostaining were also immunoreactive for GABA. Terminals of many additional double bouquet cell-like axons were identified as containing GABA by postembedding immunocytochemistry.

In the monkey visual cortex, double bouquet cells axons do not appear to stain immunocytochemically for parvalbumin, in keeping with the virtual lack of colocalization of calbindin and parvalbumin in the same cortical neurons (Hendry *et al.*, 1989). Otherwise, double bouquet cells appear to be chemically heterogeneous, since they can be immunoreactive for calbindin, somatostatin, or tachykinin (Fig. 18) (Table II), depending on the area in which they are located (de Lima and Morrison, 1989; DeFelipe *et al.*, 1990). In area 17 of monkeys where numerous calbindin-immunoreactive double bouquet cell axons are found, very few are immunoreactive for tachykinin or somatostatin, but in other regions, double bouquet axon bundles immunoreactive for the peptides are common. Autoradiographic studies involving the uptake and transport of [^{3}H]-GABA after small injections of [^{3}H]-GABA in deep layers of the cortex, result in multiple long, vertical rows of silver grains joining the injection site with retrogradely labeled somata in layers II and III (Somogyi *et al.*, 1981a, 1983a; De-Felipe and Jones, 1985). These vertical rows closely resemble the immunoreactive bundles of double bouquet cell axons (Somogyi and Cowey, 1984; DeFelipe and Jones, 1985).

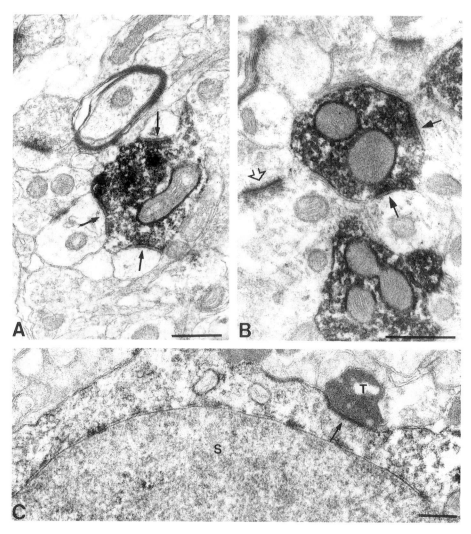

Figure 17. (A) Electron micrograph of a tachykinin-immunoreactive axon terminal, provisionally identified as arising from a double bouquet cell axon bundle, terminating in symmetric synapses (arrows) on three dendritic spines in layer III of area 17. Bar = 0.5 μm. (B) Electron micrograph of GABA-immunoreactive terminal making symmetrical synapses (arrows) in layer IVC of area 17. Bar = 0.5 μm. J. DeFelipe and E. G. Jones (unpublished). (C) Electron micrograph showing a GABA-immunoreactive cell soma (S) in layer IVC of area 17, receiving a thalamocortical axon terminal (T) identified by Wallerian degeneration consequent upon a lesion placed in the lateral geniculate nucleus 4 days earlier. Bar = 0.5 μm. S. H. C. Hendry and E. G. Jones (unpublished).

3.2.5. Large Basket Cells

Large basket cells are considered to be GABAergic on the grounds that the largest nonpyramidal cell somata (15–20 μm in diameter) invariably stain immunocytochemically for GABA and GAD. In addition, pyramidal cell somata and their proximal dendrites are commonly surrounded by GABA- and GAD-immunoreactive multiterminal endings of the basket type, a pattern of staining that has also been found in material stained immunocytochemically for parvalbumin (Houser *et al.*, 1983; Hendry *et al.*, 1983a, 1989; DeFelipe *et al.*, 1986b,

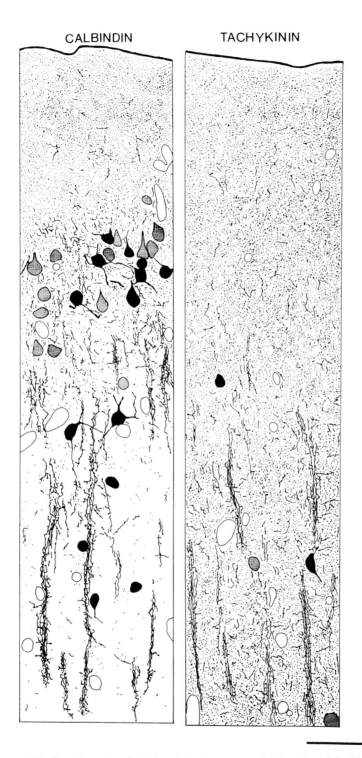

Figure 18. Calbindin (left) and tachykinin (right) immunoreactive bundles of double bouquet cell axons descending through layer III of area 18 in a macaque monkey. Larger profiles are immunoreactive cell somata. Bar = 100 μm. From DeFelipe *et al.* (1990).

1989a; Fitzpatrick *et al.*, 1987; Schiffmann *et al.*, 1988; Blümke *et al.*, 1990; Kisvárday *et al.*, 1990; Hendrickson *et al.*, 1991). Parvalbumin-immunoreactive terminal formations that closely resemble the classic basket formations of Ramón y Cajal are prominent in the primary motor cortex (Hendry *et al.*, 1989; Akil and Lewis, 1992), but appear to be less evident in areas 17 and 18 and in the prefrontal cortex (Akil and Lewis, 1992). Nevertheless, parvalbumin-immunoreactive large multipolar cells and perisomatic immunoreactive boutons on pyramidal cells are present in the monkey visual cortex (Blümke *et al.*, 1990; Hendrickson *et al.*, 1991).

Large basket cells do not stain immunocytochemically for known peptides or for calbindin (e.g., see Hendry *et al.*, 1984a, 1989). Material stained for CCK and tachykinins shows a significant number of stained perisomatic boutons on nonimmunoreactive pyramidal cells (Hendry *et al.*, 1983b; Jones *et al.*, 1988), but these probably arise from cells other than classic basket cells, since large, multipolar cells are not stained for the peptides.

3.2.6. Layer I Cells

Virtually all neurons in layer I of adult primate visual cortex stain immunocytochemically for GABA (Hendry *et al.*, 1987) (Figs. 1 and 2). In fetal primates, the additional population of neurons formed by the Cajal–Retzius cells that are not usually detectable in adults, are not GABA-immunoreactive (Huntley and Jones, 1990). The Cajal–Retzius cells can be distinguished from the GABA cells of layer I by the presence in the former of acetylcholinesterase activity and immunoreactivity for calbindin and occasionally parvalbumin (Huntley and Jones, 1990; Hendrickson *et al.*, 1991).

3.2.7. Other Forms of Aspiny Cell

The identification of other forms of aspiny cortical cell in the primate cortex as GABAergic has mainly been by indirect reasoning or on the basis of studies carried out on what appear to be the same cell types in other species, particularly rats and cats. The rather generic form of aspiny cell with axonal arcades or sparsely spinous cell forming local axonal connections is common in primate visual cortex and is undoubtedly GABAergic because similarly sized somata and dendritic configurations show GABA or GAD immunoreactivity (Houser *et al.*, 1983, 1984) and axonal terminals of Golgi-impregnated cells of this type terminate in symmetrical synapses, mainly on dendritic shafts of pyramidal neurons. Terminals ending in symmetrical synapses on apical dendritic shafts are invariably GABAergic (Hendry *et al.*, 1983a). Similar reasoning tends to identify the long, vertical bipolar or bitufted cells as GABAergic, along with the fact that they usually show immunoreactivity for one or more neuropeptides that are commonly colocalized with GABA.

3.3. Laminar Distributions of Morphologically Characterized GABA Cells

The difficulties of obtaining reliable quantitative estimates of specific cell populations in Golgi-impregnated material have made it difficult to assess any laminar differences in the distributions of specific aspiny cell classes in primate visual cortex. Certain morphological classes have been reported to be preferentially located in single layers or in border zones between two layers. Precise axonal morphologies may vary in these sites (Ramón y Cajal, 1899a; Valverde, 1971; Lund, 1973, 1987, 1990; Mates and Lund, 1983; Lund *et al.*, 1988; Werner *et al.*, 1989). The two most easily recognized classes, chandelier and basket cells, are present at most depths from layer III to layer VI. However, chandelier cells appear to be absent from layer IVCβ (Lund, 1987) and the classical type of basket cell, with profuse pericellular terminations, may be present only in layers V and VI (Lund, 1987), although this conflicts with Ramón y Cajal's (1899a) original descriptions in the human visual area. Other neurons which resemble basket cells in the size of their somata and the lateral extents of their axons are present in layer III, in layer IVB, and in other divisions of layer IV, but they appear to lack the pericellular axonal terminations characteristic of basket cells (Lund, 1987, 1990; Lund *et al.*, 1988). This may be misleading, however, given the fact that the typical pericellular baskets are usually built up by the terminations of axons of several basket cells (Ramón y Cajal, 1899a,b).

Double bouquet cell somata are characteristically situated in layers II and III and are concentrated in the superficial half of the latter (Ramón y Cajal, 1899a; DeFelipe *et al.*, 1990; DeFelipe and Jones, 1992). In our studies, the long thin bundles of axons so characteristic of this form of cell are less overt in area 17 than in other areas of primate cortex (Fig. 15), but there are no quantitative data on the relative numbers of such cells from area to area. Arachniform or neurogliaform and the generic form of aspiny stellate cell appear, from the works of Ramón y Cajal and others, to be distributed in all layers. These two seem to form the majority of the aspiny cells classified by Lund (1987, 1990) in layer IVC on the basis of laminar specific axonal distributions. Similar cells with ascending axons situated in deeper layers may be construed as a form of Martinotti cell. The GABA–peptide cells, usually characterized by a long stringy bipolar or bitufted form, appear in most cases to be concentrated in layer II, upper layer III, layer VI, and the underlying white matter (Hendry *et al.*, 1983b, 1984a,c). An exception is the GABA/tachykinin population found in layer IVC but the overall morphology of this class of GABA cell has not been determined. Similarly, the morphology of the GABA-concentrating cells whose somata are retrogradely labeled in layers IVC and V after injections of [^{3}H]-GABA in superficial layers (Fig. 14) (Somogyi *et al.*, 1981a, 1983a; DeFelipe and Jones, 1985) has not been specifically identified.

4. Innervation Patterns of GABA Neurons in Monkey Area 17

4.1. Afferent Innervation

4.1.1. General

Neurons in the monkey visual cortex receive synaptic inputs from neurons in the thalamus (Hubel and Wiesel, 1972), in the contralateral area 17 (Myers, 1962), and in several occipital and temporal areas in the ipsilateral hemisphere (Felleman and Van Essen, 1991). Noradrenergic axons from cells in the locus coeruleus (Foote and Morrison, 1984; Morrison and Foote, 1986; Morrison *et al.*, 1982) and serotoninergic axons from cells in the midbrain raphe (Kosofsky *et al.*, 1984; Morrison and Foote, 1986; Morrison *et al.*, 1982; Takeuchi and Sano, 1984) also terminate in area 17. GABA neurons in monkey area 17 are innervated by thalamocortical and commissural axons (see Sections 4.1.2 and 4.1.3). There is some evidence that serotonin fibers innervate the dendrites of aspiny nonpyramidal neurons (de Lima and Morrison, 1989), which are presumably GABAergic, but little is known of the terminations of other nonspecific afferents and no indication of the cells receiving inputs from other visual cortical areas has been presented.

4.1.2. Thalamocortical Innervation

Electron microscopic studies show direct terminations of geniculocortical axons in GABA neurons in monkey area 17 (Fig. 17). Although most thalamocortical terminals end on dendritic spines (Winfield *et al.*, 1982) and, therefore, on non-GABA cells, a certain proportion of the geniculocortical terminals also end on the somata and smooth dendrites of nonpyramidal neurons. The latter are almost exclusively GABA cells and GABA-immunostained somata and dendritic shafts represent 5–10% of the total population of geniculocortical recipient structures (Freund *et al.*, 1989). Physiologically characterized axons from the magnocellular and parvocellular layers of the LGN terminate with the same frequency on GABA and non-GABA elements (Freund *et al.*, 1989). Although a magnocellular axon may form six times the number of synapses as a parvocellular axon (Winfield *et al.*, 1982), both have proportionately equal access to inhibitory and excitatory cortical neurons.

Although GABA neurons clearly receive geniculocortical synapses (Fig. 17), not all GABA neurons in layer IVC may be innervated directly by the LGN. Examination of a peroxidase-filled, putatively GABAergic cell revealed no synapses from geniculocortical axons, even though neighboring neuronal elements were richly innervated (Kisvárday *et al.*, 1986). The relative proportions of GABA cells innervated and not innervated monosynaptically by geniculocortical axons, therefore, remain uncertain.

Thalamic inputs to monkey area 17 also arise from the inferior and lateral pulvinar nuclei (Rezak and Benevento, 1979) and from the intralaminar nuclei (Towns *et al.*, 1990). The pulvinar afferents terminate most densely in layer I and also in layer II, where they are distributed in patches of approximately the same size and spacing as the CO-rich periodicities (Ogren and Hendrickson, 1977). Pulvinar axons and those from the S and intercalated layers of the LGN (Fitzpatrick *et al.*, 1983) appear to provide thalamocortical innervation to GABA cells and other neurons in CO patches of layers II and III, respectively. No direct examination of thalamocortical terminations or other circuit elements in the CO patches has been reported.

4.1.3. Commissural Innervation

Commissural axons from area 17 of the opposite hemisphere terminate exclusively along the border between areas 17 and 18, where the vertical meridian of the visual field is represented (Myers, 1962; Zeki, 1970; Newsome and Allman, 1980). The commissural axons terminate most frequently on dendritic spines but also end on the somata and dendritic shafts of nonpyramidal cells (Fisken *et al.*, 1975), which are presumed to be GABA neurons. The axons do not end in layer IV of monkey area 17 (Fisken *et al.*, 1975) and, thus, appear to avoid the numerous GABA neurons in this layer.

4.2. Synaptic Connections of Identified GABA Cells

4.2.1. General

There have been fewer studies of the synaptic connectivity of identified nonpyramidal cells in the primate neocortex than in other species. Some cells taken for investigation in primates have come from the visual cortex, but much of the available information derives from investigations of temporal, sensory-motor, and auditory areas. As seen in Table III, the only characterized cells so far studied belong to the neurogliaform, chandelier, and double bouquet classes.

4.2.2. Neurogliaform Cells

Mates and Lund (1983) studied the synaptic connections of three Golgi-impregnated neurogliaform cells in layer IVC of monkey area 17. They reported that the axons of these cells formed symmetric synapses with cell bodies of spine-bearing stellate cells and with unidentified dendrites, but no quantitative data were given. A more detailed study of clutch cells in layer IVC by Kisvárday *et al.* (1986) involved a sample of 130 axonal boutons from three Golgi-impregnated cells. These authors found that 17 boutons formed multiple synapses (15 formed two synapses and 2 formed three synapses) while the rest made single synapses. The 159 synapses thus analyzed were all of the symmetric type and were distributed as follows (Table III): 10–17% on somata, 43.8–58.5% on dendritic shafts, and 20.8–46.3% on dendritic spines. One postsynaptic element was an axon initial segment, and five postsynaptic elements were unidentified. In addition, Kisvárday and colleagues studied 17 postsynaptic elements from three other neurogliaform cells: of these elements, eight were somata, five dendritic shafts, and four dendritic spines. These authors suggested that the majority of the postsynaptic elements engaged by neurogliaform cell axons belonged to spiny stellate cells.

4.2.3. Chandelier Cells

Chandelier cell axon terminals form symmetric synapses only with the axon initial segments of pyramidal cells (Somogyi, 1977, 1979; Fairén and Valverde, 1980; Somogyi *et al.*, 1982; Peters *et al.*, 1982; Freund *et al.*, 1983; DeFelipe *et al.*, 1985), and they are the major source of axon initial segment synapses (DeFelipe and Fariñas, 1993). Chandelier cells are, so far, the only type of aspiny non-pyramidal cell in the cerebral cortex that has been demonstrated to form synapses exclusively with a highly localized region of a single cell type; other types of cortical neurons, although sometimes targeted mainly at a particular region, tend to form synapses with a variety of postsynaptic elements. In the primate cerebral cortex the relatively few studies at the electron microscope level of chandelier cell axon terminals have been made on the visual, temporal, and sensory-motor areas. As seen in Table III, a total of 49 terminal rows of boutons

Table III. Synaptic Connections of Identified Aspiny Nonpyramidal Cells in the Primate Cerebral Cortex

Cell type	Staining[a]	Species	Cortical region	Number of cells studied[b]	Postsynaptic elements	References
Neurogliaform	Golgi	Monkey	Visual (area 17)	3	Dendritic shafts somata	Mates and Lund (1983)
				3[c]	49% dendritic shafts	Kisvárday et al. (1986)
					34% dendritic spines	
					13% somata	
					0.6% axon initial segments	
					3% unidentified	
Chandelier	Golgi	Human	Temporal	2	100% axon initial segments	Kisvárday et al. (1986)
	Golgi	Monkey	Visual (area 17)	6	100% axon initial segments	Somogyi et al. (1982)
	Golgi	Monkey	Visual (area 18)	3	100% axon initial segments	Somogyi et al. (1982)
	Golgi	Monkey	Motor (area 4)	6	100% axon initial segments	DeFelipe et al. (1985)
	PV	Monkey	Sensory-motor (areas 3b, 4)	32	100% axon initial segments	DeFelipe et al. (1989a)
Double bouquet	Golgi	Monkey	Visual (area 17)	1	60% dendritic shafts	Somogyi et al. (1982)
					40% dendritic spines	
	CaBP	Monkey	Somatic sensory (areas 3a, 1)	7	62% dendritic shafts	DeFelipe et al. (1989b)
					38% dendritic spines	
	SIRF	Monkey	Superior temporal gyrus	?	62.5% dendritic shafts	de Lima and Morrison (1989)
					37.5% dendritic spines	
	SRIF	Monkey	Inferior temporal gyrus	?	61% dendritic shafts	de Lima and Morrison (1989)
					39% dendritic spines	
	TK	Monkey	Primary auditory	9	57% dendritic shafts	DeFelipe et al. (1990)
					43% dendritic spines	

[a]Abbreviations: CaBP; calbindin; PV, parvalbumin; SRIF, somatostatin; TK, tachykinin.
[b]In the case of chandelier cells, the number refers to the number of terminal rows of boutons (candlesticks) which can belong to one, two, or more chandelier cells. In the case of double bouquet cells, the number refers to individual bundles of axons. In the monkey superior and inferior temporal gyri, the reported numbers of synapses formed by double bouquet cell axons arising from an unspecified number of bundles wer 64 and 36, respectively (de Lima and Morrison, 1989). See text for further details.
[c]Data do not include 17 postsynaptic structures from three additional cells. See text for further details.

have been studied. Of the 49 rows, 17 were identified by the Golgi method and 32 by immunocytochemistry for parvalbumin. Of the Golgi-stained rows, 2 were from the human temporal cortex, 9 from the monkey visual cortex, and 6 from the monkey motor cortex. The 32 parvalbumin-immunoreactive rows were all studied in the monkey sensory-motor cortex. In all species and in all cortical areas, the morphology and synaptic connections of the terminal rows of boutons of chandelier cell axons appear to be the same (see Somogyi *et al.*, 1982; Fairén *et al.*, 1984).

Not all pyramidal cells, however, are equally innervated by chandelier cell axons. The terminal portions of chandelier cell axons vary in their complexity (see Fairén and Valverde, 1980, and DeFelipe *et al.*, 1985) and a single chandelier cell axon can give rise to both very simple (made up of 5–10 boutons) and very complex (made up of more than 20 boutons) terminal segments. Serial reconstructions from electron micrographs in the monkey motor and somatic sensory areas revealed that the axon initial segments of pyramids in layers II–III receive, on average, more synapses than those of layer V pyramids (DeFelipe *et al.*, 1985), conforming the suggestion of Sloper and Powell (1979) that there is a higher density of axon initial segment synapses in superficial layers than in deep layers. For a given layer, the number of initial segment synapses was variable (from 2 to 52 for layers II and III and from 2 to 26 for layer V). In area 17 of the cat, different populations of pyramidal cells retrogradely labeled with horseradish peroxidase, received a characteristic and rather homogeneous number of initial segment synapses (Fariñas and DeFelipe, 1991). Layer VI pyramidal cells projecting to the thalamus received very few synapses (from 1 to 5) on their axon initial segments, compared with layer III callosal pyramidal cells (from 16 to 23) and ipsilateral corticocortical pyramidal cells (from 22 to 28). This laminar variability in the number of axon initial segment synapses may indicate a general differential laminar distribution and/or complexity of chandelier cell axons. However, there are insufficient data to generalize this to primate visual cortex.

4.2.4. Double Bouquet Cells

Double bouquet cells characteristically form symmetric synapses with dendritic shafts and spines in all cortical areas studied. Somogyi and Cowey (1981) studied the synaptic connections of one Golgi-impregnated double bouquet cell axon from monkey area 17. The soma was located in layer III and 35 Golgi-impregnated boutons arising from the axon were examined in layers III and IV. Forty percent of these boutons formed axospinous synapses, while the rest made synapses with dendritic shafts of small or medium caliber. Similar observations have been made on more extensive series of double bouquet cell axons in other cortical areas of monkeys (DeFelipe *et al.*, 1989b, 1990; de Lima and Morrison, 1989). Many of the postsynaptic dendrites were side branches of apical or basal dendrites of pyramidal cells. There is a lack of double bouquet synapses on the apical dendrites themselves, despite the strong verticality of both apical dendrites and double bouquet cell axons.

In spite of the dense packing of axons and their terminals in the bundles, and the tendency of the axons to form multiple synapses, there is very little convergence on a given postsynaptic structure (Somogyi and Cowey, 1981; DeFelipe *et al.*, 1989b). In the case of 276 terminals made by calbindin-immunoreactive double bouquet axons, most ended on different dendrites: only five dendritic shafts received two immunoreactive terminals.

Since double bouquet cells are very abundant, and the axon of each cell may have hundreds of boutons (and therefore, form hundreds of synapses), double bouquet cells can be considered as one of the most important sources of symmetric synapses on spines and distal dendrites in layer III.

4.2.5. Other Forms of GABA Cell

The circuitry entered into by basket cells, by the "generic" form of short-axon cell and by the vertical bipolar of bitufted cells, while not specifically investigated in primate visual cortex, can be deduced provisionally from investigations of these cells in other primate cortical areas or in area 17 of other species. Axons of basket cells end in multiterminal endings on the somata and proximal portions of the dendrites of pyramidal neurons (Somogyi *et al.*, 1983b; DeFelipe *et al.*, 1986b). It is not yet certain that similar multiterminal endings on the somata of basket cells themselves also emanate from basket cell axons. The aspiny cells with axons forming arcades (sparsely spinous neurons forming local axonal endings) have axons that terminate principally on shafts and dendritic spine stalks of apical and other major dendrites of pyramidal cells (Peters and Saint-Marie, 1984). The axons of the long, stringy, bipolar or bitufted cell, at least when identified by peptide immunoreactivity, terminate on a variety of postsynaptic elements: tachykinin- or CCK-immunoreactive cell axons end primarily on pyramidal cell somata and proximal dendritic shafts (Hendry *et al.*, 1983b; Jones *et al.*, 1988), while NPY- and SRIF-immunoreactive terminals end predominately on medium-sized dendritic shafts, some of which may belong to pyramidal cells (Hendry *et al.*, 1984c).

4.3. GABAergic Circuits in Primate Visual Cortex

The data outlined in preceding sections on the synaptic inputs and outputs of the various classes of aspiny neurons, whether defined by morphology, by GABA, GAD, or peptide immunoreactivity, or by both morphology and immunoreactivity, are insufficient to provide a comprehensive picture of GABAergic circuitry in the visual or any other area of primate neocortex. However, drawing together the few data that are available from the primate and extrapolating from those obtained in other species, it is possible to make two broad generalizations. The first, deriving from the evidence of the distribution of GABA axon terminals, indicates that every part of the surface of a cortical pyramidal cell soma, the various generations of dendrites, spines, and axon initial segment, are targets of GABA cell axons but that the GABA terminals at each of these locations derive in large part from a different class of GABAergic neuron.

A preliminary scheme, therefore, would have the terminations on initial segments being almost exclusively provided by chandelier cells and those on the somata and proximal portions of dendrites by basket cells, with lesser contributions from certain GABA–peptide cells. On side branches of the apical and basal dendritic systems, double bouquet cells may be the major sources of GABA terminals while other contributions may come from the generic form of aspiny cell and certain other GABA–peptide cells. In layer I, apical dendritic tufts may have their main GABA innervation from the local GABA cells. What is missing from this generalization is knowledge of the relative contributions of individual forms of GABA cell to the inhibitory input to a single pyramidal cell, the degree of convergence of inputs from members of the same GABAergic cell class, and the variations that occur in these parameters among pyramidal cells of different types, especially those with projections to different subcortical targets. These are the kinds of data that will be required before a comprehensive picture of GABAergic circuitry at the single cell level can be built up.

The second generalization, deriving from the evidence on those GABA cells that receive the terminations of thalamocortical axons, indicates that certain GABAergic cells form part of the immediate circuitry leading from the thalamic input, while others are involved in circuitry removed from this by several synaptic steps. In receiving direct thalamic synapses, basket cells and neurogliaform cells can be seen as directly involved in the disynaptic inhibitory cortical response that invariably follows the initial monosynaptic excitatory response to a thalamic input and are, therefore, presumably intricately involved in the shaping of receptive field properties of neurons at the heart of the thalamic termination layers. Double bouquet cells, chandelier cells, GABA/peptide cells, and other forms of GABA cell, in not receiving thalamocortical terminations and being situated outside the layers in which thalamic fibers terminate, are in a position to influence the subsequent steps in thalamocortical processing, in affecting receptive fields of neurons in superficial and deep layers of cortex, in modulating the flow of information between layers, and in influencing the processing leading from input to output in the cortex.

Apart from the large basket cells, all GABA neurons of the primate visual cortex have axons whose distribution is either concentrated in the vicinity of the soma or projected vertically across layers. Even the basket cells have the majority of their axon terminations in the vicinity of the dendritic field. As a manifestation of this, few basket cell somata are retrogradely labeled at any distance from an intracortical injection of [³H]-GABA (Hendry and Jones, 1981). Hence, most direct GABA actions would appear to be local or intracolumnar. This contrasts with the far-flung extent of the horizontal collaterals of pyramidal cells which may be the principal route for spread of direct excitatory influences from column to column in the visual and other areas of the primate cortex (Gilbert and Wiesel, 1983, 1989; DeFelipe *et al.*, 1986a).

Some idea of the complexity of circuitry effected by GABA neurons of area 17 can be obtained by examining the axonal distributions of GABA cells in layer IVC, the largest layer of densest thalamic terminations. From her observations on Golgi-stained material, Lund (1987, 1990) has described a very precise pattern of intra- and interlaminar connections whereby aspiny, putatively GABA cells of layer IVC (Fig. 19) enter into the circuitry leading from the thalamic input. Some of the cells have dendritic fields restricted to precise subdivisions of layer IVC, e.g., the lower half of layer IVCα, and the principal branches of the axon can be similarly restricted. It is possible that these sublaminar specific distributions are reflected in the sublaminae detected in layer IVC by *in situ* hybridization histochemistry (Fig. 7). According to Lund (1990), they may correlate with specific distribution patterns of individual thalamic axons and with the dendritic fields of particular spiny stellate cells that are probably among the major recipients of thalamic axon terminals. Other aspiny cells of layer IVC project axons upwards to terminate specifically in layers IVB, IVA, or III, or downwards to terminate in layers V and/or VI. In terms of potential GABAergic inputs to layer IVC, aspiny cells in layers IVA, V, and VI are main contributors and commonly have highly stratified axonal distributions in layer IVC (Fig. 19).

While the variety of intra- and interlaminar connections effected by aspiny,

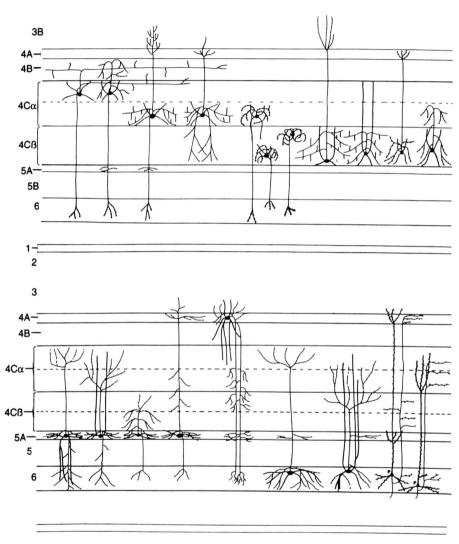

Figure 19. (Top) Some of the varieties of interneuron (probably GABAergic) found in divisions of layer 4C in the macaque monkey visual cortex. The restriction of dendritic arbors and intense local axon arbors to different divisions of layer 4C contrasts with the dendritic overlap and diffuse local axon collaterals of spiny stellate neurons in the same layer. (Bottom) Intrinsic axon relays to layer 4C. With the exception of input from the pyramidal neurons (P) of layer 6, these relays derive from interneurons of layers 6, 5A, and 4A, which are believed to be largely GABAergic. The axon terminals show very discrete patterns of distribution in terms of the various subdivisions in depth of layer 4C. From Lund (1990).

putatively GABA neurons in and projecting to layer IVC is considerable, there is a clear sense from Fig. 19 of a great deal of precision. The different populations of cells may therefore form parts of different functional streams created by the differential levels of termination of separate classes of thalamic afferents.

It is noticeable that the lateral spread of axonal distribution of all the cells depicted in Fig. 19 is rather limited, as one would expect from the general character of most classes of aspiny cortical cell (Section 3.1). Functional streaming is therefore either local or interlaminar and, thus, intracolumnar. Throughout all layers of area 17 the strong intracolumnar character of GABA-mediated inhibition is likely to be maintained by the strong, bidirectional, vertical GABAergic loop that links supra- and infragranular layers. The descending link of this loop is obviously formed by the double bouquet cells of layers II–III and the ascending link by the populations of less well characterized GABA cells located in layers IVC and V–VI and with ascending axons (Fig. 14). The effects mediated by this bidirectional, translaminar flow of GABA-mediated inhibition can only be conjectured. The synaptic relationships of the bundles of double bouquet cell axons suggest, however, that the descending link is potentially capable of engaging virtually all pyramidal cells, presumably in a column-specific manner. Whether the ascending link similarly engages side branches of potentially all apical dendrites as it ascends is not yet known. The terminal ramifications of the double bouquet cell axon bundles in layer V and apparently of the ascending axons in layers II–III would also appear to ensure that the major sources of output connections of area 17 are capable of modulating one another's activity directly. The implications of this vertical, bidirectional, GABAergic circuitry therefore impinge on all aspects of cortical columnar processing.

Long, horizontal GABAergic connections that would be in a position to mediate intercolumnar inhibitory effects appear to be relatively few and the potential cells of origin limited to the large basket cells. When identified in area 17 of monkeys, these cells have not been described as having the very long horizontal axon collaterals found on similar cells in other areas. Even in these other areas and in other species where the cells have been more fully characterized, the number of terminals made by the collaterals at a distance from the parent cell is relatively small (DeFelipe *et al.*, 1986b). Instead, there is a rather dense concentration of terminals in and close to the dendritic field of the cell. These cells, therefore, are unlikely to be mediators of long-range, GABAergic inhibition.

In seeking a circuitry whereby relatively direct inhibition might be extended over long horizontal distances in area 17, attention should be paid to the long horizontal collaterals of the pyramidal cell axons. Although undoubtedly releasing an excitatory transmitter, the terminals of these axons, if ending selectively on GABAergic interneurons, could effect a disynaptic inhibitory influence over functional columns at a distance from that in which the parent pyramidal cell lies. Unfortunately, the data which one could use to elucidate further such a conjecture are extremely limited. Two layer III pyramidal cells have been injected with horseradish peroxidase in area 17 of monkeys and subjected to intensive electron microscopic examination (McGuire *et al.*, 1991). The majority (80%) of the terminals made by local and distant collaterals of these cells were on spines and dendrites of other pyramidal cells. Only 20% of the terminals were on aspiny dendrites and, thus, likely to be on GABAergic neurons.

5. Development of GABA Neurons and GABA Receptors in Monkey Area 17

5.1. Development of GABA Immunoreactive Neurons

GABA-immunoreactive neurons have been described in area 17 of fetal monkeys as early as embryonic day 61 (E61; Hendrickson *et al.*, 1988; Meinicke and Rakic, 1989) which is approximately one-third of the way through the 165-day macaque gestation period. At this stage, most GABA neurons are present in the marginal and subplate zones, but some immunoreactive somata are also found in the cortical plate (Hendrickson *et al.*, 1988; Meinicke and Rakic, 1989). Later, at E110, GABA-immunostained somata remain concentrated in layer I and the subcortical white matter (Huntley *et al.*, 1988), which contain the neurons of the earlier marginal and subplate zones (Fig. 20). By E135, additional GABA-positive somata are present in layers II–III, but it is not until E155 that a relatively high density of immunostained somata becomes evident in layer IVC (Huntley *et al.*, 1988). *In situ* hybridization histochemistry for GAD mRNA reveals a localization of GAD-expressing neurons principally in the marginal zone early in development (E75) with the adult pattern becoming evident only in the final days of fetal development (Hendrickson *et al.*, 1988). There is some indication from immunostaining and *in situ* hybridization histochemistry that GAD gene expression may be higher in the late fetal and early postnatal periods than

GABA

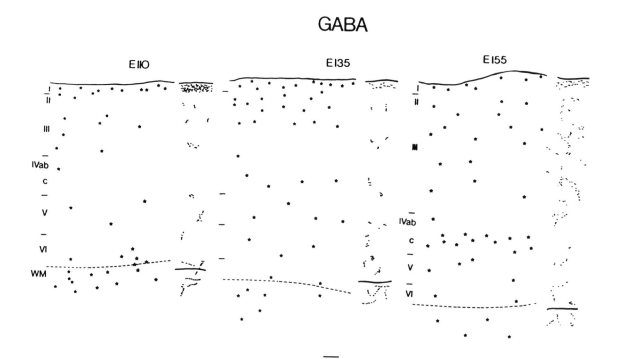

Figure 20. Distribution of GABA-immunoreactive cell somata (stars) and fiber plexuses (dots) in area 17 of rhesus monkey fetuses at 110, 135, and 155 days of gestation. Note reduction in GABA cells in white matter (WM) and increase in layer IV over time. Bar = 100 μm. From Huntley *et al.* (1988).

in adulthood (Hendrickson *et al.*, 1988). This tends to match the pattern of GABA$_A$ receptor development in fetal and neonatal monkeys (see below).

Neuropeptides that coexist with GABA in certain neurons of the adult visual cortex are expressed by both GABA and non-GABA neurons in area 17 of fetal monkeys. During the last third of fetal development, GABA cells that are NPY- and SRIF-immunoreactive exhibit adultlike patterns of distribution, density, and neuronal morphology (Huntley *et al.*, 1988; Yamashita *et al.*, 1989). However, the development of the typical plexuses of NPY-immunoreactive fibers is a more prolonged event that continues into the postnatal period (Tigges *et al.*, 1989). Unlike SRIF and NPY immunoreactivity and unlike the adult pattern, tachykininlike immunoreactivity is present in non-GABA neurons which form a dense band of somata in layer V from E110 until E155 (Huntley *et al.*, 1988). For a brief period, around E135, these neurons are also CCK-immunoreactive. It has not been determined when the adult pattern of tachykinin immunoreactivity in half the GABA cells of layer IVC (Jones *et al.*, 1988) is first expressed. Other patterns, such as immunoreactivity for the proenkephalin peptide BAM 18, appear in certain neurons only during the late fetal period and disappear at birth (Huntley *et al.*, 1988). Similarly, mRNA for prepro-NPY appears and then disappears in layer IV cells during the middle period of fetal development (S. H. C. Hendry and E. G. Jones, unpublished results). These findings indicate that the acquisition of neuropeptidelike immunoreactivity in GABA cells and its transient display in certain non-GABA cells are a general feature of the neurochemical development of monkey visual cortex. Conceivably, the neuropeptides may perform neurotrophic functions in the developing cerebral cortex, but this has not been explored.

The development of immunoreactivity for the calcium-binding proteins calbindin and parvalbumin proceeds along very different paths in monkey area 17. Two waves of intense but transient calbindin immunostaining are found in populations of pyramidal and spiny stellate cells in late fetal and early postnatal periods (Van Brederode *et al.*, 1990). The first wave begins at E113, when intense calbindin immunoreactivity appears in neurons of layers IV–VI. This continues to increase up to birth and then declines by the third postnatal month. In the second wave, calbindin immunostaining of neurons in layers II and III increases between the fourth and ninth postnatal months and then declines slowly, only reaching adult levels by the second year. Parvalbumin immunoreactivity is noticeable only in the week prior to birth (E162) and reaches adult levels by the fifth postnatal month. The initial wave of calbindin immunostaining, said to affect principally aspiny stellate neurons in layer IVC, has been correlated with the ingrowth of geniculocortical axons and the establishment of their synapses, while the delayed intense parvalbumin immunostaining of presumed GABA neurons in layer IVC has been correlated with the end of geniculocortical plasticity (Van Brederode *et al.*, 1990).

5.2. Development of GABA Receptors

GABA$_A$ receptors, localized by the binding of [^{3}H]muscimol and [^{3}H]flunitrazepam, are present in developing monkey visual cortex at E61 (Shaw *et al.*, 1991) (Fig. 21). At that age, binding sites for both ligands are present in the cortical plate and marginal zone (Fig. 21). Layer IV begins to show signs of

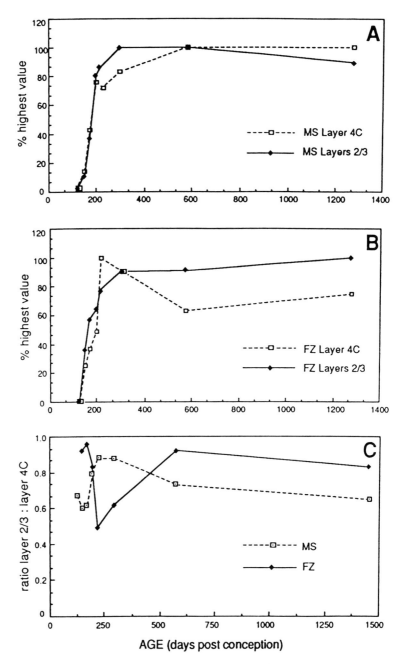

Figure 21. Changes in [³H]muscimol (MS) and in [³H]benzodiazepine (FZ) binding illustrating postnatal development of GABA$_A$ receptors in layers 4C and 2/3 of macaque monkeys. In the species used, birth occurs at approximately 165 days postconception. From Shaw *et al.* (1991).

greater binding in the last 2 months of fetal development (E119) and the adult laminar pattern is reached at or near the time of birth. However, after birth the density of muscimol-binding sites increases while the density of flunitrazepam-binding sites is reduced outside layer IV (Shaw *et al.*, 1991). Scatchard analyses indicate that the muscimol- and flunitrazepam-binding sites are most dense after the fourth postnatal month (170 and 113% of adult levels, respectively) and that they reach adult levels very late, at 8.5 years postnatal. A developmental change in the affinity of muscimol for the $GABA_A$ receptor is also evident during development, with an unusual high-affinity site dropping out at birth. These data indicate that the $GABA_A$ receptor, with its multitude of binding sites, is expressed by neurons of monkey area 17 in fetal life but that a slow maturation process, including a postnatal overabundance of receptor sites, characterizes its development. The presence of high densities of GABA neurons and GABA terminals and synaptic contacts from the late fetal period onward suggests that the receptors may be functional over this entire period. However, the absence of synaptic contacts (Rakic *et al.*, 1986; Zielinski and Hendrickson, 1990) at a time (E61) when $GABA_A$ receptors are present has led to the suggestion that the receptors may play an unconventional, possibly neurotrophic role early in cortical development (Shaw *et al.*, 1991).

$GABA_B$ receptor binding is evident relatively late in development (E126), and adopts an adultlike laminar pattern at that time (Shaw *et al.*, 1991).

6. Plasticity of GABA Neurons in Area 17 of Adult Monkeys

6.1. Changes in GABA Immunoreactivity

During a critical period in the development of the monkey visual cortex (over the first 4–5 postnatal months), manipulations of one eye can alter the geniculocortical connectivity of area 17. Monocular deprivation in the critical period causes the ocular dominance columns innervated by one set of lateral geniculate laminae (those receiving their inputs from the normal eye) to expand in size while the columns innervated by the other set of laminae (those receiving their inputs from the deprived lateral geniculate laminae) contract. After the critical period, monocular deprivation, even that caused by removal of an eye, produces little or no alteration in column size, as measured by bulk labeling of thalamocortical axons or single-unit recording (LeVay *et al.*, 1980).

Nevertheless, in adult monkeys, eye removal or the elimination of ganglion cell activity in one retina by intravitreal injections of tetrodotoxin (TTX) dramatically changes the pattern of GABA immunostaining in area 17 (Fig. 22). Instead of the normal, homogeneous distribution of immunostained neurons across layer IVC, GABA cells are stained in alternating lightly and intensely immunoreactive bands (Hendry and Jones, 1986). In sections cut parallel to the flattened pial surface of the cortex, the lightly and darkly stained bands appear as elongated stripes, each of which is 0.4–0.5 mm wide and matches a deprived or nondeprived ocular dominance stripe, respectively. The intensely immunostained stripes include a density and proportion of GABA-immunoreactive somata and immunoreactive terminals within the normal range. By contrast, the lightly immunostained stripes contain a much reduced density and proportion

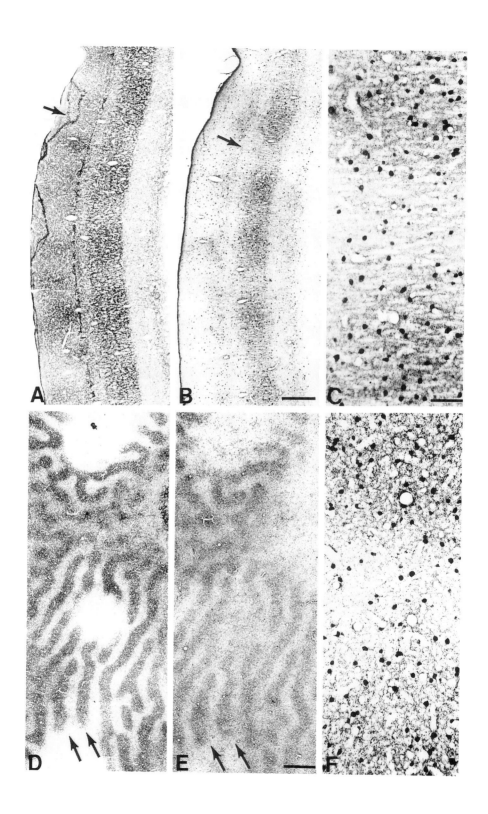

of immunoreactive somata and terminals: only half the normal number of GABA neurons is immunostained in layer IVC of the deprived-eye stripes (Figs. 23 and 24) (Hendry and Jones, 1986, 1988). Even those cells in the deprived-eye columns that continue to display GABA immunoreactivity tend to be more lightly immunostained than the GABA neurons in the adjacent normal-eye columns. These data indicate that intracellular levels of GABA in 50% of layer IVC GABAergic neurons deprived of visual input are reduced below an immunocytochemically detectable threshold. Experiments in which dot blots were stained with the same anti-GABA antisera as those used in the immunocytochemical studies indicated that a reduction in GABA concentration to levels 0.1% of those normally detected by immunocytochemistry in the cerebral cortex is necessary to eliminate immunostaining (S. H. C. Hendry and E. G. Jones, unpublished results).

The effects of abolishing all ganglion cell activity in one retina, either by enucleation or by intravitreal injection of TTX, are also evident outside layer IVC. In layer IVA, stripes of normal, intense GABA immunostaining alternate with stripes of abnormally light immunostaining (Hendry and Jones, 1986, 1988). These stripes in layer IVA line up with the intense and light stripes in layer IVC and correspond to normal-eye and deprived-eye columns, respectively. Reductions in both the density and proportion of GABA-immunostained neurons can also be detected in parts of deprived columns in layers II and III. The reductions are evident both at the centers of the deprived-eye columns, namely in the CO-dense patches, as well as in the surrounding CO-weak interpatch regions (Fig. 25). Significant reductions in GABA cell numbers also extend for 100–150 μm into the part of the CO-weak interpatch region associated with the adjacent, nondeprived-eye column. Thus, only the center of the normal-eye column, marked by the CO-dense patch, contains a normal proportion of GABA neurons (Hendry and Jones, 1988). These and other data (see below) suggest that manipulations of one eye can produce changes in both sets of ocular dominance columns. No perceptible changes in GABA immunoreactivity occur in either layer V or layer VI of monocularly deprived adult monkeys.

6.2. Time Course of GABA Plasticity

The periodic reductions of GABA immunostaining in adult monkey area 17 occur quickly following elimination of retinal activity. Within 5 days, eye removal or TTX injections produce a 50% reduction in the number of GABA-immunostained neurons in layer IVC (Fig. 23) (Hendry and Jones, 1986, 1988, and unpublished). No changes in cortical GABA immunostaining are evident at 3 days, even though reductions in CO staining occur by this survival time (see also Wong-Riley and Carroll, 1984a). The first indication of a deprivation effect

---

Figure 22. Matched pairs of sections cut perpendicular (top row) and parallel (bottom row) to the surface of area 17 and stained histochemically for CO (A, D) or immunocytochemically for GABA (B, C, E, F). A–C show reduction in GABA immunostaining in deprived eye dominance columns 2 weeks after removal of one eye. D–F show a similar reduction 11 weeks after suturing the lids of one eye closed. C and F are higher-magnification views of two normal and an intervening deprived column from B and E, respectively, and show reduced somal and neuropil staining for GABA in the deprived columns. Arrows indicate same columns. Bars = 500 μm (A, B), 100 μm (C, F), 1 mm (D, E). Hendry and Jones (1986).

on GABA immunostaining is a reduction in the immunoreactivity of individual neurons, which occurs within 4 days (90 hr) of eye removal. Thus, the reduction in immunoreactivity and eventual loss of immunostaining in half the GABA neurons has a rapid onset when retinal ganglion cell activity is eliminated. These effects have been seen in several species of Old World monkey from 2 to more than 20 years of age.

Elimination of pattern vision by suturing the eyelids together over one eye produces a loss in GABA immunostaining that is delayed and, apparently, dependent on the age of the monkey. Changes in GABA immunoreactivity can be detected in cortical neurons only after one eye has been sutured shut for more than 2 months (Hendry and Jones, 1986, 1988) (Fig. 24), in contrast to the 4 days

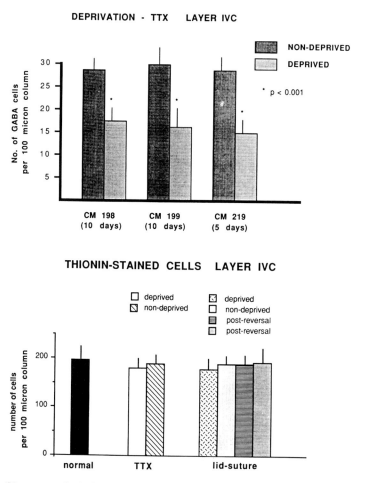

Figure 23. (Upper panel) Reduction in GABA-immunoreactive cells in 100-μm-wide columns through layer IVC at the centers of deprived ocular dominance columns in two monkeys subjected to monocular TTX injections. Statistically significant differences ($p < 0.001$) between the centers of deprived and nondeprived columns are observed. (Lower panel) Absence of significant cell death in layer IVC, demonstrated by counting the total number of Thionine-stained neurons in 100-μm-wide columns through layer IVC in normal animals, in deprived and nondeprived ocular dominance columns of lid-sutured and TTX-injected animals, and in an animal subjected to lid reopening (two sets of counts). Counts were made from three different sections in each case. Differences are statistically insignificant. From Hendry and Jones (1988).

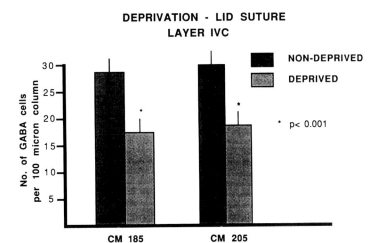

Figure 24. Reduction of GABA neurons in deprived ocular dominance columns in two monkeys that were subjected to eyelid suture for 11 and 13 weeks. From Hendry and Jones (1988).

required when one retina is removed or silenced. Furthermore, the reduced GABA immunostaining in layer IVC is most evident in monkeys that have had monocular lid suture between the ages of 1 and 2 years. It is striking that in these and older monkeys, lid suturing produces a less dramatic effect on the CO staining in layer IVC than either enucleation or TTX injection (Horton, 1984; Trusk *et al.*, 1990), but the change in GABA immunostaining is just as robust, with nearly a 50% reduction in GABA cell density (Fig. 24) (Hendry and Jones, 1988). Although it has not been extensively explored, the effect on GABA immunostaining of even longer periods of monocular deprivation by lid suturing appears less dramatic in older monkeys.

6.3. Mechanisms

6.3.1. Absence of Cell Death

Reductions in the number of GABA-immunoreactive somata and terminals are not the result of a deprivation-induced cell death in the adult visual cortex. Studies of the transneuronal effects of eye removal indicate that cell death in the primate visual cortex only becomes apparent many months or years following enucleation of one eye (Campbell, 1905; Cowan, 1970). Direct quantification of the same enucleated and TTX-injected monkeys in which 50% reductions in GABA-stained cells were detected, demonstrated that the total neuronal population is not reduced in the visual cortex, generally, and not in layer IVC of the deprived-eye columns, specifically (Fig. 23). That is, the cells in the deprived columns cease to be GABA-immunoreactive but do not die. That these cells remain able to resynthesize detectable levels of GABA if renewed visual stimuli are allowed to reach them is revealed by the following experiments. Animals were monocularly deprived by TTX injections for 1 week or by monocular lid suture for 3 months, then a biopsy was taken from area 17 of one side and the GABA-deprivation effect confirmed by immunocytochemistry. The animals recovered and TTX injections ceased or the lids were reopened. After 2 months of renewed binocular vision, the animals were killed and the cortex opposite that biopsied was sectioned and stained for CO and GABA immunoreactivity. There was no evidence of stripes in layer IVC and the density of immunostained cells had returned to normal (Fig. 26). Hence, cortical GABA neurons can be induced to modulate their levels of GABA immunoreactivity under activity-dependent conditions.

6.3.2. Decline in GAD Immunoreactivity

In enucleated, lid-sutured, or TTX-injected monkeys, the pattern of GAD immunostaining in layer IVC also consists of alternating intensely and lightly immunoreactive stripes corresponding to normal- and deprived-eye dominance col-

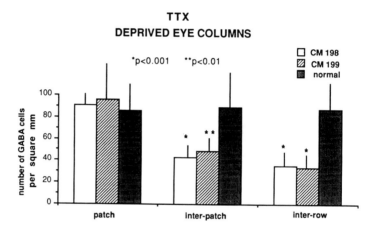

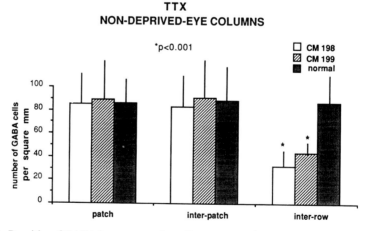

Figure 25. Densities of GABA-immunoreactive cell somata associated with CO-positive periodicities ("patch") lying in layer III above deprived (upper panel) and nondeprived (lower panel) ocular dominance columns of layer IVC, compared with the densities lying between CO periodicities of the same ("inter-patch") and adjacent ("inter-row") rows in a TTX-injected animal. Statistically significant differences indicate a reduction in the density of GABA-stained cells around and between CO periodicities lying above deprived ocular dominance columns. From Hendry and Jones (1988).

umns (Fig. 27). In the deprived stripes in layer IVC there is a reduction in GAD-immunostained somata to 50% of normal density and GAD-immunoreactive terminals virtually disappear. The activity-dependent changes in GABA immunostaining, therefore, depend on changes in the GABA-synthesizing enzyme, GAD.

6.3.3. GAD Gene Expression

Changes in immunocytochemically detectable levels of GAD and GABA in area 17 of monocularly deprived monkeys are likely to depend on an activity-dependent regulation of GAD gene expression. In an initial study aimed at detecting changes in GAD mRNA levels by *in situ* hybridization histochemistry in animals deprived by TTX injections for 7–15 days, no differences could be detected in mRNA levels in the normal- and deprived-eye columns in layer IVC,

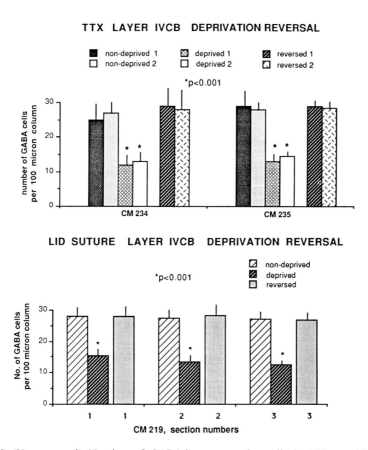

Figure 26. (Upper panel) Number of GABA-immunoreactive cells in 100-μm-wide columns through the centers of deprived and nondeprived ocular dominance columns in layer IVCβ in biopsies from two TTX-injected animals, compared with comparable counts through layer IVCβ of the contralateral visual cortex after recovery. Two sets of counts are shown in each case. (Lower panel) Number of GABA-immunoreactive cells in 100-μm-wide columns through the centers of deprived and nondeprived ocular dominance columns in layer IVCβ in a biopsy from a lid-sutured animal, compared with comparable counts through layer IVCβ of the contralateral visual cortex after reopening the eyelids. Data are from three different sections (1–3). From Hendry and Jones (1988).

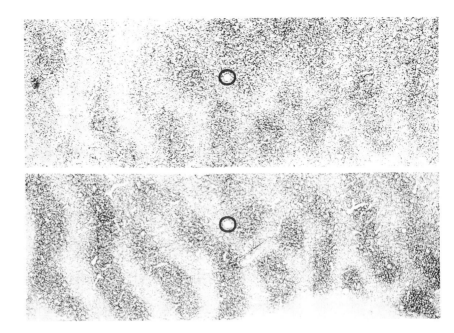

Figure 27. Adjacent sections from area 17 of the same monkey stained immunocytochemically for GAD (upper) and histochemically for CO (lower) and showing reduction of GAD immunoreactivity in deprived ocular dominance columns consequent upon monocular deprivation caused by suturing the lids of one eye closed for 11 weeks. Same blood vessel is circled. Hendry and Jones (1986).

despite large changes in mRNA levels for the alpha subunit of type II, calcium–calmodulin-dependent protein kinase (Benson *et al.*, 1991a). This led to the suggestion that the changes in GAD and GABA levels resulting from monocular deprivation depend on posttranslational changes in one or both of the two known forms of GAD. Recent experiments, however, reveal that when animals are monocularly deprived for periods of 15 days or longer, robust reductions in 67-kDa GAD mRNA appear (Fig. 28) (D. L. Benson, M. M. Huntsman, P. J. Isackson, and E. G. Jones, unpublished observations). The reasons for the reductions in GAD protein levels preceding those in GAD mRNAs are still unclear but suggest that many cellular mechanisms in the visual cortex come under the influence of changes in arriving neural activity at different times.

6.3.4. Competition in GABA Regulation

Apart from activity-dependent regulation of GAD expression (and thus of GABA levels) at transcriptional and posttranslational levels, competitive interactions between left and right eye influences may continue to play a role in GABA regulation even in adult animals. During the development of the visual system, axons driven by one or the other of the two eyes compete for synaptic territory. Geniculocortical axons placed at a competitive disadvantage by monocular deprivation during the critical period come to occupy abnormally small terminal zones unless inputs from the other eye are contemporaneously reduced or eliminated (LeVay *et al.*, 1980). A similar competitive mechanism that continues to operate even beyond the window of developmental vulnerability, appears to control levels of GABA in cortical neurons, as measured by immunoreactivity. When all geniculocortical activity entering area 17 of one hemisphere in an adult monkey is eliminated by destruction of the LGN, GABA immunoreactivity in area 17 ipsilateral to the lesion (deprived hemisphere) differs neither qualitatively nor quantitatively from the immunoreactivity contralateral to the lesion (normal hemisphere), and does not differ from that seen in area 17 of a normal monkey (S. H. C. Hendry and E. G. Jones, unpublished results). The normal pattern persists in the deprived hemisphere for at least several days beyond the time that elimination of activity in one retina will reduce GABA immunoreactivity in area 17. This suggests that in the monocularly deprived animal, the maintenance of an imbalance in geniculocortical activity may be the key element in regulating GABA levels.

In a second type of analysis along these lines, the visuotopic organization of area 17 was exploited to examine an expanse of cortex normally driven by one eye. In each hemisphere the temporal crescent of the contralateral visual field is represented deep in the calcarine fissure. Stimuli from the temporal crescent of the visual field fall only on the nasal part of the retina in the ipsilateral eye and do not impinge on any part of the contralateral retina. The cortical representation of the contralateral temporal crescent is thus strictly monocular (the monocular segment). It occupies several square millimeters of the calcarine cortex. In the monocular segment contralateral to a TTX-injected eye, the alternating intensely and lightly GABA-immunoreactive stripes of the larger binocular region give way to a uniformly and intensely immunoreactive zone. In this deprived monocular segment, the number and proportion of GABA-immunostained neurons in layer IVC remain equal to those in layer IVC of a normal monkey. These analyses of the cortex after geniculate lesions and of the deprived monocular segment strongly suggest that interactions between neurons in normally driven and deprived ocular dominance column contribute to the regulation of GABA immunoreactivity in adult area 17.

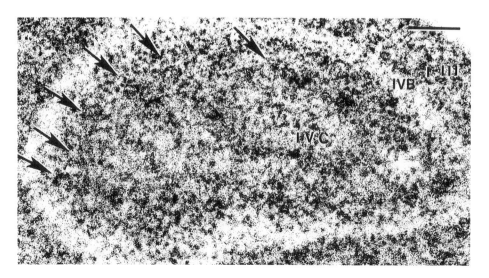

Figure 28. Reduction in GAD mRNA levels in deprived ocular dominance columns, as demonstrated by *in situ* hybridization histochemistry in a monkey monocularly deprived by intraocular injections of TTX for 15 days. Arrows indicate normal columns in layer IVC. Bar = 500 μm. D. L. Benson, M. M. Huntsman, P. J. Isackson, and E. G. Jones (unpublished).

6.4. Effects of Visual Deprivation on GABA Cells in the Lateral Geniculate Nucleus

Reductions in CO staining occur more rapidly in deprived layers of the monkey LGN than in deprived-eye dominance columns of area 17 (Wong-Riley and Carroll, 1984a). However, changes in GABA immunoreactivity in lateral geniculate neurons take place more slowly. The LGN contains a large population of GAD- and GABA-immunoreactive cells, which make up 30–35% of the neurons in the magnocellular layers and 25% of those in the parvocellular layers (Montero and Zempel, 1986). For more than 2 weeks following the elimination of input from one retina, the intensity of GABA immunostaining and the density and proportion of GABA-immunostained neurons do not change in the deprived layers (Fig. 29). Only by the third week after enucleation or TTX injections (i.e., more than 2 weeks after the first detectable changes in area 17) does an effect on GABA immunoreactivity become obvious in the LGN (Hendry, 1991). At that time, the proportion and density of GABA neurons in the deprived layers of the nucleus are reduced by one-third. These data indicate that neurochemical changes in the LGN follow rather than contribute to the rapid changes in area 17.

6.5. Changes in GABA Receptors

Both radioligand binding and immunocytochemical studies in monocularly deprived monkeys show that $GABA_A$ receptors are downregulated in deprived ocular dominance columns of layer $IVC\beta$ (Hendry *et al.*, 1990) approximately along the same time course as the reduction in GABA and GAD immunoreactivity (Fig. 30). Quantitatively, [^{3}H]muscimol and [^{3}H]flunitrazepam binding is reduced by approximately 25% in deprived-eye columns. This reduction results from a lower B_{max} (receptor number) and not from a change in K_D (receptor affinity) (J. L. Fuchs, personal communication).

Parallel immunocytochemical studies using peptide-specific antisera (Huntsman *et al.*, 1991), indicate that at least three subunits of the $GABA_A$ receptor downregulate under conditions of monocular deprivation. Under these conditions, the uniform distribution of immunoreactivity for the $\alpha1$, $\beta2/\beta3$, and $\gamma2$ subunits normally seen in layer $IVC\beta$ is replaced by alternating intensely and lightly immunostained stripes (Fig. 30). In association with this, mRNA levels for the various subunits are also reduced in the deprived stripes (M. M. Huntsman, P. J. Isackson, and E. G. Jones, unpublished results). The stripes intensely immunostained for each subunit display a normal mRNA level and correspond to the normal-eye columns, while those lightly immunostained display a reduced level and correspond to deprived-eye columns. These and the ligand binding studies indicate that the elimination of input from one retina reduces the density of $GABA_A$ receptors and their constituent subunits on the surfaces of deprived cortical neurons, rather than inducing a compensatory upregulation that might lead to receptor supersensitivity.

6.6. Changes in Proteins and Peptides That Coexist with GABA

Of the several neuropeptides that coexist with GABA in neurons of monkey area 17, only substance P (and possibly other members of the tachykinin family) display obvious changes in immunoreactivity with monocular deprivation. In area 17 of a normal monkey, half of the GABA neurons in layer IVC exhibit detectable levels of tachykininlike immunoreactivity (Jones *et al.*, 1988). Following the loss of retinal inputs from one eye, however, the immunoreactivity in the deprived-eye columns virtually disappears (Fig. 31) (Hendry *et al.*, 1988a) and only 10–20% of the normal complement of GABA–tachykinin neurons in layer IVC retain their peptide immunoreactivity. A parallel loss occurs in layers IVA

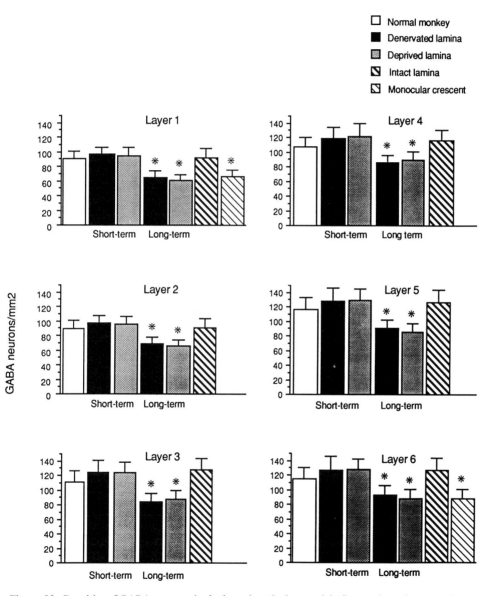

Figure 29. Densities of GABA neurons in the lateral geniculate nuclei of normal monkeys, monkeys surviving 2 weeks or less following monocular enucleation (short-term, denervated lamina), 5–17 days of TTX injections (short-term, deprived lamina), and monkeys surviving 20–30 days following the same manipulations (long-term, denervated or deprived lamina). Quantitative analyses were performed on the nuclei of both sides so that layers with normal activity or innervation (intact lamina) could be compared with equivalent laminae of the other side. For layers 1 and 6, the representation of the monocular crescent was also examined. *$p < 0.01$. From Hendry (1991).

and II–III. As with the reduction in GABA immunostaining, tachykininlike immunoreactivity downregulates very quickly and becomes undetectable in most deprived neurons within 5 days of eye removal or intravitreal injection of TTX. *In situ* hybridization histochemistry reveals that this downregulation is brought about as the result of a precipitous drop in preprotachykinin mRNA levels (Benson and Jones, 1991).

The GABA neurons in layer IVC which lose their immunoreactivity following monocular deprivation make up a specific and well-defined population. At the electron microscopic level, GABA neurons in layer IVCβ receive a variable density of synaptic contacts on their somata and proximal dendrites but most are either very densely innervated or very lightly innervated (Hendry and Jones, 1989). Examination of tachykinin-immunostained neurons in layer IVCβ (Hendry and Jones, 1989) shows that they belong almost exclusively to the lightly innervated group of GABA cells. These lightly innervated GABA–tachykinin neurons, however, receive a large proportion of their synaptic contacts from geniculocortical axons. It is they that completely lose GABA (and tachykinin) immunoreactivity in deprived-eye columns of layer IVCβ. Those remaining with reduced GABA immunoreactivity appear to belong to the more densely innervated class in which large numbers of nongeniculate inputs also converge on the somata and proximal dendrites. From these data, a picture emerges of the GABA–tachykinin neuronal population in layer IVCβ as a group that is more dependent on geniculocortical input for their synaptic drive and, probably for that reason, more susceptible to the loss of visually driven afferent activity.

Parvalbumin-immunoreactive neurons in layer IVC of the normal monkey are all GABA cells (Hendry *et al.*, 1989). Parvalbumin immunoreactivity persists

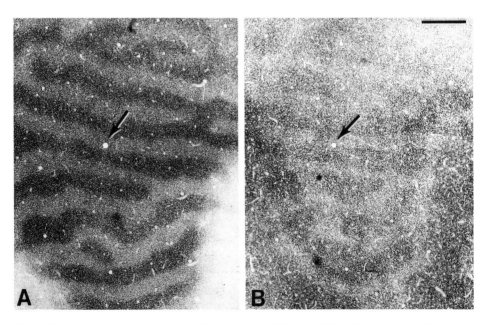

Figure 30. Adjacent tangential sections through layer IVC stained histochemically for CO (A) and immunocytochemically for β2/β3 subunits of the $GABA_A$ receptor (B), showing reduction in β2/β3 immunostaining in deprived ocular dominance stripes consequent upon monocular visual deprivation by TTX injection for 5 days. Arrows indicate same blood vessels. Similar reductions are seen in immunostaining for α1 and γ2 subunits. Bar = 1 mm. Hendry *et al.* (1990).

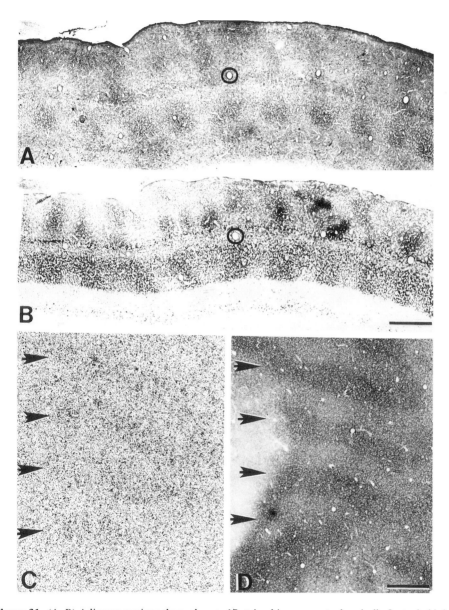

Figure 31. (A, B) Adjacent sections through area 17 stained immunocytochemically for tachykinins (A) and histochemically for CO (B), showing reduction in tachykinin immunostaining in deprived ocular dominance columns following monocular deprivation caused by TTX injections for 15 days. Same blood vessel is circled. Bar = 1 mm. Hendry *et al.* (1988a). (C, D) Adjacent tangential sections through layer IVC of a similarly deprived monkey showing reduction in β-preprotachykinin mRNA in deprived columns, as identified by *in situ* hybridization histochemistry. Arrows indicate normal columns in autoradiograph (left) and CO-stained section (right). Bar = 500 μm. D. L. Benson, M. M. Huntsman, P. J. Isackson, and E. G. Jones (unpublished).

in these cells of deprived-eye columns for as much as 2 weeks following the loss of input from one eye. Thus, from day 5 following elimination of retinal activity, when GABA immunoreactivity declines by a massive amount, until day 14 the deprived GABA neurons of layer IVC can still be localized by parvalbumin immunostaining (S. H. C. Hendry and E. G. Jones, unpublished observations). With longer survival times, the parvalbumin immunoreactivity in a large proportion of the deprived GABA neurons in layer IVC also drops below detectable levels. These findings indicate that levels of GABA and GAD are more rapidly regulated than levels of parvalbumin and that the continuing presence of parvalbumin has no effect on the neurotransmitter levels in the deprived neurons.

Of the other proteins and peptides that have been colocalized with GABA in neurons of monkey area 17, only calbindin changes with monocular deprivation (see below). Levels of NPY-, SRIF-, and CCK-like immunoreactivity remain normal, although neurons containing these are mainly found outside layer IVC and other geniculocortical termination zones, where the most robust changes in immunostaining usually occur.

6.7. Afferent-Dependent Plasticity of Other Molecules

6.7.1. Glutamate

Afferent axons from the thalamus and other areas of cortex as well as the intrinsic excitatory interneurons and the efferent pyramidal neurons use glutamate or related excitatory amino acids as a neurotransmitter (Conti *et al.*, 1987, 1988; DeFelipe *et al.*, 1988). In the monkey visual cortex, neurons immunoreactive for glutamate are densely and homogeneously distributed in layer IVC and are present in variable densities in the other layers. Monocular deprivation in adult monkeys alters this pattern of immunoreactivity (Fig. 32) (Carder *et al.*, 1991). Over the same time period (5 days) that intravitreal injections of TTX lead to reduced numbers of detectable GABA- and GAD-immunoreactive neurons in layer IVC, glutamate immunostaining is virtually lost from the deprived-eye columns but remains intense within normal-eye columns. As a result, the familiar pattern of alternating intensely and lightly immunostained bands appears in layer IVC. Because glutamate exists in both metabolic and transmitter pools, either or both might be reduced in the deprivation-induced loss of immunoreactivity. However, studies of the normal distribution of glutamate immunoreactivity indicate that it is mainly the neurotransmitter pool that is localized by immunocytochemistry, as employed to study glutamate plasticity (Conti *et al.*, 1987, 1988). If these conclusions gain support from further study, monocular deprivation would be seen to reduce levels of not only the major inhibitory neurotransmitter in monkey area 17 (GABA) but also the principal excitatory neurotransmitter (glutamate).

6.7.2. Calcium/Calmodulin-Dependent Protein Kinase

Recognized as the major postsynaptic density protein of asymmetrical synapses (Kelly and Montgomery, 1982; Kennedy *et al.*, 1983; Kelly *et al.*, 1984; Ouimet *et al.*, 1984), type II calmodulin-dependent protein kinase (CaM II kinase) makes up 1% of the total neuronal protein in the mammalian forebrain (Erondu and Kennedy, 1985). CaM II kinase is active in phosphorylating a wide variety of substrates, including cytoskeletal proteins such as MAP2 (Kelly *et al.*, 1984; Miller and Kennedy, 1985; Schulman, 1984; Vulliet *et al.*, 1984) and proteins involved in the processes of synaptic vesicle movement (McGuinness *et al.*, 1985, 1989) and transmitter release (Llinás *et al.*, 1985; Lin *et al.*, 1990). It also plays a role in the induction of long-term potentiation at certain hippocampal synapses (Malenka *et al.*, 1989; Malinow *et al.*, 1989). The α subunit of the holoenzyme is particularly enriched in the forebrain (Erondu and Kennedy, 1985), and recent studies show that gene expression for this subunit is confined to non-GABA cells (Benson *et al.*, 1991b, 1992) (and see Fig. 33). Immunostaining for the α subunit of the enzyme is particularly intense in layers II–III, IVB, and V and relatively light in layers IVA, IVC, and VI (Hendry and Kennedy, 1986). Within each sublamina of layer IV the

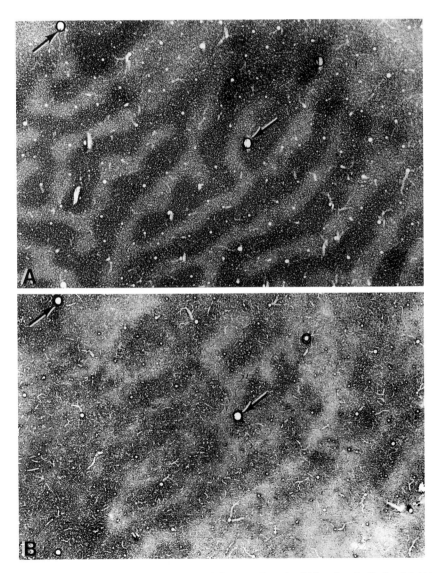

Figure 32. Adjacent tangential sections through layer IVC, stained histochemically for CO (A) and immunocytochemically for glutamate (B) showing reduction in glutamate immunoreactivity in deprived ocular dominance columns after 15 days of TTX injections in one eye. S. H. C. Hendry, R. Carder, and E. G. Jones (unpublished).

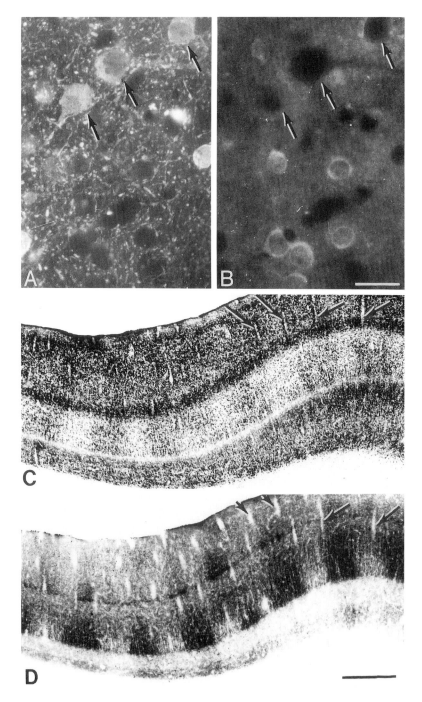

Figure 33. (A, B) Fluorescence photomicrographs of the same microscopic field, showing immuno-staining for GABA (A) and for α-CAM II kinase (B) in layer IVC of area 17 of a normal monkey. The two antigens are found in different neurons. α-CAM II kinase in this layer is probably contained in spiny stellate cells. Bar = 20 μm. S. H. C. Hendry and E. G. Jones (unpublished). (C, D) Adjacent sections through area 17 stained immunocytochemically for α-CAM II kinase (C) and histo-chemically for CO (D), showing increased immunoreactivity for α-CAM II kinase in deprived eye dominance column of layer IVC 10 days after removal of one eye. Arrows indicate same blood vessels. Bar = 1 mm. Hendry and Kennedy (1986).

immunostaining is normally homogeneous. However, monocular deprivation by enucleation or TTX injection in the adult monkey leads to a rapid change in the pattern of immunostaining particularly in layer IVC (Hendry and Kennedy, 1986). Alternating light and dark bands, each 0.4–0.6 mm in diameter and corresponding to the ocular dominance stripes, now appear (Fig. 33). Comparison of the pattern of immunostaining with the pattern of CO staining in adjacent sections shows that this change in CaM II kinase immunoreactivity reflects an enhancement in the deprived-eye columns: the intensely immunostained bands correspond to columns lightly stained for CO. Comparison of normal and deprived monkey cortex indicates that the number of kinase-positive cells in layer IVC is normal but individual cells in the deprived-eye columns are more intensely immunoreactive (Hendry and Kennedy, 1986; D. L. Benson and E. G. Jones, unpublished results).

The enhanced CaM II kinase immunostaining alone could reflect an increase in the cellular concentration of kinase or enhanced phosphorylation (Erondu and Kennedy, 1985). Either or both may therefore occur in the deprived-eye neurons. *In situ* hybridization histochemistry, however, shows that levels of mRNAs coding for the α subunit of CaM II kinase are enhanced in the deprived-eye dominance columns of layer IVC (Fig. 34) and that this enhancement extends into other layers, particularly to the CO-rich patches of layers II–III, V, and VI, as though following the vertical lines of intracortical connectivity (Benson *et al.*, 1991a). These findings are the converse of the effects of deprivation on GAD and preprotachykinin mRNA levels and indicate that activity-dependent regulation of gene transcription or mRNA stability can occur in the positive as well as negative direction.

6.7.3. Cholinesterases

Deprivation-induced changes can also be revealed in adult monkey visual cortex by histochemical staining for acetylcholinesterase and butyrylcholinesterase. Staining for these enzymes is elevated in deprived-eye columns (Graybiel and Ragsdale, 1982). Neurons in the LGN also display cholinesterase activity, but there is a mismatch between the lateral geniculate and cortical staining patterns since the most intense staining normally occurs in the parvocellular layers of the nucleus and in the magnocellular-recipient layer IVCα of the cortex (Graybiel and Ragsdale, 1982; Fitzpatrick and Diamond, 1980). Horton (1984) attributes the deprivation-induced increase in acetylcholinesterase staining to the shrinkage of neuronal somata and other intracortical elements in the deprived columns (see Haseltine *et al.*, 1979), leaving a normal level of stained fibers behind and producing an appearance of increased density in those columns. However, a *reduction* in acetylcholinesterase staining is also induced by monocular deprivation in the regions of deprived-eye CO-dense patches in layers II and III (Horton, 1984). This suggest that a shrinkage effect cannot be responsible for all changes in cholinesterase activity in area 17.

6.8. Deprivation Effects on Normal-Eye Columns

Monocular deprivation also produces changes in columns dominated by the normal eye. The changes are particularly evident in and around the CO patches in layers II and III. The CO periodicities are centered on the ocular dominance columns and form parallel rows that are aligned with the ocular dominance stripes and run orthogonal to the border between areas 17 and 18, becoming progressively less orderly in the cortical representation of the fovea (Horton, 1984; Hevner and Wong-Riley, 1990). Elimination of retinal activity causes patches at the centers of the deprived-eye columns to shrink and to show reduced CO staining (Horton, 1984; Wong-Riley and Carroll, 1984a; Hevner and Wong-Riley, 1990). These effects occur as rapidly as the deprivation effects on layer IV but the deprived-eye patches persist in this shrunken state for at least several months following eye removal in monkeys (Trusk *et al.*, 1990) and for

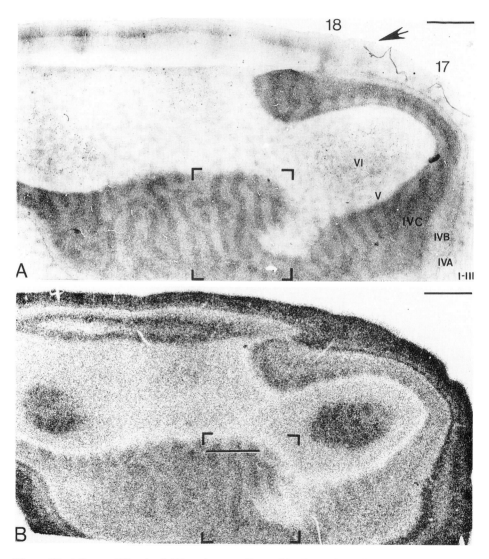

Figure 34. Adjacent CO-stained (A) and autoradiographic (B) sections showing pattern of CO-stained normal and deprived ocular dominance columns matched to *in situ* hybridization with a cRNA probe to α-CAM II kinase. Stripes of enhanced *in situ* hybridization in layers IV and VI correlate with the deprived eye dominance columns. Arrow indicates border between areas 17 and 18. Boxed area was used for quantitative measurements. Bars = 2 mm. From Benson *et al.* (1991a).

more than 20 years following eye removal in humans (Horton and Hedley-White, 1984).

In a monocularly deprived monkey, the CO staining of patches at the centers of the intact-eye columns becomes increased and the patches expand and join with their neighbors of the same intact-eye row to resemble a string of beads (see Fig. 2 of Trusk *et al.*, 1990). CO-stained bridges between individual patches are occasionally found in normal monkeys (Livingstone and Hubel, 1984; Ts'o and Gilbert, 1988) but are unusual. In the intact-eye columns of monocularly deprived monkeys they are the rule.

Expansion of the members of rows of intact-eye patches takes place when one eye is enucleated or silenced by TTX and also when it is chronically defocused. Removal of the lens from one eye of an adult monkey produces progressive expansion of the CO patches in rows related to the intact eye and progressive shrinkage of those in rows related to the aphakic eye (Hendry *et al.*, 1988a). Three months after the removal of one lens, the bridges between the intact-eye patches have filled in so that layers II and III now contain wide, homogeneously stained stripes alternating with shrunken rows of aphakic-eye patches (Fig. 35).

The increase of CO staining in the rows of intact-eye patches is accompanied by increased immunoreactivity for the neuropeptide, substance P. In the normal monkey, somata immunostained for substance P and other members of the tachykinin family are evenly distributed in layers II and III, but there is enhanced immunostaining of terminals in the CO patches in comparison with the interpatch and interrow regions (Fig. 35). The tachykinin immunostaining, thus, mirrors the immunostaining for GAD (Hendrickson *et al.*, 1981; Fitzpatrick *et al.*, 1987). After 3 months of monocular aphakia in adulthood, tachykinin immunostaining expands within the patches of intact-eye rows and shrinks within those of the deprived-eye rows. The expanded, intact-eye pattern results from an increased density of immunostained somata in the patches and in the bridges between the patches (Hendry *et al.*, 1988a). No changes in GABA immunostaining are evident in these regions in the same monkeys (i.e., the number and proportion of GABA-immunostained cells does not increase). This suggests either that tachykinin levels are increasing in GABA cells in which tachykinin immunoreactivity could not previously be demonstrated or that tachykinin expression is being induced in a new population of cells.

Calbindin immunoreactivity is normally found in GABA cells in and around the CO patches of layer II but only in the interpatch regions and interrow regions in layer III (Celio *et al.*, 1986; Van Brederode *et al.*, 1990). In layers II and III of a monocularly deprived monkey, calbindin immunostaining of rows of patches related to the deprived-eye columns is uniformly low while immunostaining of rows related to normal-eye columns is uniformly high (S. H. C. Hendry and R. K. Carder, unpublished observations). These data suggest that under normal conditions only the part of the population of GABA–calbindin cells found in the interpatch and interrow regions contains a concentration of calbindin sufficiently high to be detected immunocytochemically. Under conditions of monocular deprivation, levels rise in the remainder. The ability of deprivation of one eye to produce changes in the neurochemical characteristics of cells dominated by the other, normal eye is undoubtedly related to a functional interaction between inputs from the two eyes that exists under normal conditions. An anatomical substrate for binocular interactions across ocular dominance domains does exist in area 17. Neurons of layer IVC, although tending to maintain dendritic fields in only one ocular dominance column, send axons across columnar borders (Katz *et al.*, 1989). Many of these cross-columnar connections link layer IVC neurons dominated by one eye with layer III neurons dominated by the other eye. It is not known whether these axons innervate preferentially CO patches or interpatch regions. Intracortical connections within layers II and III separately link patches with patches and interpatch regions with other interpatch regions (Livingstone and Hubel, 1984; Gilbert and Wiesel, 1989). If these projections cross ocular dominance columns, they would supply an anatomical basis for an interaction between eye inputs. Therefore, both connections from layer IV and connections within layers II and III are in a position

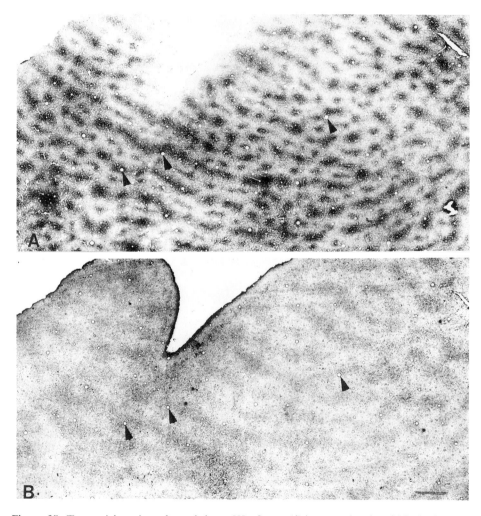

Figure 35. Tangential sections through layer III of area 17 in a monkey in which the lens was removed from one eye 3 months previously. A shows pattern of alternating rows of shrunken and expanded CO-stained patches centered on the aphakic and normal ocular dominance columns. B shows enhanced tachykinin immunostaining in the expanded normal-eye rows. Arrowheads indicate same blood vessels. From Hendry *et al.* (1988a).

to mediate the transfer of normal activity and of deprivation effects across ocular dominance columns.

6.9. Summary

Evidence for activity-dependent changes in non-GABA-related molecules reviewed here shows that: (1) the effects of monocular deprivation in the adult monkey extend beyond changes in the levels of GABA- and GAD-immunoreactivity, related proteins and peptides and receptors to include regulation of excitatory neurotransmitters, second messenger-linked enzymes, and certain other enzymes. (2) Reductions in immunoreactivity occur without a general reduction in neuronal protein synthesis. The immunocytochemical and *in situ* hybridization histochemical data indicate that the level of at least one protein, CaM II kinase, is increased in deprived neurons. Immunoreactivity for most peptides and other molecules, including the synaptic vesicle protein synapsin I (Hendry and Jones, 1986), and housekeeping enzymes such as neuron-specific enolase display no change with monocular deprivation. (3) Different neuronal mechanisms may serve to regulate levels of immunoreactivity (Benson *et al.*, 1991a), but most of the changes revealed to date seem to depend on an activity-dependent regulation of mRNA levels. Whether this results from changes in transcription or in mRNA stability remains to be determined.

7. Significance of Activity-Dependent Regulation of GABA-Related Gene Expression

7.1. Effects in the Critical Period

Neural activity, by which we imply the initiation and conduction of action potentials and the induction of membrane conductance and polarization changes, plays an important role in the development and early postnatal maturation of the visual system (reviewed recently in Shatz, 1990). In the case of the primate visual cortex, competitive interactions between the axons of thalamic cells innervated by each of the two eyes ensure that monocular deprivation during a critical period of early postnatal development will result in axons driven by the undeprived eye gaining access to more "synaptic space" in area 17 (Hubel *et al.*, 1977; LeVay *et al.*, 1980). The responsivity of many cortical cells is accordingly modified in favor of the nondeprived eye, putatively for the remainder of the lifetime of the individual (LeVay *et al.*, 1980). It is widely believed that the adult visual cortex shows little or no capacity for afferent-dependent plasticity beyond the critical developmental period. The results reviewed in Section 6 which show the important role of neural activity in the regulation of gene expression for transmitters and other neuroactive molecules cast some doubt on the received dogma. Moreover, it is likely that the mechanisms that underlie the activity-dependent plasticity seén in adulthood are fundamentally similar to those operating during the establishment and stabilization of connections during the normal development of the visual cortex, mechanisms that are probably compromised by perturbations of visual experience during the critical period. Although similar effects have not yet been demonstrated in the maturing visual cortex of monkeys, a number of observations point to the significance of GABA in critical-period-dependent plasticity in kittens and these have obvious parallels with the activity-dependent regulation of GABA observed in adult monkeys.

GABA-mediated inhibition is present early in infant cats (Albus and Wolf, 1984; Tsumoto and Sato, 1985; Wolf *et al.*, 1986) and chronic infusion or iontophoretic application of the $GABA_A$ antagonist, bicuculline, in the visual cortex of kittens during the critical period leads to reductions in orientation selectivity, unusually large receptive fields, loss of directional selectivity, and certain other receptive field changes in visual cortical neurons (Wolf *et al.*, 1986; Ramoa *et al.*, 1988). Chronic intracortical infusion of the GABA agonist, muscimol, during monocular deprivation in the critical period causes cells in the visual cortex to prefer the deprived, rather than the nondeprived eye, as customarily occurs. That is, the more active, undeprived-eye inputs fail to consolidate (Reiter and Stryker, 1988). After monocular deprivation in the critical period when a certain percentage of visual cortical cells become driven only by the undeprived eye, iontophoretic application of bicuculline later in life uncovers suppressed, deprived-eye inputs to the cells (Sillito *et al.*, 1981). All of these effects seem to indicate that GABA-mediated synaptic influences can be modified or compromised under activity-dependent conditions during critical-period-dependent plasticity. Are any comparable phenomena demonstrable in adult animals subjected to monocular deprivation leading to downregulation of GABA receptor and transmitter function?

7.2. Effects in Adults

The number of physiological and behavioral data available on the visual cortex of monkeys subjected to monocular deprivation after the critical period is rather small. In monkeys monocularly deprived or made strabismic during the critical period and examined behaviorally 3 or more years later, there were decreases in spatial modulation sensitivity, temporal modulation sensitivity, and increment threshold spectral sensitivity (Haweth *et al.*, 1983, 1986). The extent to which similar changes occur in animals deprived in adulthood is not clear.

In one animal from which an eye had been removed in adulthood, there was some expansion of normal layer IVC eye dominance columns, as measured by single-unit recording (LeVay *et al.*, 1980). Anatomical tracing in a second adult indicated that this physiological expansion is not accompanied by expansion of terminations of thalamocortical fibers related to the normal eye. In the adult deprived animal, there were continuous responses to the normal eye in the upper cortical layers without changes in orientation selectivity. Our own investigations confirm these observations (K. A. C. Martin, H. D. Schwark, R. A. Warren, G. W. Huntley and E. G. Jones, unpublished observations, Fig. 36).

Observations of this kind have usually been taken to imply that deprived-eye inputs simply "drop out" and that most cells in superficial layers become responsive to the nondeprived eye due to the presence normally of binocular inputs to cells in these layers. The presence of a greater degree of functional plasticity under these conditions is, however, revealed by studies in which instead of total monocular deprivation, small retinal lesions have been used to deprive only a small part of the representation in one visual cortex. In the first of these studies, in cats, Kaas *et al.* (1990) removed one eye in order to obtain an ipsilateral cortex receiving inputs only from the contralateral normal eye. Lesions 5–10° in extent were then placed in the retina of this eye just above the area centralis. Two to six months later, neurons in portions of areas 17 and 18 normally representing the lesioned area and covering 4–8 mm in extent, were found to be responsive to visual stimuli and to have receptive fields in parts of the retina adjacent to the lesion. In comparable experiments in 5- to 9-year-old monkeys, Heinen and Skavenski (1991) placed bilateral foveal lesions 600 μm in extent and recorded 2.5 months later in the foveal representation of area 17. They found that approximately half the cells recorded in the area that should have been deprived of retinal inputs showed responses to stimuli applied to parts of the retina outside the lesioned area. Cells responding in the affected area showed weaker responses than normal and had larger, abnormally shaped receptive fields with no antagonistic subregions. The changes in the cortex affected a region representing more than

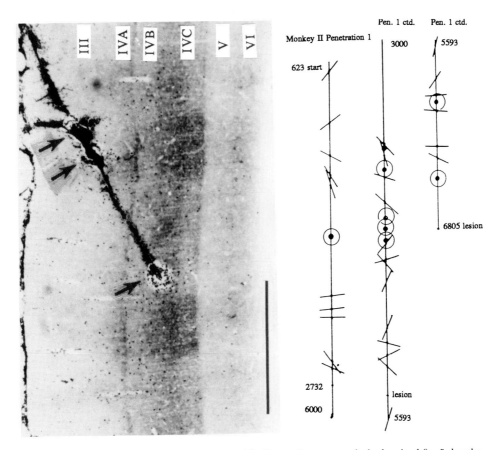

Figure 36. (Left) CO-stained section from area 17 of a monkey monocularly deprived for 5 days by TTX injections, showing an oblique electrode track and marking lesions (arrows). Bar = 1 mm. (Right) Sequences of orientation selectivities (bars) of single neurons encountered as electrode entering along track at left crossed successive orientation columns. Circles indicate nonoriented cells. Sequences of change in orientations as electrode crosses deprived and undeprived ocular dominance columns appear normal. K. A. C. Martin, H. D. Schwark, R. A. Warren, G. W. Huntley, and E. G. Jones (unpublished).

1° of visual field, that is to say, a region corresponding to several millimeters of cortical extent.

The rapidity with which visual inputs can be revealed in an anticipated silenced zone of area 17 was shown by Gilbert and Wiesel (1992) who recorded single-unit responses both before and after making lesions in homotopic regions of the two retinas in cats and monkeys (Fig. 37). Within minutes of making the lesions, it was found that cortical neurons at the edges of the silenced area had acquired enlarged receptive fields that had expanded into the portions of the retina adjacent to the lesions. After 2 months, neurons in the silenced area proper had acquired receptive fields in loci adjacent to the lesion. This was a 4- to 5-mm topographical shift in the normal cortical representation. The receptive fields were larger than normal, although the enlargement diminished somewhat

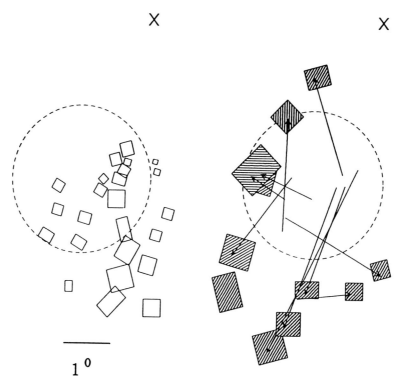

Figure 37. Receptive field maps in a region of area 17 in a monkey deafferented by a retinal lesion, immediately before the lesion was made (left) and 2 months following the lesion (right). A large shift in receptive field position is seen, with the predicted cortical scotoma (dashed outline) having entirely filled in. All recorded sites in the predicted scotoma region of the cortex had cells with receptive fields (hatched rectangles) located outside the lesioned retinal area, although a few fields overlay the scotoma, probably owing to incomplete destruction at the edge of the lesion in the retina. Overall the shifts maintained a crude retinotopic order, with fields that were originally located in the lower part of the scotoma shifting down, and those located in the upper part shifting up. Note that for one site where the receptive field was initially located outside the lesion, the field shifted horizontally. This result was observed in several experiments, and indicates that the effects of the perturbation caused by the lesion are propagated beyond the deafferented area of cortex. There was also receptive field enlargement, though less than those observed in short-term experiments: the field areas averaged $0.036\ \text{deg}^2$ ($\pm0.022\ \text{deg}^2$) before the lesion and $0.100\ \text{deg}^2$ ($\pm0.025\ \text{deg}^2$) 2 months later ($p \ll 0.01$). The X's mark the position of the fovea. From Gilbert and Wiesel (1992).

with time; the neurons with new receptive fields displayed normal orientation selectivity, directionality, and binocularity implying that the fundamental pattern of intracortical connectivity had been preserved or modified to adapt to the new conditions.

The rapidity with which the receptive field shifts occur and the rapidity with which representational changes occur in other sensory and motor cortical areas under comparable conditions (Merzenich *et al.*, 1983a,b, 1987; Stryker *et al.*, 1987; Wall *et al.*, 1986; Jenkins *et al.*, 1990; Clark *et al.*, 1988; Sanes *et al.*, 1988) make it unlikely that these effects are caused by axonal sprouting and the formation of new connections. Instead, a rapid unmasking or previously silent or subthreshold inputs seems more likely. That this unmasking is primarily cortical is revealed by the lack of a corresponding shift in the representation in the LGN of Gilbert and Wiesel's cats. A large silent area remained, even after 2 months.

There are two potential sources of preexisting widespread connections whose existence is normally unrevealed in conventional mapping studies. These are the thalamocortical axons and the horizontal, excitatory collaterals of cortical pyramidal cells. Although the former have quite extensive domains of cortical termination, a span of several ocular dominance columns in the case of cat Y axons (Martin, 1984; Freund *et al.*, 1985), it is generally felt that these are insufficient to account for the shifts of several millimeters recorded in the three studies reviewed above. The number of investigations pertaining to the spread of single geniculocortical axons is, however, quite small (Gilbert and Wiesel, 1979; Blasdel and Lund, 1983) and most single axons from which data might be obtained have been labeled by injections in the cortex or immediately subjacent white matter, not in the LGN or deep in the optic radiation which would be required to demonstrate early branching.

At present, the most attractive candidates for providing a preexisting set of connections whose presence may be revealed by loss of geniculocortical inputs to a focal region of the cortex are the horizontal intracortical collaterals. Pyramidal cells projecting to other cortical areas or to subcortical sites, particularly those in layers III and V, have extensive systems of intracortical collaterals that can extend horizontally for 6 mm or more within the visual and other areas (Gilbert and Wiesel, 1979; DeFelipe *et al.*, 1986a). In layer III of area 17 in monkeys, these horizontal collaterals interconnect focal aggregations of neurons with the same orientation selectivity (Gilbert and Wiesel, 1989). In motor cortex, there is evidence that these connections are normally subthreshold under experimental conditions (Huntley and Jones, 1991). The preservation of orientation selectivity in neurons with rapidly shifted receptive fields and located in a focal region of cortex deprived of retinal inputs, suggests that it is these horizontal, excitatory connections that have been unmasked. A potential mechanism for such unmasking would seem to be in the overt downregulation of GABA and $GABA_A$ receptors revealed by our immunocytochemical and gene expression studies. Although we have not conducted investigations on animals deprived by subtotal lesions of the retina, there is reason to believe that these would have effects identical to those seen with full monocular deprivation, only restricted to the part of area 17 representing the lesioned area. The time course of the effects on GABA mechanisms is sufficiently short to play a major role in the rapid receptive field shifts. Under these circumstances, we may anticipate an effect similar to that diagrammed in Fig. 38. A zone of cortex in which GABA transmission is

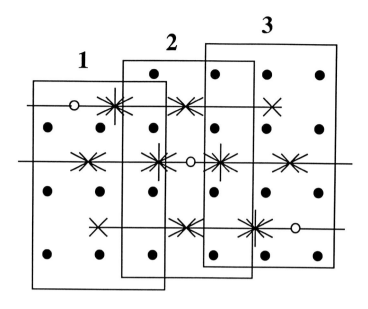

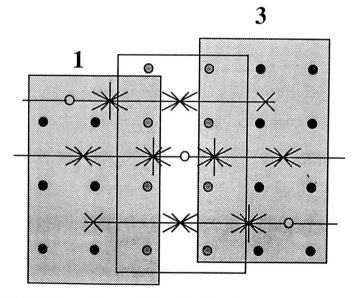

Figure 38. Predicted effects of downregulation of GABA and GABA$_A$ receptors in a region of area 17 deprived of afferent input. Upper diagram indicates putative normal pattern of three overlapping zones representing adjacent portions of the visual field, each containing approximately the same number of GABA cells (black) and the same balance of excitation and inhibition that sets up boundary constraints where inputs to the three zones tend to overlap. Pyramidal cells with long horizontal intracortical collaterals crossing from one zone into adjacent zones are indicated by unfilled circles and focal terminations.

When input to zone 2 is compromised (lower diagram), GABA mechanisms are reduced in zone 2 allowing overlapping afferents to zones 1 and 3 and horizontal excitatory connections of the pyramidal cells to be revealed, thus causing apparent spread of regions represented in zones 1 and 3 into functionally deafferented zone 2.

downregulated in an activity-dependent manner should permit the functional expression of subthreshold inputs coming into it via the long-range horizontal collaterals from adjacent normally active areas and, thus, the apparent acquisition of receptive fields proper to the surrounding regions in the GABA-deficient zone. The same diagram could also be used to explain the functional expression of normally suppressed thalamocortical inputs in the deprived zone, should it be revealed that geniculocortical fiber branching and terminations in the cortex are more extensive than previously anticipated.

Possibly the most obvious functional manifestation of the representational shift induced in the striate cortex in the presence of subtotal lesions of the retina is the "filling in" phenomenon experienced by subjects with visual field scotomata (e.g., Ramachandran and Gregory, 1991). Under these conditions, a large area of visual field loss commonly goes unreported and is only revealed by perimetry. Effects of total monocular deprivation that may be attributed to downregulation of visual cortical GABA mechanisms in the human are uncertain. One potential effect is the well-known Charles Bonnet syndrome in which patients with long-standing cataracts experience complex visual hallucinations in the absence of psychosis (Weinberger and Grant, 1940; Bartlet, 1951; McNamara *et al.*, 1982; Damas-Mora *et al.*, 1982; White, 1980). Conceivably, these perceptual illusions are manifestations of high-order cortical areas receiving a distorted input from an area 17 in which the normal balance of excitation and inhibition has been profoundly disturbed. The transient period of readaptation to the visual environment after removal of a cataract or after restoration of vision lost by other pathology (Gregory, 1966) may be manifestations of the return of gene expression for molecules related to GABA transmission, comparable to that revealed by our recovery experiments in monkeys.

8. Conclusions

At least one-fifth of the neurons in area 17 of the adult primate are GABAergic and, thus, can be expected to be major components of cortical circuitry. Both these cells and the $GABA_A$ receptor which is also enriched in area 17, although found in all layers, show lamina-specific concentrations that seem to correlate mainly with the layers in which the major densities of geniculocortical terminations occur. There is now ample evidence to indicate that all GABA cells are aspiny interneurons and that all major classes of aspiny, nonpyramidal cortical neurons are GABAergic. Many of these types, in addition to being characterized morphologically, can be organized in subcategories on the basis of colocalized neuropeptides or other polypeptides, particularly the calcium-binding proteins calbindin and parvalbumin. Although evidence is growing regarding the synaptic connections formed by these individual classes of aspiny, GABA cell, much of this derives from investigations on other species and it is still not feasible to place each variety of GABA cell into a framework of intracortical circuitry. Each morphological type, however, appears to send its axon to terminate on a restricted and reproducible location on the initial axon segment, soma, spines, or dendrites of the pyramidal cells. However, GABA neurons, while forming both very local and strong, bidirectional, vertical, and thus intracolumnar connections between cortical layers, appear to contribute little to long-range intralaminar and thus intercolumnar connections.

One of the most significant recent findings on GABA neurons in area 17 of primates is the facility with which levels of immunoreactive GABA, of its synthesizing enzyme, GAD, and of the major subunits of the GABA$_A$ receptor, along with certain other peptides contained in GABA neurons, are regulated by activity entering the visual cortex from the eyes, even in adult monkeys. In the case of GAD and certain other molecules, this regulation is mediated by an activity-dependent control of mRNA levels. Activity-dependent plasticity of this kind has not previously been thought of as a phenomenon involving adult visual cortex. Its existence suggests the presence of a finely tuned adaptive system through which cortical transmitter and receptor levels are regulated by ongoing visual experience throughout the lifetime of the individual. An adaptive mechanism of this kind may contribute to the rapid unmasking of silent connections that appear to underlie the afferent-dependent plasticity of retinotopic organization in the visual cortex.

ACKNOWLEDGMENTS. Personal work described in the text was supported by grants EY06432, EY01793, and NS21377 from the National Institutes of Health, United States Public Health Service, by the Frontier Research Program in Brain Mechanisms of Mind and Behavior, RIKEN, Japan, and by the Irvine Research Unit in Molecular Neurobiology. We thank Dr. Paul Isackson and Ms. Molly Huntsman for numerous contributions.

9. References

Akil, M., and Lewis, D. A., 1992, Differential distribution of parvalbumin-immunoreactive pericellular clusters of terminal boutons in developing and adult monkey neocortex, *Exp. Neurol.* **115**:239–249.

Albus, K., and Wolf, W., 1984, Early post-natal development of neuronal function in the kitten's visual cortex: A laminar analysis, *J. Physiol. (London)* **348**:153–185.

Bartlet, J. E. A., 1951, A case of organized visual hallucinations in an old man with cataract, and their relation to the phenomenon of the phantom limb, *Brain* **74**:363–373.

Benson, D. L., and Jones, E. G., 1991, Differential intraneuronal regulation of GAD and β-preprotachykinin mRNAs in monkey visual cortex following monocular deprivation, *Soc. Neurosci. Abstr.* **17**:115.

Benson, D. L., Isackson, P. J., and Jones, E. G., 1990, In situ hybridization reveals VIP precursor mRNA-containing neurons in monkey and rat neocortex, *Mol. Brain Res.* **9**:169–174.

Benson, D. L., Isackson, P. J., Gall, C. M., and Jones, E. G., 1991a, Differential effects of monocular deprivation on glutamic acid decarboxylase and type II calcium-calmodulin-dependent protein kinase gene expression in the adult monkey visual cortex, *J. Neurosci.* **11**:31–47.

Benson, D. L., Isackson, P. J., Hendry, S. H. C., and Jones, E. G., 1991b, Differential gene expression for glutamic acid decarboxylase and type II calcium-calmodulin-dependent protein kinase in basal ganglia, thalamus and hypothalamus of the monkey, *J. Neurosci.* **11**:1540–1564.

Benson, D. L., Isackson, P. J., Gall, C. M., and Jones, E. G., 1992, Contrasting patterns in the localization of glutamic acid decarboxylase and Ca^{2+}/calmodulin protein kinase gene expression in the rat central nervous system, *Neuroscience* **46**:825–850.

Blasdel, G. G., and Lund, J. S., 1983, Termination of afferent axons in macaque striate cortex, *J. Neurosci.* **3**:1389–1413.

Blasdel, G. G., Lund, J. S., and Fitzpatrick, D., 1985, Intrinsic connections of macaque striate cortex: Axonal projections of cells outside lamina 4C, *J. Neurosci.* **5**:3350–3369.

Blümke, I., Hof, P. R., Morrison, J. H., and Celio, M. R., 1990, Distribution of parvalbumin immunoreactivity in the visual cortex of Old World monkeys and humans, *J. Comp. Neurol.* **301**:417–432.

Blümke, I., Hof, P. R., Morrison, J. H., and Celio, M. R., 1991, Parvalbumin in the monkey striate cortex: A quantitative immunoelectron-microscopy study, *Brain Res.* **554**:237–243.

Bu, D. R., Erlander, M. G., Hitz, B. C., Tillakaratne, N. J. K., Kaufman, D. L., Wagner-McPherson, C. B., Evans, G. A., and Tobin, A. J., 1992, Two human glutamate decarboxylases, 65-kDa GAD and 67-kDa GAD, are each encoded by a single gene, *Proc. Natl. Acad. Sci. USA* **89**:2115–2119.

Campbell, A. W., 1905, *Histological Studies on the Localization of Cerebral Function,* Cambridge University Press, London.

Campbell, M. J., Lewis, D. A., Benoit, R., and Morrison, J. H., 1987, Regional heterogeneity in the distribution of somatostatin-28- and somatostatin-28 1–12 immunoreactive profiles in monkey neocortex, *J. Neurosci.* **7**:1133–1144.

Carder, R. K., Jones, E. G., and Hendry, S. H. C., 1991, Distribution of glutamate neurons and terminals in striate cortex of normal and monocularly deprived monkeys *Soc. Neurosci. Abstr.* **17**:115.

Celio, M. R., 1986, Parvalbumin in most γ-aminobutyric acid-containing neurons of the rat cerebral cortex, *Science* **231**:995–997.

Celio, M. R., Scharer, L., Morrison, J. H., Norman, A. W., and Bloom, F. E., 1986, Calbindin immunoreactivity alternates with cytochrome c-oxidase-rich zones in some layers of the primate visual cortex, *Nature* **323**:715–717.

Clark, S. A., Allard, T., Jenkins, W. M., and Merzenich, M. M., 1988, Receptive fields in the body-surface map in adult cortex defined by temporally correlated inputs, *Nature* **332**:444–445.

Conti, F. A., Rustioni, A., Petrusz, P., and Towle, A. C., 1987, Glutamate-positive neurons in the somatic sensory cortex of rats and monkeys, *J. Neurosci.* **7**:1887–1901.

Conti, F., Fabri, M., and Manzoni, T., 1988, Immunocytochemical evidence for glutamatergic cortico-cortical connections in monkeys, *Brain Res.* **462**:148–153.

Cowan, W. M., 1970, Anterograde and retrograde transneuronal degeneration in the central and peripheral nervous system, in: *Contemporary Research Methods in Neuroanatomy* (W. J. H. Nauta and S. O. E. Ebbesson, eds.), Springer, Berlin, pp. 217–249.

Damas-Mora, J., Skelton-Robinson, M., and Jenner, F. A., 1982, The Charles Bonnet syndrome in perspective, *Psychol. Med.* **12**:251–261.

DeFelipe, J., and Fariñas, I., 1993, The pyramidal neuron of the cerebral cortex. Morphological and chemical characteristics of the synaptic inputs, *Prog. Neurobiol.,* in press.

DeFelipe, J., and Jones, E. G., 1985, Vertical organization of γ-aminobutyric acid-accumulating intrinsic neuronal systems in monkey cerebral cortex, *J. Neurosci.* **5**:3246–3260.

DeFelipe, J., and Jones, E. G., 1988, *Cajal on the Cerebral Cortex,* Oxford University Press, London.

DeFelipe, J., and Jones, E. G., 1991, Parvalbumin immunoreactivity reveals layer IV of the monkey cerebral cortex as a mosaic of microzones of thalamic afferent terminations, *Brain Res.* **562**:39–47.

DeFelipe, J., and Jones, E. G., 1992, High resolution light and electron microscopic immunocytochemistry of co-localized GABA and calbindin D-28k in somata and double bouquet cell axons of monkey somatosensory cortex, *Eur. J. Neurosci.* **4**:46–60.

DeFelipe, J., Hendry, S. H. C., Jones, E. G., and Schmechel, D., 1985, Variability in the terminations of GABAergic chandelier cell axons on initial segments of pyramidal cell axons in the monkey sensory-motor cortex, *J. Comp. Neurol.* **231**:364–384.

DeFelipe, J., Conley, M., and Jones, E. G., 1986a, Long-range focal collateralization of axons arising from cortico-cortical cells in monkey sensory-motor cortex, *J. Neurosci.* **6**:3749–3766.

DeFelipe, J., Hendry, S. H. C., and Jones, E. G., 1986b, A correlative electron microscopic study of basket cells and large GABAergic neurons in the monkey sensory-motor cortex, *Neuroscience* **7**:991–1009.

DeFelipe, J., Conti, F., Van Eyck, S. L., and Manzoni, T., 1988, Demonstration of glutamate-positive axon terminals forming asymmetric synapses in cat neocortex, *Brain Res.* **455**:162–165.

DeFelipe, J., Hendry, S. H. C., and Jones, E. G., 1989a, Visualization of chandelier cell axons by parvalbumin immunoreactivity in monkey cerebral cortex, *Proc. Natl. Acad. Sci. USA* **86**:2093–2097.

DeFelipe, J., Hendry, S. H. C., and Jones, E. G., 1989b, Synapses of double bouquet cells in monkey cerebral cortex visualized by calbindin immunoreactivity, *Brain Res.* **503**:49–54.

DeFelipe, J., Hendry, S. H. C., Hashikawa, T., Molinari, M., and Jones, E. G., 1990, A microcolumnar structure of monkey cerebral cortex revealed by immunocytochemical studies of double bouquet cell axons, *Neuroscience* **37:**655–673.

de Lima, A. D., and Morrison, J. H., 1989, Ultrastructural analysis of somatostatin-immunoreactive neurons and synapses in the temporal and occipital cortex of the macaque monkey, *J. Comp. Neurol.* **283:**212–227.

Demeulemeester, H., Archens, L., Vandesande, F., Orban, G. A., Heizmann, C. W., and Pochet, R., 1991, Calcium binding proteins and neuropeptides as molecular markers of GABAergic interneurons in the cat visual cortex, *Exp. Brain Res.* **84:**538–544.

DeYoe, E. A., Hockfield, S., Garren, H., and Van Essen, D. C., 1990, Antibody labeling of functional subdivisions in visual cortex: Cat-301 immunoreactivity in striate and extrastriate cortex of the macaque monkey, *Visual Neurosci.* **5:**67–81.

Erondu, N. E., and Kennedy, M. B., 1985, Regional distribution of type II Ca^{2+}/calmodulin-dependent protein kinase in rat brain, *J. Neurosci.* **5:**3270–3277.

Fairén, A., and Valverde, F., 1980, A specialized type of neuron in the visual cortex of the cat: A Golgi and electron microscope study of chandelier cells, *J. Comp. Neurol.* **194:**761–779.

Fairén, A., DeFelipe, J., and Regidor, J., 1984, Nonpyramidal neurons. General account, in: *Cerebral Cortex,* Volume 1 (A. Peters and E. G. Jones, eds.), Plenum Press, New York, pp. 201–253.

Fariñas, I., and DeFelipe, J., 1991, Patterns of synaptic input on cortico-cortical and cortico-thalamic cells in the cat visual cortex. II. The axon initial segment, *J. Comp. Neurol.* **304:**70–77.

Felleman, D. J., and Van Essen, D. C., 1991, Distributed hierarchical processing in the primate cerebral cortex, *Cereb. Cortex.* **1:**1–47.

Ferrer, I., Tuñon, T., Soriano, E., del Rio, A., Iraizoz, I., Fonseca, M., and Guionnet, N., 1992, Calbindin immunoreactivity in normal human temporal cortex, *Brain Res.* **572:**33–41.

Ferster, D., 1986, Orientation selectivity of synaptic potentials in neurons of cat primary visual cortex, *J. Neurosci.* **6:**1284–1301.

Ferster, D., 1987, Origin of orientation-selective EPSPs in simple cells of cat visual cortex, *J. Neurosci.* **7:**1780–1791.

Ferster, D., 1988, Spatially opponent excitation and inhibition in simple cells of the cat visual cortex, *J. Neurosci.* **8:**1172–1180.

Fisken, R. A., Garey, L. J., and Powell, T. P. S., 1975, The intrinsic, association and commissural connections of area 17 of the visual cortex, *Philos. Trans. R. Soc. London Ser. B* **272:**487–536.

Fitzpatrick, D., and Diamond, I. T., 1980, Distribution of acetylcholinesterase in the geniculostriate system of *Galago senegalensis* and *Aotus trivirgatus:* Evidence for the origin of the reaction product in the lateral geniculate body, *J. Comp. Neurol.* **194:**703–720.

Fitzpatrick, D., Itoh, K., and Diamond, I. T., 1983, The laminar organization of the lateral geniculate body and the striate cortex in the squirrel monkey (*Samiri sciureus*), *J. Neurosci.* **3:**673–702.

Fitzpatrick, D., Lund, J. S., and Blasdel, G. G., 1985, Intrinsic connections of macaque striate cortex: Afferent and efferent connections of lamina 4C, *J. Neurosci.* **5:**3329–3349.

Fitzpatrick, D., Lund, J. S., Schmechel, D. E., and Towles, A. C., 1987, Distribution of GABAergic neurons and axon terminals in the macaque striate cortex, *J. Comp. Neurol.* **264:**73–91.

Foote, S. L., and Morrison, J. H., 1984, Postnatal development of laminar innervation patterns by monoaminergic fibers in monkey (*Macaca fascicularis*) primary visual cortex, *J. Neurosci.* **4:**2667–2680.

Freund, T. F., Martin, K. A. C., Smith, A. D., and Somogyi, P., 1983, Glutamate decarboxylase-immunoreactive terminals of Golgi-impregnated axoaxonic cells and of presumed basket cells in synaptic contact with pyramidal neurons of cat's visual cortex, *J. Comp. Neurol.* **221:**263–278.

Freund, T. F., Martin, K. A. C., and Whitteridge, D., 1985, Innervation of cat visual areas 17 and 18 by physiologically identified X- and Y-type thalamic afferents. I. Arborization patterns and quantitative distribution of postsynaptic elements, *J. Comp. Neurol.* **242:**263–274.

Freund, T. F., Martin, K. A. C., Soltesz, I., Somogyi, P., and Whitteridge, D. I., 1989, Arborization pattern and postsynaptic targets of physiologically identified thalamocortical afferents in striate cortex of the macaque monkey, *J. Comp. Neurol.* **289:**315–336.

Gaspar, P., Berger, B., Febvret, A., Vigny, A., Krieger-Poulet, M., and Borri-Voltattorni, C., 1987, Tyrosine hydroxylase-immunoreactive neurons in the human cerebral cortex: A novel catecholaminergic group, *Neurosci. Lett.* **80:**257–262.

Gilbert, C. D., and Wiesel, T. N., 1979, Morphology and intracortical projections of functionally characterized neurones in the cat visual cortex, *Nature* **280:**120–125.

Gilbert, C. D., and Wiesel, T. N., 1983, Clustered intrinsic connections in cat visual cortex, *J. Neurosci.* **3:**1116–1133.

Gilbert, C. D., and Wiesel, T. N., 1989, Columnar specificity of intrinsic horizontal and cortico-cortical connections in cat visual cortex, *J. Neurosci.* **9**:2432–2442.

Gilbert, C. D., and Wiesel, T. N., 1992, Receptive field dynamics in adult primary visual cortex, *Nature* **356**:150–152.

Graybiel, A. M., and Ragsdale, C. W., Jr., 1982, Pseudocholinesterase staining in the primary visual pathway of the macaque monkey, *Nature* **299**:439–442.

Gregory, R. L., 1966, *Eye and Brain: The Psychology of Seeing*, Weidenfeld & Nicholson, London, pp. 188–219.

Haseltine, E. C., DeBruyn, E. J., and Casagrande, V. A., 1979, Demonstration of ocular dominance columns in Nissl-stained sections of monkey visual cortex following enucleation, *Brain Res.* **176**:153–158.

Haweth, R. S., Smith, E. L., III, Crawford, M. L. J., and Von Noorden, G. K., 1983, Effects of enucleation of the deprived eye on stimulus deprivation amblyopia in monkeys, *Invest. Ophthalmol. Vis. Sci.* **25**:10–17.

Haweth, R. S., Smith, E. L., III, Duncan, G. C., Crawford, M. L. J., and Von Noorden, G. K., 1986, Effects of enucleation of the fixating eye on strabismic amblyopia in monkeys, *Invest. Ophthalmol. Vis. Sci.* **27**:246–254.

Hayes, T. L., and Lewis, D. A., 1992, Nonphosphorylated neurofilament protein and calbindin immunoreactivity in layer III pyramidal neurons of human neocortex, *Cereb. Cortex* **2**:56–67.

Heinen, S. J., and Skavenski, A. A., 1991, Recovery of responses in foveal V_1 neurons following bilateral foveal lesions in adult monkey, *Exp. Brain Res.* **83**:670–674.

Hendrickson, A. E., 1982, The orthograde axoplasmic transport autoradiographic technique and its implications for additional neuroanatomical analysis of the striate cortex, in: *Cytochemical Methods in Neuroanatomy* (V. Chan-Palay and S. L. Palay, eds.), Liss, New York, 1–16.

Hendrickson, A. E., Wilson, J. R., and Ogren, M. P., 1978, The neuroanatomical organization of pathways between dorsal lateral geniculate nucleus and visual cortex in old and new world primates, *J. Comp. Neurol.* **182**:123–136.

Hendrickson, A. E., Hunt, S. P., and Wu, J. L., 1981, Immunocytochemical localization of glutamic acid decarboxylase in monkey striate cortex, *Nature* **292**:605–607.

Hendrickson, A. E., Ogren, M., Vaughn, J. E., Barber, R. P., and Wu, J., 1983, Light and electron microscopic immunocytochemical localization of glutamic acid decarboxylase in monkey geniculate complex: Evidence for GABAergic neurons and synapses, *J. Neurosci.* **3**:1245–1262.

Hendrickson, A. E., Mehra, R., and Tobin, A., 1988, In situ hybridization and immunocytochemical labeling GABA neurons during development of monkey visual cortex *Soc. Neurosci. Abstr.* **14**:188.

Hendrickson, A. E., Van Brederode, J. F. M., Mulligan, K. A., and Celio, M. R., 1991, Development of the calcium-binding proteins parvalbumin and calbindin in monkey striate cortex, *J. Comp. Neurol.* **307**:626–646.

Hendry, S. H. C., 1991, Delayed reduction in GABA and GAD immunoreactivity of neurons in the adult monkey dorsal lateral geniculate nucleus following monocular deprivation or enucleation, *Exp. Brain Res.* **86**:47–59.

Hendry, S. H. C., and Carder, R. K., 1993, Laminar specific compartmentation of calbindin immunoreactivity in monkey and human visual cortex, *Vis. neurosci.*, in press.

Hendry, S. M. C., and Jones, E. G., 1981, Sizes and distribution of intrinsic neurons incorporating tritiated GABA in monkey sensory-motor cortex, *J. Neurosci.* **1**:390–405.

Hendry, S. H. C., and Jones, E. G., 1986, Reduction in number of GABA immunostained neurons in deprived-eye dominance columns of monkey area 17, *Nature* **320**:750–753.

Hendry, S. H. C., and Jones, E. G., 1988, Activity-dependent regulation of GABA expression in the visual cortex of adult monkeys, *Neuron* **1**:701–712.

Hendry, S. H. C., and Jones, E. G., 1989, Synaptic organization of GABA and GABA/tachykinin immunoreactive neurons in layer IVCβ of monkey area 17, *Soc. Neurosci. Abstr.* **14**:1123.

Hendry, S. H. C., and Kennedy, M. B., 1986, Immunoreactivity for a calmodulin-dependent protein kinase in selectively increased in macaque striate cortex after monocular deprivation, *Proc. Natl. Acad. Sci. USA* **83**:1536–1540.

Hendry, S. H. C., Houser, C. R., Jones, E. G., and Vaughn, J. E., 1983a, Synaptic organization of immunocytochemically identified GABA neurons in the monkey sensory-motor cortex, *J. Neurocytol.* **12**:639–660.

Hendry, S. H. C., Jones, E. G., and Beinfeld, M. C., 1983b, Cholecystokinin-immunoreactive neurons in rat and monkey cerebral cortex make symmetric synapses and have intimate associations with blood vessels, *Proc. Natl. Acad. Sci. USA* **80:**2400–2404.

Hendry, S. H. C., Jones, E. G., DeFelipe, J., Schmechel, D., Brandon, C., and Emson, P. C., 1984a, Neuropeptide-containing neurons of the cerebral cortex are also GABAergic, *Proc. Natl. Acad. Sci. USA* **81:**6526–6530.

Hendry, S. H. C., Hockfield, S., Jones, E. G., and McKay, R., 1984b, Monoclonal antibody that identifies subsets of neurons in the central visual system of monkey and cat, *Nature* **307:**267–269.

Hendry, S. H. C., Jones, E. G., and Emson, P. C., 1984c, Morphology, distribution and synaptic relations of somatostatin- and neuropeptide Y-immunoreactive neurons in rat and monkey neocortex, *J. Neurosci.* **4:**2497–2517.

Hendry, S. H. C., Jones, E. G., Schwark, H. D., and Yan, J., 1987, Numbers and proportions of GABA-immunoreactive neurons in different areas of monkey cerebral cortex, *J. Neurosci.* **7:**1503–1519.

Hendry, S. H. C., Jones, E. G., and Burstein, N., 1988a, Activity-dependent regulation of tachykinin-like immunoreactivity in neurons of the monkey primary visual cortex, *J. Neurosci.* **8:**1225–1238.

Hendry, S. H. C., Jones, E. G., Hockfield, S., and McKay, D. G., 1988b, Neuronal populations stained with the monoclonal antibody Cat-301 in the mammalian cerebral cortex and thalamus, *J. Neurosci.* **8:**518–542.

Hendry, S. H. C., Jones, E. G., Emson, P. C., Lawson, D. E. M., Heizmann, C. W., and Streit, P., 1989, Two classes of cortical GABA neurons defined by differential calcium binding protein immunoreactivities, *Exp. Brain Res.* **76:**467–472.

Hendry, S. H. C., Fuchs, J., de Blas, A. L., and Jones, E. G., 1990, Distribution and plasticity of immunocytochemically localized GABA receptors in adult monkey visual cortex, *J. Neurosci.* **10:**2438–2450.

Hevner, R. F., and Wong-Riley, M. T. T., 1990, Regulation of cytochrome oxidase protein levels by functional activity in the macaque monkey visual system, *J. Neurosci.* **10:**1331–1340.

Hockfield, S., and Sur, M., 1990, Monoclonal antibody Cat-301 identifies Y-cells in the dorsal lateral geniculate nucleus of the cat, *J. Comp. Neurol.* **300:**320–331.

Hockfield, S., McKay, R. D., Hendry, S. H. C., and Jones, E. G., 1984, A surface antigen that identifies ocular dominance columns in the visual cortex and laminar features of the lateral geniculate nucleus, *Cold Spring Harbor Symp. Quant. Biol.* **35:**877–889.

Hof, P. R., Cox, K., Young, W. G., Celio, M. R., Rogers, J., and Morrison, J. H., 1991, Parvalbumin-immunoreactive neurons in the neocortex are resistant to degeneration in Alzheimer's disease, *J. Neuropathol. Exp. Neurol.* **50:**451–462.

Hornung, J. P., Török, I., and De Tribolet, N., 1989, Morphology of tyrosine hydroxylase-immunoreactive neurons in the human cerebral cortex, *Exp. Brain Res.* **76:**12–20.

Horton, J. C., 1984, Cytochrome oxidase patches: A new cytoarchitectonic feature of monkey visual cortex, *Philos. Trans. R. Soc. London Ser. B* **304:**199–253.

Horton, J. C., and Hedley-White, E. T., 1984, Mapping of cytochrome oxidase patches and ocular dominance columns in human visual cortex, *Philos. Trans. R. Soc. London Ser. B* **304:**255–272.

Horton, J. C., and Hubel, D. H., 1981, Regular patchy distribution of cytochrome oxidase staining in primary visual cortex of macaque monkey, *Nature* **292:**762–764.

Horton, J. C., Dagi, L. R., McCrane, E. P., and de Monasterio, F. M., 1990, Arrangement of ocular dominance columns in human visual cortex, *Arch. Ophthalmol.* **180:**1025–1031.

Houser, C. R., Hendry, S. H. C., Jones, E. G., and Vaughn, J. E., 1983, Morphological diversity of GABA neurons demonstrated immunocytochemically in monkey sensory-motor cortex, *J. Neurocytol.* **12:**617–638.

Houser, C. R., Vaughn, J. E., Hendry, S. H. C., Jones, E. G., and Peters, A., 1984, GABA neurons in the cerebral cortex, in: *Cerebral Cortex*, Volume 2 (E. G. Jones and A. Peters, eds.), Plenum Press, New York, pp. 63–90.

Hubel, D. H., and Wiesel, T. N., 1968, Receptive fields and functional architecture of monkey striate cortex, *J. Physiol. (London)* **195:**215–243.

Hubel, D. H., and Wiesel, T. N., 1972, Laminar and columnar distribution of geniculo-cortical fibers in the macaque monkey, *J. Comp. Neurol.* **146:**421–450.

Hubel, D. H., and Wiesel, T. N., 1977, Functional architecture of macaque monkey visual cortex, *Proc. R. Soc. London Ser. B* **198:**1–59.

Hubel, D. H., Wiesel, T. N., and LeVay, S., 1977, Plasticity of ocular dominance columns in monkey striate cortex, *Philos. Trans. R. Soc. London Ser. B* **278:**377–409.

Humphrey, A. L., and Hendrickson, A. E., 1983, Background and stimulus-induced patterns of high metabolic activity in the visual cortex (area 17) of the squirrel and macaque monkey, *J. Neurosci.* **3**:345–358.

Huntley, G. W., and Jones, E. G., 1990, Cajal–Retzius neurons in developing monkey neocortex show immunoreactivity for calcium binding proteins, *J. Neurocytol.* **19**:200–212.

Huntley, G. W., and Jones, E. G., 1991, Relationship of intrinsic connections to forelimb movement representations in monkey motor cortex: A correlative anatomic and physiological study, *J. Neurophysiol.* **66**:390–413.

Huntley, G. W., Hendry, S. H. C., Killackey, H. P., Chalupa, L. M., and Jones, E. G., 1988, Temporal sequence of neurotransmitter expression by developing neurons of fetal monkey visual cortex, *Dev. Brain Res.* **43**:69–96.

Huntsman, M. M., Jones, E. G., Möhler, H., and Hendry, S. H. C., 1991, Distribution of immunocytochemically localized GABA receptor subunits in monkey and human visual cortex, *Soc. Neurosci. Abstr.* **17**:115.

Itaya, S. K., Itaya, P. W., and Van Hoesen, G. W., 1984, Intracortical termination of the retinogeniculo-striate pathway studies with transsynaptic tracer (wheat germ agglutinin–horseradish peroxidase) and cytochrome oxidase staining in the macaque monkey, *Brain Res.* **304**:303–310.

Jenkins, W. M., Merzenich, M. M., Ochs, M. T., Allard, T., and Guic-Robles, E., 1990, Functional reorganization of primary somatosensory cortex in adult owl monkeys after behaviorally controlled tactile stimulation, *J. Neurophysiol.* **63**:82–104.

Jones, E. G., 1975, Varieties and distribution of non-pyramidal cells in the somatic sensory cortex of the squirrel monkey, *J. Comp. Neurol.* **160**:205–268.

Jones, E. G., 1984, Neurogliaform or spiderweb cells, in: *Cerebral Cortex*, Volume 1 (A. Peters and E. G. Jones, eds.), Plenum Press, New York, pp. 409–418.

Jones, E. G., 1990, The role of afferent activity in the maintenance of primate neocortical function, *J. Exp. Biol.* **153**:155–176.

Jones, E. G., and Hendry, S. H. C., 1984, Basket cells, in: *Cerebral Cortex*, Volume 1 (A. Peters, and E. G. Jones, eds.), Plenum Press, New York: pp. 309–336.

Jones, E. G., and Hendry, S. H. C., 1986, Co-localization of GABA and neuropeptides in neocortical neurons, *Trends Neurosci.* **10**:71–76.

Jones, E. G., DeFelipe, J., Hendry, S. H. C., and Maggio, J. E., 1988, A study of tachykinin-immunoreactive neurons in monkey cerebral cortex, *J. Neurosci.* **8**:1206–1224.

Kaas, J. H., Krubitzer, L. A., Chino, Y. M., Langston, A. L., Polley, E. H., and Blair, N., 1990, Reorganization of retinotopic cortical maps in adult mammals after lesions of the retina, *Science* **248**:229–231.

Katz, L. C., Gilbert, C. D., and Wiesel, T. N., 1989, Local circuits and ocular dominance in monkey striate cortex, *J. Neurosci.* **9**:1389–1399.

Kaufman, D. L., Houser, C. R., and Tobin, A. J., 1989, Two forms of glutamate decarboxylase (GAD), with different N-terminal sequences, have distinct intraneuronal distributions, *Soc. Neurosci. Abstr.* **15**:487.

Kelly, P. T., and Montgomery, P. R., 1982, Subcellular localization of the 52,000 molecular weight major postsynaptic density protein, *Brain Res.* **223**:265–286.

Kelly, P. T., and Vernon, P., 1985, Changes in the subcellular distribution of calmodulin kinase II during brain development, *Dev. Brain Res.* **18**:221–224.

Kelly, P. T., McGuinness, T. L., and Greengard, P., 1984, Evidence that the major postsynaptic density protein is a component of a Ca^{++}/calmodulin dependent protein kinase, *Proc. Natl. Acad. Sci. USA* **81**:945–949.

Kennedy, M. B., Bennett, M. K., and Erondu, N. E., 1983, Biochemical and immunochemical evidence that the "major postsynaptic density protein" is a subunit of a calmodulin-dependent protein kinase, *Proc. Natl. Acad. Sci. USA* **80**:7357–7361.

Kisvárday, Z. F., Cowey, A., and Somogyi, P., 1986, Synaptic relationships of a type of GABA-immunoreactive neuron (clutch cell), spiny stellate cells and lateral geniculate nucleus afferents in layer IVC of the monkey striate cortex, *Neuroscience* **19**:741–761.

Kisvárday, Z. F., Martin, K. A. C., Friedlander, M. J., and Somogyi, P., 1987, Evidence for interlaminar inhibitory circuits in the striate cortex of the cat, *J. Comp. Neurol.* **260**:1–19.

Kisvárday, Z. F., Gulyas, A., Beroukas, D., North, J. B., Chubb, I. W., and Somogyi, P., 1990, Synapses, axonal and dendritic patterns of GABA-immunoreactive neurons in human cerebral cortex, *Brain* **113**:793–812.

Kobayashi, K., Emson, P. C., and Mountjoy, C. Q., 1989, *Vicia villosa* lectin-positive neurones in human cerebral cortex. Loss in Alzheimer-type dementia, *Brain Res.* **498:**170–174.

Kosofsky, B. E., Molliver, M. E., Morrison, J. H., and Foote, S. L., 1984, The serotonin and norepinephrine innervation of primary visual cortex in the cynomolgus monkey (*Macaca fascicularis*), *J. Comp. Neurol.* **230:**168–178.

Kuljis, R. O., and Rakic, P., 1989a, Distribution of neuropeptide Y-containing perikarya and axons in various neocortical areas in the macaque monkey, *J. Comp. Neurol.* **280:**383–392.

Kuljis, R. O., and Rakic, P., 1989b, Neuropeptide Y-containing neurons are situated outside cytochrome oxidase puffs in macaque visual cortex, *Visual Neurosci.* **2:**57–62.

LeVay, S., Wiesel, T. N., and Hubel, D. H., 1980, The development of ocular dominance columns in normal and visually deprived monkeys, *J. Comp. Neurol.* **191:**1–51.

Lewis, D. A., and Lund, J. S., 1990, Heterogeneity of chandelier neurons in monkey neocortex: Corticotropin-releasing factor- and parvalbumin-immunoreactive populations, *J. Comp. Neurol.* **293:**599–615.

Lewis, D. A., Foote, S. L., and Cha, C. I., 1989, Corticotropin-releasing factor immunoreactivity in monkey neocortex: An immunohistochemical analysis, *J. Comp. Neurol.* **290:**599–613.

Lin, J. W., Sugimore, M., Llinás, R. R., McGuinness, T. L., and Greengard, P., 1990, Effects of synapsin I and calcium/calmodulin-dependent protein kinase II on spontaneous neurotransmitter release in the squid giant synapse, *Proc. Natl. Acad. Sci. USA* **87:**8257–8261.

Livingstone, M. S., and Hubel, D. H., 1982, Thalamic inputs to cytochrome oxidase-rich regions in monkey visual cortex, *Proc. Natl. Acad. Sci. USA* **79:**6098–6101.

Livingstone, M. S., and Hubel, D. H., 1984, Anatomy and physiology of a color system in the primate visual cortex, *J. Neurosci.* **4:**309–356.

Llinás, R., McGuinness, T. L., Leonard, C. S., Sugimori, M., and Greengard, P., 1985, Intraterminal injection of synapsin I or calcium/calmodulin dependent protein kinase II alters neurotransmitter release at the squid giant synapse, *Proc. Natl. Acad. Sci. USA* **82:**3035–3039.

Lund, J. S., 1973, Organization of neurons in the visual cortex, area 17, of the monkey (*Macaca mulatta*), *J. Comp. Neurol.* **147:**455–496.

Lund, J. S., 1987, Local circuit neurons of macaque striate cortex: I. Neurons of laminae 4C and 5A, *J. Comp. Neurol.* **257:**60–92.

Lund, J. S., 1990, Excitatory and inhibitory circuiting and laminar mapping strategies in the primary visual cortex of the monkey, in: *Signal and Sense: Local and Global Order in Perceptual Maps* (G. M. Edelman, W. E. Gall, and W. M. Cowan, eds.), Wiley–Liss, New York, pp. 51–82.

Lund, J. S., and Yoshioka, T., 1991, Local circuit neurons of macaque monkey striate cortex: III. Neurons of laminae 4B, 4A, and 3B, *J. Comp. Neurol.* **311:**234–258.

Lund, J. S., Henry, G. H., McQueen, C. L., and Harvey, A. R., 1979, Anatomical organization of the visual cortex of the cat: A comparison with area 17 of the macaque monkey, *J. Comp. Neurol.* **184:**559–618.

Lund, J. S., Hawken, M. J., and Parker, A. J., 1988, Local circuit neurons of macaque monkey striate cortex: II. Neurons of laminae 5B and 6, *J. Comp. Neurol.* **276:**1–29.

McGuinness, T. L., Lai, Y., and Greengard, P., 1985, Ca^{2+}/calmodulin-dependent protein kinase II. Isozymic forms from rat forebrain and cerebellum, *J. Biol. Chem.* **260:**1696–1704.

McGuinness, T. L., Brady, S. T., Gruner, J. A., Sugimori, M., Llinás, R., and Greengard, P., 1989, Phosphorylation-dependent inhibition by synapsin I of organelle movement in squid axoplasm, *J. Neurosci.* **9:**4138–4149.

McGuire, B. A., Gilbert, C. D., Rivlin, P. K., and Wiesel, T. N., 1991, Targets of horizontal connections in macaque primary visual cortex, *J. Comp. Neurol.* **305:**370–392.

McGuire, P. K., Hockfield, S., and Goldman-Rakic, P. S., 1989, Distribution of Cat-301 immunoreactivity in the frontal and parietal lobes of the macaque monkey, *J. Comp. Neurol.* **288:**280–296.

McNamara, M. E., Heros, R. C., and Boller, F., 1982, Visual hallucinations in blindness: The Charles Bonnet syndrome, *Int. J. Neurosci.* **17:**13–15.

Malenka, R. C., Kauer, J. A., Perkel, D. J., Mauk, M. D., Kelly, P. T., Nicoll, R. A., and Waxham, M. N., 1989, An essential role for postsynaptic calmodulin and protein kinase activity in long-term potentiation, *Nature* **340:**554–557.

Malinow, R., Schulman, H., and Tsien, R. W., 1989, Inhibition of postsynaptic PKC or CaMKII blocks induction but not expression of LTP, *Science* **245:**862–866.

Marin-Padilla, M., 1969, Origin of the pericellular baskets of the pyramidal cells of the human motor cortex: A Golgi study, *Brain Res.* **14:**633–646.

Marin-Padilla, M., 1987, The chandelier cell of the human visual cortex: A Golgi study, *J. Comp. Neurol.* **256:**61–70.

Martin, K. A. C., 1984, Neuronal circuits in cat striate cortex, in: *Cerebral Cortex,* Volume 2 (E. G. Jones and A. Peters, eds.), Plenum Press, New York, pp. 241–283.

Mates, S. L., and Lund, J. S., 1983, Neuronal composition and development in lamina 4C of monkey striate cortex, *J. Comp. Neurol.* **221:**60–90.

Meinicke, D. L., and Rakic, P., 1989, The temporal relationship of GABA and GABA-A/benzodiazepine receptor expression in neurons of the visual cortex of developing monkeys, *Soc. Neurosci. Abstr.* **15:**1335.

Merzenich, M. M., Kaas, J. H., Wall, J., Nelson, R. J., Sur, M., and Felleman, D., 1983a, Topographic reorganization of somatosensory cortical areas 3B and 1 in adult monkeys following restricted deafferentation, *Neuroscience* **8:**33–56.

Merzenich, M. M., Kaas, J. H., Wall, J. T., Sur, M., Nelson, R. J., and Felleman, D. J., 1983b, Progression of change following median nerve section in the cortical representation of the hand in areas 3b and 1 in adult owl and squirrel monkeys, *Neuroscience* **10:**639–666.

Merzenich, M. M., Nelson, R. J., Kaas, J. H., Stryker, M. P., Jenkins, W. M., Zook, J. M., Cynader, M. S., and Schoppmann, A., 1987, Variability in hand surface representations in areas 3b and 1 in adult owl and squirrel monkeys, *J. Comp. Neurol.* **258:**281–296.

Miller, S. G., and Kennedy, M. B., 1985, Distinct forebrain and cerebellar isozymes of type II Ca^{2+}/calmodulin-dependent protein kinase associate differently with the post-synaptic density fraction, *J. Biol. Chem.* **260:**9039–9046.

Montero, V. M., and Zempel, J., 1986, The proportion and size of GABA-immunoreactive neurons in the magnocellular and parvicellular layers of the lateral geniculate nucleus of the rhesus monkey, *Exp. Brain Res.* **62:**215–223.

Morrison, J. H., and Foote, S. L., 1986, Noradrenergic and serotoninergic innervation of cortical, thalamic and tectal visual structures in Old and New World monkeys, *J. Comp. Neurol.* **243:**117–138.

Morrison, J. H., Foote, S. L., Molliver, M. E., Bloom, F. E., and Lidov, H. G. W., 1982, Noradrenergic and serotonergic fibers innervate complementary layers in monkey primary visual cortex: An immunohistochemical study, *Proc. Natl. Acad. Sci. USA* **79:**2401–2405.

Mulligan, K. A., van Brederode, J. F. M., and Hendrickson, A. E., 1989, The lectin *Vicia villosa* labels a distinct subset of GABAergic cells in macaque visual cortex, *Visual Neurosci.* **2:**63–72.

Myers, R. E., 1962, Commissural connections between occipital lobes of the monkey, *J. Comp. Neurol.* **118:**1–16.

Naegle, J. R., and Barnstable, C. J., 1989, Molecular determinants of GABAergic local-circuit neurons in the visual cortex, *Trends Neurosci.* **12:**28–34.

Nakagawa, F., Schulte, B. A., and Spicer, S. S., 1986, Selective cytochemical demonstration of glycoconjugate-containing terminal N-acetylgalactosamine on some brain neurons, *J. Comp. Neurol.* **243:**280–290.

Newsome, W. T., and Allman, J. M., 1980, Interhemispheric connections of visual cortex in the owl monkey, *Aotus trivirgatus,* and the bushbaby, *Galago senegalensis, J. Comp. Neurol.* **194:**209–233.

Ogren, M. P., and Hendrickson, A. E., 1977, The distribution of pulvinar terminals in visual areas 17 and 18 of the monkey, *Brain Res.* **137:**343–350.

O'Kusky, J., and Colonnier, M., 1982, A laminar analysis of the number of neurons, glia and synapses in the visual cortex (area 17) of adult macaque monkeys, *J. Comp. Neurol.* **210:**278–290.

Ouimet, C. C., McGuinness, T. L., and Greengard, P., 1984, Immunocytochemical localization of calcium/calmodulin-dependent protein kinase II in rat brain, *Proc. Natl. Acad. Sci. USA* **81:**5604–5608.

Peters, A., and Saint-Marie, R. L., 1984, Smooth and sparsely spinous cells forming local axonal plexuses, in: *Cerebral Cortex,* Volume 1 (A. Peters and E. G. Jones, eds.), Plenum Press, New York, pp. 419–446.

Peters, A., Proskauer, C. C., and Ribak, C. E., 1982, Chandelier cells in rat visual cortex, *J. Comp. Neurol.* **206:**397–416.

Rakic, P., Bourgeois, J., Eckenhoff, M. F., Zecevic, N., and Goldman-Rakic, P., 1986, Concurrent overproduction of synapses in diverse regions of the primate cerebral cortex, *Science* **232:**232–235.

Rakic, P., Goldman-Rakic, P. S., and Gallagher, D., 1988, Quantitative autoradiography of major neurotransmitter receptors in the monkey striate and extrastriate cortex, *J. Neurosci.* **8:**3670–3690.

Ramachandran, V. S., and Gregory, R. L., 1991, Perceptual filling in of artificially induced scotomas in human vision, *Nature* **350**:699–702.

Ramoa, A. S., Paradiso, M. A., and Freeman, R. D., 1988, Blockade of intracortical inhibition in kitten striate cortex: Effects on receptive field properties and associated loss of ocular dominance plasticity, *Exp. Brain Res.* **73**:285–298.

Ramón y Cajal, S., 1899a, Estudios sobre la corteza cerebral humana I: Corteza visual, *Rev. Trim. Micrograf. Madrid.* **4**:1–63. Translated in: DeFelipe, J., and Jones, E. G., 1988, *Cajal on the Cerebral Cortex*, Oxford University Press, London, pp. 147–187.

Ramón y Cajal, S., 1899b, Estudios sobre la corteza cerebral humana II: Estructura de la corteza motriz del hombre y mamíferos superiores, *Rev. Trim. Micrograf. Madrid* **4**:117–200. Translated in: DeFelipe, J., and Jones, E. G., 1988, *Cajal on the Cerebral Cortex*, Oxford University Press, London, pp. 188–250.

Reiter, H. O., and Stryker, M. P., 1988, Neural plasticity without postsynaptic action potentials: Less-active inputs become dominant when kitten visual cortical cells are pharmacologically inhibited, *Proc. Natl. Acad. Sci. USA* **85**:3623–3627.

Rezak, M., and Benevento, L. A., 1979, A comparison of the organization of the projections of the dorsal lateral geniculate nucleus, the inferior pulvinar and adjacent lateral pulvinar to primary visual cortex (area 17) in the macaque monkey, *Brain Res.* **167**:19–40.

Ribak, C. E., 1978, Aspinous and sparsely-spinous stellate neurons in the visual cortex of rats contain glutamic acid decarboxylase, *J. Neurocytol.* **7**:461–478.

Roberts, E., 1975, GABA in nervous system function—An overview, in: *The Nervous System*, Volume 1 (D. B. Tower, ed.), New York, Raven, pp. 541–552.

Rockel, A. J., Hiorns, R. W., and Powell, T. P. S., 1980, The basic uniformity in structure of the neocortex, *Brain* **103**:221–244.

Sanes, J. N., Suner, S., Lando, J. F., and Donoghue, J. P., 1988, Rapid reorganization of adult rat motor cortex somatic representation patterns after motor nerve injury, *Proc. Natl. Acad. Sci. USA* **85**:2003–2007.

Schiffmann, S., Campistrom, G., Tugendhaft, P., Brotchi, J., Flament-Durand, J., Geffard, M., and Vanderhaeghen, J. J., 1988, Immunocytochemical detection of GABAergic nerve cells in the human temporal cortex using a direct γ-aminobutyric acid antiserum, *Brain Res.* **442**:270–278.

Schmechel, D. E., Vickrey, B. G., Fitzpatrick, D., and Elde, R. P., 1984, GABAergic neurons of mammalian cerebral cortex: Widespread subclass defined by somatostatin content, *Neurosci. Lett.* **47**:227–232.

Schulman, H., 1984, Phosphorylation of microtubule-associated proteins by a Ca^{2+}/calmodulin-dependent protein kinase, *J. Cell Biol.* **99**:11–19.

Shatz, C. J., 1990, Impulse activity and the patterning of connections during CNS development, *Neuron* **5**:745–756.

Shaw, C., and Cynader, M. C., 1986, Laminar distribution of receptors in monkey (*Macaca fascicularis*) geniculostriate system, *J. Comp. Neurol.* **248**:301–312.

Shaw, C., Cameron, L., March, D., Cynader, M. C., and Hendrickson, A., 1991, Pre- and postnatal development of GABA receptors in *Macaca* monkey visual cortex, *J. Neurosci.* **11**:3943–3959.

Shields, S. M., Ingebritsen, T. S., and Kelly, P. T., 1985, Identification of protein phosphate I in synaptic junctions: Dephosphorylation of endogenous calmodulin-dependent kinase II and synaptic-enriched phosphoproteins, *J. Neurosci.* **5**:3414–3422.

Sillito, A. M., 1974, Modification of the receptive field properties of neurons in the visual cortex by bicuculline, a GABA antagonist, *J. Physiol. (London)* **239**:36–37P.

Sillito, A. M., 1975a, The contribution of inhibitory mechanisms to the receptive field properties of neurones in the striate cortex of the cat, *J. Physiol. (London)* **250**:305–329.

Sillito, A. M., 1975b, The effectiveness of bicuculline as an antagonist of GABA and visually evoked inhibition in the cat's striate cortex, *J. Physiol. (London)* **250**:287–304.

Sillito, A. M., 1984, Functional considerations of the operation of GABAergic inhibitory processes in the visual cortex, in: *Cerebral Cortex*, Volume 2 (E. G. Jones, and A. Peters, eds.), Plenum Press, New York, pp. 91–117.

Sillito, A. M., Kemp, J. A., and Blakemore, C., 1981, The role of GABAergic inhibition in the cortical effects of monocular deprivation, *Nature* **291**:318–320.

Sloper, J. J., and Powell, T. P. S., 1979, A study of the axon initial segment and proximal axon of neurons in the primate motor and somatic sensory cortices, *Philos. Trans. R. Soc. London Ser. B* **285**:173–197.

Somogyi, P., 1977, A specific "axo-axonal" interneuron in the visual cortex of the rat, *Brain Res.* **136:**345–350.

Somogyi, P., 1979, An interneuron making synapses specifically on the axon initial segment (AIS) of pyramidal cells in the cerebral cortex of the cat, *J. Physiol. (London)* **296:**18–19.

Somogyi, P., and Cowey, A., 1981, Combined Golgi and electron microscopic study on the synapses formed by double bouquet cells in the visual cortex of the cat and monkey, *J. Comp. Neurol.* **195:**547–566.

Somogyi, P., and Cowey, A., 1984, Double bouquet cells, in: *Cerebral Cortex*, Volume 1 (A. Peters and E. G. Jones, eds.), Plenum Press, New York, pp. 337–360.

Somogyi, P., Cowey, A., Halasz, N., and Freund, T. F., 1981a, Vertical organization of neurones accumulating ^{3}H-GABA in visual cortex of rhesus monkey, *Nature* **294:**761–763.

Somogyi, P., Freund, T. F., Helasz, N., and Kisvárday, Z. F., 1981b, Selectivity of neuronal [^{3}H]GABA accumulation in the visual cortex as revealed by Golgi staining of the labeled neurons, *Brain Res.* **225:**431–436.

Somogyi, P., Freund, T. F., and Cowey, A., 1982, The axo-axonic interneuron in the cerebral cortex of the rat, cat and monkey, *Neuroscience* **7:**2577–2607.

Somogyi, P., Cowey, A., Kisvárday, Z. F., Freund, T. F., and Szentágothai, J., 1983a, Retrograde transport of γ-amino[^{3}H]butyric acid reveals specific interlaminar connections in the striate cortex of monkey, *Proc. Natl. Acad. Sci. USA* **80:**2385–2389.

Somogyi, P., Kisvárday, Z. F., Martin, K. A. C., and Whitteridge, D., 1983b, Synaptic connections of morphologically identified and physiologically characterized large basket cells in the striate cortex of cat, *Neuroscience* **10:**261–294.

Somogyi, P., Freund, T. F., Hodgson, A. J., Somogyi, J., Beroukas, D., and Chubb, I. W., 1985, Identified axo-axonic cells are immunoreactive for GABA in the hippocampus and visual cortex of cats, *Brain Res.* **332:**143–149.

Stryker, M. P., Jenkins, W. M., and Merzenich, M. M., 1987, Anesthetic state does not affect the map of the hand representation within area 3b somatosensory cortex in owl monkey, *J. Comp. Neurol.* **258:**208–303.

Sur, M., Frost, D. O., and Hockfield, S., 1988, Expression of a surface-associated antigen on Y-cells in the cat lateral geniculate nucleus is regulated by visual experience, *J. Neurosci.* **8:**874–882.

Takeuchi, Y., and Sano, Y., 1983, Immunohistochemical demonstration of serotonin nerve fibers in the neocortex of the monkey (*Macaca fuscata*), *Anat. Embryol.* **166:**155–168.

Takeuchi, Y., and Sano, Y., 1984, Serotonin nerve fibers in the primary visual cortex of the monkey quantitative and immunoelectron-microscopical analysis, *Anat. Embryol.* **169:**1–8.

Tigges, M., Tigges, J., McDonald, J. K., Slattery, M., and Fernandes, A., 1989, Postnatal development of neuropeptide Y-like immunoreactivity in area 17 of normal and visually deprived rhesus monkeys, *Visual Neurosci.* **2:**315–328.

Tootell, R. B. H., Hamilton, S. L., and Switkes, E., 1988a, Functional anatomy of macaque striate cortex. IV. Contrast and mango-parvo streams, *J. Neurosci.* **8:**1594–1609.

Tootell, R. B. H., Silverman, M. S., Hamilton, S. L., and De Valois, R. L., 1988b, Functional anatomy of macaque striate cortex. V. Spatial frequency, *J. Neurosci.* **8:**1610–1624.

Tootell, R. B. H., Silverman, M. S., Hamilton, S. L., De Valois, R. L., and Switkes, E., 1988c, Functional anatomy of macaque striate cortex. III. Color, *J. Neurosci.* **8:**1569–1593.

Towns, L. C., Tigges, J., and Tigges, M., 1990, Termination of thalamic intralaminar nuclei afferents in visual cortex of squirrel monkey, *Visual Neurosci.* **5:**151–154.

Trottier, S., Geffard, M., and Evrard, B., 1989, Co-localization of tyrosine hydroxylase and GABA immunoreactivities in human cortical neurons, *Neurosci. Lett.* **106:**76–82.

Trusk, T. C., Kaboord, W. S., and Wong-Riley, M. T. T., 1990, Effects of monocular enucleation, tetrodotoxin, and lid suture on cytochrome-oxidase reactivity in supragranular puffs of adult macaque striate cortex, *Visual Neurosci.* **4:**185–204.

Ts'o, D. Y., and Gilbert, C. D., 1988, The organization of chromatic and spatial interactions in the primate striate cortex, *J. Neurosci.* **8:**1712–1727.

Ts'o, D. Y., Gilbert, C. D., and Wiesel, T. N., 1986, Relationships between horizontal interactions and functional architecture in cat striate cortex as revealed by cross-correlation analysis, *J. Neurosci.* **6:**1160–1170.

Tsumoto, T., and Sato, H., 1985, GABAergic inhibition and orientation selectivity of neurons in the kitten visual cortex at the time of eye opening, *Vision Res.* **25:**383–388.

Tsumoto, T., Masui, H., and Sato, H., 1986, Excitatory amino acid transmitters in neuronal circuits of the cat visual cortex, *J. Neurophysiol.* **55:**469–483.

Tsumoto, T., Hagihara, K., Sato, H., and Hata, Y., 1987, NMDA receptors in the visual cortex of kittens are more effective than those of adult cats, *Nature* **327:**513–514.

Valverde, F., 1971, Short axon neuronal subsystems in the visual cortex of the monkey, *Int. J. Neurosci.* **1:**181–197.

Van Brederode, J. F. M., Mulligan, K. A., and Hendrickson, A. E., 1990, Calcium-binding proteins as markers for subpopulations of GABAergic neurons in monkey striate cortex, *J. Neurosci.* **298:**1–22.

Vulliet, P. R., Woodgett, J. R., and Cohen, P., 1984, Phosphorylation of tyrosine hydroxylase by calmodulin-dependent multiprotein kinase, *J. Biol. Chem.* **259:**13,680–13,683.

Wall, J. T., Kaas, J. H., Sur, M., Nelson, R. J., Felleman, D. J., and Merzenich, M. M., 1986, Functional reorganization in somatosensory cortical areas 3b and 1 of adult monkeys after median nerve repair: Possible relationships to sensory recovery in humans, *J. Neurosci.* **6:**218–233.

Weinberger, L. M., and Grant, F. C., 1940, Visual hallucinations and their neuro-optical correlates, *Arch. Ophthalmol.* **23:**166–199.

Werner, L., Winkelmann, E., Koglin, A., Neser, J., and Rodewohl, H., 1989, A Golgi deimpregnation study of neurons in the rhesus monkey visual cortex (areas 17 and 18), *Anat. Embryol.* **180:**583–597.

White, N. J., 1980, Complex visual hallucinations in partial blindness due to eye disease, *Br. J. Psychiatry* **136:**284–286.

Williams, S. M., Goldman-Rakic, P. S., and Leranth, C., 1992, The synaptology of parvalbumin-immunoreactive neurons in the primate prefrontal cortex, *J. Comp. Neurol.* **320:**353–369.

Winfield, D. A., Rivera-Dominguez, M., and Powell, T. P. S., 1982, The termination of geniculocortical fibres in area 17 of the visual cortex in the macaque monkey, *Brain Res.* **231:**19–32.

Wolf, W., Hicks, T. P., and Albus, K., 1986, The contribution of GABA-mediated inhibitory mechanisms to visual response properties of neurons in the kitten's striate cortex, *J. Neurosci.* **6:**2779–2795.

Wong-Riley, M. T. T., and Carroll, E. W., 1984a, Effect of impulse blockage on cytochrome oxidase activity in monkey visual system, *Nature* **307:**262–264.

Wong-Riley, M. T. T., and Carroll, E. W., 1984b, Quantitative light and electron microscopic analysis of cytochrome oxidase-rich zones in V II prestriate cortex of the squirrel monkey, *J. Comp. Neurol.* **222:**18–37.

Wong-Riley, M. T. T., and Norton, T. T., 1988, Histochemical localization of cytochrome oxidase activity in the visual system of the tree shrew: Normal patterns and the effect of retinal impulse blockage, *J. Comp. Neurol.* **272:**562–578.

Yamashita, A., Hayashi, M., Shimizu, K., and Oshima, K., 1989, Ontogeny of somatostatin in cerebral cortex of macaque monkey: An immunohistochemical study, *Dev. Brain Res.* **45:**103–111.

Zeki, S. M., 1970, Interhemispheric connections of prestriate cortex of monkey, *Brain Res.* **19:**63–75.

Zezula, J., Cortea, R., Probst, A., and Palacios, J. M., 1988, Benzodiazepine receptor sites in the human brain: Autoradiographic mapping, *Neuroscience* **25:**771–796.

Zielinski, B., and Hendrickson, A., 1990, Development of synapses in macaque monkey striate cortex shows an "inside-out" pattern, *Soc. Neurosci. Abstr.* **14:**494.

Zinner-Feyerabend, M., and Braak, E., 1991, Glutamic acid decarboxylase (GAD)-immunoreactive structures in the adult human lateral geniculate nucleus, *Anat. Embryol.* **183:**111–117.

4

Primate Visual Cortex
Dynamic Metabolic Organization and Plasticity Revealed by Cytochrome Oxidase

MARGARET T. T. WONG-RILEY

1. Introduction

Few regions in the brain have received as much attention and scrutiny as the visual cortex, whose structural and functional organization provides an ideal model for understanding cerebral cortex in general. Binocularity in the visual system further permits experimental manipulations of a single eye input with the other eye serving as a useful internal reference point. In the last century, the visual cortex has consistently been used as a fertile testing ground for virtually every new neurobiological technique and innovation. As a result, much of its anatomical, neurochemical, and functional organizations have been examined. Of special significance is the discovery by Hubel and Wiesel (1968) of ocular dominance columns, orientation columns, and the exquisite system of functional modules in the primate striate cortex. Our understanding of the visual system has reached new heights over the last decade, with new techniques based on brain metabolism, enzyme histochemistry, immunohistochemistry, voltage-sensitive dyes, and brain imaging having been applied rigorously to the study of the visual cortex (e.g., Kennedy *et al.*, 1976; Wong-Riley, 1979b; Horton and Hubel, 1981; Hendrickson *et al.*, 1981; Tootell *et al.*, 1982, 1988a–e; Hockfield *et al.*, 1983; Horton, 1984; Carroll and Wong-Riley, 1984; Wong-Riley and Carroll,

MARGARET T. T. WONG-RILEY • Department of Cellular Biology and Anatomy, Medical College of Wisconsin, Milwaukee, Wisconsin 53226.

Cerebral Cortex, Volume 10, edited by Alan Peters and Kathleen S. Rockland. Plenum Press, New York, 1994.

1984a,b; Livingstone and Hubel, 1984a; Hendry and Jones, 1986; Blasdel and Salama, 1986; Wong-Riley *et al.*, 1989a,b; Ts'o *et al.*, 1990; Beaulieu *et al.*, 1992). In this chapter, I will concentrate on cytochrome oxidase, and review what this endogenous metabolic marker has revealed about the primate visual cortex. The species on which most of the studies are based is the macaque monkey, but other primate species including man will be described when appropriate. A comparative study with other mammalian species has been reported previously (Wong-Riley, 1988).

2. Significance of Cytochrome Oxidase as a Metabolic Marker for Neuronal Activity

Cytochrome c oxidase (cytochrome aa_3, ferrocytochrome c, oxygen oxidoreductase, EC 1.9.3.1; abbreviated here as CO) is an integral transmembrane protein of the inner mitochondrial membrane and catalyzes the last step of oxidative metabolism, generating ATP in the process (Wikstrom *et al.*, 1981). It is particularly critical in the brain, where neuronal energy is derived almost exclusively from the oxidative pathway (Erecinska and Silver, 1989). ATP is used for a variety of neuronal functions, including fast axoplasmic transport, synthesis of macromolecules and neurotransmitters, and maintenance of the resting membrane potential by energy-dependent ion pumping mechanisms. Of these, only minute amounts of ATP are consumed in neurotransmitter metabolism, while the bulk of the energy is used to actively drive Na^+ and other cations out of the cell once they enter down their concentration and electrical gradients (Sokoloff, 1974; Bachelard, 1975; Lowry, 1975). Such ion pumping activities occur largely along dendritic membranes, which represent the major receptive sites for depolarizing inputs. Thus, dendritic membrane depolarization, which consists largely of local graded potentials, dominates slow adjustments of the background excitation level of the brain (Rall, 1962), and dendritic membrane repolarization by the energy-dependent ATPase system (notably Na^+, K^+-ATPase) makes the largest single contribution to the metabolic activity of the brain (Lowry *et al.*, 1954). It is, therefore, not surprising that the level of CO is consistently high in dendrites and up to 60% of mitochondria in the cortical neuropil examined reside in dendrites (Wong-Riley, 1989; Wong-Riley *et al.*, 1989b). Neurons also receive hyperpolarizing inputs, but repolarization after hyperpolarization is mainly passive and requires little energy. Neuronal cell bodies that receive exclusively inhibitory (hyperpolarizing) synapses have very low levels of CO (Mjaatvedt and Wong-Riley, 1988). Likewise, action potentials *per se* consume very little energy, and Creutzfeldt (1975) calculated that only about 0.3–3% of the human cortical energy consumption can be accounted for by spiking activity of cortical neurons. Axon trunks and axon bundles in the white matter that propagate action potentials do exhibit low levels of CO activity. However, spiking activity is triggered by the spatial and temporal integration of graded depolarization and, in that sense, may indirectly correlate with how metabolically active the neurons are. Moreover, subthreshold depolarization may be offset by graded hyperpolarization, and the energy consumed by these suppressed depolarizing events should not be overlooked when spiking activity is being considered. Axon terminals may vary

143

CYTOCHROME
OXIDASE
STUDY OF
PRIMATE
VISUAL
CORTEX

in their energy demand depending on the frequency of their membrane depolarization and the frequency with which they extrude both Na^+ and Ca^{2+} ions, possibly explaining why terminals vary in their CO content (Wong-Riley, 1989; Wong-Riley *et al.*, 1989b). Thus, the entire neuron need not be metabolically homogeneous; indeed, most neurons are metabolically heterogeneous among their various compartments (Wong-Riley, 1989). Neuronal activity, therefore, includes both local potential changes and action potentials, but the former consume far more energy than the latter. There is tight coupling between energy metabolism and neuronal activity, but neuronal activity controls energy expenditure (Lowry, 1975) as well as energy-generating enzymes such as CO (reviewed in Wong-Riley, 1989). CO can, therefore, serve not only as a metabolic marker denoting cellular capacity for aerobic metabolism, but also as an indicator of the relative levels of energy-dependent neuronal activity averaged over a period of time. Histochemical changes in CO have been detected within 8 to 14 hr following functional alterations (Mawe and Gershon, 1986; Liu *et al.*, 1990; Trusk *et al.*, 1992; see also Section 6.3 and Fig. 18). Shorter time periods have not yet been examined.

To be sure, CO is but one such marker in neurons. However, its attributes surpass in usefulness those of a number of oxidative enzymes, such as succinate dehydrogenase (SDH). First, CO can withstand a wide range of pH, temperature, and limited aldehyde fixation, permitting good tissue preservation in the presence of adequate enzyme activity (Seligman *et al.*, 1968; Wong-Riley, 1979b; Wikstrom *et al.*, 1981). Aldehyde fixation is particularly critical for ultrastructural localization of reaction product within mitochondria of defined neuronal compartments (e.g., dendrites, axonal terminals, somata, and axonal trunks). Second, CO is an integral membrane protein of the inner mitochondrial membrane. As such, it is much more stable than soluble proteins in the cytoplasm. Third, being the terminal enzyme of the electron transport chain, there is little need for complex posttranslational modifications for it to function. Fourth, the consistency with which CO activity can be revealed in a simple histochemical reaction has facilitated its usefulness in routine laboratory settings. Fifth, CO can be studied in a variety of ways in the nervous system, such as biochemical assay for its molecular activity (Liu *et al.*, 1991; Hevner *et al.*, 1993), histochemical reaction for its relative enzyme activity (Wong-Riley, 1979b), optical densitometric measurements for semiquantitation of products of its reaction with cytochrome c (Wong-Riley and Kageyama, 1986), immunohistochemistry for its relative protein amount (Hevner and Wong-Riley, 1989), electron microscopy for the cellular and subcellular localization of its reaction product (Carroll and Wong-Riley, 1984), *in situ* hybridization for its mRNA localization (Hevner and Wong-Riley, 1990), and various molecular techniques for analyzing its bigenomic composition, mitochondrial and nuclear (Attardi and Schatz, 1988). Such versatility has broadened its applicability as a neuronal metabolic marker. Sixth, CO histochemistry or immunohistochemistry can be effectively used in conjunction with other techniques, such as anterograde or retrograde axonal tracing (Mjaatvedt and Wong-Riley, 1988), high-affinity uptake of radioactive tracers (Livingstone and Hubel, 1982, 1983; Carroll and Wong-Riley, 1985, 1987a), 2-deoxyglucose autoradiography (Tootell *et al.*, 1985; 1988a–e), immunohistochemistry (Hendry and Jones, 1986, 1988), double and triple immunolabeling (Luo *et al.*, 1989; Liu *et al.*, 1990), and even classical techniques such as Golgi (Malach, 1992).

Finally, CO labels metabolically active dendritic and axonal terminals that elude routine electrophysiological recordings (Carroll and Wong-Riley, 1984; Wong-Riley and Carroll, 1984a; Wong-Riley *et al.*, 1989b). Since CO is regulated at a very local level, sites of high CO activity can be regarded as sites of intense energy demand and, most likely, of constant membrane depolarization. Thus, dendrites of almost all neurons consistently have high levels of CO. Active depolarization of certain classes of neuronal cell bodies and presynaptic terminals also imposes elevated levels of CO (DiRocco *et al.*, 1989). CO, then, provides an additional means of charting metabolically active sites in the brain that generate energy to fuel neuronal activity.

If neuronal activity governs the regulation of CO, then the level of this enzyme should covary with energy demands of neurons under normal and altered states of functioning. Studies done in many laboratories over the last 15 years have helped to unravel some of these relationships in various parts of the brain (see review by Wong-Riley, 1989). They revealed the exquisite sensitivity of this enzyme to changing functional needs. I shall restrict my discussion in this chapter to studies of the primate visual cortex.

3. Primary Visual Cortex of Primates

The pioneering work of Hubel and Wiesel (1968, 1974, 1977) has demonstrated the basic functional building blocks of the primate visual cortex: a system of ocular dominance columns superimposed on sets of orientation columns. Each set of 180° orientation columns or a pair of left- and right-eye columns, in turn, forms a hypercolumn. Analysis of single points in the visual field would require the neural substrate to analyze all orientations and all colors by both eyes. Such a functional module would include both ocular dominance and orientation hypercolumns, and would occupy a surface area of about 4 mm² (Livingstone and Hubel, 1984a).

Such functional modules are readily inferred from electrophysiological recordings, but are missed by conventional anatomical approaches. It was not until the advent of metabolic markers such as 2-deoxyglucose and optical imaging methods that acute images of modular components were captured (Hubel *et al.*, 1978; Blasdel and Salama, 1986; Ts'o *et al.*, 1990). A stabilized metabolic map that distinguishes functionally active from less active zones can consistently be generated from the primate visual cortex by means of CO histochemistry or immunohistochemistry (Wong-Riley, unpublished observations, 1978; Horton and Hubel, 1981; Carroll and Wong-Riley, 1982, 1984; Wong-Riley and Carroll, 1984b; Hevner and Wong-Riley, 1990). This map bears a remarkable repeating pattern that appears intimately related to the spacing and geometry of various functional columns, such as orientation, ocular dominance, color, and, perhaps, spatial frequency (Horton and Hubel, 1981; Tootell *et al.*, 1982, 1988b; Horton, 1984; Livingstone and Hubel, 1984a). At the cores of the repeating units are "puffs" or supragranular "blobs" (see Section 3.1.1a) which, in Old World monkeys and man, are centered on ocular dominance columns (Horton, 1984; Horton and Hedley-Whyte, 1984; Wong-Riley and Carroll, 1984b) and as such, are ideal candidates for organizing centers of functional modules.

3.1. Dynamic Metabolic Map of V1 Revealed by Cytochrome Oxidase

145

CYTOCHROME
OXIDASE
STUDY OF
PRIMATE
VISUAL
CORTEX

Anyone who has compared a Nissl-stained section with a CO-reacted section of the primate striate cortex can attest to the striking difference between the two. Nissl staining reveals a homogeneous distribution of cells across individual laminae, whereas CO brings out the intrinsic heterogeneity, much like the development of a latent image from a photographic paper in the darkroom (Fig. 1A–D). This image is stable for normal animals (even after a week of physiological recordings under anesthesia; unpublished observations by DeYoe, Trusk, and Wong-Riley) and can serve as a useful reference point against which other functional and anatomical patterns can be compared (Lund, 1988). When the densi-

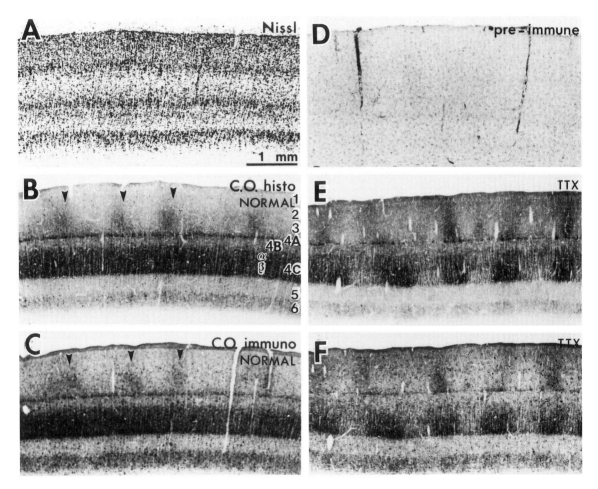

Figure 1. Serial coronal sections from normal macaque monkeys (A–D) and adjacent sections from TTX-treated monkeys (E, F). Sections were stained for Nissl (A) or reacted for CO histochemistry (B and E), CO immunohistochemistry (C and F), or with control preimmune serum (D). Layers of the cortex are identified in B. Note the exact alignment between puffs stained histochemically and immunohistochemically (arrowheads in B and C) in the normal cortex, and the alignment of dark and light ocular dominance columns in cortex after 3 weeks of monocular TTX treatment (E and F). Modified from Hevner and Wong-Riley (1990).

ties of reaction product are semiquantitated by optical densitometry, values are found to differ significantly between cortical laminae (Fig. 2). Layers 4C and 4A, for example, have optical density values that are 14, 33, and 11% higher than those of 4B, 5, and 6, respectively, and values in supragranular puffs are 15% higher than those in interpuffs ($p < 0.01$ for all). On the other hand, there is no significant difference in optical densities between puffs, 4A, and 4C. What is more important, the level of CO is highly responsive to altered functional demands of neurons. Monocular retinal impulse blockade, for example, induces a reduction in cortical neuronal activity (Trusk *et al.*, 1992) that is paralleled by a decrease in CO levels within affected ocular dominance columns (Wong-Riley and Carroll, 1984b; Wong-Riley *et al.*, 1989a, b). The decreases in CO activity (demonstrable histochemically in Figs. 1E and 4C) and CO protein (shown immunohistochemically in Figs. 1F and 4D) are most prominent in layer 4C, but are also visible in layers 2, 3, 4A, 5, and 6 (see also Section 6 and related figures).

The basic pattern in the striate cortex highlights laminae 2/3, 4, and 6 (Horton, 1984; Carroll and Wong-Riley, 1984; Hendrickson, 1985). The well-

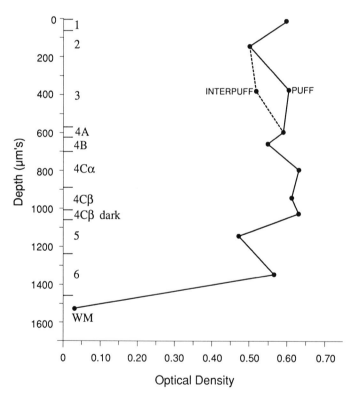

Figure 2. Laminar distribution of cytochrome oxidase densities as measured by optical densitometry in the macaque striate cortex. Each point represents the mean of at least 25 readings. The standard deviations are too minute to be plotted. Note that the values for puffs, layer 4A, and 4C are significantly higher than interpuffs and the other layers. The intense staining in layer 1 results mainly from artifactual accumulation of reaction product along edges of sections. WM, white matter. Modified from Wong-Riley (1988). A bar graph of optical density values from layer 4A to the white matter is shown in Fig. 20A.

147

CYTOCHROME
OXIDASE
STUDY OF
PRIMATE
VISUAL
CORTEX

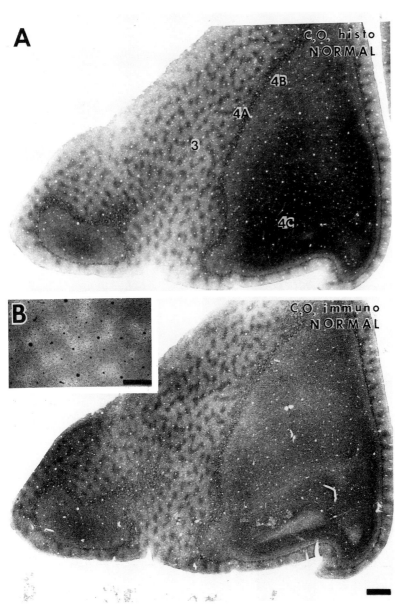

Figure 3. Adjacent tangential sections of normal macaque striate cortex reacted for CO histochemistry (A) or CO immunohistochemistry (B). Layers of the cortex are indicated in A. The inset in B shows a section of cortical layers 2 and 3 stained by indirect immunofluorescence. Note that the puffs of layers 2 and 3 (visible as dark spots, or as light spots in the immunofluorescence micrograph) and the reticulated pattern of layer 4A form closely matched patterns histochemically and immunohistochemically. Bar = 1 mm for A and B, 0.5 mm for inset in B. Modified from Hevner and Wong-Riley (1990).

known polka-dot-like distribution of CO-rich zones visible in tangential sections decorates the supragranular layers 2 and 3 (Fig. 3A,B). Sometimes they appear to form bridges linking them in rows. Layer 4A takes on a honeycomb appearance (Fig. 4A,B), while layer 4C is subdivided into a slightly more reactive 4Cα above the less reactive 4Cβ, with the lowermost edge of 4Cβ again assuming a darker shade of staining (Fig. 5). We have called this darker band 4Cβ dark (Hiltgen and Wong-Riley, 1986). Layer 6 is moderately active, while layers 4B and 5 are less reactive but significantly higher than the white matter. In cross sections, one can detect a faint pillar of higher CO activity extending from layer 1 to 6 and centered on the supragranular CO-rich puffs. Such pillars render a periodicity not only to layers 2/3 and 6, but also to 4B and 5, which are otherwise rather pale (Figs. 1B,C, and 5).

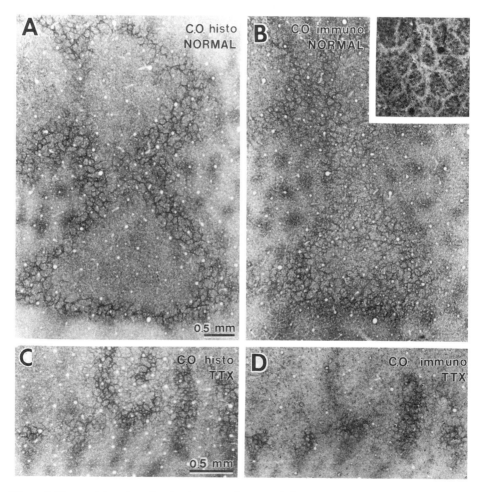

Figure 4. Tangential sections showing honeycomb pattern in layer 4A stained for CO histochemistry (A and C) or CO immunohistochemistry (B and D). The pattern is continuous in normal cortex (A and B) but is broken up by ocular dominance columns in an animal that has received intravitreal TTX for 3 weeks (C and D). The inset in B shows the honeycomb pattern with immunofluorescence for CO. Modified from Hevner and Wong-Riley (1990).

149

CYTOCHROME
OXIDASE
STUDY OF
PRIMATE
VISUAL
CORTEX

3.1.1. Supragranular Puffs

One of the most striking features of the primate striate cortex is the existence of an extremely regular array of CO-rich zones in the supragranular layers 2 and 3. Over the past decade, a flurry of investigations has concentrated on the morphology, functional significance, connectional patterns, and plastic changes of these zones. They have been regarded as the "fundamental cytoarchitectonic unit" of the primate visual cortex (Horton and Hedley-Whyte, 1984).

3.1.1a. Nomenclature. Just as the primary visual cortex has multiple names, such as striate cortex, area 17, or V1, so do the CO-rich supragranular zones. Their numerous designations are reminiscent of characters in a Russian novel, except that they are decidedly easier to pronounce and remember. The term "puffs" was adopted in our laboratory from inception (Wong-Riley, unpublished observation, 1978; Carroll and Wong-Riley, 1982, 1984) because these

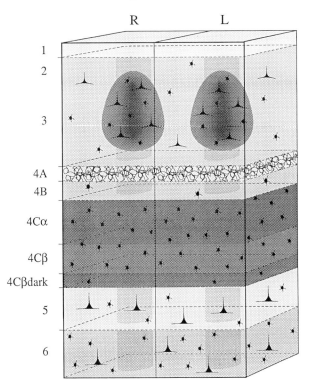

CYTOCHROME OXIDASE PATTERN
IN MACAQUE STRIATE CORTEX

Figure 5. Schematic diagram showing laminar and cellular patterns of cytochrome oxidase labeling in two adjacent ocular dominance columns (R and L), forming a postulated functional module. Labeling is intense in puffs of layers 2–3, honeycomblike distribution in 4A, and laminar distribution in 4C, with $4C\alpha$ and $4C\beta$ dark being slightly more reactive than the upper portion of $4C\beta$. Layer 6 is moderately reactive for CO. In addition, there is a pillarlike distribution of CO through the cores of ocular dominance columns. CO-reactive nonpyramidal neurons are distributed throughout the cortical gray, while CO-reactive pyramidal cells are found in the supra- and infragranular layers.

structures are globular-shaped and contain a nonhomogeneous reticulated interior, like a puff of cotton. The heterogeneity is an important concept, as neurons within puffs do not share uniform staining and are not likely to have identical functional attributes. The neuropil is likewise heterogeneous. Horton and Hubel (1981) called these zones "patches," but Livingstone and Hubel (1982) converted them to "blobs" and extended the definition to layers 4B, 5, and 6. Other terms such as "spots" (Tootell *et al.*, 1983) and "dots" (Tigges *et al.*, 1984; Hendrickson, 1985) have also been used. In recent years, the name "blob" has enjoyed greater popularity; however, authors usually do not specify the laminar location and it is not always clear whether "blob" is applied in every case to layer 2/3 only, to 4B, to 5 and 6, or all of the above.

For consistency and to avoid using "layer 2/3" every time, I shall use the terms "puffs" and "interpuffs" in V1 to mean CO-rich and CO-poor zones housed exclusively in layers 2/3, as originally intended by this author, and "blobs" and "interblobs" to mean CO-rich and CO-poor pillars through layers 2, 3, 4B, 5, and 6, as originally intended by Livingstone and Hubel (1982, 1983).

3.1.1b. Dimensions. The average diameter of puffs in the squirrel monkey is 210 μm (Carroll and Wong-Riley, 1984). In the macaque monkey, they measure approximately 262 by 377 μm (Wong-Riley and Carroll, 1984b; however, Horton in 1984 reported lower values of 150 by 250 μm), and in man, they are about 230 by 450 μm (Wong-Riley *et al.*, 1993a; again, slightly lower values of 250 by 400 μm were reported by Horton and Hedley-Whyte in 1984).

Trusk *et al.* (1990) reconstructed 50 puffs in normal adult macaque monkeys and found them to be vase-shaped structures with a larger base in layer 3B. Puff areas decrease from upper layer 3B, where they are 0.046–0.050 mm², to 3A above as well as to 4A below. In upper layer 3A and throughout layer 2, the number of visible puffs gradually declines, although the mean cross-sectional areas of puffs remain fairly constant (0.040 mm²). The volume of individual puffs ranges from 0.0071 to 0.0193 mm³, with an average of 0.0127 ± 0.0027 mm³.

In the squirrel monkey, where ocular dominance columns (ODCs) have not been demonstrated anatomically, the center-to-center spacing of puffs is about 500 μm (Carroll and Wong-Riley, 1984). In the macaque, puffs form rows in register with ODCs in layer 4C. The center-to-center spacing of puffs within rows is about 450 μm [Horton (1984) reported 550 μm] and between rows is about 350 μm (Wong-Riley and Carroll, 1984b). In man, these values are 650 and 800 μm, respectively (Horton and Hedley-Whyte, 1984; Wong-Riley *et al.*, 1993a). Rows tend to radiate orthogonally from the 17–18 border, which represents the vertical meridian (Fig. 6). Although the center-to-center spacing of puffs within the same ocular dominance row is known to vary with eccentricity (Livingstone and Hubel, 1984a), the spacing between rows tends to remain stable, as ODCs retain relatively constant widths at eccentricities of 0–15° in the macaque (Hubel and Wiesel, 1972; LeVay *et al.*, 1985). At the foveal representation, the distance between the centers of adjacent puffs corresponds to a visual angle of about 4' of arc. At 8° eccentricity, this distance increases to about ¹/₂° (Tootell *et al.*, 1988b).

Puffs are present throughout the entire primary visual cortex, and Horton (1984) estimated about one puff per 0.22 mm² in the squirrel monkey and one puff per 0.2 mm² in the macaque. Near the foveal representation of human, the density of puffs is about a third of that in the macaque, or about 1 puff per 0.6–

151

CYTOCHROME
OXIDASE
STUDY OF
PRIMATE
VISUAL
CORTEX

0.8 mm² (Horton and Hedley-Whyte, 1984). Within 1 mm² of the macaque striate cortex, there are at least 5 puffs (Purves and LaMantia, 1990). Horton (1984) calculated about 5.6 puffs/mm² next to the optic disk representation, 9 puffs/mm² near the monocular crescent, and 5.5 puffs/mm² within the monocular crescent. Our own unpublished datum on the opercular surface is 6.08 puffs/mm², which is closer to the theoretical value of 6.56 puffs/mm² reported by Schein and de Monasterio (1987). The density is about 2–3 puffs/mm² in man (Wong-Riley *et al.*, 1993a) [Horton and Hedley-Whyte (1984) reported 1.25–1.67 puffs/mm²].

On the basis of an estimated surface area of 1200 to 1400 mm² for the macaque striate cortex, Horton (1984) postulated that there are at least 6000 puffs, and perhaps as many as 7000 to 9000 puffs, since the density of puffs increases in the peripheral binocular cortex. If the human striate cortex has a surface area of 3000–3600 mm², it would contain approximately the same total number of puffs as the macaque (Horton and Hedley-Whyte, 1984). However, if the surface area is 2130 mm² (Stensaas *et al.*, 1974), then there may only be about 3000 puffs (Tootell *et al.*, 1988b), a value that has been ascribed to the squirrel monkey (Horton and Hedley-Whyte, 1984).

3.1.1c. Specialized Regions. The cortical optic disk representation is an oval region located in the roof of the calcarine fissure with visual input only from the intact, ipsilateral eye (Horton, 1984). Even though ODCs are absent here, puffs are present and are larger, rounder, and more widely spaced than the surrounding region (Horton, 1984). The density of puffs there is reported to be

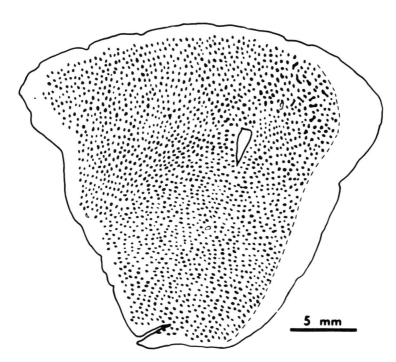

Figure 6. Sketch of CO-rich puffs in the macaque striate cortex showing their distribution into rows and intersecting the 17/18 border (dashed line at the right). 1917 "patches" are present in 373 mm³ (one patch per 0.2 mm²). Figure kindly provided by Horton, and taken from Horton (1984).

3.1 puffs/mm², which is almost half the normal value. This suggests that rows of puffs corresponding to the contralateral eye may simply be absent in this portion of the striate cortex.

Another specialized region is the monocular representation of the visual field far periphery, where the cortex receives exclusively contralateral eye input (Wiesel *et al.*, 1974). Again, puffs here appear larger, rounder, and more widely spaced than the binocular region of the cortex (Horton, 1984).

3.1.1d. Neurons within Puffs and Interpuffs. Quantitative analysis of supragranular neurons confirms the qualitative impression from Nissl stains that the overall density of neurons within puffs is equal to that in interpuffs (Carroll and Wong-Riley, 1984; Trusk *et al.*, 1990), and range between 65 and 135 cells per 20,000 μm² in the macaque, with a great deal of variability between monkeys. What, then, contributes to the heightened metabolic activity within puffs?

While much of the CO activity is attributable to the neuropil, there are consistent differences between puff and interpuff neurons. First, the number of CO-rich neurons in puffs is significantly higher than that in interpuffs (16.4 versus 12.1 CO-rich neurons per 20,000 μm² of cortex). Second, the number of CO-poor neurons tends to be significantly higher in interpuffs than in puffs (88 versus 76.5 CO-poor cells per 20,000 μm²) (Trusk *et al.*, 1990).

Thus, neurons within puffs are metabolically nonhomogeneous and the mean size of CO-rich neurons is significantly higher than that of CO-poor neurons. At least three major neuronal types have been analyzed at the EM level (Carroll and Wong-Riley, 1984; Wong-Riley *et al.*, 1989a): types A, B, and C. About 57% of puff neurons in the macaque belong to type A, which are small (52.54 ± 0.65 μm²), lightly reactive nonpyramidal (and perhaps some small pyramidal) cells whose somata receive exclusively symmetric synapses (Fig. 7A). Roughly 18% of neurons are type B cells, comprising moderately reactive medium and large pyramidal cells (83.09 ± 6.08 μm²) whose cell bodies also receive only symmetric synapses (Fig. 8). The remaining 25% of neurons are made up of medium-sized (75.73 ± 4.99 μm²), darkly reactive, nonpyramidal type C cells which receive both symmetric and asymmetric synapses on their perikarya (Fig. 10). Rare type D cells are small, darkly reactive nonpyramidal neurons whose cell bodies can give rise to somatodendritic synapses.

What factors possibly contribute to the greater metabolic demands of type C cells? It is not cell size, because type C's are intermediate in size between types A and B and more reactive than either of them. It is unlikely to be the type of neurotransmitter *per se;* type C cells are immunoreactive for GABA (Nie and Wong-Riley, 1993), but GABAergic neurons can be either darkly or lightly reactive for CO (Luo *et al.*, 1989). A major factor is that type C neurons receive asymmetric axosomatic synapses not present on the other two cell types. Asymmetric synapses have often been associated with excitatory synapses, and neuronal cell bodies that are postsynaptic to this type of synapse tend to be more CO-reactive than those receiving symmetric, presumed inhibitory synapses (Mjaatvedt and Wong-Riley, 1988). It remains to be proven whether type C cells are more tonically active than the other metabolic cell types.

Not only do puff neurons differ under normal conditions, they also do not respond alike to a common functional insult (Wong-Riley *et al.*, 1989a). With monocular retinal impulse blockade, type A cells are the least affected, since they show no change in somal size and remain lightly reactive for CO (Fig. 7B);

153

**CYTOCHROME
OXIDASE
STUDY OF
PRIMATE
VISUAL
CORTEX**

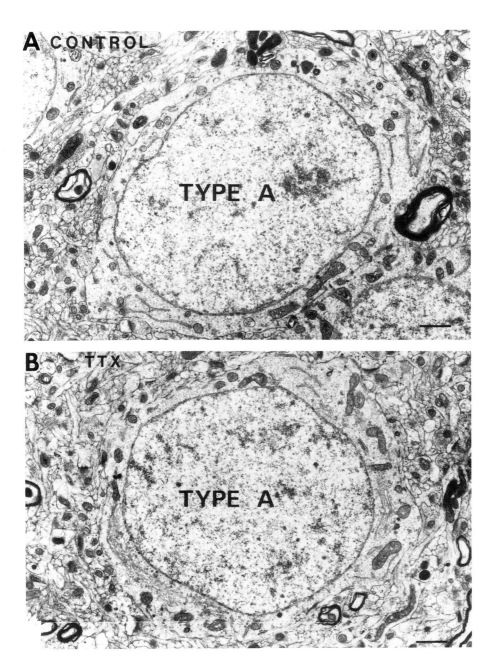

Figures 7–11. Electron micrographs of macaque V1 puffs reacted cytochemically for CO.

Figure 7. The most common type of neurons found in normal puffs are type A cells with a thin rim of cytoplasm containing mitochondria with little CO reaction product (A). They are nonpyramidal and small pyramidal neurons that receive exclusively symmetrical asoxomatic synapses. Four weeks of retinal TTX treatment did not bring about significant change in this neuronal type (B). Bars = 1 μm. From Wong-Riley *et al.* (1989a).

type B cells undergo a slight reduction in somal size and their mitochondria are converted from moderately reactive to mainly the lightly reactive variety (Fig. 9). Type C cells, on the other hand, suffer a significant shrinkage in somal size, a drastic reduction in the packing density of their mitochondria, and a striking conversion of mitochondria from mainly the darkly reactive type to primarily the lightly reactive variety (Fig. 11). Thus, metabolically most active neurons appear most vulnerable to functional insults induced in the adult (Table I). This issue will be discussed further in Section 6.3.

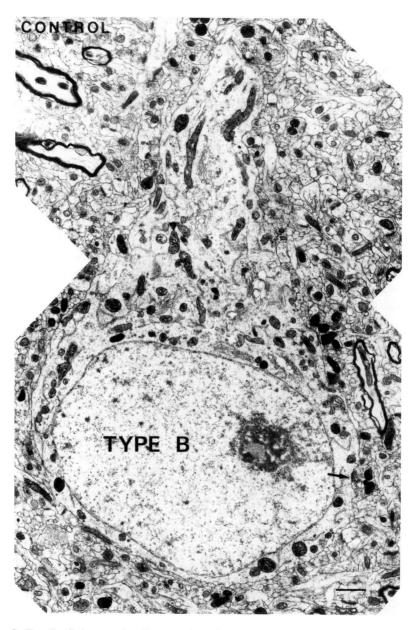

Figure 8. Type B cells in control puffs are moderately reactive pyramidal cells with an assortment of mitochondria that are mostly moderately reactive for CO. They also receive exclusively symmetrical synapses on their somata (arrow). Bar = 1 μm. From Wong-Riley *et al.* (1989a).

Neuronal compositions within interpuffs are qualitatively similar to those in puffs and the same three major neuronal types are present. However, there are some quantitative differences: All three cell types contain fewer darkly reactive mitochondria than the comparable cell types in puffs, the numerical and areal densities of mitochondria are lower in types B and C of interpuffs than those in puffs, and type C cells in interpuffs have a smaller mean size than those in puffs (Carroll and Wong-Riley, 1984; Wong-Riley *et al.*, 1992). These ultrastructural differences are consistent with light microscopic observations that interpuffs have lower levels of CO activity than puffs. Two weeks of TTX blockade induce a significant downward shift in the size and CO reactivity of mitochondria in type C, and less so in types B and A of interpuffs (Wong-Riley *et al.*, 1992). This topic will be discussed further in Section 6.3.

3.1.1e. Neuropil of Puffs and Interpuffs. The neuropil of puffs is distinctly more reactive than that of interpuffs (Fig. 2) but the same types of profiles, such as dendrites, axons, and glia, are found in both. The single largest

155

CYTOCHROME
OXIDASE
STUDY OF
PRIMATE
VISUAL
CORTEX

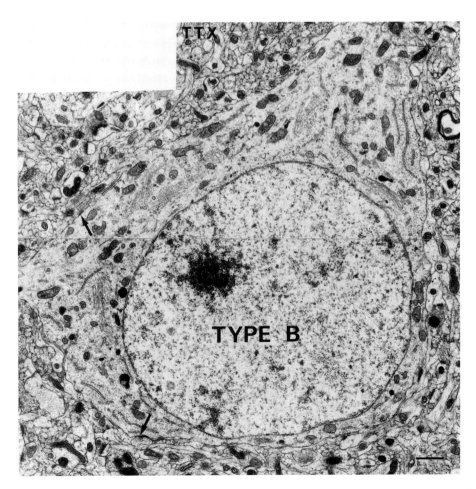

Figure 9. After 4 weeks of retinal TTX treatment, mitochondria in type B neurons have converted to mainly the lightly reactive variety. However, the ultrastructure remains intact and synapses are still present on the cell body (arrows). Bar = 1 μm. From Wong-Riley *et al.* (1989a).

group of CO-rich profiles in the neuropil are the dendrites, which probably belong to both local and infragranular neurons (Carroll and Wong-Riley, 1984; Wong-Riley *et al.*, 1989b). In fact, 60% of the total mitochondrial population in macaque puffs reside in dendrites (Fig. 12), and the majority are highly reactive for CO. While conventional microelectrodes do not routinely monitor dendritic activities, the sizable amount of graded potentials generated within a major component of the neuropil should not be ignored.

As for the rest of the neuropil, the distribution of CO is not homogeneous and is not proportional to profile size or numerical densities (Fig. 12). The

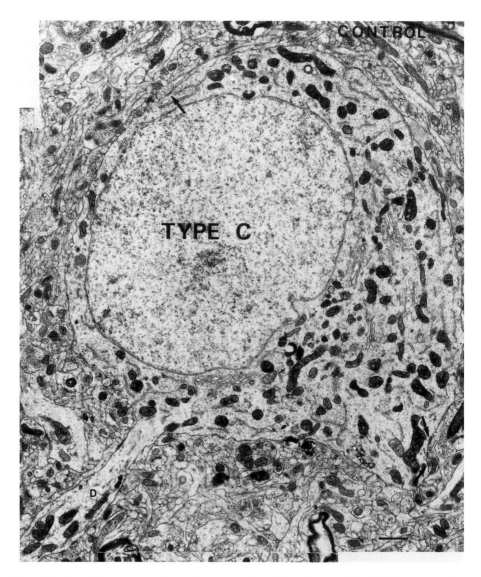

Figure 10. Type C neurons constitute the most reactive cell type in control puffs. They are medium-sized nonpyramidal neurons whose perikarya and dendrites (D) contain a rich supply of mitochondria darkly reactive for CO. This is also the only cell type found thus far to receive both asymmetrical (arrow) and symmetrical synapses on their somata. Bar = 1 μm. From Wong-Riley *et al.* (1989a).

157

CYTOCHROME
OXIDASE
STUDY OF
PRIMATE
VISUAL
CORTEX

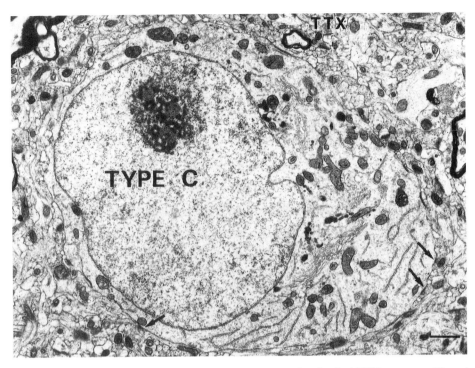

Figure 11. Type C cells exhibit the greatest change after 4 weeks of retinal TTX treatment. There is a significant reduction in cell size and a conversion of mitochondria from darkly reactive to light and moderately reactive types. Two asymmetrical synapses on the right (arrows) and one symmetrical synapse on the left (arrow) still contact this neuron in this plane of section. Bar = 1 μm. From Wong-Riley *et al.* (1989a).

Table I. Percent Distribution, Size, and Mitochondrial Packing Density of Neurons within Macaque Puffs

	Treatment	Type A	Type B	Type C
Percent distribution	Control	57%	18%	25%
	2 week TTX	67%	15%	18%
	4 week TTX	52%	15%	33%
Average size (±SE)	control	52.54 ± 0.65	83.09 ± 6.08	75.73 ± 4.99
	2 week TTX	52.95 ± 1.36	76.21 ± 5.51	66.37 ± 4.72*
	4 week TTX	52.65 ± 4.61	71.83 ± 7.02	66.29 ± 4.99*
Mito count per 100	Control	111 ± 4	140 ± 9	143 ± 7
μm^2 cytoplasm	2 week TTX	84 ± 5***	89 ± 10**	107 ± 8**
(±SE)	4 week TTX	113 ± 4	131 ± 7	132 ± 7
Mito area per 100	Control	13.2 ± 0.4	15.1 ± 0.7	16.9 ± 0.6
μm^2 cytoplasm	2 week TTX	7.6 ± 0.4***	8.3 ± 1.0***	10.2 ± 0.9***
(±SE)	4 week TTX	12.1 ± 0.5	12.8 ± 0.7**	14.4 ± 0.5***

[a]Changes in average cell size (cross-sectional area in μm^2) were assessed by Student's *t*-test. Changes in mitochondrial packing density (count and area) were tested by one-way analysis of variance and Dunnett's test (*$p < 0.05$; **$p < 0.01$; ***$p < 0.001$). From Wong-Riley *et al.* (1989a).

majority of axon terminals give rise to asymmetric synapses (axons asymmetric or AA), and they contain either no mitochondria or a few very lightly reactive ones. Axon terminals forming symmetric synapses (axons symmetric or AS), on the other hand, usually possess darkly reactive mitochondria and are immunoreactive for GABA (Beaulieu *et al.*, 1992; Nie and Wong-Riley, 1993). This implies that AS terminals may be more tonically active than AA ones. Not surprisingly, both myelinated and unmyelinated axonal trunks display low levels of CO activity, as the propagation of action potentials consumes a minimal amount of energy (Creutzfeldt, 1975). Glial cells, in general, also contain very few mitochondria and most of those are of the lightly reactive type. Oligodendroglia are routinely low in CO activity, whereas an occasional astroglia may be more reactive for CO. Overall, glial mitochondria contribute only a minute percentage to the total mitochondrial population in puffs.

3.1.1f. Puffs versus Interpuffs. Qualitatively, puffs and interpuffs are very similar in their neuronal and neuropil compositions, but quantitatively they differ in at least three aspects. First, the mean size of neurons population in interpuffs is less than that in puffs, and much of this difference reflects the larger CO-reactive neurons in puffs, especially the metabolically active type C cells (Carroll and Wong-Riley, 1984; Trusk *et al.*, 1990; Wong-Riley *et al.*, 1992). Second, the levels of CO activity in both neurons and neuropil are much lower in interpuffs than in puffs. Thus, even though the numerical densities of mitochondria do not differ between the two regions, the number of darkly reactive ones in interpuffs is only half as much as that in puffs.

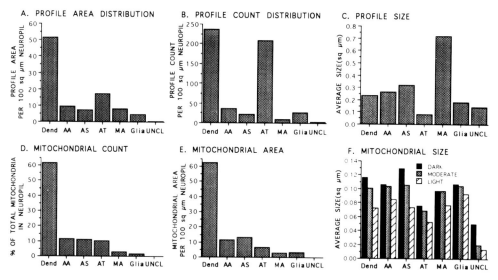

Figure 12. Areal (A) and numerical (B) distributions and average size (C) of neuropil profiles located in control puffs (Dend: dendrites; AA: axon terminals making asymmetrical synapses, AS: axon terminals making symmetrical synapses; AT: unmyelinated axon trunks; MA: myelinated axons; UNCL: unclassified profiles). The relative distributions of mitochondrial count and area within profiles of the neuropil are shown in D and E. A positive relationship exists between average mitochondrial size and CO reactivity (dark, moderate, and light) within each profile type (F). However, the mean size of mitochondria in each category differs among neuropil profiles. From Wong-Riley *et al.* (1989b).

159

CYTOCHROME
OXIDASE
STUDY OF
PRIMATE
VISUAL
CORTEX

Another striking difference lies in the synaptic compositions of the neuropil of these two regions. Extensive analysis of over 30,000 synapses in macaque interpuffs and a comparable number in macaque puffs (Wong-Riley *et al.*, 1989b, 1992; Nie and Wong-Riley, 1993; and unpublished observations) revealed that the ratio of asymmetric to symmetric synapses in puffs is 3 to 1, while that in the interpuffs is only slightly less than 2 to 1 (however, see Beaulieu *et al.*, 1992). Virtually all of the symmetric synapses in the macaque striate cortex are found to be GABA(+) while the asymmetric ones are GABA(−) (Beaulieu *et al.*, 1992; Nie and Wong-Riley, 1993). If the symmetric synapses prove to be inhibitory and the asymmetric ones excitatory, then puffs are likely to be dominated by excitatory inputs and interpuffs may be more strongly influenced by inhibitory interactions. This would be consistent with a greater energy demand and higher CO activity for puffs than for interpuffs, since repolarization after depolarization consumes much more energy than that after hyperpolarization, as discussed above (Section 2). Moreover, several functional attributes of interpuffs, such as orientation specificity and high-spatial-frequency tuning (see below), may be molded by inhibitory synapses. In the cat striate cortex, inhibitory mechanisms have been implicated in the generation of certain receptive field properties, including orientation specificity (Sillito, 1975; Sillito *et al.*, 1980).

While synaptic difference is prominent between puffs and interpuffs in macaque monkeys, it is not remarkable in squirrel monkeys (Carroll and Wong-Riley, 1984) where the ratio of asymmetric to symmetric synapses is only about 1.2 to 1 in both regions. However, levels of CO activity are much higher in terminals of puffs than of interpuffs, indicating that depolarizing activities in terminals may be more robust in puffs than in interpuffs. The higher proportion of symmetric synapses in puffs of squirrel monkeys may be related to the fact that puffs in this species are largely binocular, as are interpuffs, whereas puffs in the macaque are dominated by a single eye (Tigges *et al.*, 1984; Livingstone and Hubel, 1984a; Ts'o and Gilbert, 1988).

3.1.1g. Functional Significance. Why do puffs possess a higher level of metabolic activity? The answer is not immediately obvious, but the pioneering work of Livingstone and Hubel (1984a) and subsequent studies by others (Tootell *et al.*, 1983, 1988c–e; Ts'o and Gilbert, 1988; Silverman *et al.*, 1989; Ts'o *et al.*, 1990) have shown that puff neurons do have functional attributes that are different from interpuff neurons.

In the macaque, where the bulk of physiological experiments has been conducted, puff units tend to be selective for color but nonselective for stimulus orientation; they tend to be monocularly driven, sustain a high level of spontaneous activity, and are activated by stimuli of low-spatial-frequency gratings. The last four attributes would be consistent with a greater demand for energy metabolism, and long-term activity rate should determine the density of CO. A higher level of spontaneous activity would require greater energy expenditure; less orientation specificity would imply a higher probability of activation by stimuli of any orientation; low-spatial-frequency cells have broader spatial frequency and orientation bandwidths than those tuned to high spatial frequencies, and they would be expected chronically to receive more stimulation (Silverman *et al.*, 1989); and likewise, monocular activation signifies less binocular competition and less interference from the other eye. Interpuff neurons, on the other hand, are less spontaneously active and are less likely to be activated unless specific stimuli are presented with a particular orientation and within an appropriate range of high spatial frequencies. They may not respond vigorously until inputs from the two eyes have been coactivated.

The case for color preference is less clear. Color-varying stimuli produce much greater 2-DG uptake in V1 than do luminance-varying stimuli (Tootell *et al.*, 1988c). Yet, greater color selectivity for puff neurons means that cells respond more vigorously only when conditions are optimal, thereby lowering their frequency of activation. Broadband units, on the other hand, tend to be activated by all three types of cone input, are less discriminatory, and therefore may require greater metabolic capacity to sustain their activity. Interpuffs also contain broadband units, and the cells there are not color-blind. Even though they appear to lose the information concerning the colors that form a border or the sign of the contrast of the border, they still may respond to color-contrast borders in which the two colors are equiluminant (Livingstone and Hubel, 1988). Moreover, nocturnal primates such as owl monkeys also possess a full complement of puffs (Horton, 1984); thus, it is unclear at present as to how much color processing contributes to an overall enhancement of metabolic demand that helps to shape puffs.

Allman and Zucker (1990) proposed that puffs are involved in intensity coding over a very broad dynamic range, requiring neurons to have the energetic capacity, and thus the high concentration of CO, to sustain the range of activity levels. They further postulated that a sizable proportion of puff neurons may be very nonselective and respond simply to a graded input encoding contrast. Interpuff neurons, on the other hand, code the strength of match between geometric variables and the image. They may use contrast adaptation to extend their operating range from dim to bright light. Their preference for higher spatial frequencies (and possibly binocular disparities) further reduces the statistical probability of their activation, as discussed above. However, these postulates need to be rigorously tested.

What are the relationships between the functional map and the metabolic map? If one considers an individual globular puff with its surrounding interpuff region as a functional unit, akin to the "blob domain" of Tootell *et al.* (1988b), then from the center to the periphery of such a unit one may find a continuum from metabolically most active to least active. Indeed, CO density as measured by optical densitometry also appears to fall on a continuum, with highest values in the center to lowest in the periphery of puffs, and still lower in the interpuffs (Trusk *et al.*, 1990). Even dendrites of many border zone puff neurons extend freely into interpuff territory, indicating a smooth transition from one compartment to the next (Malach, 1992). A continuum of peak tuning of cells also appears to exist in a radial array, with cells tuned to the lowest spatial frequencies in the center of puffs and those tuned to increasingly higher frequencies in rings around this center (Silverman *et al.*, 1989). Whether similar continua exist for other functional properties is as yet unknown. It is possible that orientation selectivity may lie along such a continuum, so that cells in the centers of puffs may be least selective and those in the centers of interpuffs most selective. By the same token, we find morphologically similar cell types in puffs and interpuffs, but their degree of CO activity is higher in puffs than in interpuffs (Carroll and Wong-Riley, 1984; Wong-Riley *et al.*, 1989a, 1992). Within each region and subregion, those with the highest degree of selectivity, such as color-opponent cells in puffs and orientation-selective units in interpuffs, may be least metabolically

161

CYTOCHROME
OXIDASE
STUDY OF
PRIMATE
VISUAL
CORTEX

active, and may correspond to CO-poor, type A cells. On the other hand, cells with the lowest degree of selectivity, such as broadband units in puffs and inter-puffs, may be metabolically most active and may correspond to CO-rich, type C or type B cells. These hypothetical correlations may one day be testable, when single-unit recordings are combined with CO labeling at the EM level. A critical factor is that the intracellular label should not interfere with endogenous enzymatic activity.

To be sure, selectivity also entails suppression by inhibitory interactions, which is intimately dependent on the activation of inhibitory interneurons. Thus, while color-opponent or orientation-selective cells themselves may have relatively low energy demand, their properties are molded by sustained activity of inhibitory interneurons whose metabolic needs are high. Type C cells, for example, are immunoreactive for GABA (Nie and Wong-Riley, 1993) and have an intense level of CO (Wong-Riley *et al.*, 1989a). Generally, neurons are more reactive in puffs than in interpuffs, and one would expect a higher frequency of depolarizing activities in puffs than in interpuffs, whether the activities are spontaneous or synaptically elicited, and whether they excite non-GABAergic or GABAergic postsynaptic neurons. If GABAergic neurons are activated in puffs, however, it remains an open question whether their axons terminate entirely in puffs or invade interpuffs as well.

The CO-rich type C cells may comprise a heterogeneous population. Although they may all be GABAergic, they may contain different neuropeptides whose exact functions are still unclear (Hendry *et al.*, 1984). The fact that type C cells constitute about a fourth of the puff population (Wong-Riley *et al.*, 1989a) makes them candidates for the non-color-coded units that constitute a similar proportion of the recording sample for puff neurons (Livingstone and Hubel, 1984b; Ts'o and Gilbert, 1988). On the other hand, Allman and Zucker (1990) hypothesized that the type C cells encode contrast over a broad dynamic range and serve as the basis for a cortical brightness constancy system. In nocturnal primates, they reasoned that there may be a larger proportion of type C cells. When primates became diurnal, they may have elaborated this system to encompass the color-specific lightness for determining color constancy.

Maguire and Baizer (1982) found in awake monkeys that nonoriented cells responded in a graded fashion to luminance variation spanning 4 log units from threshold to saturation. At saturation, some of these cells sustained firing rates of 300 spikes per second. Since these cells would require a relatively high energetic capacity to sustain their intense activity, they may correspond to CO-rich type C cells in puffs. These neurons may also correlate with "luxotonic" units that have a maintained discharge to diffuse light and/or have a discharge that varied monotonically with changes in luminance intensity over a range of at least 3 log units (Bartlett and Doty, 1974; Kayama *et al.*, 1979).

3.1.1h. Connections of Puffs. At least two parallel visual channels reach supragranular striate cortex from the retina (reviewed in Lund, 1988). In one channel, large retinal ganglion cells project to the magnocellular layers of the lateral geniculate nucleus (LGN), where the neurons are highly sensitive to changes in the contrast and temporal content of visual stimuli, but exhibit low spatial resolution and are insensitive to wavelength. The information is relayed mainly to layer 4Cα of V1, but some of it may reach CO-rich puffs (Livingstone and Hubel, 1984b). The second channel relays information from the smaller retinal ganglion cells to the parvocellular layers of the LGN, where neurons have high spatial resolution and are responsive to color and high-spatial-frequency gratings; but they exhibit low contrast sensitivity. The information is relayed mainly to layer 4Cβ and from there to layers 2–3 of V1 (Fitzpatrick *et al.*, 1985). The functional segregation into puffs and interpuffs is as described above.

Thus, whether direct or by way of layer 4, puffs receive input from both the magno- and parvocellular laminae of the LGN (Livingstone and Hubel, 1982; Fitzpatrick *et al.*, 1983; Weber *et al.*, 1983). The broadband cells there may receive input either from the magnocellular cells or from the broadband parvocellular cells, but the physiological properties of many of them would be more consistent with input from the magnocellular system. Color information may be relayed to puffs from parvocellular interlaminar leaflets (Fitzpatrick *et al.*, 1983), where color-opponent center-only ("COCO") or type 2 cells are found (Livingstone and Hubel, 1982, 1984a; Hubel and Livingstone, 1990), and/or from layer 4Cβ (Michael, 1988). Interpuff neurons, on the other hand, probably receive inputs from color-coded parvocellular cells and, although they are not distinctly color-selective, they nonetheless respond to color contrast (Livingstone and Hubel, 1988).

Individual puffs are probably connected to other puffs of similar receptive field type, color opponency, and ocular dominance (Livingstone and Hubel, 1984b; Ts'o and Gilbert, 1988). Since there may be three times as many puffs that specialize in red/green opponency than blue/yellow opponency (Ts'o and Gilbert, 1988), there may be far more interconnections between the former than there are between the latter. Interactions also exist between interpuff regions (Livingstone and Hubel, 1984b) and cells with matching orientation and eye preference are probably joined by clustered horizontal connections (Ts'o and Gilbert, 1988).

In addition, V1 also receives feedback fibers from V2 (Tigges *et al.*, 1973; Wong-Riley, 1979a; Lund *et al.*, 1981; see Rockland, this volume) but it is not known if they target puffs and/or interpuffs. More recently, Cusick and Kaas (1988) injected the retrograde/anterograde tracer (WGA/HRP into area 18 of squirrel monkeys and found that the projection patterns within area 17 were "latticelike" and "dotlike." Unfortunately, the injections were not confined to specific subregions of area 18 and the exact parceling of feedback connections remains undefined.

3.1.1i. Development of Puffs. In macaque monkeys, puffs appear as early as 3.5 weeks before birth (Horton, 1984; Wong-Riley *et al.*, 1988), indicating that their genesis is largely a response to an intrinsic signal independent of visual experience. These intrinsic factors may include spontaneous activity, housekeeping functions, and intracortical synaptic interactions. Cells that exhibit synchronized or correlated activities may interact and form groupings with similar metabolic properties. In time, puff neurons with correlated activities may segregate in an anatomical, functional, and metabolic fashion from the network of interpuff neurons.

Human visual cortex appears to lag behind that of macaque monkeys in the full expression and organization of CO-rich puffs. Puffs probably exist in a rudimentary fashion at birth, consolidate toward the end of the first month, and only become well organized by the fourth month (Wong-Riley *et al.*, 1993a). This means that visual experience could potentially influence the ultimate differen-

163

CYTOCHROME
OXIDASE
STUDY OF
PRIMATE
VISUAL
CORTEX

tiation of puffs in human. Thus, even though cortical neurons in human have all been laid down and are well laminated, functional modules may not have been completed, and the necessary circuits for organizing the puff system and other processing streams may not have matured.

Purves and LaMantia (1990) initially found that the density of puffs in newborns was not different from those in adult monkeys. They therefore postulated that puffs were added gradually during postnatal development. A later study by the same authors (1993) uncovered a decrease in puff density from an average of $5.2/mm^2$ at birth to $4.3/mm^2$ in maturity. This difference of 18% is comparable to a 16% increase in the average area of the primary visual cortex from 919 mm^2 in newborns to 1069 mm^2 in adult monkeys. Thus, they concluded that the number of puffs remained approximately the same during maturation, at about 4800/hemisphere. Substantial variability in puff density probably exists among animals.

What are the effects of prenatal bilateral enucleation on the development of CO pattern in the visual cortex? Kuljis and Rakic (1990) performed a bilateral retinal ablation at midgestation (E81 and E120), before contacts were established between photoreceptors and other retinal neurons, and before layers 2 and 3 of the cortex were generated. Surprisingly enough, they found normal size and spacing of CO puffs. This strongly implies that the development of puffs in the macaque is genetically determined, independent of retinal photoreceptor input and independent of visual experience.

Dehay *et al.* (1988) found that enucleation at E81 and E110 reduced the striate cortex by 14–40%, whereas enucleation at E59 and E68 led to an areal reduction of more than 70%. However, there was only an 8% linear reduction of interpuff spacing (Kennedy *et al.*, 1990). The latter authors proposed two levels of specifications. The first one is dependent on the presence of the sensory periphery (retina) during early stages of development and determines the areal extent of area 17. The second is independent of the retina and specifies the periodicity of CO puffs. Further experimentation is needed to clarify these issues.

3.1.1j. Comparative Studies among Species. The question of species variation in the possession of puffs was addressed by Horton (1984). In seven nonprimate species (mouse, rat, ground squirrel, cat, rabbit, mink, and tree shrew), none appeared to show a nonhomogeneous staining pattern in the supragranular layers. On the other hand, of the seven primate species (squirrel monkey, rhesus monkey, cynomolgus monkey, owl monkeys, bushbaby, baboon, and human), all exhibited a patchy pattern. These findings strongly indicate that puffs are strictly a primate entity. This view remained unchallenged until recently, when Van Sluyters's group (Murphy *et al.*, 1991) reported the presence of puffs in the cat striate cortex. This finding awaits confirmation by other investigators.

Although puffs are in register with ODCs in macaque monkeys and man, they are nonetheless present in the monocular crescent where ODCs are absent, and in owl and squirrel monkeys where ODCs are only weakly developed (Carroll and Wong-Riley, 1984; Horton, 1984; Wong-Riley and Carroll, 1984b; Horton and Hedley-Whyte, 1984). However, these findings by themselves do not rule out the possibility that puffs are strongly dominated by input from a single eye and that they form the functional organizing centers for modules.

As discussed above, color processing may be an important attribute of puffs but may not be the decisive factor for the existence of puffs. Nocturnal primates such as galagos, lorises, and owl monkeys live in dim light but possess puffs (Horton, 1984; Tootell *et al.*, 1985; Allman and Zucker, 1990). However, owl monkeys are considered protanomalous trichromats (Jacobs, 1977) with a fully functional photopic visual system, which provides them with some color vision and reasonably acute spatial resolution. There are two diurnal primates, *Hapalemur griseus* and *Propithecus verrauxi* (Allman and Zucker, 1990), and a diurnal nonprimate with well-developed color vision, the tree shrew (Horton, 1984; Wong-Riley and Norton, 1988), that lack puffs. Other mammals with cone-dominated retinae, such as ground squirrels, also do not possess puffs (Horton, 1984). A comparative study of CO distribution patterns in visual cortices of several mammalian species has been given elsewhere (Wong-Riley, 1988).

3.1.2. Layer 4

Layer 4 is the major recipient of geniculocortical projections. Just as the supragranular layers are heterogeneous, so is layer 4 subdivided into several sublaminae, each with distinct connections and functional properties. These sublaminae are vividly portrayed by CO histochemistry as well as CO immunohistochemistry (Figs. 1B,C and 5) (see also Horton, 1984; Carroll and Wong-Riley, 1984; Wong-Riley and Carroll, 1984b; Hevner and Wong-Riley, 1990).

3.1.2a. Laminar and Cellular Compositions of Layer 4. Layer 4A receives its major input from parvocellular geniculate laminae (Lund, 1988). Its neuropil, which contains both afferent terminals and postsynaptic dendritic processes, assumes a curious honeycomblike configuration that is highly reactive and immunoreactive for CO (Horton, 1984; Carroll and Wong-Riley, 1984; Hevner and Wong-Riley, 1990). This pattern is quite striking in tangential sections of V1 in macaque and squirrel monkeys, and is disrupted by monocular impulse blockade (Fig. 4A–D). However, layer 4A is not clearly discernible in human striate cortex (Horton and Hedley-Whyte, 1984; Wong-Riley *et al.*, 1993a) and the question arises as to whether 4A is simply less metabolically active, whether it is more sensitive to postmortem changes, or whether it is absent in the human. Burkhalter and Bernardo (1989) recognized a thin band between layers 3 and 4B that was devoid of feedback fibers from V2 (their Fig. 4B), raising the possibility that this thin band may represent a geniculate-recipient layer (layer 4A) that normally does not receive direct V2 projections.

Layer 4B lacks geniculate input and the level of CO is low except for periodic "blobs" in register with puffs in layers 2 and 3 (Livingstone and Hubel, 1982). 4B derives afferents predominantly from layer 4Cα (Lund and Boothe, 1975), and both of these layers are very sensitive to changes in luminance, but not to color-varying stimuli (Tootell *et al.*, 1988c). 4B is functionally heterogeneous, and stereo-tuned cells are selectively located under supragranular interpuffs (Hubel and Livingstone, 1990).

Layer 4C is rich in both CO activity and enzyme amount (Hevner and Wong-Riley, 1990). However, laminae 4Cα and 4Cβ do not always have the same enzyme levels. Layer 4Cα receives input mainly from the magnocellular layers of the LGN, in which color sensitivity is weak, while 4Cβ derives its afferents from parvocellular LGN, which exhibits distinct color opponency (Blasdel and Lund, 1983; Livingstone and Hubel, 1984b). Interestingly, both 4Cα and magnocellular geniculate laminae have higher levels of CO activity than the bulk of 4Cβ and

165

CYTOCHROME
OXIDASE
STUDY OF
PRIMATE
VISUAL
CORTEX

parvocellular laminae in the macaque (Liu and Wong-Riley, 1990; Wong-Riley and Carroll, 1984b). Layers 4Cα and 4Cβ also have different intracortical afferent and efferent connections: 4Cα projects mainly to 4B, 5A, and 6, and receives input from the latter two sources; while 4Cβ sends fibers chiefly to 4A, 3B, and has a reciprocal connection with layer 6 (Fitzpatrick *et al.*, 1985).

Layer 4Cβ, in turn, is not homogeneous (Carroll and Wong-Riley, 1984; Fitzpatrick *et al.*, 1985; Wong-Riley and Carroll, 1984b). The bulk of 4Cβ is packed with cells and so stains intensely for Nissl but less intensely for CO. The base of 4Cβ is more cell sparse and forms a thin, CO-rich band known as "4Cβ dark" (Hiltgen and Wong-Riley, 1986). This dichotomy is much more marked in the newborn macaque or human (Fitzpatrick *et al.*, 1985; Kennedy *et al.*, 1985; Wong-Riley *et al.*, 1993a) or after monocular enucleation in the adult macaque (Horton, 1984; Trusk *et al.*, 1990), when upper 4Cβ is much paler for CO than is 4Cβ dark. It is not surprising, then, that the two groups of neurons respond with different latencies to thalamic fiber stimulation (Mitzdorf and Singer, 1979). They have different mitochondrial and synaptic compositions (Hiltgen and Wong-Riley, 1986), unique efferent cortical projections, with upper 4Cβ terminating mainly in layer 3B and lower 4Cβ contributing mainly to layer 4A (Fitzpatrick *et al.*, 1985), and, perhaps, receive input from different populations of parvocellular geniculate neurons. One possibility is that the CO-rich 4Cβ dark is contacted mainly by the darkly reactive lamina 6 or moderately reactive lamina 5 neurons of the LGN, while the paler portion of 4Cβ may be the termination site for lamina 3 or 4 neurons of the LGN, which are less reactive than the dorsal laminae (Liu and Wong-Riley, 1990).

There are only two major neuronal types in layer 4C (Lund, 1973). Most of them are small spiny stellates whose somata exhibit relatively low levels of CO (Hiltgen and Wong-Riley, 1986). The remaining 20–25% of the population are medium-sized aspinous stellates with relatively high levels of CO and are reminiscent of type C cells in layers 2–3. Much of CO labeling, however, resides in dendrites of both cell types as well as in terminals forming symmetrical synapses (Hiltgen and Wong-Riley, 1986). Thus, layer 4C shares some common metabolic and synaptic features with the supragranular layers (see Section 3.1.1).

3.1.2b. Functional Significance. Gratings of very high spatial frequencies produce much higher 2-DG uptake in 4Cβ than in 4Cα, whereas gratings of low spatial frequencies produce the converse, i.e., higher uptake in 4Cα than in 4Cβ. Furthermore, the cells in layer 4C lying beneath puffs of a given color opponency tend to have identical color opponency. Tootell *et al.* (1988c) found that long and short wavelengths (reds and blues, being the most saturated) produce robust 2-DG uptake, and middle wavelengths (yellows and greens, being the least saturated) produce little or no uptake. According to these authors, the lack of uptake may be caused by an insignificant change in the firing of the predominant cell type, the red–green opponent cells.

Besides the laminar segregation of functions, layer 4C is obviously partitioned into vertical ocular columns (or slabs) subserved by the left or the right eye (Hubel and Wiesel, 1977). Trusk *et al.* (1990) measured ODCs made visible by transneuronal transport of [³H]proline in a normal macaque and found them to be 350–360 μm wide (Fig. 13D). Bands innervated by the contralateral eye were slightly, but not significantly, wider than bands innervated by the ipsilateral eye. In addition to ODCs, there appears to be a columnar organization in layer

4B for spatial frequency. Gratings of relatively high spatial frequency (5–7 cycles/deg) of varied orientations induce 2-DG uptake in 4B that lies in register with labeled interpuffs, while low-spatial-frequency gratings (1–1.5 cycles/deg) label 4B directly underneath the puffs (Tootell *et al.*, 1988e).

3.1.2c. Development of Layer 4. At birth, the CO pattern differs slightly from the adult but layer 4C appears quite similar between macaque monkeys and human (Kennedy *et al.*, 1985; Wong-Riley *et al.*, 1993a). The basic pattern does not appear to be influenced by eccentricity in the newborn macaque (Kennedy *et al.*, 1985). The parvo-recipient 4Cβ proper is simply paler than that in the adult, thereby highlighting the thin band of CO-rich 4Cβ dark, while the reactive band in 4Cα is quite robust and even spills slightly into 4B. This implies that the maturation of the magno-recipient zone may precede that of the parvocellular one, and that 4Cβ dark may be more active than the rest of the sublamina. Evidence exists that magnocellular geniculate layers may mature earlier than parvocellular ones (Rakic, 1976; Gottlieb *et al.*, 1985), that 4Cβ may develop later than 4Cα (Blasdel and Lund, 1983), and that the critical period for plasticity may end earlier in 4Cα than in 4Cβ (LeVay *et al.*, 1980). The existence of 4Cβ dark implies that this sublamina may mature earlier and may receive a set of inputs that is different from the rest of 4Cβ (Fitzpatrick *et al.*, 1985). Thus, three sets of channels may exist that differ in their developmental rates, connections, metabolic properties, and functional characteristics: the more active magno pathway to 4Cα, the less active parvo pathway to the upper part of 4Cβ, and the more active pathway from an unspecified portion of the LGN to 4Cβ dark. Future studies using more refined tracing or recording techniques may provide more insight into this issue.

3.1.3. Layers 5 and 6

Of the two infragranular layers, layer 5 receives no geniculate input and is rather pale, whereas layer 6 is a geniculate-recipient layer with moderate levels of CO (Figs. 1B,C and 5) (see also Horton, 1984; Carroll and Wong-Riley, 1984; Wong-Riley and Carroll, 1984b). In addition, layer 5A projects to and receives input from 4Cα, while layer 6 is reciprocally connected with both 4Cα and 4Cβ (Fitzpatrick *et al.*, 1985).

The Meynert cells at the border between layers 5 and 6 are highly reactive for CO. Their cell bodies lie mainly below CO-poor interpuffs of layers 2–3 (Payne and Peters, 1989). They may be involved in the detection of movement, since they project to the superior colliculus (Fries and Distel, 1983) and/or cortical area MT known as V5 (Lund and Boothe, 1975; Fries *et al.*, 1985). However, it is not clear whether the dendritic field of Meynert cells lies within or outside the domains of puffs. Other CO-reactive cells found in the infragranular layers include both pyramidal and medium-sized nonpyramidal neurons (Carroll and Wong-Riley, 1984; Horton, 1984; Wong-Riley and Carroll, 1984b).

Besides the laminar arrangement, the infragranular layers actively participate in ocular dominance and orientation columns as originally described by Hubel and Wiesel (1968, 1974, 1977). More recently, spatial frequency columns have also been postulated to involve layers 5 and 6 (Tootell *et al.*, 1988e). The CO-rich blobs found in these two layers are probably intimately associated with

167

CYTOCHROME
OXIDASE
STUDY OF
PRIMATE
VISUAL
CORTEX

vertically aligned blobs in 4B and 2/3 to form the organizing centers of functional modules in the striate cortex.

3.2. Comparisons with Patterns Shown by Other Markers

The fact that CO consistently reveals a pattern of puffs and interpuffs, laminae 4A, 4C, and 6 has been exploited by researchers who use CO pattern as the reference point against which other markers/techniques/patterns are compared. These other approaches include 2-DG, voltage-sensitive dyes, high-resolution optical imaging, microvascular pattern, GABA, various neuropeptides, anterograde and retrograde tracings, and additional histological stains.

3.2.1. 2-Deoxyglucose Labeling

2-Deoxy-D-[^{14}C]glucose (2-DG) has been effectively used as a marker for activity-driven glucose uptake and utilization in the CNS (Sokoloff *et al.*, 1977). This approach labels acute changes in neuronal activity, but when the simulation is prolonged, the acute patterns disappear. CO, on the other hand, is an integral protein of the inner mitochondrial membrane and requires times to be synthesized, assembled, degraded, and renewed. The level of CO at any given time represents the stabilized state of cellular capacity for aerobic metabolism, which is needed to sustain a more chronic level of neuronal activity. This metabolic capacity is regulated at a very local level so that it differs among brain regions, laminae, cell groups, and even between compartments of a single neuron (Kageyama and Wong-Riley, 1982; Wong-Riley, 1989).

When the condition of acute visual stimulation matches that which presumably chronically activates the puffs, such as monocular exposure to a complex geometric pattern or stripes of all orientations, 2-DG labels every other row of dots or patches that lie in register with CO puffs (Horton and Hubel, 1981; Humphrey and Hendrickson, 1983; Horton, 1984). If stripes of a single orientation are presented binocularly, 2-DG labels a lattice that includes the CO puffs but is more extensive (Horton and Hubel, 1981). This is consistent with Livingstone and Hubel's (1984a) finding that puff neurons are not selective for orientation but interpuff neurons are. Likewise, Tootell's (1988a, c–e) group has elegantly demonstrated the relationship between 2-DG patterns elicited with a variety of visual stimuli and the more stable pattern of CO in the macaque visual cortex.

Since glucose can be metabolized via either the oxidative or the glycolytic pathway, 2-DG potentially labels not only neurons but also glia, which depend more on glycolytic metabolism for their energy supply (Siegel *et al.*, 1981). CO tends to label mainly neurons, since the vast majority of mitochondria in the brain reside in neurons. Certain glial elements, such as reactive astrocytes and glial limitans, may have CO levels higher than regular oligodendroglia and astroglia, although this point needs to be further investigated. 2-DG has been postulated to label mainly axons, axon terminals, and perhaps dendrites (reviewed in DiRocco *et al.*, 1989), while CO level is high in most dendrites, some neuronal somata, and those axon terminals that are presumably tonically active (Wong-Riley, 1989).

3.2.2. Optical Imaging Techniques

Ts'o *et al.* (1990) have effectively used high-resolution optical imaging techniques (activity-dependent intrinsic signals) to visualize the functional architecture of visual cortex, such as ocular dominance and orientation columns. By deriving a map of monocularity, they found an excellent match between the most monocular regions and CO puffs. In V2, they found a close match between the optical imaging maps, which showed regions of increased activity resulting from stimulation of either eye, and the thick and thin stripes revealed by CO. Thus, they concurred that "the presence of elevated concentrations of cytochrome oxidase probably indicates regions of long-term elevated activity."

3.2.3. Microvascular Pattern

Regions of high CO activity denote loci of high metabolic activity, and one would expect these regions to have more abundant microvessels. Indeed, Zheng *et al.* (1991) found that puffs were more richly vascularized than interpuffs, and the mean total length of microvessel profiles was 42% greater in puffs than in interpuffs. The distribution of microvessels was densest in lamina 4C of V1, where CO activity was also the highest. Likewise, the total microvessel length per unit area was 27% greater in CO-rich stripes than in the interstripes of V2. When comparing V1 and V2, these authors found that the microvessel length per unit area in V1 was 26% greater than in V2. They concluded that the modular, laminar, and regional organization of the primate visual cortex is reflected in the underlying distribution of cortical microvessels.

3.2.4. Na^+, K^+-ATPase

CO represents a key energy-generating enzyme whose yield of ATP in neurons is used largely to fuel ATPase activity, and Na^+, K^+-ATPase is a key energy-consuming enzyme whose ion pumping function is critical in maintaining ionic gradients for the generation of neuronal membrane potentials. It is, therefore, reasonable to assume that the levels of CO and Na^+, K^+-ATPase should be positively correlated. This is indeed the case for a number of brain regions, including the primate visual cortex (Hevner *et al.,* 1992). By the same token, the levels of both enzymes are regulated by neuronal activity, the deprivation of which induces a downregulation of the two enzymes in corresponding portions of the macaque striate cortex (Fig. 13A–D) (see also Hevner *et al.,* 1992).

3.2.5. GABA

Hendrickson *et al.* (1981) first noted that antibodies against the synthesizing enzyme for GABA, glutamic acid decarboxylase, more intensely labeled puffs than interpuffs. They suggested that CO staining might be largely localized to GABAergic neurons. Recent studies, however, indicate that the number of GABA-immunoreactive neurons is comparable between puffs and interpuffs (Fitzpatrick *et al.,* 1987; Hendry *et al.,* 1988; Beaulieu *et al.,* 1992), and the percent distribution of CO-rich type C cells, being immunoreactive for GABA (Nie and Wong-Riley, 1993), is also comparable between the two zones (Wong-Riley *et al.,* 1989a, 1992, and unpublished observations). The distribution of GABA-immunoreactive symmetric terminals is reported to be either the same

169

CYTOCHROME
OXIDASE
STUDY OF
PRIMATE
VISUAL
CORTEX

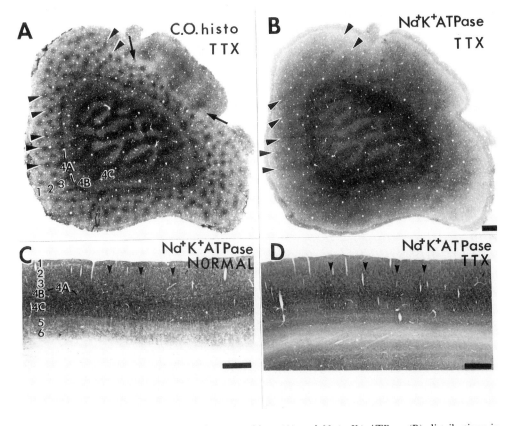

Figure 13. Comparison between cytochrome oxidase (A) and Na+, K+-ATPase (B) distributions in adjacent tangential sections of the macaque striate cortex subsequent to 3 weeks of monocular TTX treatment. (A) Note the well-defined ocular dominance stripes in layer 4C; the lightly stained stripes correspond to the TTX-injected eye. In the supragranular layers 2–3, rows of large, darkly stained puffs (arrowheads) are seen alternating with rows of small, lightly stained puffs. Arrows indicate the border of V1 and V2, where rows of normal, larger pufflike regions are seen. (B) Ocular dominance stripes in layer 4C form the same pattern with ATPase as for CO. Dark stripes extend into the supragranular layers (arrowheads), where they lie in register with the rows of darkly CO-reactive puffs. C and D represent cross sections of striate cortex from normal and TTX-treated monkeys, respectively, showing the laminar distribution of ATPase. Note that layers 4A and 4C are darkly stained, and there is some suggestion of puffs in layers 2–3 (arrowheads in C). Ocular dominance stripes are faintly visible after TTX treatment (arrowheads in D). Bars for A and B = 1 mm and for C and D = 0.5 mm. A–C, modified from Hevner *et al.* (1992); D, unpublished observation of Hevner and Wong-Riley.

between the two regions (Beaulieu *et al.*, 1992), or more prevalent in interpuffs than in puffs (Nie and Wong-Riley, 1993). However, type C cells and symmetric terminals are more CO-reactive in puffs (Wong-Riley *et al.*, 1989a, 1992), raising the possibility that they may also contain more GAD enzymes there. Nevertheless, CO is enriched not only in GABAergic type C cells and symmetric terminals, but also in non-GABAergic type B pyramidal cells and in dendrites of most if not all neurons (Carroll and Wong-Riley, 1984; Wong-Riley *et al.*, 1989a,b; Nie and Wong-Riley, 1993).

Many GABAergic neurons also exhibit immunoreactivity to a number of neuropeptides (Hendry *et al.*, 1984). For example, virtually all the small tachykinin (such as substance P and substance K)-immunoreactive neurons also display GABA immunoreactivity (Jones *et al.*, 1988). Thus, GABAergic neurons comprise a heterogeneous population whose metabolic activity may prove to be heterogeneous as well.

3.2.6. AChE, LDH, SDH, and Myelin

Horton (1984) first described a positive correlation between CO staining pattern and acetylcholinesterase (AChE), lactic dehydrogenase (LDH), and succinate dehydrogenase (SDH) in macaque and squirrel monkeys. AChE is probably localized in fibers and synaptic terminals in both monkeys and human (Horton and Hedley-Whyte, 1984); however, the honeycomb pattern so vividly shown in layer 4A by CO does not stain as such by AChE (Horton, 1984). Moreover, monocular enucleation induces a stripelike pattern with AChE or butyrylcholinesterase (BuChE) stain in the monkey striate cortex, but the AChE-rich and BuChE-rich stripes correspond instead to CO-poor stripes (Graybiel and Ragsdale, 1982). Thus, even though BuChE is decreased in the deafferented layers of the LGN, it is not decreased in cortical dominance columns corresponding to the removed eye.

The reports on myelin have been contradictory. Tootell *et al.* (1983) and Horton (1984) found that V1 puffs and V2 stripes were more densely myelinated than surrounding regions, while Krubitzer and Kaas (1989) attributed myelin-rich zones to CO-poor interpuffs of V1 and interstripes of V2. At any rate, myelinated axons in both the gray matter and the white matter have low levels of CO (Carroll and Wong-Riley, 1984; Wong-Riley and Carroll, 1984a; Wong-Riley *et al.*, 1989b), and so a positive relationship cannot be established between CO content and myelin density, although the two can certainly coexist.

3.2.7. Cat-301

The pattern created by the monoclonal antibody Cat-301 is quite different from that of CO. Cat-301 tends to highlight the magnocellular channel, such as layer 4B of V1, V2 thick stripes, V3, and MT (Hockfield *et al.*, 1983; DeYoe *et al.*, 1990). It is also strong in at least parts of areas V3a, the MST complex, and the posterior parietal complex, but not in area V4 or inferotemporal cortex. In addition to layer 4B, Cat-301 also labels somata and neuropil of layer 6 and small cellular clusters in layers 2–3. These clusters sometimes coincide with cores of CO puffs and appear to be centered on ODCs. Cat-301-positive cells may also represent a subset of GABAergic neurons (Hendry *et al.*, 1988).

3.2.8. Calbindin and Neuropeptide Y

Not all neurochemical markers bear a positive relationship with CO. The pattern conferred by the vitamin D-induced calcium-binding protein calbindin-D-28K (calbindin) is complementary to that of CO (Celio *et al.*, 1986). Likewise, neurons immunoreactive against the neurotransmitter candidate neuropeptide Y are preferentially located in interpuffs (Kuljis and Rakic, 1989). These findings further substantiate a chemical and functional distinction between puffs and interpuffs.

4. Secondary Visual Cortex

171

CYTOCHROME
OXIDASE
STUDY OF
PRIMATE
VISUAL
CORTEX

The secondary visual cortex or V2 is known to border on V1 and to establish reciprocal connections with V1. Its laminar organization is relatively unremarkable with the Nissl stain, but the pattern with CO is quite striking and bears a significant relationship to functional subdivisions in this area (Livingstone and Hubel, 1982; Tootell *et al.*, 1983; Wong-Riley and Carroll, 1984a). Again, differences in CO levels are likely to reflect differences in levels of neuronal activity (Livingstone, 1990).

4.1. Dynamic Metabolic Map of V2 Revealed by Cytochrome Oxidase

CO-rich zones in V2 extend mainly from lower layer 2 to upper layer 4, with a hint of a columnar pattern through layers 4, 5A, and between layers 5 and 6, where CO levels are slightly higher than adjacent areas (Fig. 14; see also Wong-Riley and Carroll, 1984a). The subdivision into alternating thin and thick CO-rich stripes is much more prominent in the squirrel monkeys (Figs. 14 and 15A) than it is in the macaque (Fig. 15B) or human (Hubel and Livingstone, 1987; Burkhalter and Bernardo, 1989; Tootell and Hamilton, 1989; Ts'o *et al.*, 1990; Wong-Riley *et al.*, 1993a). In the macaque, for example, the use of Cat-301 and tract tracers can improve the definition of thick and thin bands where the CO pattern may be indistinct or irregular (DeYoe *et al.*, 1990). These stripes radiate orthogonally from the 17–18 border (Livingstone and Hubel, 1982; Tootell *et al.*, 1983; Horton, 1984; Wong-Riley and Carroll, 1984a) and measure about 0.8 mm wide, with a center-to-center spacing of about 1.1–1.6 mm. Within laminae 2 to upper 4 of each stripe, however, CO activity is *not* homogeneous but forms globular zones (called "V2 puffs" by Wong-Riley and Carroll, 1984a) that are much larger and less regular than those in V1 (Fig. 14C). V2 puffs in the thick stripes of squirrel monkeys range between 700 and 1100 μm in diameter, while those in the thin stripes are between 400 and 890 μm in diameter. These globular zones may form the basis for further functional subdivisions within the stripes.

Ultrastructural features in V2 puffs bear a remarkable similarity to those in V1 puffs (Wong-Riley and Carroll, 1984a). CO-rich neurons include medium-sized nonpyramidals and medium to large pyramidal cells, while CO-poor neurons comprise mainly the small stellates. The majority of mitochondria in the neuropil again reside in dendrites, and terminals that form symmetric synapses are more CO-reactive than those that give rise to asymmetric synapses. CO-poor zones in V2 have qualitatively similar neuronal compositions, but they have significantly fewer darkly reactive mitochondria and proportionally fewer asymmetric synapses than CO-rich zones. These findings are consistent with the notion that CO-rich zones are under greater excitatory influence than CO-poor ones.

4.2. Functional Significance

The modular organization of V2 was revealed by CO and encompasses CO-rich thin and thick stripes and CO-poor intervening interstripes (Living-

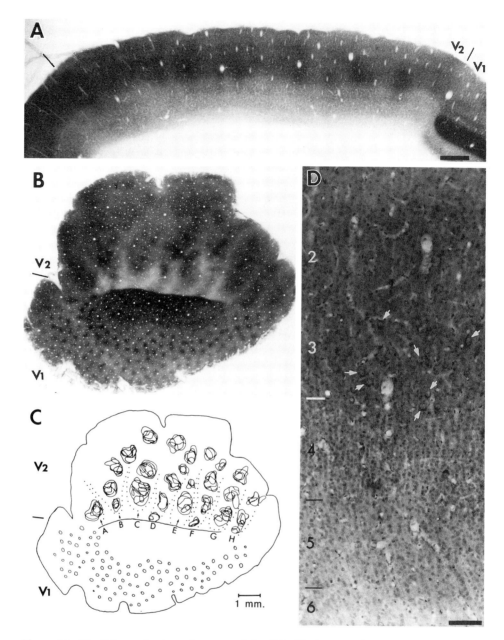

Figure 14. V2 of the squirrel monkey reacted for CO. (A) Cross section of the V1/V2 border showing the abrupt transition of CO pattern. In V2, periodic zones or puffs of high CO activity appear in layers 2 and 3, extending into the upper edge of 4. There is a hint of a columnar pattern of staining, with layers 4 and 5 below the puffs being slightly more reactive than the adjacent region. A thin band of staining can also be seen in upper 5 (5A) and another one between 5 and 6. The black lines above the section demarcate the approximate boundaries of V2 where distinct supragranular puffs can be seen in periodic array. (B) Tangential section showing the small and closely spaced puffs in V1 and larger puffs forming thick and thin stripes in V2. (C) Camera lucida tracing of puffs in V2 through seven tangential sections spaced 120 μm apart. Note that the puffs remain globular with some irregular boundaries, and that they form alternating wide (A, C, E, G) and narrow (B, D, F, H) stripes. The puffs in the thin stripes tend to be smaller and begin right at the V1/V2 border. (D) CO-reacted and Nissl-counterstained coronal section through V2, showing the location of the puff

173

CYTOCHROME
OXIDASE
STUDY OF
PRIMATE
VISUAL
CORTEX

stone and Hubel, 1982; Tootell *et al.*, 1983; Horton, 1984; Wong-Riley and Carroll, 1984a). Functional cell types tend to predominate in different regions, with unoriented wavelength-selective cells in thin stripes, oriented, disparity-sensitive, and directionally selective cells in thick stripes, and orientation-selective cells in interstripes (Hubel and Livingstone, 1985; Shipp and Zeki, 1985; DeYoe and Van Essen, 1985, 1988; Hubel and Livingstone, 1987; Livingstone and Hubel, 1988; Felleman and Van Essen, 1991). Units that respond mainly to low-spatial-frequency gratings are found in CO-rich stripes, while units favoring high spatial frequencies tend to aggregate in CO-poor interstripes (Tootell and Hamilton, 1989). Ts'o *et al.* (1990) used high-resolution optical imaging techniques and found no indication of ocular dominance in V2. Instead, the left and right eye maps tend to overlap and to coincide with the thick and thin CO stripes. They also found larger iso-orientation modules with uniform orientation tuning in V2 than in V1.

These functional properties are molded by segregated inputs from V1. Thus, V1 puffs project to thin stripes, V1 interpuffs to interstripes, and layer 4B of V1 connects with thick stripes (Livingstone and Hubel, 1983, 1987a). The thin and interstripes, in turn, project to V4, and the thick stripes send fibers to MT (DeYoe and Van Essen, 1985; Shipp and Zeki, 1985). These selective connections form the anatomical basis of functionally distinct streams in the primate visual cortex.

5. Functional Streams in the Primate Visual Cortex

From the pioneering work of Hubel and Wiesel (1968, 1974, 1977) to studies of the past decade, the functional architecture of the primate visual cortex has been carefully scrutinized and its exquisite organization gradually elucidated. Sets of ODCs and orientation columns each form a hypercolumn, and hypercolumns, in turn, form functional modules in V1. CO contributes structural and metabolic identities to such modules by delineating vertical columns centered on CO-rich supragranular puffs (the "blob" system) from those that lie in register with the CO-poor supragranular interpuffs (the "interblob" system) (Horton, 1984; Carroll and Wong-Riley, 1984; Livingstone and Hubel, 1984a, 1988). These two major systems plus the channel that involves CO-poor layer 4B prove to have distinct yet interrelated functions as well as specific connections (Livingstone and Hubel, 1987a,b, 1988). Moreover, the CO-rich thin and thick stripes and the CO-poor interstripes of V2 have also been shown to have remarkable structural and functional characteristics as well as distinct connections (Tootell *et al.*, 1983; Livingstone and Hubel, 1983; Wong-Riley and Carroll, 1984a; DeYoe and Van Essen, 1985). These form the V1 and V2 links to the tripartite computational streams that have been postulated to process color, form, and position/motion/depth in a relatively independent, parallel fashion (Zeki, 1978; Van Essen and Maunsell, 1983; Lennie, 1984; Shipp and Zeki, 1985; Maunsell and Newsome, 1987; Livingstone and Hubel, 1984b, 1987b, 1988;

Figure 14. *(Continued).* (upper half) in layers 2 and 3, extending into the upper limit of layer 4. CO-reactive neurons are prominent within layers 3 and upper 4 (arrows). From Wong-Riley and Carroll (1984a).

SQUIRREL MONKEY

A

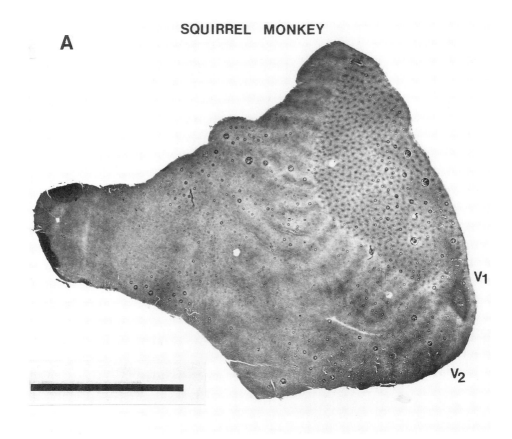

V1

V2

MACAQUE MONKEY

B

V2

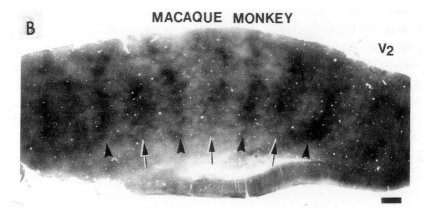

Figure 15. (A) Flat-mounted section from the lateral surface of squirrel monkey cortex treated to reveal CO. Anterior in the brain is toward the left, and dorsal is toward the top. The semicircular region on the right (containing an array of spots) is central V1; the adjoining region (showing parallel stripes) is central V2. Section is taken from layer 3. Bar = 1 cm. Section kindly provided by R. Tootell and is similar to Fig. 1 of Tootell *et al.* (1983). (B) Tangential section through macaque V2 reacted for CO. Note the CO-rich globular zones forming thick (arrows) and thin stripes (arrowheads). Bar = 1 mm. From Wong-Riley *et al.* (unpublished observations).

175

CYTOCHROME
OXIDASE
STUDY OF
PRIMATE
VISUAL
CORTEX

DeYoe and Van Essen, 1988; Zeki and Shipp, 1988; Felleman and Van Essen, 1991). Such parcellations had been suspected from anatomical physiological, and psychophysical data (Hubel and Wiesel, 1972; Lund *et al.*, 1975; Leventhal *et al.*, 1981; Fitzpatrick *et al.*, 1985; Livingstone and Hubel, 1988). Metabolic markers such as CO help to reveal the hidden functional channels in a striking, global manner. Thus, separate channels have been found to go through separate compartments defined by CO.

The three functional streams are known as P-B (parvocellular-blob), P-I (parvocellular-interblob), and M (magnocellular) streams (Maunsell and Newsome, 1987; DeYoe and Van Essen, 1988; Livingstone and Hubel, 1988; Zeki and Shipp, 1988). They are derived from the two major parallel channels: parvocellular and magnocellular, which differ at the LGN level in their color selectivity, contrast sensitivity, temporal resolution, and spatial resolution. Most parvocellular neurons are color opponent and have low contrast sensitivity, whereas magnocellular cells are largely non-color-selective and have higher luminance contrast sensitivity (Wiesel and Hubel, 1966; De Valois *et al.*, 1977; Schiller and Malpeli, 1978; Shapley *et al.*, 1981; Derrington and Lennie, 1984; Derrington *et al.*, 1984; Kaplan *et al.*, 1987; Hubel and Livingstone, 1990). The P-B stream probably contributes strongly to the processing of color, and involves type B retinal ganglion cells (P-like) and parvocellular geniculate (P) neurons, layer 4Cβ, 4A, 6, and puffs/blobs of V1, V2 thin stripes, and V4. This system has a lower acuity than the interblob system by a factor of 3 or 4 (Livingstone, 1990). However, since V1 puffs and V2 thin stripes are prominent in nocturnal primates with poor color vision, such as owl monkeys and galagos, the P-B stream is probably not involved exclusively in the hue discrimination. Moreover, puffs probably receive input from both the parvocellular and magnocellular pathways. The P-I stream involves type B retinal ganglion cells (P-like) and geniculate P neurons, 4Cβ, 4A, 6, interpuffs/interblobs of V1, and V2 interstripes, and is responsible mainly for high-resolution form perception. As discussed above, cells in interpuff regions are orientation-selective and most of them are complex cells. They respond to both luminance- and color-contrast borders, but not to color as such, and they are not wavelength or contrast-sign selective (Hubel and Livingstone, 1990). They also show less contrast sensitivity than 4B and 4Cα cells. However, some interpuff neurons have sensitivities intermediate between magno- and parvocellular geniculate cells, indicating a possible contribution from both systems (Hubel and Livingstone, 1990). Both P-B and P-I streams ultimately project to inferotemporal cortex (IT) and contribute to object recognition (the "what" of objects) (Mishkin *et al.*, 1983). The M stream is heavily involved in the processing of motion and stereoscopic depth, and encompasses type A retinal ganglion cells (M-like) and magnocellular geniculate (M) neurons, layers 4Cα, 4B, and 6 of V1, V2 thick stripes, middle temporal (MT), and posterior parietal (PP) areas. This system has low acuity, a very high sensitivity to luminance contrast, and fast temporal resolution, but is nonselective for color. Cells in 4Cα and 4B show high contrast sensitivity and their responses are markedly decreased at or near equiluminance of color-contrast stimuli (Blasdel and Fitzpatrick, 1984; Hawken *et al.*, 1987; Tootell *et al.*, 1988d; Hubel and Livingstone, 1990). Layer 4B, however, appears to be heterogeneous and segregated, with the disparity-tuned cells selectively located under interpuffs (Hubel and Livingstone, 1990). Cells in V2 thick stripes and MT show movement-direction selectivity or disparity selectivity, or both (Zeki, 1974; Maunsell and

Van Essen, 1983; Hubel and Livingstone, 1987). The M stream also contributes to spatial discrimination learning as well as visual tracking and attention (the "where" of objects) (Mishkin *et al.*, 1983).

Although these functional streams are dissociable under specific testing conditions, there are extensive anatomical and functional interconnections between them (reviewed in Felleman and Van Essen, 1991). Some strategies will be handled in parallel by more than one stream or will involve significant crossover between streams. The McCulloch effect strongly suggests an intimate relationship between orientation-selective cells and hue perception. The perception of three-dimensional form, for example, can be derived not only from object contours (involving orientation-selective cells; P-I stream), but from shading (shape-from-shading-brightness coding; P-B stream) and from motion (M-stream). Thus, the P-I stream is not uniquely involved in form perception (DeYoe and Van Essen, 1988) and normal perception involves simultaneous processing of color, orientation, direction, stereopsis, and spatial resolution.

By the same token, neurons in different CO compartments may share similar functional properties. Indeed, similar morphological types of neurons are found in puffs and interpuffs (see Sections 3.1.1d to 3.1.1f) except that puff neurons tend to be more CO-reactive than interpuff neurons. While CO compartments often are related to functional streams, there are examples for non-correspondence between physiological properties and CO compartments (DeYoe and Van Essen, 1985). The correlation between CO pattern and functional subdivisions may also be more relevant during earlier stages of visual cortical processing, such as V1 and V2, than in later stages, when more extensive integration of functions and pathways occur. These regions, including MT, VP/VA, DX, and DL in the New World primate and extrastriate areas in the human, exhibit topographic inhomogeneities with CO (Tootell *et al.*, 1985; Cusick and Kaas, 1988; Wong-Riley *et al.*, 1993a), but the relationship between CO and functional attributes has not been established. The visual system of higher primates contains as many as 32 visual areas with more than 300 complex interconnections involving both feedforward and feedback projections (Felleman and Van Essen, 1991). Much more work is needed to establish the actual relationship between CO-defined compartments and visual functions in these areas.

6. Metabolic Plasticity in the Adult Primate Visual Cortex

The exquisite parcellation of left and right eye inputs to the striate cortex provides a fertile ground for experimental manipulations. It permits direct comparison between deprived and nondeprived neural elements within the same animal. One of the classical experiments was to remove one eye in adult monkeys and observe the shrinkage of deafferented lateral geniculate neurons (Matthews *et al.*, 1960). However, changes in the visual cortex were either unremarkable or showed only faint bands of gliosis (Hazeltine *et al.*, 1979; Hendrickson and Tigges, 1985). The assumption was that mature cortical neurons had become refractory to change if the cortex and its geniculate afferents were not physically damaged. This assumption was challenged by the advent of 2-DG autoradiography and CO histochemistry. Functional deprivation leads presumably to decreased neuronal activity that, in turn, reduces energy demand and the level of metabolism in the affected neurons. Various types of manipulations have been tried, including monocular enucleation, lid suture, and retinal impulse blockade; all have induced a downregulation of CO in the affected ODCs of adult monkeys (Horton and Hubel, 1981; Horton, 1984; Wong-Riley and Carroll, 1984b; Hendrickson and Tigges, 1985; Hendry and Jones, 1986; Wong-Riley et al., 1989a,b; Trusk et al., 1990). Such functionally induced metabolic plasticity can be reliably demonstrated by CO histochemistry or immunohisto-chemistry. Only those paradigms involving the adult visual system will be reviewed here, as plasticity during development can be readily demonstrated with more classical techniques.

6.1. Monocular Lid Suture

This is the mildest form of visual deprivation tested in the adult, since the lids only diminish light input by 1–4 log units (Crawford and Marc, 1976), and the loss is mainly in pattern discrimination of high spatial frequencies. Retinal ganglion cells retain their spontaneous activity and lateral geniculate neurons can still respond to diffuse light through a lid sutured for at least 6 months (Horton, 1984). There is neither detectable neuronal atrophy nor changes in the level of several enzymes, including CO, AChE, LDH, SDH, citrate dehydrogenase, and NADH diaphorase in the LGN of adult monkeys (Cotlier et al., 1965; Horton, 1984). However, some faint banding pattern is detectable with CO histochemistry within layer 4 of V1 (Horton, 1984; Trusk et al., 1990). The bands are clearer in juveniles lid-sutured for short term (11 weeks) than in adults deprived for the same period of time (compare Fig. 16A with 16B,C). The effects become more severe with longer periods of deprivation (1–3 years) (Fig. 16D,E). However, there is a great deal of interanimal variation with long-term lid suture, suggesting that either the thickness of the lids varies among animals, or that some functional recovery may occur in some animals, or both (Trusk et al., 1990). There is no indication of sprouting or contraction of geniculate afferents, as transneuronal transport of label showed normal-sized columns in deprived animals (LeVay et al., 1980; Horton, 1984); yet, the paler bands corresponding to the deprived eye appear slightly wider than the darker bands, similar to the wide pale bands seen with the Liesegang stain (LeVay et al., 1980). Thus, monocular lid suture probably affects CO activity along border zones of nondeprived columns, where afferent segregation may not be as complete as the cores of columns. While the cores are strictly monocular, the border zone neurons may extend some of their dendrites into the neighboring column and a decrease in those inputs may cause a decrease in CO activity within postsynaptic dendrites, even though the presynaptic terminals may stay within their boundaries. This situation is quite different from those of monocular enucleation or TTX, as we shall see below.

Above 4C, alternate rows of puffs show lighter CO staining and these rows are centered on the light CO columns in layer 4 (Horton, 1984; Trusk et al., 1990). Puff volume is significantly reduced after 11 weeks of monocular lid suture ($p < 0.001$), and, as expected, the decrease is greater in the juvenile than it is in the adult (38% shrinkage versus 33%) (Trusk et al., 1990). A reduction in puff volume does not imply a shrinkage of cortical tissue, but simply a decrease in the volume that is CO-reactive. Long-term lid suture (48 weeks to 2.5 years)

177

CYTOCHROME
OXIDASE
STUDY OF
PRIMATE
VISUAL
CORTEX

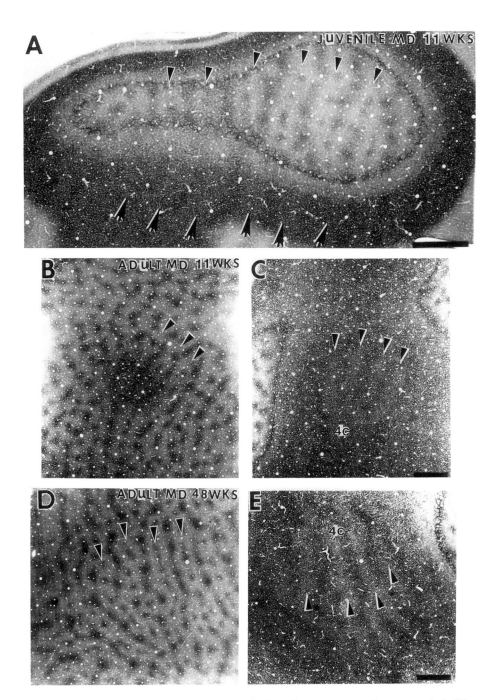

Figure 16. (A) Tangential section through V1 of a juvenile macaque monkey monocularly lid sutured (MD) for 11 weeks. Note the alternating rows of normal puffs and shrunken puffs (small arrowheads) and the faint ocular dominance bands in layer 4C (large arrowheads) revealed by CO histochemistry. The effect of 11 weeks of monocular lid suture is much milder in the adult V1, where pale puffs are difficult to discern (arrowheads in B) and pale bands in layer 4C are relatively faint (arrowheads in C). Long-term MD can lead to more distinct banding pattern in supragranular puffs (arrowheads in D) and ocular dominance columns in 4C (arrowheads in E). However, there are substantial interanimal differences in their response to long-term MD. Bars = 1 mm. Unpublished observations of Trusk and Wong-Riley, comparable to findings of Trusk *et al.* (1990).

reduces puff volume up to 53% ($p < 0.001$), although there is a great deal of interanimal variability (Trusk *et al.*, 1990). In all cases, the reduction in the upper portion of puffs (layers 2 and 3A) is greater than that in layer 3B, which does not commence until after 48 weeks or more. Nevertheless, the cross-sectional area of deprived puffs is significantly smaller than that of nondeprived puffs at all depths of puffs in all monkeys examined.

Below 4C, light and dark patchy staining is present in layers 5 and 6, indicating that deprivation affects cells through the entire extent of deprived columns. In most cases examined, the effect of lid suture appears more pronounced in the supra- and infragranular layers than it is in layer 4 (Horton, 1984; Trusk *et al.*, 1990).

6.2. Monocular Enucleation

The removal of one eye deprives the system of light input, afferent neuronal activity, as well as possible neurotropic factors. The induction of transneuronal degeneration in the mature postsynaptic LGN neurons is well known (Matthews *et al.*, 1960). Cellular changes in the striate cortex are more subtle, but bands of increased cellular density signifying gliosis can be demonstrated in layer 4C of Nissl-stained sections (Hazeltine *et al.*, 1979; Hendrickson and Tigges, 1985).

A much more dramatic effect is shown by CO histochemistry. The metabolic integrity of deprived-eye columns is severely affected and is demonstrable in the cat, monkey, and man (Wong-Riley, 1979b; Horton and Hubel, 1981; Horton, 1984; Horton and Hedley-Whyte, 1984; Hess and Edwards, 1987; Trusk *et al.*, 1990). A striped pattern of low and high CO staining is evident in layer 4C as well as in the supra- and infragranular layers (Fig. 17A,C). Within 4C, the effect is more prominent in 4Cβ than in 4Cα, so that along the enucleated eye columns 4Cβ appears paler than 4Cα (Horton, 1984). The reasons for the milder effect on 4Cα are not entirely clear, but magnocellular geniculate neurons may be more resistant to transneuronal shrinkage (Matthews *et al.*, 1960), and their terminals may be less segregated in 4Cα than parvocellular input to 4Cβ (Horton, 1984). The light labeling in 4Cβ also helps to highlight the thin band at the base of 4C (4Cβ dark) (Hiltgen and Wong-Riley, 1986), which appears to "resist" the effect of enucleation more than the rest of 4Cβ. Although this band may represent a binocular subregion, existing autoradiographic or lesion studies do not indicate any mixing of right and left eye afferents at the base of 4C (Horton, 1984). Another more likely possibility is that this band is normally darker than the rest of 4Cβ and an equivalent decrease in enzyme activity in both sublayers would still result in an uneven staining intensity.

The combination of functional impairment and cellular atrophy probably contributes to the severe reduction in energy demand of cortical neurons and, hence, in their level of CO activity. In sharp contrast to monocular lid suture, the darkly stained columns in layer 4Cβ are significantly wider than the lightly stained ones in monkeys surviving monocular enucleation for 2, 16.5, 28.5, 56, and 60 weeks (Fig. 17D) (Trusk *et al.*, 1990). Again, there is no sign of expansion or shrinkage of geniculocortical afferents in layer 4C (LeVay *et al.*, 1980). This strongly suggests that primary afferent denervation sends a strong signal to deprived cortical neurons along border zones to strengthen their interactions

179

CYTOCHROME
OXIDASE
STUDY OF
PRIMATE
VISUAL
CORTEX

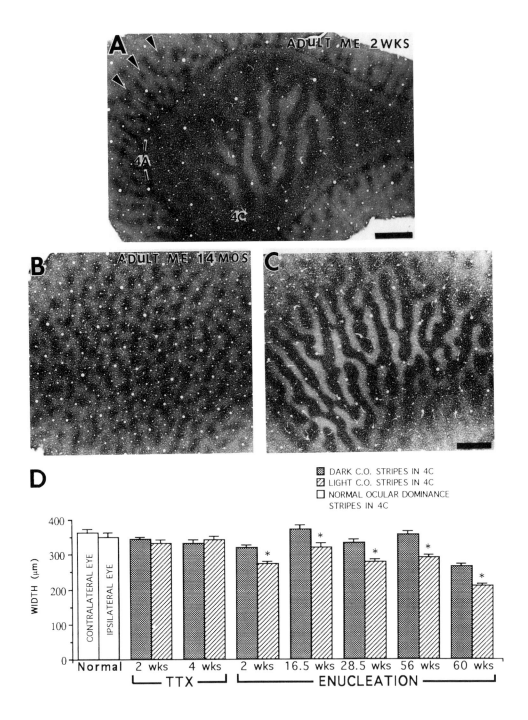

Figure 17. Tangential sections through V1 of adult macaque monkeys monocularly enucleated (ME) for 2 weeks (A) or 14 months (B and C). Alternate dark and light CO bands can be seen in layer 4C (A and C). Changes in every other row of puffs are more discernible after 2 weeks (A) than after 14 months (B). Presumably, long-term ME induces a severe shrinkage of puffs so that the cortex appears to have been taken over by puffs subserving the remaining eye. (D) The mean (± SE) widths of CO-

181

**CYTOCHROME
OXIDASE
STUDY OF
PRIMATE
VISUAL
CORTEX**

with the neighboring nondeprived eye columns, perhaps through dendrites that have already existed in those columns but were previously functionally dormant, or through terminal dendritic branchings that have invaded the columns after the deprivation.

In the supragranular layers, alternating rows of puffs become paler and smaller than their counterparts (Fig. 17A), and they lie in register to pale ODCs in 4C. Likewise, puffs are paler in the representations of the optic disk area (Horton, 1984). Within the intact eye columns, puffs remain darkly reactive and appear more confluent (Horton, 1984). This could either mean that the puffs are expanding or that the adjacent interpuffs have increased their CO activity, perhaps as a result of competitive advantage over the deprived interpuffs, as interpuff neurons tend to be binocular yet dominated by a single eye.

Changes were quantified by Trusk *et al.* (1990). Deprived puffs suffer a considerable shrinkage in their cross-sectional areas, with a greater reduction in layer 3B than in layers 2 and 3A, a reverse of the situation in lid suture (see above). Puff volumes are likewise reduced by 34% after 2 weeks and by 53% after 60 weeks of enucleation ($p < 0.001$ for both). The decrease is so severe after long-term enucleation that deprived puffs often become inconspicuous, while puffs corresponding to the intact eye appear larger and more confluent (Fig. 17B). The distance between adjacent rows of puffs in layer 3B is consistently equal to the width of dark CO-stained columns in 4B, and significantly greater than the widths of lightly stained columns. These findings indicate that long-term enucleation leads to actual shrinkage of cortical tissues previously driven by the removed eye.

The pattern of light and dark banding extends from layer 4C to layer 4A above and to layers 5 and 6 below (Horton, 1984). The effect in the cortex rapidly follows that in the LGN and involves mainly postsynaptic dendrites and cell bodies whose energy demands are markedly curtailed by the removal of a potent excitatory drive. Long-term enucleation (28.5 and 60 weeks) also induces a transneuronal shrinkage of CO-reactive cells in deprived puffs. By comparison, interpuff neurons in deprived interpuffs show little change in size (Trusk *et al.*, 1990). The milder effect on deprived interpuff neurons could be related to several factors. First, interpuff neurons tend to be metabolically less active and are therefore less vulnerable to functional deprivation. Second, interpuff neurons may normally be under stronger inhibitory influence and may be released from some of these influences after enucleation. Third, since interpuff neurons tend to be binocular, some of the activity may be sustained by input from the intact eye.

Figure 17. (*Continued*) reacted ocular dominance columns (ODCs) in layer 4Cβ are shown for monkeys surviving enucleation for varying periods and for monkeys monocularly injected with TTX for 2 or 4 weeks. ODCs in the normal monkey were measured in autoradiograms following transneuronal transport of [³H]proline from the contralateral eye. In the normal and TTX-treated monkeys, ODCs from each eye are of equal width in layer 4Cβ. The widths of ODCs served by the enucleated eye (striped bars) are significantly lower than those of ODCs served by the spared eye (shaded bars) in each of the enucleated monkeys (*$p < 0.01$). A–C, unpublished observations of Trusk and Wong-Riley, comparable to findings of Trusk *et al.* (1990). Bars = 1 mm. D, Modified from Trusk *et al.* (1990).

6.3. Monocular Retinal Impulse Blockage with TTX

Intravitreal injections of tetrodotoxin, a specific blocker of voltage-dependent sodium channels, provide several advantages over enucleation. First, within a tested range of dosage, TTX does not cause cell death or interfere with such cellular functions as fast axoplasmic transport (Hille, 1968; Wong-Riley and Riley, 1983). Second, the vitreous body serves as a natural reservoir for prolonged, slow, and safe release of TTX. Thus, a single injection of 19 μg of TTX in 10 μl of sterile water effectively blocks retinal impulse activity for at least 4 days in adult monkeys (Wong-Riley and Carroll, 1984b; Wong-Riley *et al.,* 1989a,b). Third, the efficacy of the drug can be readily assessed by visually monitoring the disappearance and return of the pupillary light reflex. Fourth, the eyelids remain open so that the natural stimulus, light, can still enter both the injected and noninjected eye. Fifth, the process is reversible so that there is no permanent damage to the system, and the return of enzymatic activity can denote the return of neuronal functional activity (Wong-Riley and Riley, 1983; Wong-Riley *et al.,* 1989a). Finally, the removal of excitatory inputs by TTX results in decreased metabolic demand, CO levels, and average spike rates that can be analyzed and compared experimentally (Trusk *et al.,* 1992) without the confounding effect of denervation.

In the primate, monocular retinal impulse blockade induces a drastic reduction in cortical activity in affected ODCs (Trusk *et al.,* 1992) followed by a dramatic decrease in their CO levels (Wong-Riley and Carroll, 1984b; Wong-Riley *et al.,* 1989a,b, 1992). As early as 14 hr after TTX injection, pale ocular dominance bands are faintly visible in layer 4C, while puffs and infragranular blobs remain relatively normal in appearance (Fig. 18). With longer periods of inactivation, the bands in 4C and alternating rows of pale shrunken puffs become increasingly more prominent (Fig. 19). The decrease in CO activity is invariably accompanied by parallel changes in CO protein content, and is most prominent in layer 4C, followed by layers 2/3 puffs, 4A, 5, and 6 (Figs. 1E,F, 4C,D, 19A–D). Unlike the case of monocular enucleation, the widths of dark and light ODCs in layer 4C are equal and are comparable to those in normal monkeys (Fig. 17D and Trusk *et al.,* 1990). This suggests that without denervation, the cessation of retinal impulse alone is not sufficient to cause synaptic dominance by the nondeprived eye along border zones of deprived columns. In terms of CO activity as measured by optical densitometry, there is a significant reduction in enzyme levels in all of the layers affected by TTX. Figure 20 illustrates the values in granular and infragranular layers under normal conditions (Fig. 20A) and the percentage change from control after 1, 2, and 4 weeks of monocular TTX treatment (Fig. 20B). All changes are statistically significant ($p < 0.01$) except for those in layers 4B, 5, and lower 6 after 2 weeks of TTX. The reduction is most severe in 4Cβ proper, where it approaches 25% after 4 weeks of TTX. Layers 4Cβ dark and 4A are more severely affected than 4Cα, 5, and upper 6. Layers 4B and lower 6 are the least affected (Carroll and Wong-Riley, unpublished observations).

Two weeks of TTX induce a 17% reduction in the volume of deprived puffs, a loss that is less severe than that of enucleation but nonetheless significant ($p < 0.001$) (Trusk *et al.,* 1990). The decrease is mainly in layers 3A and upper 3B. Extending the TTX treatment to 4 weeks leads to a drastic 57% reduction in the volume of CO-reactive puffs ($p < 0.001$) that is most prominent in layer 3B.

This result is very similar to that of 60-week enucleation, suggesting that a plateau level may be reached after 1 month of severe interruption of retinal input. However, unlike enucleation, there is no evidence of cortical tissue shrinkage, but merely a decrease of CO activity that can be reversed if functional recovery should occur (see Section 6.4).

183

CYTOCHROME
OXIDASE
STUDY OF
PRIMATE
VISUAL
CORTEX

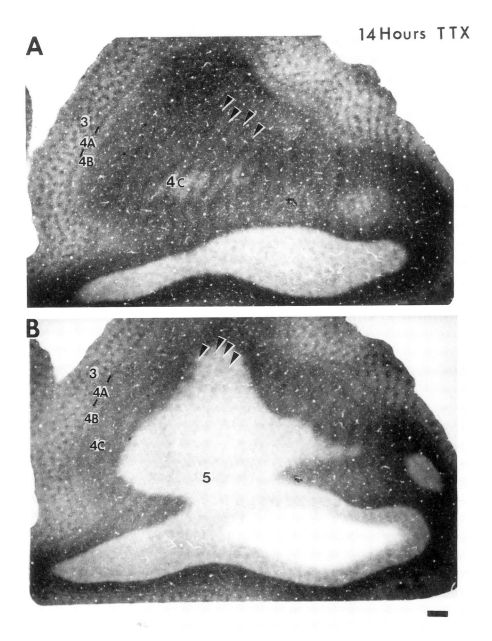

Figure 18. As early as 14 hr after monocular TTX injection, pale ocular dominance bands are faintly visible in layer 4C of the macaque striate cortex (arrowheads in A). However, CO patterns in the other layers, such as puffs in layer 3 (A and B) and blobs in layer 5 (arrowheads in B), are relatively normal. Bar = 1 mm. Unpublished observations of Trusk, Wong-Riley, and DeYoe (1993).

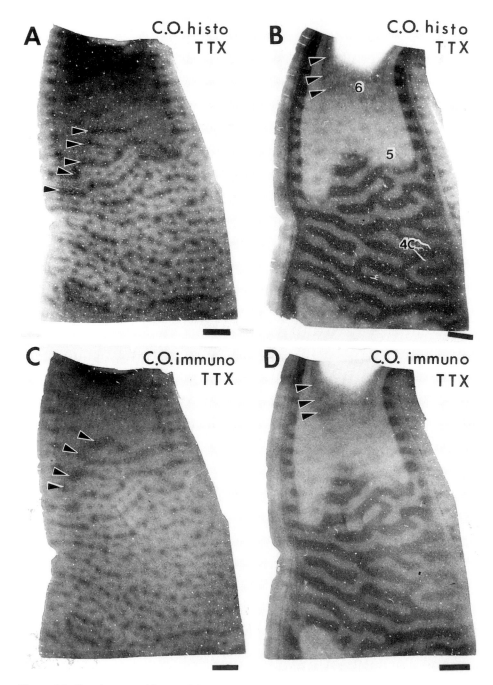

Figure 19. Cytochrome oxidase activity (A and B) and amount (C and D) in striate cortex of macaque monkeys monocularly treated with TTX for 3 weeks. Note the alternating rows of small, pale puffs and large, dark puffs, visible in adjacent sections reacted histochemically (A) and immunohistochemically (C). Dark and pale ocular dominance columns are also clearly visible in layer 4C (B and D), as well as in 4A (arrowheads in A and C) and 6 (arrowheads in B and D). Bars = 1 mm. From Hevner and Wong-Riley (1990).

185

CYTOCHROME
OXIDASE
STUDY OF
PRIMATE
VISUAL
CORTEX

As discussed above, not all mature neurons respond alike to the same functional insult (Figs. 7B, 9, and 11; Table I). Those with the highest baseline level of CO activity are invariably the most severely affected by chronic intravitreal TTX treatment (Wong-Riley *et al.*, 1989a). In cortical puffs, the most vulnerable neurons are the reactive nonpyramidals or type C cells, whose rich supply of darkly reactive mitochondria is transformed by TTX into a reduced collection of mostly moderate and lightly reactive ones (compare Figs. 10 and 11). In addition to a dramatic decrease in the numerical and areal densities of mitochondria, type C cells are the only neurons in puffs that show a significant reduction in mean cell size after 2 or 4 weeks of TTX (Table I). Their vulnerability is not related to size, since type C cells are intermediate in size between types A and B (Table I), nor is it related to neurotransmitter type *per se,* as neurons with the same transmitter type can have decidedly different levels of metabolic capacity (Luo *et al.*, 1989; see also Section 3.1.1d). Rather, type C cells receive both asymmetric and symmetric axosomatic synapses; thus, direct depolarizing input normally drives energy production upward in the cell bodies, but weakening of such input by TTX substantially reduces energy demand, resulting in a downregulation of CO.

The moderately reactive type B pyramidal cells are less affected than type C neurons. Even though the packing density of their mitochondria is severely reduced after 2 weeks of TTX and remain low after 4 weeks of TTX, their

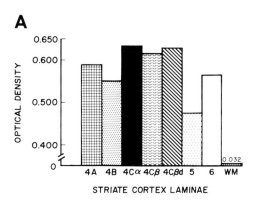

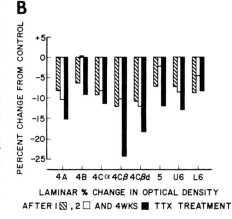

Figure 20. Optical density values of various granular and infragranular layers in normal macaque striate cortex (A; see also Fig. 2). Each value represents the mean of at least 25 readings, and the standard deviations are too minute to plot. Values in layer 4C are significantly higher than those in 4B, 5, and 6 ($p < 0.01$) but only slightly higher than those in 4A. The percentage change from control after 1, 2, and 4 weeks of monocular TTX treatment is shown in B. The reduction is most severe in 4Cβ proper, followed by 4Cβ dark and 4A; it is less severe in 4Cα and the other layers. All reductions are significant ($p < 0.01$) except for those in 4B, 5, and lower 6 (L6) after 2 weeks of TTX. These layers are normally low in CO and changes there may not be detectable in the animals examined. Carroll and Wong-Riley (unpublished observations).

shrinkage in cell size is statistically not significant (compare Figs. 8 and 9; Table I). Unlike type C cells, pyramidal cells receive only symmetric (putatively inhibitory) synapses on their somata, and their reduced energy demand after TTX treatment probably results more from a decrease in depolarizing input on their dendrites than a possible suppression of inhibitory synapses on their cell bodes, since the latter should lead to disinhibition.

The least affected cell type is the relatively abundant but metabolically least active type A neurons (compare Figs. 7A and 7B). There is virtually no change in the size of these neurons even after 4 weeks of TTX. However, there is a significant decrease in the numerical and areal densities of their mitochondria after 2 weeks of TTX.

In all three metabolic cell types, there is a dramatic reduction in the numerical density of their mitochondria after 2 weeks of TTX and a return to control levels following 4 weeks of TTX (Table I). This implies that chronically quiescent neurons are able to regain a "set point" established under normal conditions for the ratio of mitochondrial number to tissue volume. However, deprived neurons are unable to regain normal levels of mitochondrial volume density, because many mitochondria are converted to the smaller, less-reactive variety.

Recent evidence indicates that types B and C in deprived interpuffs are also more adversely affected by monocular TTX than type A neurons (Wong-Riley *et al.*, 1992). By the same token, afferent blockade is more detrimental to processes of higher metabolic activity in the neuropil, such as dendrites and terminals that form symmetric synapses, than to processes of lower CO activity (Wong-Riley *et al.*, 1989b, 1993b). In general, then, metabolically most active neurons and neuronal processes are also metabolically most plastic. They maintain a high level of oxidative capacity to meet their active functional needs, and they are able to rapidly adjust their local enzyme levels if their energy demand is drastically reduced.

Another significant aspect of striate cortical plasticity is that nondeprived interpuffs respond differently from nondeprived puffs. The difference appears to be related to the fact that puffs are largely monocular while interpuffs are mainly binocular (Livingstone and Hubel, 1984a; Ts'o and Gilbert, 1988). While cells in nondeprived puffs are not different from their normal controls (Wong-Riley *et al.*, 1989a), cells in nondeprived interpuffs appear to be operating at a competitive advantage by gaining in size and in the absolute number of mitochondria. Type A cells also show increased levels of CO. This is offset, however, by a net loss of darkly reactive mitochondria in type B and C cells, indicating that these cells may be compromised somewhat by a loss of excitatory drive from the nondominant, TTX-inactivated eye (Wong-Riley *et al.*, unpublished observations). The loss is especially prominent during the acute phase of TTX blockade, when ongoing spiking activity in both left- and right-eye dominated interpuffs decreases up to threefold in less than 1 hr (Trusk *et al.*, 1992).

In addition to metabolic plasticity, the mature visual cortex is also capable of ultrastructural and synaptic reorganization following functional inactivation (Wong-Riley *et al.*, 1989b). For example, quantitative EM analyses suggest that a subpopulation of excitatory afferents that are silenced by TTX may undergo retraction, while other intrinsic inhibitory terminals may remain active and sprout collaterals to replace the receding terminals. At the same time, dendrites deprived of their normal rates of excitatory input may shrink in size and some of their spines may be resorbed (Fig. 21). These minute adjustments indicate that

187

CYTOCHROME
OXIDASE
STUDY OF
PRIMATE
VISUAL
CORTEX

synaptic competition, initiated during development, persists to a milder degree well into adulthood, and point to the dynamic ability of mature neurons to respond to changing functional demands.

6.4. TTX Recovery

A distinct advantage of the TTX paradigm is that the detrimental effect of retinal impulse blockade is largely, if not entirely, reversible (Wong-Riley and Riley, 1983; Carroll and Wong-Riley, 1987b; Wong-Riley *et al.*, 1989a). The pupillary light reflex returns within 1 week, but CO activity requires a much longer time period for full recovery. Figure 22 illustrates the detrimental effect of TTX (Fig. 22A) followed by the return of a virtually normal pattern of CO labeling in layer 4C and puffs (Fig. 22B); likewise, optical density measurements of central to peripheral portions (zones 1, 2, and 3) of puffs also return to control values (Fig. 22C), and the size of puffs eventually attains control values as well. As shown in Fig. 22D, the longer the interval of inactivation, the longer it takes for

SYNAPTIC REMODELING

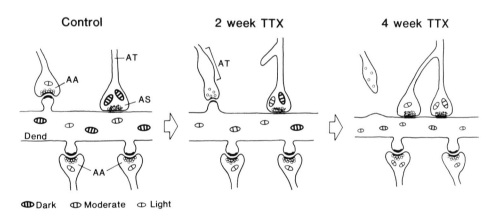

Figure 21. A conceptual diagram of one hypothesized model for synaptic remodeling in cortical puffs after retinal impulse blockade. In the left panel, there is normally a 3 to 1 ratio of asymmetrical (AA) to symmetrical (AS) synapses on dendrites (Dend) and dendritic spines. After 2 weeks of TTX (middle panel), a subpopulation of AA terminals, possibly the major afferents that have been silenced the most, may initiate retraction. At the same time, some AS terminals that remain active may respond by sending out collaterals to replace the receding AA terminal. This reorganization would explain the increase in size and areal density of axon trunks (which include the swollen, retracted portion of preterminal axon, AT). Following 2 additional weeks of TTX (right panel), a subpopulation of AA terminals has retracted completely from the synaptic site, causing resorption of the postsynaptic spines, while other AA terminals may have grown back. The AA terminal to the upper left of this panel represents one that may be retracting or advancing. Some of these newly formed terminals may establish synapses (not shown). Simultaneously, collaterals of AS have sprouted and formed new synapses on the dendrites. These events could explain the increase in numerical densities of both AS and AT profiles as well as the rise in symmetrical synapses coupled with a fall in asymmetrical synapses in that region. Presumably, the mean sizes of dendrite and AS decrease in response to reduced impulse activity. Likewise, mitochondria within these profiles progressively shift from predominantly darkly reactive to mainly moderately and lightly reactive types. From Wong-Riley *et al.* (1989b).

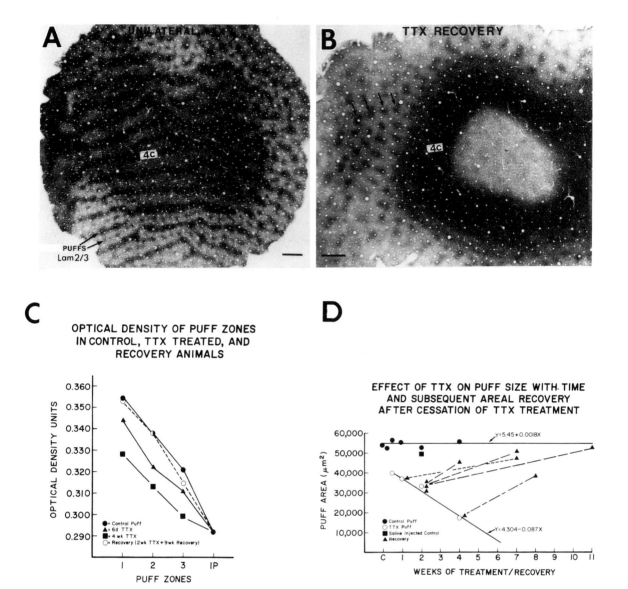

Figure 22. (A) Tangential section through the striate cortex of an adult macaque that had received 4 weeks of monocular TTX injections. Alternating bands of high and low CO activity are present in layer 4C and alternating dark and pale puffs (arrows) occur in supragranular layers. (B) Tangential cortical section from an adult monkey allowed to recover for 6 weeks following 1 week of TTX administration into one eye. The staining pattern is indistinguishable from that of normal monkeys, indicating complete reversal of the TTX effect and metabolic recovery of cortical neurons. Complete recovery is also observed following longer TTX treatment. Bars = 0.5 mm. (C) Optical density values of puffs and interpuffs (IP) in striate cortex of normal macaque monkeys and of monkeys monocularly injected with TTX for 6 days, 4 weeks, or for 2 weeks followed by a 9-week recovery. Note that the values normally decrease from the central (zone 1) to the peripheral (zone 3) portions of puffs and are lowest in the interpuffs for all animals. There is a parallel reduction in optical densities in all three zones with TTX, so that by the 4th week the value in zone 3 is comparable to that of interpuffs, and the puffs appear optically to be severely "shrunken" in size. Changes for 2 and 4 weeks of TTX are significant ($p < 0.01$) while values for the recovered animal are comparable to those of the control. (D) The time course of areal recovery of puffs from TTX treatment is plotted in this graph. The sizes of control puffs are taken from three groups of monkeys: untreated normals (black circle at "C"), saline injected for 2 weeks (black square), or TTX-injected (remaining black circles). There is no significant difference between sizes of all the control puffs (straight line at the

189

CYTOCHROME
OXIDASE
STUDY OF
PRIMATE
VISUAL
CORTEX

full metabolic recovery to occur. Presumably, new enzyme molecules are synthesized and assembled into functional holoenzyme in the mitochondria, and new synapses are established and/or strengthened. Recovery may also involve an expanded influence of the nondeprived eye, which has been reported to exhibit supernormal vernier acuity in humans after long-term monocular deficits (Freeman and Bradley, 1980; see also Moran and Gordon, 1982). On the other hand, recovery may simply reflect a reconsolidation of input from the previously inactivated eye.

6.5 Laser Retinal Lesions

In monocularly deprived striate cortex, the level of CO activity often appears to be elevated in nondeprived interpuffs such that they form more prominent bridges between puffs. Such an "increase" in CO levels is difficult to substantiate when the baseline metabolic level of the animal in question is unknown.

One paradigm that can provide an internal basis of comparison is to make focal retinal lesions with a blue-green argon laser and compare the laser-affected cortical areas with remote, non-laser-affected zones within the same animal (Trusk and Wong-Riley, 1990; Trusk et al., 1993). Such lesions destroyed the photoreceptors but left the ganglion cells intact. Surprisingly, even though all ganglion cells at the lesioned sites retained virtually normal CO levels, a dramatic dark and pale CO banding pattern was seen from supra- to infragranular layers in the relevant areas of the striate cortex, with distant unaffected areas serving as an internal control (Fig. 23). Optical densitometric measurements verified the significant reduction in CO levels within deprived puffs and layer $4C\beta$ as compared with unaffected control regions of the same animal (Fig. 24A,C). Puffs and layer $4C\beta$ in the companion nondeprived rows had density values close to those of the control. On the other hand, deprived interpuffs showed little difference from the control while nondeprived interpuffs within the lesioned sites upregulated their CO above control levels (Fig. 24B). A transitional area had values intermediate between lesioned and normal regions. Thus, substantial metabolic plasticity occurs not only in cortical areas deprived of their normal afferent input, but also in interpuff regions.

The laser paradigm also reveals that the removal of photoreceptors is as detrimental to the cortex as removal of the entire eye. These findings could have significant clinical relevance, as focal laser retinal burns can result from therapeutic procedures as well as accidental exposures.

Figure 22. (*Continued*) top). Values for puffs affected by 3 days, 1, 2, and 4 weeks of TTX are shown as open circles. Another group of animals were tested for the time course of recovery and received varying periods of TTX treatment followed by varying periods of recovery (1 week TTX/6 weeks recovery; 2 weeks/2 weeks; 2 weeks/5 weeks; 2 weeks/9 weeks; and 4 weeks/4 weeks). The recovery in puff size (black triangles) is longer with longer periods of treatment, but reaches normal values with the 2-week TTX/9-week recovery paradigm. A and B, from Wong-Riley (1989) and Wong-Riley et al. (1989a); C and D, unpublished observations of Carroll and Wong-Riley.

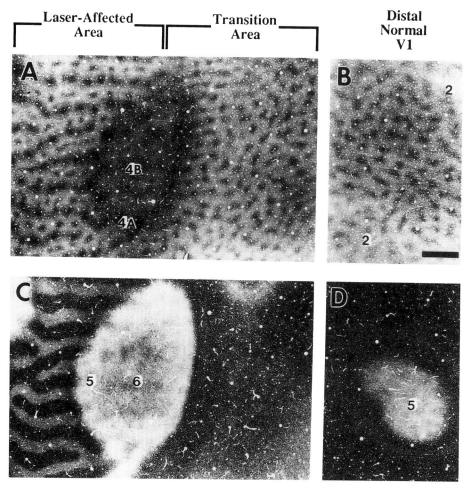

Figure 23. The effect of focal laser (600-μm spot size and 150- to 400-mW blue-green argon laser) retinal lesions (centered on the vertical midline above but not including the fovea) on the macaque striate cortex is shown in tangential sections reacted for CO. Within the laser-affected region, CO activity is lower in alternating rows of layer 3 puffs (left half of A) that are in register with lightly reacted ocular dominance stripes in layer 4C (left half of C). In cortical regions serving undamaged retina surrounding the lesion (transition area), puffs have no apparent banding pattern (right half of A), and CO reactivity in layer 4C is homogeneous (right half of C) but it is at a level lower than that of distal normal V1 (D) and the darkly reactive stripes of the laser-affected area (left half of C). The normal patterns of CO reactivity in puffs (B) and 4C (D) are seen in an isoeccentric region approximately 10 mm away. Note that interpuffs within the darkly stained puff rows of the laser-affected region (left half of A) have a higher level of CO activity than interpuffs in transitional (right half of A) and distal normal layer 3 (B) (see also Fig. 24). Bar = 1 mm. From Trusk and Wong-Riley (1990) and Trusk *et al.* (1993).

191

CYTOCHROME
OXIDASE
STUDY OF
PRIMATE
VISUAL
CORTEX

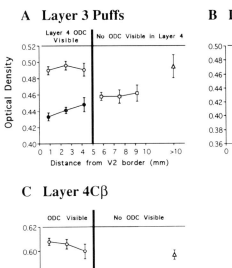
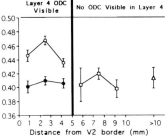

A Layer 3 Puffs

B Layer 3 Interpuff

C Layer 4Cβ

- Laser Eye ODC
- Normal Eye ODC
- Transitional Area
- Distal Normal V1

Figure 24. Optical densities of CO reaction product in puffs, interpuffs, and layer 4Cβ within laser-affected, transitional, and normal striate cortex (same cases as in Fig. 23). Values are the average optical densities (± SE) of 10–20 sites. (A) CO reactivity is greatest in puffs above darkly stained ocular dominance columns (ODC) in layer 4 (corresponding to regions of the untreated eye homonymous to lesioned retina) and in puffs more than 10 mm away from the laser-affected region. CO reactivity is significantly reduced in puffs above lightly stained ODC in layer 4 (associated with lesioned retina), and in puffs within cortex adjacent to the laser-affected region. No ODCs are visible in layer 4 of the transitional region. (B) CO reactivity is at equivalent levels within interpuffs of distal normal cortex, transitional cortex, and in interpuffs above lightly stained ODC. However, density is significantly greater in interpuffs above darkly reactive ODC, suggesting that they have increased their CO activity above that of the normal. (C) CO reactivity is equally high in distal normal cortex and in darkly reactive ODCs of the laser-affected region; however, it is significantly reduced throughout transitional layer 4Cβ and in stripes associated with lesioned portions of the retina. From Trusk and Wong-Riley (1990) and Trusk *et al.* (1993).

7. Regulation of Cytochrome Oxidase by Neuronal Activity

It is clear from studies performed in the past decade that the level of CO activity is very much controlled by the level of neuronal activity (see above). CO activity, in turn, largely reflects local adjustment of CO protein amount (Hevner and Wong-Riley, 1989, 1990). Furthermore, neuronal activity plays an important role in the regulation of CO subunit mRNA levels as well as the copy number of mitochondrial DNA, which encodes the largest 3 of the 13 subunits of CO (Hevner and Wong-Riley, 1991). As shown in Fig. 25, the mRNA level of one of the major subunits of CO (CO I) is drastically reduced within the same ODCs as those that exhibit decreased levels of CO activity (compare A, C, and E with B, D, and F). Thus, changing functional activities of neurons impose local adjustment

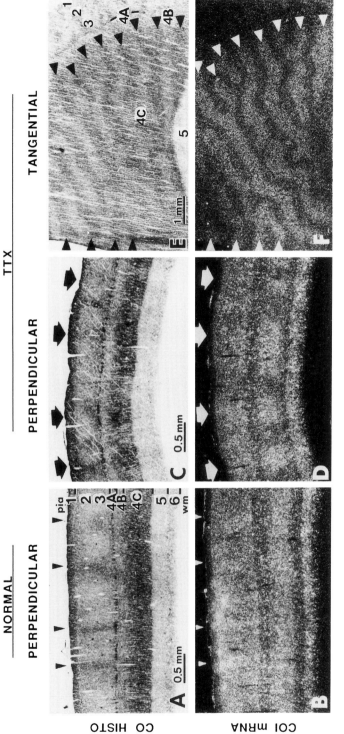

Figure 25. Regulation of mRNA encoding CO subunit I (COI) by functional activity in the macaque visual system. In normal V1, CO activity (A) and COI mRNA levels (B) are essentially constant within a cortical layer, except in layers 2–3 where the puffs (arrowheads in A and B) contain more CO activity and COI mRNA than the interpuff regions. In cortex from animals treated monocularly with TTX for 7 days (C and D, cross sections) or 3 days (E and F, tangential sections), CO activity (C, E) and COI mRNA levels (D, F) were normal in ocular dominance columns corresponding to the untreated eye (arrows in C and D, arrowheads in E and F) and decreased in columns related to the treated eye. Bright-field optics were used for A, C, and E, and dark-field optics for B, D, and F. Pia, pia mater; wm, white matter. From Hevner and Wong-Riley (1991).

of energy production and consequent up- or downregulation of CO gene expression.

193

CYTOCHROME
OXIDASE
STUDY OF
PRIMATE
VISUAL
CORTEX

8. Conclusions

CO has provided a window through which we may glimpse at the metabolic properties of neurons and synapses as well as the basis for functional modules and streams within the primate visual cortex.

The adult visual cortex is still capable of dynamic change, and CO levels reflect metabolic plasticity resulting directly from alterations in energy demand as neurons readjust their functional load. The adjustment is fine-tuned, and CO allows us to observe these changes even in the adult, when nothing short of denervation was thought to cause any structural or functional modifications in the visual cortex (Hubel *et al.*, 1977). It also allows us to differentiate between responses to different types of functional insults (such as lid suture, enucleation, and TTX). Moreover, it permits us to examine changes that occur not only globally between adjacent ODCs, but also between subregions of the same laminae (puffs versus interpuffs in the supragranular layers; 4Cα versus 4Cβ in the granular layer). Within each subregion, one can also compare changes among neurons of different types, among compartments of the same neuron, and among mitochondria of different CO contents. All of these analyses have been done (see above), and all of them point to the dynamic metabolic plasticity of mature neurons to finely tune responses geared toward different insults, to vary adjustments among different types of neurons, and to exquisitely regulate energy generation at a local, subcellular level.

It has been a colorful and rewarding expedition, with promise of more discoveries yet to come.

ACKNOWLEDGMENTS. It is a pleasure to thank Z. Huang and S. Tjepkema-Burrows for preparing the plates, C. Snyder for the artwork, and Dr. E. DeYoe for critically reading the manuscript. The work described in the author's laboratory was supported by research grants EY05439 and NS18122 from the National Institutes of Health.

9. References

Allman, J., and Zucker, S., 1990, Cytochrome oxidase and functional coding in primate striate cortex: A hypothesis, *Cold Spring Harbor Symp. Quant. Biol.* **55:**979–982.

Attardi, G., and Schatz, G., 1988, Biogenesis of mitochondria, *Annu. Rev. Cell Biol.* **4:**289–333.

Bachelard, H. S., 1975, Energy utilized by neurotransmitters, in: *Brain Work, Alfred Benzon Symposium, VIII* (D. H. Inguar and N. A. Lassen, eds.), Academic Press, New York, pp. 79—81.

Bartlett, J., and Doty, R., 1974, Response of units in striate cortex of squirrel monkeys to visual and electrical stimuli, *J. Neurophysiol.* **37:**621–641.

Beaulieu, C., Kisvárday, Z., Somogyi, P., Cynader, M., and Cowey, A., 1992, Quantitative distribution of GABA-immunopositive and -immunonegative neurons and synapses in the monkey striate cortex (area 17), *Cereb. Cortex* **2:**295–309.

Blasdel, G. G., and Fitzpatrick, D., 1984, Physiological organization of layer 4 in macaque striate cortex, *J. Neurosci.* **4:**880–895.

Blasdel, G. G., and Lund, J. S., 1983, Termination of afferent axons in macaque striate cortex, *J. Neurosci.* **3:**1389–1413.

Blasdel, G. G., and Salama, G., 1986, Voltage-sensitive dyes reveal a modular organization in monkey striate cortex, *Nature* **321**:579–585.

Burkhalter, A., and Bernardo, K. L., 1989, Organization of corticocortical connections in human visual cortex, *Proc. Natl. Acad. Sci. USA* **86**:1071–1075.

Carroll, E. W., and Wong-Riley, M. T. T., 1982, Light and EM analysis of cytochrome oxidase-rich zones in the striate cortex of squirrel monkeys, *Soc. Neurosci. Abstr.* **8**:706.

Carroll, E. W., and Wong-Riley, M. T. T., 1984, Quantitative light and electron microscopic analysis of cytochrome oxidase-rich zones in the striate cortex of the squirrel monkey, *J. Comp. Neurol.* **222**:1–17.

Carroll, E. W., and Wong-Riley, M., 1985, Correlation between cytochrome oxidase staining and the uptake and laminar distribution of triturated aspartate, glutamate, γ-aminobutyrate and glycine in the striate cortex of the squirrel monkey, *Neuroscience* **15**:959–976.

Carroll, E. W., and Wong-Riley, M., 1987a, Neuronal uptake and laminar distribution of triturated aspartate, glutamate, gamma-aminobutyrate and glycine in the prestriate cortex of squirrel monkeys: Correlation with levels of cytochrome oxidase activity and their uptake in area 17, *Neuroscience* **22**:395–412.

Carroll, E. W., and Wong-Riley, M., 1987b, Recovery of cytochrome oxidase activity in the adult macaque visual system after termination of impulse blockage due to tetrodotoxin, *Soc. Neurosci. Abstr.* **13**:1046.

Celio, M. R., Scharer, L., Morrison, J. H., Norman, A. W., and Bloom, F. E., 1986, Calbindin immunoreactivity alternates with cytochrome c-oxidase-rich zones in some layers of the primate visual cortex, *Nature* **323**:715–717.

Cotlier, E., Lieberman, T. W., and Gray, A. J., 1965, Dehydrogenases and diaphorases in monkey lateral geniculate body, *Arch. Neurol.* **12**:295–299.

Crawford, M. L. J., and Marc, R. E., 1976, Light transmission of cat and monkey eyelids, *Vision Res.* **16**:323–324.

Creutzfeldt, O. D., 1975, Neurophysiological correlates of different functional states of the brain, in: *Brain Work: Alfred Benzon Symposium, VIII* (D. H. Ingvar and N. A. Lassen, eds.), Academic Press, New York, pp. 21–46.

Cusick, C. G., and Kaas, J. H., 1988, Cortical connections of area 18 and dorsolateral visual cortex in squirrel monkeys, *Visual Neurosci.* **1**:211–237.

DeHay, C., Horsburgh, G., Berland, M., Killackey, H., and Kennedy, H., 1989, Maturation and connectivity of the visual cortex in monkey is altered by prenatal removal of retinal input, *Nature* **337**:265–267.

Derrington, A. M., and Lennie, P., 1984, Spatial and temporal contrast sensitivities of neurones in lateral geniculate nucleus of macaque. *J. Physiol. (London)* **357**:219–240.

Derrington, A. M., Krauskopf, J., and Lennie, P., 1984, Chromatic mechanisms in lateral geniculate nucleus of macaque, *J. Physiol. (London)* **357**:241–265.

De Valois, R. L., Snodderly, D. M., Yund, E. W., and Hepler, N. K., 1977, Responses of macaque lateral geniculate cells to luminance and color figures, *Sens. Processes* **1**:244–259.

DeYoe, E. A., and Van Essen, D. C., 1985, Segregation of efferent connections and receptive field properties in visual area V2 of the macaque, *Nature* **317**:58–61.

DeYoe, E. A., and Van Essen, D. C., 1988, Concurrent processing streams in monkey visual cortex, *Trends Neurosci.* **11**:219–226.

DeYoe, E. A., Hockfield, S., Garren, H., and Van Essen, D. C., 1990, Antibody labeling of functional subdivisions in visual cortex: Cat-301 immunoreactivity in striate and extrastriate cortex of the macaque monkey, *Visual Neurosci.* **5**:67–81.

DiRocco, R. J., Kageyama, G. H., and Wong-Riley, M. T. T., 1989, The relationship between CNS metabolism and cytoarchitecture: A review of ^{14}C-deoxyglucose studies with correlation to cytochrome oxidase histochemistry, *Comput. Med. Imag. Graph.* **13**:81–92.

Erecinska, M., and Silver, I. A., 1989, ATP and brain function, *J. Cereb. Blood Flow Metab.* **9**:2–19.

Felleman, D. J., and Van Essen, D. C., 1991, Distributed hierarchical processing in the primate cerebral cortex, *Cereb. Cortex* **1**:1–47.

Fitzpatrick, D., Itoh, K., and Diamond, I. T., 1983, The laminar organization of the lateral geniculate body and the striate cortex in the squirrel monkey (*Saimiri sciureus*), *J. Neurosci.* **3**:673–702.

Fitzpatrick, D., Lund, J. S., and Blasdel, G. G., 1985, Intrinsic connections of macaque striate cortex. Afferent and efferent connections of lamina 4C, *J. Neurosci.* **5**:3329–3349.

Fitzpatrick, D., Lund, J. S., Schmechel, D. E., and Towles, A. C., 1987, Distribution of GABAergic neurons and axon terminals in the macaque striate cortex, *J. Comp. Neurol.* **264**:73–91.

195

CYTOCHROME
OXIDASE
STUDY OF
PRIMATE
VISUAL
CORTEX

Freeman, R. D., and Bradley, A., 1980, Monocularly deprived humans: Nondeprived eye has super-normal vernier acuity, *J. Neurophysiol.* **43:**1645–1653.

Fries, W., and Distel, H., 1983, Large layer VI neurons of monkey striate cortex (Meynert cells) project to the superior colliculus, *Proc. R. Soc. London Ser. B* **219:**53–59.

Fries, W., Keizer, K., and Kuypers, H. G. J. M., 1985, Larger layer VI cells in macaque striate cortex (Meynert cells) project to both superior colliculus and prestriate visual area V5, *Exp. Brain Res.* **58:**613–616.

Gottlieb, M. D., Pasik, P., and Pasik, T., 1985, Early postnatal development of the monkey visual system. I. Growth of the lateral geniculate nucleus and striate cortex, *Dev. Brain Res.* **17:**53–62.

Graybiel, A. M., and Ragsdale, C. W., Jr., 1982. Pseudocholinesterase staining in the primary visual pathway of the macaque monkey, *Nature* **299:**439–442.

Hawken, M. J., Parker, A. J., and Lund, J. S., 1987, Contrast sensitivity and laminar distribution of direction selective neurons in monkey striate cortex, *Invest. Ophthalmol. Vis. Sci. (Suppl.)* **28:**197.

Hazeltine, E. C., De Bruyn, E. J., and Cassagrande, V. A., 1979, Demonstration of ocular dominance columns in Nissl-stained sections of monkey visual cortex following enucleation, *Brain Res.* **176:**153–158.

Hendrickson, A. E., 1985, Dots, stripes and columns in monkey visual cortex, *Trends Neurosci.* **8:**406–410.

Hendrickson, A. E., and Tigges, M., 1985, Enucleation demonstrates ocular dominance columns in Old World macaque but not in New World squirrel monkey visual cortex, *Brain Res.* **333:**340–344.

Hendrickson, A. E., Hunt, S. P., and Wu, J.-Y., 1981, Immunocytochemical localization of glutamic acid decarboxylase in monkey striate cortex, *Nature* **292:**605–607.

Hendry, S. H. C., and Jones, E. G., 1986, Reduction in number of immunostained GABAergic neurones in deprived-eye dominance columns of monkey area 17, *Nature* **320:**750–753.

Hendry, S. H. C., and Jones, E. G., 1988, Activity-dependent regulation of GABA expression in the visual cortex of adult monkeys, *Neuron* **1:**701–712.

Hendry, S. H. C., Jones, E. G., DeFelipe, J., Schmechel, D. E., Brandon, C., and Emson, P. C., 1984, Neuropeptide-containing neurons of the cerebral cortex are also GABAertic, *Proc. Natl. Acad. Sci. USA* **81:**6526–6530.

Hendry, S. H. C., Jones, E. G. Hockfield, S., and McKay, R. D. G., 1988, Neuronal populations stained with the monoclonal antibody Cat-301 in the mammalian cerebral cortex and thalamus, *J. Neurosci.* **8:**518–542.

Hess, D. T., and Edwards, M. A., 1987, Anatomical demonstration of ocular segregation in the retinogeniculocortical pathway of the New World capuchin monkey (*Cebus apella*), *J. Comp. Neurol.* **264:**409–420.

Hevner, R. F., 1990, Cytochrome oxidase in mammalian brain: Regulation of its activity, protein amount, and mRNA by neural functional activity, Doctoral dissertation, Medical College of Wisconsin.

Hevner, R. F., and Wong-Riley, M. T. T., 1989, Brain cytochrome oxidase: Purification, antibody generation, and immunohistochemical/histochemical correlations in the CNS, *J. Neurosci.* **9:**3884–3898.

Hevner, R. F., and Wong-Riley, M. T. T., 1990, Regulation of cytochrome oxidase protein levels by functional activity in the macaque monkey visual system, *J. Neurosci.* **10:**1331–1340.

Hevner, R. F., and Wong-Riley, M. T. T., 1991, Neuronal expression of nuclear and mitochondrial genes for cytochrome oxidase (CO) subunits analyzed by *in situ* hybridization: Comparison with CO activity and protein, *J. Neurosci.* **11:**1942–1958.

Hevner, R. F., Duff, R. S., and Wong-Riley, M. T. T., 1992, Coordination of ATP production and consumption in brain: Parallel regulation of cytochrome oxidase and Na^+, K^+-ATPase, *Neurosci. Lett.* **138:**188–192.

Hevner, R. F., Liu, S., and Wong-Riley, M. T. T., 1993, An optimized method for determining cytochrome oxidase activity in brain tissue homogenates, *J. Neurosci. Methods*, in press.

Hille, B., 1968, Pharmacological modifications of the sodium channels of frog nerve, *J. Gen. Physiol.* **51:**199–219.

Hiltgen, G., and Wong-Riley, M., 1986, Quantitative EM analysis of the effect of retinal impulse blockage on cytochrome oxidase activity in lamina IVC of macaque striate cortex, *Soc. Neurosci. Abstr.* **12:**130.

Hockfield, S., McKay, R. D., Hendry, S. H. C., and Jones, E. G., 1983, A surface antigen that identifies ocular dominance columns in the visual cortex and laminar features of the lateral geniculate nucleus, *Cold Spring Harbor Symp. Quant. Biol.* **48:**877–889.

Horton, J. C., 1984, Cytochrome oxidase patches: A new cytoarchitectonic feature of monkey visual cortex, *Philos. Trans. R. Soc. London Ser. B* **304**:199–253.

Horton, J. C., and Hedley-Whyte, E. T., 1984, Mapping of cytochrome oxidase patches and ocular dominance columns in human visual cortex, *Philos. Trans. R. Soc. London Ser. B* **304**:255–272.

Horton, J. C., and Hubel, D. H., 1981, Regular patchy distribution of cytochrome oxidase staining in primary visual cortex of macaque monkey, *Nature* **292**:762–764.

Hubel, D. H., and Livingstone, M. S., 1985, Complex-unoriented cells in a subregion of primate area 18, *Nature* **315**:325–327.

Hubel, D. H., and Livingstone, M. S., 1987, Segregation of form, color, and stereopsis in primate area 18, *J. Neurosci.* **7**:3378–3415.

Hubel, D. H., and Livingstone, M. S., 1990, Color and contrast sensitivity in the lateral geniculate body and primary visual cortex of the macaque monkey, *J. Neurosci.* **10**:2223–2237.

Hubel, D. H., and Wiesel, T. N., 1968, Receptive fields and functional architecture of monkey striate cortex, *J. Physiol (London)* **195**:215–243.

Hubel, D. H ., and Wiesel, T. N., 1972, Laminar and columnar distribution of geniculocortical fibers in the macaque monkey, *J. Comp. Neurol.* **146**:421–450.

Hubel, D. H., and Wiesel, T. N., 1974, Sequence regularity and geometry of orientation columns in the monkey striate cortex, *J. Comp. Neurol.* **158**:267–294.

Hubel, D. H., and Wiesel, T. N., 1977, Ferrier lecture: Functional architecture of macaque monkey visual cortex, *Proc. R. Soc. London Ser. B* **198**:1–59.

Hubel, D. H., Wiesel, T. N., and LeVay, S., 1977, Plasticity of ocular dominance columns in monkey striate cortex, *Philos. Trans. R. Soc. London Ser. B* **278**:131–163.

Hubel, D. H., Wiesel, T. N., and Stryker, M. P., 1978, Anatomical demonstration of orientation columns in macaque monkey, *J. Comp. Neurol.* **177**:361–380.

Humphrey, A. L., and Hendrickson, A. E., 1983, Background and stimulus-induced patterns of high metabolic activity in the visual cortex (area 17) of the squirrel and macaque monkey, *J. Neurosci.* **3**:345–358.

Jacobs, G. H., 1977, Visual capacities of the owl monkey (*Aotus trivirgatus*). I. Spectral sensitivity and colour vision, *Vision Res.* **17**:811–820.

Jones, E. G., DeFelipe, J., Hendry, S. H. C., and Maggio, J. E., 1988, A study of tachykinin-immunoreactive neurons in monkey cerebral cortex, *J. Neurosci.* **8**:1206–1224.

Kageyama, G. H., and Wong-Riley, M. T. T., 1982, Histochemical localization of cytochrome oxidase in the hippocampus: Correlation with specific neuronal types and afferent pathways, *Neuroscience* **7**:2337–2361.

Kaplan, E., Purpura, K., and Shapley, R. M., 1987, Contrast affects the transmission of visual information through the mammalian lateral geniculate nucleus, *J. Physiol. (London)* **391**:267–288.

Kayama, Y., Riso, R., Bartlett, J., and Doty, R., 1979, Luxotonic responses of units in macaque striate cortex, *J. Neurophysiol.* **42**:1495–1517.

Kennedy, C., Des Rosiers, M. H. Sakurada, O., Shinohara, M., Reivich, M., Jehle, H. W., and Sokoloff, L., 1976, Metabolic mapping of the primary visual system of the monkey by means of the autoradiographic [^{14}C] deoxyglucose technique, *Proc. Natl. Acad. Sci. USA* **73**:4230–4234.

Kennedy, H., Bullier, J., and Dehay, C., 1985, Cytochrome oxidase activity in the striate cortex and lateral geniculate nucleus of the newborn and adult macaque monkey, *Exp. Brain Res.* **61**:204–209.

Kennedy, H., Dehay, C., and Horsburgh, G., 1990, Striate cortex periodicity, *Nature* **348**:494.

Krubitzer, L. A., and Kaas, J. H., 1989, Cortical integration of parallel pathways in the visual system of primates, *Brain Res.* **478**:161–165.

Kuljis, R. O., and Rakic, P., 1989, Neuropeptide Y-containing neurons are situated predominantly outside cytochrome oxidase puffs in macaque visual cortex, *Visual Neurosci.* **2**:57–62.

Kuljis, R. O., and Rakic, P., 1990, Hypercolumns in primate visual cortex can develop in the absence of cues from photoreceptors, *Proc. Natl. Acad. Sci. USA* **87**:5303–5306.

Lennie, P., 1984, Recent development in the physiology of color vision, *Trends Neurosci.* **7**:243–248.

LeVay, S., Wiesel, T. N., and Hubel, D. H., 1980, The development of ocular dominance columns in normal and visually deprived monkeys, *J. Comp. Neurol.* **191**:1–51.

LeVay, S., Connolly, M., Houde, J., and Van Essen, D. C., 1985, The complete pattern of ocular dominance stripes in the striate cortex and visual field of the macaque monkey, *J. Neurosci.* **5**:486–501.

197

CYTOCHROME
OXIDASE
STUDY OF
PRIMATE
VISUAL
CORTEX

Leventhal, A. G., Rodieck, R. W., and Dreher, B., 1981, Retinal ganglion cell classes in the Old World monkey: Morphology and central projections, *Science* **213**:1139–1142.

Liu, S., and Wong-Riley, M., 1990, Quantitative light and electron microscopic analysis of cytochrome oxidase distribution in neurons of the lateral geniculate nucleus of the adult monkey, *Visual Neurosci.* **4**:269–287.

Liu, S., Wilcox, D. A., Sieber-Blum, M., and Wong-Riley, M., 1990, Developing neural crest cells in culture: Correlation of cytochrome oxidase activity with SSEA-1 and dopamine-β-hydroxylase immunoreactivity, *Brain Res.* **535**:271–280.

Liu, S., Hevner, R. F., and Wong-Riley, M. T. T., 1991, Metabolic map of the normal rat brain as revealed by cytochrome oxidase histochemistry and biochemistry, *Soc. Neurosci. Abstr.* **17**:864.

Livingstone, M., 1990, Segregation of form, color, movement, and depth processing in the visual system: Anatomy, physiology, art, and illusion, in: *Vision and the Brain* (B. Cohen and I. Bodis-Wollner, eds.), Raven Press, New York, pp. 119–138.

Livingstone, M. S., and Hubel, D. H., 1982, Thalamic inputs to cytochrome oxidase-rich regions in monkey visual cortex, *Proc. Natl. Acad. Sci. USA* **79**:6098–6101.

Livingstone, M. S., and Hubel, D. H., 1983, Specificity of cortico-cortical connections in monkey visual system, *Nature* **304**:531–534.

Livingstone, M. S., and Hubel, D. H., 1984a, Anatomy and physiology of a color system in the primate visual cortex, *J. Neurosci.* **4**:309–356.

Livingstone, M. S., and Hubel, D. H., 1984b, Specificity of intrinsic connections in primate primary visual cortex, *J. Neurosci.* **4**:2830–2835.

Livingstone, M. S., and Hubel, D. H., 1987a, Connections between layer 4B of area 17 and thick cytochrome oxidase stripes of area 18 in the squirrel monkey, *J. Neurosci.* **7**:3371–3377.

Livingstone, M. S., and Hubel, D. H., 1987b, Psychophysical evidence for separate channels for the perception of form, color, movement, and depth, *J. Neurosci.* **7**:3416–3468.

Livingstone, M. S., and Hubel, D. H., 1988, Segregation of form, color, movement, and depth: Anatomy, physiology, and perception, *Science* **240**:740–749.

Lowry, O. H., 1975, Energy metabolism in brain and its control, in: *Brain Work, Alfred Benzon Symposium VIII* (D. H. Ingvar and N. A. Lassen, eds.), Academic Press, New York, pp. 48–64.

Lowry, O. H., Roberts, N. R., Leiner, K. Y., Wu, M.-L., Farr, A. L., and Albers, R. W., 1954, The quantitative histochemistry of brain. III. Ammon's horn, *J. Biol. Chem.* **207**:39–49.

Lund, J. S., 1973, Organization of neurons in the visual cortex, area 17, of the monkey (*Macaca mulatta*), *J. Comp. Neurol.* **147**:455–496.

Lund, J. S., 1988, Anatomical organization of macaque striate cortex, *Annu. Rev. Neurosci.* **11**:253–288.

Lund, J. S., and Boothe, R. G., 1975, Interlaminar connections and pyramidal neuron organization in the visual cortex, area 17, of the macaque monkey, *J. Comp. Neurol.* **159**:305–334.

Lund, J. S., Lund, R. D., Hendrickson, A. E., Bunt, A. H., and Fuchs, A. F., 1975, The origin of efferent pathways from the primary visual cortex, area 17, of the macaque monkey as shown by retrograde transport of horseradish peroxidase, *J. Comp. Neurol.* **164**:287–304.

Lund, J. S., Hendrickson, A. E., Ogren, M. P., and Tobin, E. A. 1981, Anatomical organization of primate visual area V-II, *J. Comp. Neurol.* **202**:19–45.

Luo, X. G., Hevner, R. F., and Wong-Riley, M. T. T., 1989, Double labeling of cytochrome oxidase and gamma aminobutyric acid in central nervous system neurons of adult cats, *J. Neurosci. Methods* **30**:189–195.

Maguire, W. M., and Baizer, J. S., 1982, Luminance coding of briefly presented stimuli in area 17 of the rhesus monkey, *J. Neurophysiol.* **47**:128–137.

Malach, R., 1992, Dendritic sampling across processing streams in monkey striate cortex, *J. Comp. Neurol.* **315**:303–312.

Matthews, M. R., Cowan, W. M., and Powell, T. P. S., 1960, Transneuronal cell degeneration in the lateral geniculate nucleus of the macaque monkey, *J. Anat.* **94**:145–169.

Maunsell, J. H. R., and Newsome, W. T., 1987, Visual processing in monkey extrastriate cortex, *Annu. Rev. Neurosci.* **10**:363–401.

Maunsell, J. H. R., and Van Essen, D. C., 1983, Functional properties of neurons in middle temporal visual area of the macaque monkey. II. Binocular interactions and sensitivity to binocular disparity, *J. Neurophysiol.* **49**:1148–1167.

Mawe, G. M., and Gershon, M. D., 1986, Functional heterogeneity in the myenteric plexus: Demonstration using cytochrome oxidase as a verified cytochemical probe of the activity of individual enteric neurons, *J. Comp. Neurol.* **249**:381–391.

Michael, C. R. 1988, Retinal afferent arborization patterns, dendritic field orientations, and the segregation of function in the lateral geniculate nucleus of the monkey, *Proc. Natl. Acad. Sci. USA* **85**:4914–4918.

Mishkin, M., Ungerleider, L. G., and Macko, K. A. 1983, Object vision and spatial vision: Two cortical pathways, *Trends Neurosci.* **6**:414–417.

Mitzdorf, U., and Singer, W., 1979, Excitatory synaptic ensemble properties in the visual cortex of the macaque monkey: A current source density analysis of electrically evoked potentials, *J. Comp. Neurol.* **187**:71–84.

Mjaatvedt, A. E., and Wong-Riley, M. T. T., 1988, The relationship between synaptogenesis and cytochrome oxidase activity in Purkinje cells of the developing rat cerebellum, *J. Comp. Neurol.* **277**:155–182.

Moran, J., and Gordon, B., 1982, Long-term visual deprivation in a human, *Vision Res.* **22**:27–36.

Murphy, K. M., Van Sluyters, R. C., and Jones, D. G., 1991, The organization of cytochrome-oxidase blobs in cat visual cortex, *Soc. Neurosci. Abstr.* **17**:1088.

Nie, F., and Wong-Riley, M., 1993, Double labeling of cytochrome oxidase (CO) and GABA in neurons and synapses of macaque puffs and interpuffs: A quantitative study, *Soc. Neurosci. Abstr.* **19**:334.

Payne, B. R., and Peters, A., 1989, Cytochrome oxidase patches and Meynert cells in monkey visual cortex, *Neuroscience* **28**:353–363.

Purves, D., and LaMantia, A.-S., 1990, Numbers of "blobs" in the primary visual cortex of neonatal and adult monkeys, *Proc. Natl. Acad. Sci. USA* **87**:5764–5767.

Purves, D., and LaMantia, A., 1993, Development of blobs in the visual cortex of macaques, *J. Comp. Neurol.* **334**:169–175.

Rakic, P., 1976, Prenatal genesis of connections subserving ocular dominance in the rhesus monkey, *Nature* **261**:467–471.

Rall, W., 1962, Electrophysiology of a dendritic neuron model, *Biophys, J.* 145–167.

Schein, S. J., and de Monasterio, F. M., 1987, Mapping of retinal and geniculate neurons onto striate cortex of macaque, *J. Neurosci.* **7**:996–1009.

Schiller, P. H., and Malpeli, J. G., 1978, Functional specificity of lateral geniculate nucleus laminae of the rhesus monkey, *J. Neurophysiol.* **41**:788–797.

Seligman, A. M., Karnovsky, M. J., Wasserkrug, H. L., and Hanker, J. S., 1968, Nondroplet ultra-structural demonstrations of cytochrome oxidase activity with a polymerizing osmiophilic re-agent, diaminobenzidine (DAB), *J. Cell Biol.* **38**:1–14.

Shapley, R. M., Kaplan, E., and Soodak, R., 1981, Spatial summation and contrast sensitivity of X and Y cells in the lateral geniculate nucleus of the macaque, *Nature* **292**:543–545.

Shipp, S., and Zeki, S., 1985, Segregation of pathways leading from area V2 to area V4 and V5 of macaque monkey visual cortex, *Nature* **315**:322–325.

Siegel, G. J., Albers, R. W., Agranoff, B. W., and Katzman, R., 1981, *Basic Neurochemistry,* Little, Brown, Boston.

Sillito, A. M., 1975, The contribution of inhibitory mechanisms to the receptive field properties of neurons in the striate cortex of the cat, *J. Physiol. (London)* **250**:305–329.

Sillito, A. M., Kemp, J. A., Wilson, J. A., and Berardi, N., 1980, A re-evaluation of the mechanisms underlying simple cell orientation selectivity, *Brain Res.* **194**:517–520.

Silverman, M. S., Grosof, D. H., De Valois, R. L., and Elfar, S. D., 1989, Spatial-frequency organization in primate striate cortex, *Proc. Natl. Acad. Sci. USA* **86**:711–715.

Sokoloff, L., 1974, Changes in enzyme activities in neural tissues with maturation and development of the nervous system, in: *The Neurosciences: Third Study Program* (F. O. Schmitt and F. G. Worden, eds.), MIT Press, Cambridge, Mass., pp. 885–898.

Sokoloff, L., Reivich, M., Kennedy, C., Des Rosiers, M. H., Patlak, C. S., Pettigrew, K. D., Sakurada, O., and Shinohara, M., 1977, The [^{14}C] deoxyglucose method for the measurement of local cerebral glucose utilization: Theory, procedure, and normal values in the conscious and anesthetized albino rat, *J. Neurochem.* **28**:897–916.

Stenaas, S. S., Eddington, D. K., and Dobelle, W. H., 1974, The topography and variability of the primary visual cortex in man, *J. Neurosurg.* **40**:747–755.

Tigges, J. Spatz, W. B., and Tigges, M., 1973, Reciprocal point-to-point connections between para-striate and striate cortex in the squirrel monkey (Saimiri), *J. Comp. Neurol.* **148**:481–490.

Tigges, M., Hendrickson, A. E., and Tigges, J., 1984, Anatomical consequences of long-term mon-ocular eyelid closure on lateral geniculate nucleus and striate cortex in squirrel monkey, *J. Comp. Neurol.* **227**:1–13.

199

CYTOCHROME
OXIDASE
STUDY OF
PRIMATE
VISUAL
CORTEX

Tootell, R. B. H., and Hamilton, S. L., 1989. Functional anatomy of the second visual area (V2) in the macaque, *J. Neurosci.* **9:**2620–2644.

Tootell, R. B. H., Silverman, M. S., Switkes, E., and De Valois, R. L., 1982, Deoxyglucose analysis of retinotopic organization in primate striate cortex, *Science* **218:**902–904.

Tootell, R. B. H., Silverman, M. S., De Valois, R. L., and Jacobs, G. H., 1983, Functional organization of the second cortical visual area in primates, *Science* **220:**737–739.

Tootell, R. B. H., Hamilton, S. L., and Silverman, M. S., 1985, Topography of cytochrome oxidase activity in owl monkey cortex, *J. Neurosci.* **5:**2786–2800.

Tootell, R. B. H., Hamilton, S. L., Silverman, M. S., and Switkes, E., 1988a, Functional anatomy of macaque striate cortex. I. Ocular dominance, binocular interactions, and baseline conditions, *J. Neurosci.* **8:**1500–1530.

Tootell, R. B. H., Switkes, E., Silverman, M. S., and Hamilton, S. L., 1988b, Functional anatomy of macaque striate cortex. II. Retinotopic organization, *J. Neurosci.* **8:**1531–1568.

Tootell, R. B. H., Silverman, M. S., Hamilton, S. L., De Valois, R. L., and Switkes, E., 1988c, Functional anatomy of macaque striate cortex. III. Color, *J. Neurosci.* **8:**1569–1593.

Tootell, R. B. H., Hamilton, S. L., and Switkes, E., 1988d, Functional anatomy of macaque striate cortex. IV. Contrast and magno-parvo streams, *J. Neurosci.* **8:**1594–1609.

Tootell, R. B. H., Silverman, M. S., Hamilton, S. L., Switkes, E., and De Valois, R. L., 1988e, Functional anatomy of macaque striate cortex. V. Spatial frequency, *J. Neurosci.* **8:**1610–1624.

Trusk, T. C., and Wong-Riley, M. T. T., 1990, Focal laser retinal lesions reveal differential cytochrome oxidase reactivity in four isoeccentric regions of the macaque striate cortex, *Soc. Neurosci. Abstr.* **16:**708.

Trusk, T. C., Kaboord, W. S., and Wong-Riley, M. T. T., 1990, Effects of monocular enucleation, tetrodotoxin, and lid suture on cytochrome oxidase reactivity in supragranular puffs of adult macaque striate cortex, *Visual Neurosci.* **4:**185–204.

Trusk, T. C., Wong-Riley, M., and DeYoe, E. A., 1992. Changes in cytochrome oxidase and neuronal activity in V1 induced by monocular TTX blockade in macaque monkeys, *Soc. Neurosci. Abstr.* **18:**298.

Trusk, T. C., Jaffe, G. J., and Wong-Riley, M. T. T., 1993, Laser retinal lesions: Effects on the metabolic integrity of the primate visual system, submitted for publication.

Ts'o, D. Y., and Gilbert, C. D., 1988, The organization of chromatic and spatial interactions in the primate striate cortex, *J. Neurosci.* **8:**1712–1727.

Ts'o, D. Y., Frostig, R. D., Lieke, E. E., and Grinvald, A., 1990, Functional organization of primate visual cortex revealed by high resolution optical imaging, *Science* **249:**417–420.

Van Essen, D. C., and Maunsell, J. H. R., 1983, Hierarchical organization and functional streams in the visual cortex, *Trends Neurosci.* **6:**370–375.

Weber, J. T., Huerta, M. F., Kaas, J. H., and Harting, J. K., 1983, The projection of the lateral geniculate nucleus of the squirrel monkey: Studies of the interlamina zones and the S layers, *J. Comp. Neurol.* **213:**135–145.

Wiesel, T. N., and Hubel, D. H., 1966, Spatial and chromatic interactions in the lateral geniculate body of the rhesus monkey, *J. Neurophysiol.* **29:**1115–1156.

Wiesel, T. N., Hubel, D. H., and Lam, D. M. K., 1974, Autoradiographic demonstration of ocular dominance columns in the monkey striate cortex by means of transneuronal transport, *Brain Res.* **79:**273–279.

Wikstrom, M., Krab, K., and Saraste, M., 1981, *Cytochrome Oxidase: A Synthesis*, Academic Press, New York.

Wong-Riley, M. T. T., 1979a, Columnar cortico-cortical interconnections within the visual system of the squirrel and macaque monkeys, *Brain Res.* **162:**201–217.

Wong-Riley, M., 1979b, Changes in the visual system of monocularly sutured or enucleated cats demonstrable with cytochrome oxidase histochemistry. *Brain Res.* **171:**11–28.

Wong-Riley, M. T. T., 1988, Comparative study of the mammalian primary visual cortex with cytochrome oxidase histochemistry, in: *Vision: Structure and Function* (D. T. Yew, K. F. So, and D. S. C. Tsang, eds.), World Scientific Press, New Jersey, pp. 450–486.

Wong-Riley, M. T. T., 1989, Cytochrome oxidase: An endogenous metabolic marker for neuronal activity, *Trends Neurosci.* **12:**94–101.

Wong-Riley, M. T. T., and Carroll, E. W., 1984a, Quantitative light and electron microscopic analysis of cytochrome oxidase-rich zones in V II prestrate cortex of the squirrel monkey, *J. Comp. Neurol.* **222:**18–37.

Wong-Riley, M., and Carroll, E. W., 1984b, The effect of impulse blockage on cytochrome oxidase activity in the monkey visual system, *Nature* **307**:262–264.

Wong-Riley, M. T. T., and Kageyama, G. H., 1986, Localization of cytochrome oxidase in the spinal cord and dorsal root ganglia, with quantitative analysis of ventral horn cells in the monkey, *J. Comp. Neurol.* **245**:41–61.

Wong-Riley, M. T. T., and Norton, T. T., 1988, Histochemical localization of cytochrome oxidase activity in the visual system of the tree shrew: Normal patterns and the effect of retinal impulse blockage, *J. Comp. Neurol.* **272**:562–578.

Wong-Riley, M., and Riley, D. A., 1983, The effect of impulse blockage on cytochrome oxidase activity in the cat visual system, *Brain Res.* **261**:185–193.

Wong-Riley, M., Trusk, T., and Hoppe, D., 1988, Localization of cytochrome oxidase in macaque striate cortex during prenatal development, *Soc. Neurosci. Abstr.* **14**:743.

Wong-Riley, M. T. T., Tripathi, S. C., Trusk, T. C., and Hoppe, D. A., 1989a, Effect of retinal impulse blockage on cytochrome oxidase-rich zones in the macaque striate cortex. I. Quantitative EM analysis of neurons, *Visual Neurosci.* **2**:483–497.

Wong-Riley, M. T. T., Trusk, T. C., Tripathi, S. C., and Hoppe, D. A., 1989b, Effect of retinal impulse blockage on cytochrome oxidase-rich zones in the macaque striate cortex. II. Quantitative EM analysis of neuropil, *Visual Neurosci.* **2**:499–514.

Wong-Riley, M., Trusk, T., Kaboord, W., and Huang, Z., 1992, Interpuffs in the macaque striate cortex: Quantitative EM analysis of neurons before and after unilateral retinal impulse blockade, *Soc. Neurosci. Abstr.* **18**:299.

Wong-Riley, M. T. T., Hevner, R. F., Cutlan, R., Earnest, M., Egan, R., Frost, J., and Nguyen, T., 1993a, Cytochrome oxidase in the human visual cortex: Distribution in the developing and the adult brain, *Visual Neurosci.* **10**:41–58.

Wong-Riley, M., Trusk, T., and Huang, Z., 1993b, Interpuffs in the macaque striate cortex: Quantitative EM analysis of neuropil before and after unilateral retinal impulse blockade, *Soc. Neurosci. Abstr.* **19**:334.

Zeki, S. M., 1974, Cells respond to changing image size and disparity in the cortex of the rhesus monkey, *J. Physiol. (London)* **242**:827–841.

Zeki, S. M., 1978, Functional specialisation in the visual cortex of the rhesus monkey, *Nature* **274**:423–428.

Zeki, S., and Shipp, S., 1988, The functional logic of cortical connections, *Nature,* **355**:311–317.

Zheng, D., LaMantia, A.-S., and Purves, D., 1991, Specialized vascularization of the primate visual cortex, *J. Neurosci.* **11**:2622–2629.

The Afferent, Intrinsic, and Efferent Connections of Primary Visual Cortex in Primates

VIVIEN A. CASAGRANDE and JON H. KAAS

1. Introduction

Because of its distinctive architecture, connections, and functions, primary visual cortex, area 17 or V1 of primates, can be easily identified in most mammals (Kaas, 1987). V1 (also referred to as striate cortex) is particularly distinctive in primates, and, as a result, it was the first cortical area identified histologically (see Gennari, 1782, in Fulton, 1937). V1 of most, if not all, primates has a number of conspicuous features that distinguish this structure from its homologue in other mammals. Unlike carnivores, such as cats and ferrets, almost all of the visual input relayed from the lateral geniculate nucleus (LGN) of primates terminates in V1 (Benevento and Standage, 1982; Bullier and Kennedy, 1983; see Henry, 1991, for review), and lesions of V1 produce a severe deficit known as cortical blindness (e.g., Cowey and Stoerig, 1989). In addition, visual cortex of all primates is activated by physiologically and morphologically distinguishable streams, or channels, of inputs that are relayed from the retina to V1 in a

VIVIEN A. CASAGRANDE • Department of Cell Biology, Vanderbilt University Medical School, Nashville, Tennessee 37232. JON H. KAAS • Department of Psychology, Vanderbilt University, Nashville, Tennessee 37232.
Cerebral Cortex, Volume 10, edited by Alan Peters and Kathleen S. Rockland. Plenum Press, New York, 1994.

manner unique to primates (Kaas and Huerta, 1988; Casagrande and Norton, 1991). Furthermore, the intrinsic connections of V1 in primates exhibit both vertical (laminar) and areal (modular) distinctions that appear designed to create new output channels from input channels via features of internal circuitry. Finally, the output streams project to visual areas that seem to be organized in a manner unique to primates. In particular, the major cortical target of V1, the second visual area, V2, is composed of three morphologically distinct modules that are differentially activated from V1, and at least one other major target of V1, the middle temporal visual area or MT, appears to be a unique specialization of primates (Kaas and Preuss, 1993). These common features of visual cortex in primates are of particular interest because these specializations relate to vision in humans as well as other primates. In this review, we focus on common features that have been described for V1 across a variety of primate species, and therefore are most likely to be present in most or all primates. In addition, we describe differences in V1 organization across primate groups, since these differences may relate to functional specializations and adaptations in the greatly varied primate order. Features that vary across taxa, when related to behavioral niches, may provide clues as to the significance of variations. Finally, this review briefly compares V1 in primates with V1 in some nonprimates to emphasize the distinctiveness of V1 in primates.

2. Architecture: Defining Layers and Compartments

Our understanding of the functional subdivisions, connections, and micro-architecture of V1 has increased enormously in the past 20 years. However, a difficulty in discussing this understanding is that all published papers do not relate to a common anatomical frame of reference. In order to review the connections of V1, it is useful to consider the general issue of how this area has been subdivided into layers and modules, or compartments. We begin with a discussion of traditional concepts and controversies concerning cortical lamination in primates. We then consider how the landscape of V1 is divided based on staining for the mitochondrial enzyme cytochrome oxidase.

2.1. Cortical Lamination

A description of laminar patterns of afferents in primate V1 is currently complicated by the use of different interpretations of cytoarchitectonically defined layers in Nissl stains for cell bodies.* There has been the widespread adoption of the framework of six layers as proposed by Brodmann (1909) over schemes with more layers, but variations of the six-layer framework exist (for review, see Garey, 1971; Billings-Gagliardi et al., 1974; Braak, 1984; Henry, 1991; and Peters, this volume). In particular, Hässler (1967) and others (e.g., Weller and Kaas, 1982; Diamond et al., 1985; Lachica et al., 1993) have argued that sublayers defined by Brodmann as part of layer IV in primate V1 are actually parts of layer III. Such differences in interpretation obviously compli-

---

*We use the terms area 17 or V1 and area 18 or V2 interchangeably throughout this chapter.

cate comparisons of laminar differences and similarities across primate and nonprimate taxa, as well as comparisons across cortical areas within primates. Descriptions of laminar patterns of neuron types, neurotransmitter receptor distributions (e.g., Shaw and Cynader, 1986), and connections can be misleading if homologous layers and sublayers are not correctly identified.

The architectonic observation that leads to ambiguity is that two sheets of cells in V1 of some simian primates have the appearance of layer IV in that the cells are small and densely packed. These two sheets are separated by a zone of less densely packed, larger neurons (Fig. 1). Brodmann interpreted the two sheets of small granular cells as upper and lower tiers of layer IV (termed IVA and IVC), separated by a middle tier of larger cells (termed IVB). These three "sublayers" of Brodmann's layer IV were thought to merge into a thinner, and more typical layer IV at the border of visual area 18 (V2), and indeed they often appear to do so. This view was clearly summarized by Clark (1925) with his statement that "two layers of granules have been derived from an original single layer . . . that, at the junction between the visuo-sensory and visuo-psychic areas . . . run together and connect up to form a single lamina granularis interna." On the other hand, from careful analysis of serial sections in several planes of cut, it is clear that sublayer IVB of Brodmann's area 17 merges with sublayer IIIC of area 18 (e.g., Colonnier and Sas, 1978), with IVA having no clear equivalent in area 18. This observation is more consistent with Hässler's interpretation that only IVC of Brodmann is equivalent to layer IV in "higher" primates, and that sublayers IVA and IVB of Brodmann are, instead, sublayers of layer III.

While it may appear difficult to resolve the issue of defining layers and sublayers in area 17 of primates, the bulk of the evidence clearly supports Hässler's interpretation. First, if the lamination patterns of V1 are compared across primates, it becomes apparent that layers IVA and IVB of Brodmann are less developed sublayers of layer III in most New World monkeys and hardly apparent as sublayers of layer III in prosimian primates such as galagos. Hässler (1967) used such comparisons across primate groups to support his theory of cortical lamination, and further comparisons fortify his view (e.g., Weller and Kaas, 1982; Diamond *et al.*, 1985; Lachica *et al.*, 1993). Second, layer IIIC neurons, according to Hässler's concept of layers, project to extrastriate cortex in all primate taxa examined (see below), as do layer III cells in nonprimates. According to Brodmann's scheme, these projections would originate in layer IVB of simians (e.g., macaque monkeys) and layer IIIC of prosimians (e.g., lorisids and lemurs). Clearly, Hässler's interpretation of layers allows for a simpler explanation of laminar patterns of connections, while Brodmann's interpretation calls for explanations of how layer IV became a source of projections to extrastriate cortex in higher primates, and how a major difference in prosimian and simian primates evolved. Third, large pyramidal cells are found in Brodmann's layer IVB of monkeys (e.g., Lund *et al.*, 1979), and large pyramidal cells are not typically found in layer IV. Furthermore, as Colonnier and Sas (1978) stress, this pyramidal cell layer of area 17 merges with the IIIC pyramidal cell layer of area 18. Thus, we use a modified version of Hässler's designations for layers in the present report. For convenience we have compared the laminar designations used in this chapter with those proposed by Brodmann (parentheses) in Fig. 1. The most relevant differences between Brodmann's designations and those used here are as follows: IIIBβ (IVA), IIIC (IVB), IVα (IVCα), and IVβ (IVCβ).

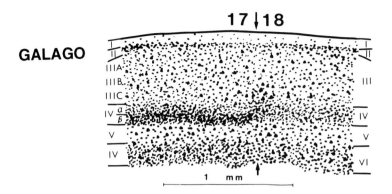

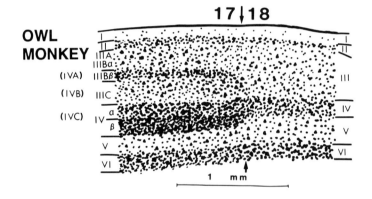

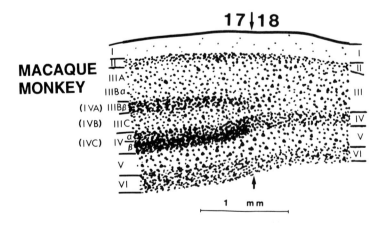

2.2. Cytochrome Oxidase Modules

Primary visual cortex of primates is distinguished not only by characteristic laminar patterns in Nissl-stained sections, but also by the presence of a distinct, periodic pattern of light and dark staining perhaps best demonstrated using the mitochondrial enzyme cytochrome oxidase (CO) (see Fig. 2). This staining pattern was first recognized in 1978 when Margaret Wong-Riley (see Horton, 1984) noted that CO staining was darker in some layers than in others (e.g., geniculate recipient layer IV; see also below) and that there were "puffs" of increased CO activity centered in layer III. Subsequently, it became apparent that these "puffs"—which have also been called dots, patches, spots, and splotches, but which are now popularly referred to as CO blobs (Livingstone and Hubel, 1984a)—marked functionally distinct modules or subdivisions of primate V1 (see Condo and Casagrande, 1990, for review). We describe the organization and variation in the appearance of CO blobs here since these blobs are ubiquitous enough to be considered a basic feature of primate cortex (see also Wong-Riley, this volume). Blobs appear to exist in area 17 of most, although possibly not all, primates (see McGuiness et al., 1986; however, see Preuss et al., 1993). Moreover, there is evidence from macaque monkeys, squirrel monkeys, and galagos linking CO blobs and interblobs with differences in receptive field properties and connections (see DeBruyn et al., 1993, for review; see also below).

CO blobs have generally not been found in nonprimates, although two preliminary reports have suggested that similar structures can be revealed in striate cortex of cats and ferrets using special fixation procedures (Cresho et al., 1992; Murphy et al., 1991; however, see Kageyama and Wong-Riley, 1986). In all mammals, however, CO appears to stain those layers within area 17 that receive direct thalamic input more darkly than layers that do not receive such direct input. Thus, layer IV and to a lesser extent layer VI always exhibit more dense staining than layers III and V (Horton, 1984). In primates the smallest class of LGN cells (W cells) also provide a patchy input that colocalizes with the CO blobs. It may be that high CO activity in blobs generally corresponds to zones of dense LGN input. In fact, the borders of layer IV defined in a CO stain appear to match the full extent of LGN arbor terminals in layer IV. This relationship has led some investigators to define cortical laminar borders based on CO stain rather than Nissl stain; borders of layers defined in a Nissl stain do not exactly match those defined in a CO stain (e.g., layer IV appears narrower in a Nissl stain) (see Fitzpatrick et al., 1985). Other aspects of the relative distribution of CO, however, such as an uneven CO staining in layer IVα in some primates, suggest that relative levels of CO do not simply reflect thalamic input (Carroll and Wong-Riley, 1984; Condo and Casagrande, 1990). More important, defining

---

Figure 1. The laminar organization of neurons in areas 17 and 18. The sublayers of layer III are more differentiated in monkeys than galagos. The layers and sublayers are numbered according to Hässler (1967), but Brodmann's (1909) numbers are given in parentheses. The use of Brodmann's terminology assumes that a broad region of sublayers in area 17 merges at the 17/18 border (arrows) to form a narrow layer IV in area 18. The use of Hässler's terminology assumes instead that only layer IVC of Brodmann is continuous with layer IV of area 18. The transition between areas is simpler in galagos, suggesting that Hässler's view is more valid. Because of such comparative and other evidence (see text), we use Hässler's terminology. Figure adapted from Weller and Kaas (1982) with permission.

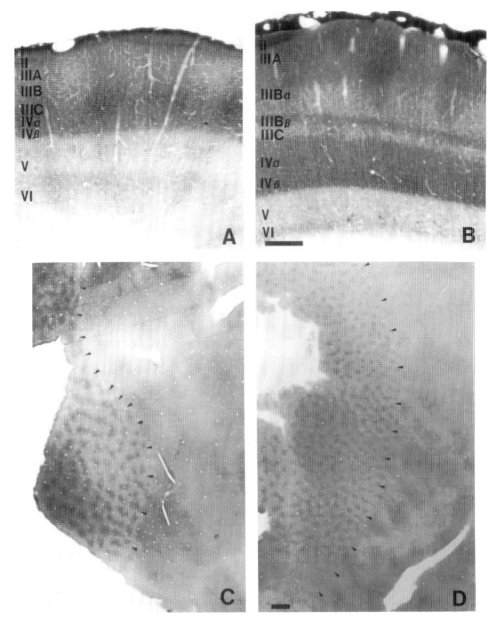

Figure 2. Cytochrome oxidase (CO)-stained sections of the striate cortex in galagos (A,C) and squirrel monkeys (B,D). Sections in A and B were cut coronal to the cortex and sections in C and D were cut tangential to the cortex after it had been unfolded and flattened. The top panels show the appearance of CO blobs in relationship to the cortical layers indicated by Roman numerals. The bottom panels show the overall distribution of the CO blobs in layer IIIB. Arrowheads indicate the border between striate cortex (i.e., area 17 and V1) and area 18 (V2). Bars = 250 μm (A, B) and 500 μm (C, D). From Lachica *et al.* (1993) with permission from the publisher.

layers with CO and/or the myriad of other markers has the potential for adding enormous confusion to the existing problem of comparing cortical layers in the same species as well as across species.

Across primates the basic pattern of CO staining is largely similar. In all primates, layer IV is densely CO positive, and CO-positive blobs appear centered in layer IIIB; weaker periodic staining aligned with these blobs occurs directly above and below the blobs as well as within the infragranular layers (Horton, 1984). A few phyletic differences in details of CO staining have also been reported. Some differences appear to be a direct reflection of differences in LGN input. Thus, as discussed in more detail below, some simian primates (e.g., macaque monkeys and squirrel monkeys) exhibit direct input from the LGN layers to a subdivision of layer III, IIIBβ. The CO staining in IIIBβ exhibits a unique honeycomb pattern that exactly matches the variation in thalamic input to this layer in these species (Humphrey and Hendrickson, 1983; see also Peters and Sethares, 1991a, for review of the structure of IIIBβ). Primates that do not have thalamic input to IIIBβ (e.g. galagos and owl monkeys) likewise show no selective increase in CO staining of this sublayer (Tootell *et al.*, 1985; Condo and Casagrande, 1990; see also Fig. 3). It is likely that humans also lack thalamic input to IIIBβ, since there is no evidence of differential CO staining of IIIBβ in human striate cortex (Horton and Hedley-Whyte, 1984; Wong-Riley *et al.*, 1993). CO blobs also appear to be centered on ocular dominance columns, in those species that exhibit ocular segregation, and variations in patterns of ocular segregation correlate with the arrangement of CO blobs (Horton and Hubel, 1981; Hess and Edwards, 1987; Rosa *et al.*, 1991). Thus, in macaque monkeys the CO blobs form elliptical patches elongated with the long axis of ocular dominance columns; in squirrel monkeys, which lack ocular dominance columns, the blobs appear round and are not arranged in elongated groups (Horton and Hubel, 1981; Humphrey and Hendrickson, 1983). Other evolved differences in CO

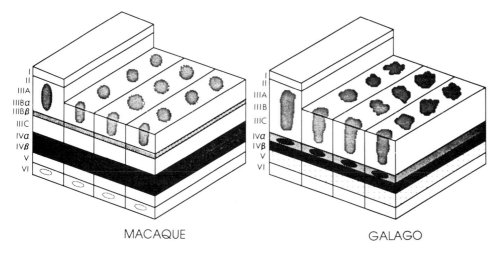

MACAQUE   GALAGO

Figure 3. Schematic drawings showing the differences between the organization of CO activity in the striate cortex of the macaque monkey (left) and galago (right). Blocks indicate ocular dominance columns. Darkened layers and patches indicate high CO activity. From Condo and Casagrande (1990) with permission from the publisher. See text for details.

staining patterns are less clearly linked to LGN input. In macaque monkeys and galagos the relative size and density of CO blobs vary with eccentricity (Livingstone and Hubel, 1984a; Condo and Casagrande, 1990); blobs are both larger and less frequent in the area of V1 representing central vision than in zones representing peripheral vision. Surprisingly, the opposite has been reported for owl monkeys (Tootell *et al.*, 1985). Phyletic differences have also been reported in the size and number of CO blobs. In nocturnal galagos CO blobs appear to occupy relatively more tangential cortical space in the region of central vision (39 versus 35%) than in macaque monkeys (Condo and Casagrande, 1990). This result does not fit with arguments that blobs play a unique role in color vision (Livingstone and Hubel, 1988); galagos are nocturnal and thus typically operate under conditions where color is not useful.

The number of blobs increases roughly linearly with estimated size of striate cortex. Thus, species with a larger V1 also have more CO blobs (Condo and Casagrande, 1990). There are exceptions, however, in that humans appear to have fewer and larger, more widely spaced blobs than do macaque monkeys (see Horton and Hedley-Whyte, 1984; Fig. 11 of Condo and Casagrande, 1990; Wong-Riley *et al.*, 1993). Beyond these quantitative evolved differences, other qualitative phyletic differences in CO patterns have been reported. For example, in owl monkeys, galagos, and humans, layer IVα shows periodic staining which aligns with the blobs in layer III; periodic staining in IVα has not been reported in squirrel monkeys (Carroll and Wong-Riley, 1984; Horton, 1984). Moreover, weak periodic staining of large pyramidal cell bodies in layer V has been observed in macaque monkeys but not in other species, and periodicity in the neuropil of layer VI has been seen in several species (see Condo and Casagrande, 1990, for review).

3. Thalamic Control: Subcortical Inputs

The major inputs to area V1 in primates and other mammals are from the (dorsal) LGN and, to a lesser extent, the nuclei of the pulvinar complex (see Kaas and Huerta, 1988). The activation of V1 appears to completely depend on the inputs from the LGN, since inactivation of the geniculate neurons also blocks visually evoked responses in V1 neurons (Malpeli *et al.*, 1981). The significance of the inputs from the pulvinar complex is unknown, but pulvinar inputs may modulate the activites of V1 neurons and alter receptive field properties relative to attention and other behavioral states (see Desimone *et al.*, 1990). Relationships with the pulvinar complex are considered briefly in a later section. Other modulating inputs originate from a number of subcortical structures reviewed recently by Tigges and Tigges (1985), and they will not be considered in detail here. Briefly, these inputs include direct serotonergic inputs from the raphe nuclei, noradrenergic connections from the locus coeruleus of the brain stem, cholinergic projections from nucleus basalis of Meynert and nucleus of the diagonal band of Broca, and inputs from a few neurons in the hypothalamus, nucleus basalis lateralis of the amygdala, and intralaminar nuclei of the thalamus. In addition, a circumscribed portion of the claustrum is reciprocally connected with V1, and activation of claustral inputs has been shown to reduce spontaneous activity and alter neuron response characteristics of V1 neurons in cats (e.g., Sherk and LeVay, 1983). This section focuses on the major input patterns from the LGN.

3.1. Geniculostriate Termination Patterns

In important ways, the geniculostriate projection pattern of primates reflects a more general mammalian pattern. In most or all mammals investigated, the major geniculostriate terminations are in layer IV, although sparser terminations have been described in the supra- and infragranular layers (e.g., cat: LeVay and Gilbert, 1976; Ferster and LeVay, 1978; mouse: Dräger, 1974; rat: Ribak and Peters, 1975; squirrel: Weber *et al.*, 1977; opossum: Sanderson *et al.*, 1980; tree shrew: Harting *et al.*, 1973; Casagrande and Harting, 1975; Hubel, 1975; Conley *et al.*, 1984; Raczkowski and Fitzpatrick, 1990). The layer IV projections originate from medium- to large-sized neurons, the physiologically defined X and Y neurons, respectively, of cats and the parvocellular, P (X-like), and the magnocellular, M (Y-like), neurons of primates. Terminations in layers I, II, and III originate from small geniculate neurons located either in layers of small cells, the koniocellular (K) layers or in intercalated (I) layers, layers of mixed small and larger cells, the superficial (S) layers, or interlaminar zones (see Casagrande and Norton, 1991, for review). These smaller LGN relay neurons typically receive inputs from the superior colliculus as well as the retina (Harting *et al.*, 1978; Lachica and Casagrande, 1993). In cats these small cells have been classified physiologically as W cells. This category is a heterogeneous population of cells grouped together mainly because of their tendency to have slower conduction velocities from the retina and larger receptive field sizes. Similar, but not identical, W-like cells have been identified in several mammalian species including tree shrews, opossums, and ferrets (see Stone, 1983; Casagrande and Norton, 1991, for review). In primates these smaller cells have only been studied in galagos, where they exhibit W-like receptive field properties (Norton and Casagrande, 1982; Irvin *et al.*, 1986; Norton *et al.*, 1988). Based on similarities in connections, it is likely that this class of small LGN cells will also be found to have W-like properties in other primate species (see Lachica and Casagrande, 1993). In part for convenience, we refer to the classes of geniculate relay neurons in primates as P (X-like), M (Y-like), and K (W-like). This nomenclature has the further advantage of neutrality in the difficult issue of whether cats, tree shrews, primates, and perhaps other mammals inherited X, Y, and W classes from a common ancestor, or evolved similar classes of retinal and geniculate neurons independently (see Kaas, 1986).

3.2. Laminar Terminations

The laminar patterns of terminations in V1 of primates have been revealed in several ways, including the use of eye injections of tracers to indirectly label geniculocortical axons via transneuronal transport, direct geniculate injections of various tracers, and by injections of horseradish peroxidase (HRP) in the white matter just beneath the cortex to label single axons and small groups of axons. The methods complement each other and provide a reasonably detailed picture of termination patterns in prosimian galagos, New World squirrel and owl monkeys, and Old World macaque monkeys. Only limited information is available for hominoids (apes and humans), but the patterns seem to be similar across primate taxa, so that supportable inferences can be made for hominoid primates. We begin here with a description of the general pattern of LGN projections to layers of cortex, and in a later section consider the details of axon arbor morphology.

The basic laminar patterns of geniculate terminations for prosimians and New and Old World simians are illustrated in Fig. 4. In galagos, geniculate axons terminate in layers IV, VI, III, and I (Glendenning *et al.*, 1976; Casagrande and DeBruyn, 1982; Florence *et al.*, 1983; Diamond *et al.*, 1985; Florence and Casagrande, 1987, 1990; Lachica and Casagrande, 1992). The projections to layer IV are the most obvious. They include M cell inputs to layer IVα where they spill over somewhat into inner layer IIIC, perhaps to terminate on the dendrites of neurons in layer IVα that extend into layer IIIC, or directly on layer IIIC neurons. P cells terminate in layer IVβ and K cells terminate in the CO blobs of layer III and in layer I. Some M and P cells also send a minor projection to layer VI.

In New World monkeys, laminar patterns of geniculate inputs into V1 have been studied in squirrel monkeys (e.g., Tigges *et al.*, 1977; Hendrickson *et al.*, 1978; Rowe *et al.*, 1978; Livingstone and Hubel, 1982; Fitzpatrick *et al.*, 1983; Weber *et al.*, 1983), cebus and spider monkeys (Hendrickson *et al.*, 1978; Florence *et al.*, 1986), owl monkeys (Kaas *et al.*, 1976; Rowe *et al.*, 1978; Diamond *et al.*, 1985; Pospical *et al.*, 1993), and marmosets (Spatz, 1979, 1989; DeBruyn and Casagrande, 1981). Laminar patterns of terminations appear to be quite similar in all of these primates. In squirrel monkeys, which have been studied more extensively, geniculate inputs are most dense in layer IV. Sparser terminations

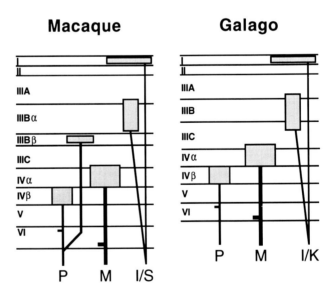

Figure 4. Schematic illustrations of the patterns of LGN axon terminations within the cortical layers of a diurnal Old World simian (macaque monkey) and a nocturnal prosimian (galago). The cortical layers are indicated using a modification of Hässler's nomenclature. The LGN relay cell pathways to cortex are indicated by: K, koniocellular [this designation includes the intercalated interlaminar (I), and superficial (S) layer cells]; M, magnocellular; P, parvocellular.

are coextensive with the CO blobs in layer III, and in layer I. As in galagos, the P geniculate cells terminate in layer IVβ, the M cells in layer IVα, and the K (interlaminar) cells in the CO blobs of layer III and layer I. In addition, and unlike galago, there is a clear projection of P cells to the inner part of layer IIIB (referred to here as layer IIIBβ). Terminations in layer VI have not been conclusively demonstrated or ruled out.

Geniculate projection patterns in other diurnal New World monkeys have not been fully studied, but they appear to be similar to those in squirrel monkeys. These projections have been more extensively investigated in the nocturnal owl monkeys, where there is again the pattern of P cell inputs to layer IVβ, M cell inputs to layer IVα, and K cell inputs to layer III blobs and layer I. A difference, however, is that there is no obvious P cell input to layer IIIBβ, which also does not appear as clear as a sublayer in Nissl preparations as it does in diurnal squirrel monkeys (Diamond *et al.*, 1985). Thus, the P cell terminations in layer IIIBβ are a feature of some, but not all, New World monkeys.

Most of what is known about geniculate termination patterns in Old World monkeys depends on studies of macaque monkeys (e.g., Hubel and Wiesel, 1972; Wiesel *et al.*, 1974; Hendrickson *et al.*, 1978; Livingstone and Hubel, 1982; Blasdel and Lund, 1983; Freund *et al.*, 1989). Again, M cells terminate in layer IVα, P cells in layer IVβ, and K cells in layer III blobs and layer I. In addition, P cells terminate densely in layer IIIBβ and some M cells and a few P cells appear to produce collateral branches that terminate sparsely in layer VI. As in galagos, the majority of direct input to layer VI appears to come from the M cell pathway.

While it has been difficult to obtain information on geniculate terminations in hominoid primates, Tigges and Tigges (1979) managed to make an eye injection and use transneuronal transport methods to determine laminar patterns of geniculate terminations in a chimpanzee that had suffered a massive stroke and had to be euthanized. Dense projections were found in layer IV, and some labeling was present in layer VI. No terminations were apparent in layers III or I, which could reflect technical difficulties, and thus the existence of such inputs in chimpanzees is still uncertain. In humans, thalamocortical terminations have been revealed in layer IV of area 17 by using silver stains in the brains of patients who died after lesions of the thalamus (Miklossy, 1992). Taken together, these results reinforce the view that there are basic similarities in the geniculate termination patterns across primate taxa. Yet, in hominoids there is no certain evidence that M and P cells terminate in different subdivisions of layer IV, or that terminations consistently exist in any of the other layers.

3.3. Ocular Dominance Columns

In many primates (see Florence *et al.*, 1986; Florence and Kaas, 1992, for review) geniculocortical inputs activated by one eye are largely separate from those activated by the other eye, forming bandlike termination zones that have been referred to as ocular dominance columns. These zones of input are more appropriately called ocular dominance bands in keeping with their shapes as viewed from the brain surface (Fig. 5). Ocular dominance bands do not occur in all primates. Moreover, they do not reflect the ancestral condition, since they are not found in other archono mammals such as tree shrews (Casagrande and Harting, 1975; Hubel, 1975; Kaas and Preuss, 1993) or most other mammals, including those with well-developed visual systems such as squirrels (Weber *et al.*, 1977). However, ocular dominance bands are found in carnivores such as cats (e.g., Löwel and Singer, 1987; Anderson *et al.*, 1988), ferrets (Redies *et al.*, 1990), and mink (LeVay *et al.*, 1987), and weakly in sheep (Clarke and Whitteridge, 1976; Clarke *et al.*, 1976; Pettigrew *et al.*, 1984) and they may exist in other unexamined taxa. The general lack of ocular dominance bands in mammals, and the presence of such bands in members of the primate, carnivore, and artiodactyl orders indicates that bands have evolved independently in at least three groups. It also appears from the distribution of such bands across primates that distinct bands have evolved, perhaps from a weak tendency, in both New World and Old World monkeys.

Ocular dominance bands are clearly expressed in all Old World monkeys examined. They have been most extensively studied in macaque monkeys, where they were described originally using physiological techniques (Hubel and Wiesel, 1968). Subsequently, they have been revealed by a number of techniques including transneuronal transport of ^{3}H-labeled amino acids injected in the eye, fiber stains and changes in the relative density of CO staining following blockade of activity from one eye (see Florence and Kaas, 1992, for review). Most recently, they have been revealed by looking at the down-regulation of immediate early gene expression following enucleation (Chaudhuri *et al.*, 1992) and by various

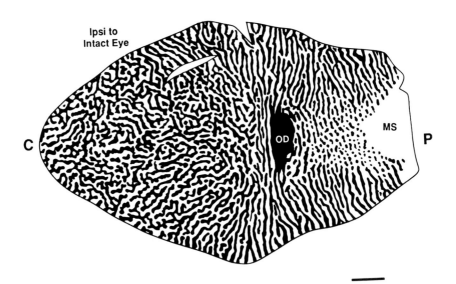

Figure 5. The complete pattern of ocular dominance bands in area 17 of a macaque monkey (*Macaca fascicularis*). The dark cytochrome oxidase regions relate to the intact ipsilateral eye (black) and light CO regions (white) relate to the suppressed contralateral eye. The figure is based on a complete reconstruction of area 17 from artifically flattened cortex cut parallel to the surface. The black oval corresponds to the projection of ipsilateral retina matched by the retina-free optic disk in the contralateral eye. The large, white area on the right corresponds to the monocular field with input only from suppressed contralateral eye. Central vision is on the left. Bar = 1 mm. Modified from Florence and Kaas (1992) with permission.

optical imaging techniques (Frostig, 1993). Although the patterns are not revealed in equal detail by each technique, the basic features revealed are very similar. Most of these features are clearly appreciated on a reconstructed representation of a flattened surface view of layer IV of V1 (e.g., LeVay *et al.*, 1985; Fig. 5). The segregation of ocular inputs in these preparations demonstrates a number of typical features: (1) The segregations occur in short bandlike segments that fuse, branch, and terminate in an irregular pattern that is similar, but nevertheless varies in detail from hemisphere to hemisphere in the same animal, and across animals. (2) The segregation involves both M and P pathways to IVα and IVβ, and the P pathway to IIIBβ. Since CO blobs are aligned with ocular dominance bands (see earlier), K inputs to blobs may demonstrate ocular segregation as well. (3) Bands are more nearly parallel and branchless in parts of V1 representing central rather than peripheral vision. The bands break down into a pattern of dots for the ipsilateral eye and larger surrounds for the contralateral eye in cortex devoted to peripheral vision. (4) Bands vary in width from central to peripheral vision with a slight decrease in average band width, and a progression from equal bands to larger bands and then surrounds for the contralateral eye. (5) Bands vary in width in different species, with the smallest bands for the small Old World talapoin monkeys and the largest bands for humans. This would suggest a relationship between body, or brain size, and band width, but since galagos (which have smaller brains) have larger ocular bands on average than talapoin monkeys, the relationship does not appear to hold across all species (see also above discussion on CO blobs, and Condo and Casagrande, 1990).

A frequent, although debatable assumption is that band width is related to the sizes of functionally significant processing units in cortex, such as hypercolumn size (see Florence and Kaas, 1992, for review). One might thus infer that processing units would vary in size across visual cortex and across species. On the other hand, this is still only inference because the arrangement of bands may be related to an original balance of developmental factors that do not relate directly to adult function (Kaas, 1988; Purves and LaMantia, 1990).

Ocular dominance bands seem to be a basic feature of all higher primates. They have been reported for humans (Hitchcock and Hickey, 1980; Horton and Hedley-Whyte, 1984; Wong-Riley *et al.*, 1993), chimpanzees (Tigges and Tigges, 1979), and all Old World monkeys examined (see Florence and Kaas, 1992, for review). There are no anatomical signs of ocular dominance bands in normal squirrel monkeys, although weak ocular periodicity has been reported from a physiological study in normal adult squirrel monkeys (Hubel and Wiesel, 1968) and there appears to be a slight tendency for anatomically defined ocular segregation after monocular rearing in this species (Tigges *et al.*, 1984). Owl monkeys and marmosets have only a weak tendency for ocular segregation (Kaas *et al.*, 1976; Rowe *et al.*, 1978; DeBruyn and Casagrande, 1981; Diamond *et al.*, 1985; Spatz, 1989), while larger New World monkeys such as *Cebus* (Hess and Edwards, 1987; Rosa *et al.*, 1988) and *Ateles* (Florence *et al.*, 1986) have obvious ocular dominance bands. Among prosimian primates, only galagos have been studied, and only a weak ocular periodicity has been demonstrated anatomically (Glendenning *et al.*, 1976; Casagrande and DeBruyn, 1982). However, recent physiological investigation of galago V1 suggests a much stronger ocular segregation (DeBruyn *et al.*, 1993). This difference raises a note of caution given that many studies of ocular segregation have used transneuronal transport of tracers that could diffuse across LGN layers, especially in species with narrow

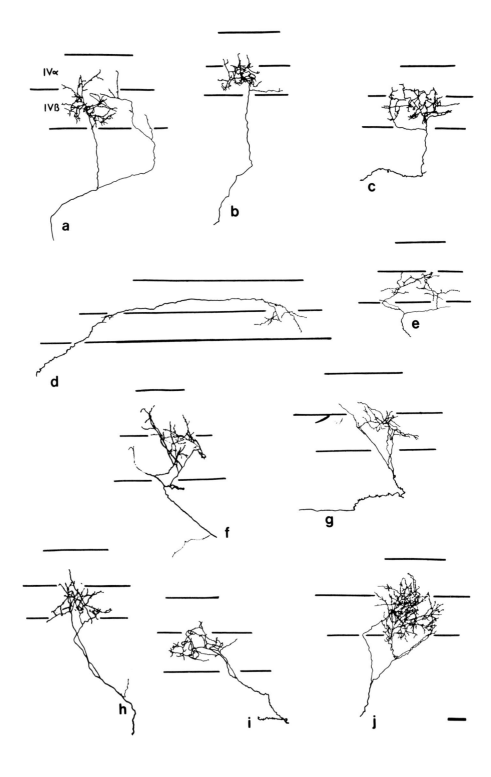

LGN layers. Such technical artifacts probably do not account for major phyletic differences in ocular segregation, but more subtle differences should be verified with more than one method. The tendency for LGN axons to segregate into ocular dominance bands may be directly related to brain size (see Rosa *et al.*, 1988) or other factors indirectly related to size such as eye separation and ocular disparity (see Florence *et al.*, 1986). This issue is complicated by a surprising lack of data on what ocular dominance columns actually contribute to visual function, and the fact, as mentioned above, that several species, such as tree shrews and squirrels, have excellent visual performance without ocular dominance columns.

3.4. Axon Arbors

In general, across mammals, axon arbors that terminate in layer IV constitute the majority of inputs to area V1. They originate from the larger geniculate cells with thicker axons, and they branch profusely in layer IV forming a dense array of synaptic swellings or boutons. In cats (Humphrey *et al.*, 1985b) and tree shrews (Muly and Fitzpatrick, 1992; Usrey *et al.*, 1992), these larger LGN cells can generally be subdivided in two groups that exhibit either small, single, more compact arbors (the X class in cats) or larger sprawling arbors that sometimes terminate in separate patches (the Y class in cats). These arbors may extend into layer III, and both classes may have branches that terminate in layer VI. In cats the arbors of X and Y cells overlap extensively in layer IV (Humphrey *et al.*, 1985a). Moreover, in cats Y axons project outside of area 17, particularly to area 18 (Humphrey *et al.*, 1985a); M axon arbors in primates do not appear to innervate extrastriate areas (however, see Benevento and Yoshida, 1981). Axons from the smallest LGN relay cell class (the W cells in cats) are thinner. As in primates, some of these thin axons terminate in layers III and I. However, the details of terminations of these W cells in area 17 in cats and tree shrews differ from those found in primates in several ways which are described in more detail below in the context of intrinsic cortical connectivity.

The main characteristics of geniculostriate axon arbors of the primate visual system are similar to those of cat and tree shrew. Arbors of individual axons have been described for several species including prosimian galago (Florence and Casagrande, 1987; Lachica and Casagrande, 1992), New World owl monkey (Pospical *et al.*, 1993), and Old World macaque monkey (Blasdel and Lund, 1983; Freund *et al.*, 1989). Detailed analysis of arbors of all three LGN classes has only been done in galagos. Other studies, however, have reported that, at least for P and M cells, the basic morphological characteristics of the arbors are similar. In galagos, the P cell axons produce small, single, dense terminal fields largely or exclusively restricted to layer IVβ. In both macaques and galagos, differences

---

Figure 6. Composite of P axons serially reconstructed from the calcarine fissure of striate cortex in lesser galagos showing the differences in their morphology. These axons ramify primarily in layer IVβ and occasionally (see f and h) project to layer VI. The axon designated d is unusual in that the axon trunk runs parallel to the cortical layers for a considerable distance before branching at the terminal focus. Comparison of the terminal arbor size of these axons with M axons in the same cortical area (Fig. 7) demonstrates a significant size difference. From Florence and Casagrande (1987) with permission.

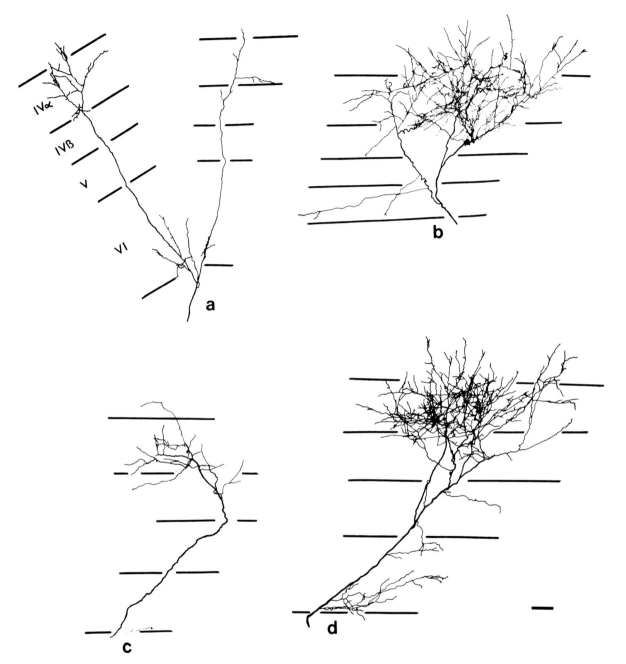

Figure 7. Composite of M arbors serially reconstructed from the dorsal surface of striate cortex showing the extent of their variation in lesser galagos. Solid lines indicate borders of cortical layers; Roman numerals identify layers according to Hässler's nomenclature. Note that all axons arborize primarily in IVα but also project to layer III and to layer VI. In some cases, the course of the axon after reaching the white matter has not been illustrated to conserve space. Bar = 50 μm. From Florence and Casagrande (1987) with permission.

have been reported in P cell arbors terminating in layer IV, but these differences probably reflect within-class variation (see Fig. 6). In macaque monkey, however, a separate type of P axon terminates in IIIBβ (Blasdel and Lund, 1983). This type of axon, which has a distinct morphology with a narrowly focused tangentially spreading arbor, is unique to those species which exhibit geniculate terminations in layer IIIBβ (Blasdel and Lund, 1983). In both galago and macaque monkey, the M arbors exhibit a more variable morphology than the P arbors and are, on average, significantly larger in area (Blasdel and Lund, 1983; Florence and Casagrande, 1987). Some M arbors spread broadly and extend branches into lower IIIC as well as occasionally into IVβ. Other M arbors are more narrowly confined to IVα. Still others bifurcate and terminate in two patches which are generally smaller in size than M arbors that terminate in a single patch (see Fig. 7). It is unclear whether these differences reflect actual subclasses or variation within a class (see Lund, 1988). Some M and occasional P axon arbors extend a few branches into layer VI. These branches appear to be restricted such that M arbors terminate primarily on cells in the lower half of layer VI, whereas P collaterals, when present, are restricted to the upper half of layer VI. Note that, unlike galago and macaque monkey, no collaterals were found to project to layer VI in owl monkey (Pospical *et al.*, 1993). However, in galago and macaque monkey the arrangement of the projections to layer VI fits with the observation that projections to P and M layers tend to arise from cells in the upper and lower divisions of layer VI, respectively (Lund *et al.*, 1975; Lachica *et al.*, 1987).

Quantitative comparisons of the distributions of M and P axons in galago cortex show that both arbor types are significantly larger in the area of layer IV innervated in the zone of area V1 representing central vision than in cortex representing the visual periphery (Florence and Casagrande, 1987). Although such quantitative comparisons of axon area across layer VI are not available for other primate species, these size differences fit with the proposed proportional increase in magnification of the representation of central vision over peripheral vision in V1 versus in the LGN (see Florence and Casagrande, 1987, for discussion). These differences are also reflected in the relative sizes of ocular dominance columns and the average diameter of M arbors, although P arbors are on average much smaller than the width of an ocular dominance column. In addition, M and P arbors tend to be elongated parallel to ocular dominance columns in galagos. This relationship suggests that their shape may be constrained by binocular interactions in species that have ocular dominance columns. However, given that P arbors are much smaller than ocular dominance columns, anisotropies in shapes of arbors may simply reflect anisotropies in the topography of V1 (Tootell *et al.*, 1982; Van Essen *et al.*, 1984).

The K arbors have been described only in one primate, galago (Lachica and Casagrande, 1992). In this primate, all K axons terminate within CO blobs in layer III; no axons are found that terminate within interblob zones. This relationship can be appreciated by comparing the overall pattern of input from the K cells with the location of CO blobs on adjacent stained sections (Fig. 8). In addition to inputs to CO blobs, K arbors extend collateral branches to layer I where arbors spread tangentially over a broad zone. Branches in layer I clearly extend beyond the boundaries of underlying parent arbors in the CO blobs in layer III. Thus, K arbors could have transcompartmental influence via contacts with apical dendrites extending into layer I. Within a CO blob column, the focus of most K arbors is in cortical layer IIIB. However, some K arbors are centered

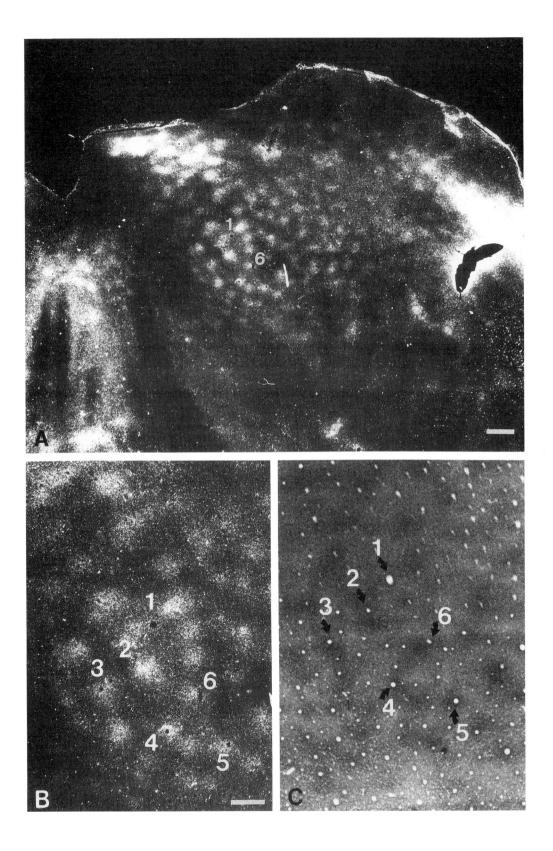

in either IIIA or IIIC (see Fig. 9). Thus, the K pathway is in a position to directly influence cells in the major pathways that project to extrastriate cortex that exit via layers IIIA or IIIC (see also below). The distribution of K arbors in cortex also reinforces the view that blob and interblob columns have different functions and that these differences may extend to cells in layers above and below the blobs. This view is further supported by differences in vertical intrinsic connections in blob and interblob cortex.

4. Intrinsic Connections

The physiological properties of cells in V1 are very different from those seen at the level of the retina and LGN. Moreover, many properties seen in V1 for the first time can be identified in higher-order visual areas to which V1 projects. Therefore, it is clear that a key to understanding the functional significance of visual cortical organization lies in determining how inputs to V1 are transformed into new output streams which subserve different extrastriate visual areas. One way to begin to understand how inputs are transformed into outputs in V1 is to examine the details of its internal circuitry, or wiring. Presumably, aspects of such circuitry that are basic to all primates (as well as to other mammals) are of fundamental importance to visual cortical function. In keeping with the purposes of this chapter we begin with a review of these basic primate features of V1 circuitry. We then compare this organization with that seen in other mammals. Finally we examine features of intrinsic circuitry that appear to differ between primate groups.

4.1. Background

The intrinsic circuitry of V1 of cortex has been examined in only a few primate species. The most extensive work has been done in macaque monkeys (e.g., Lund, 1988, 1990; see also Lund, this volume). In fact, many details of V1 cell morphology and connections have been described only in macaque monkeys. Recent studies, however, have begun to provide information on intrinsic connections in three other species, namely galagos, squirrel monkeys, and owl monkeys, and some information is now available for humans (i.e., Burkhalter and Bernardo, 1989; Miklossy, 1992). Several features of intracortical circuitry which relate V1 inputs to outputs appear to exist in all four of these primates. We consider these consistent features here.

Figure 8. (A) Low-magnification (caudal to the top, medial to the left), dark-field photomicrograph of striate cortex that has been flattened, cut tangential to the surface, and reacted for TMB histochemistry to reveal the patchy pattern of K-cell geniculostriate terminations in layer III. A higher-magnification photomicrograph of a small field marked in A (by Nos. 1 and 6) is shown in B. Blood vessels in B are identified by number so that they may be easily matched with blood vessels in C, which shows an adjacent section stained for cytochrome oxidase. Bars = 1 mm (A) and, 500 μm (B, C). From Lachica and Casagrande (1992) with permission.

As described earlier, the main inputs to V1 arise from the LGN via three distinct pathways: the parvocellular (P), magnocellular (M), and the koniocellular (K) or its equivalent (Fig. 4). The P and M pathways terminate principally in the upper and lower tiers of granular layer IV (IVα and IVβ), and to a lesser extent, in the lower and upper tiers of layer VI, respectively. The K pathway terminates only within the CO blobs in layer III and in layer I.

The output pathways are considered in detail in a separate section below, but there is some necessary background to any consideration of the internal wiring of V1. There are three pathways to extrastriate cortical areas that have been consistently identified across primate taxa. The first output pathway arises primarily from supragranular layer IIIC and projects both directly to the middle temporal visual area (MT) and to distinct compartments in V2, which then projects to MT; the second and third pathways arise from layer IIIA CO blobs

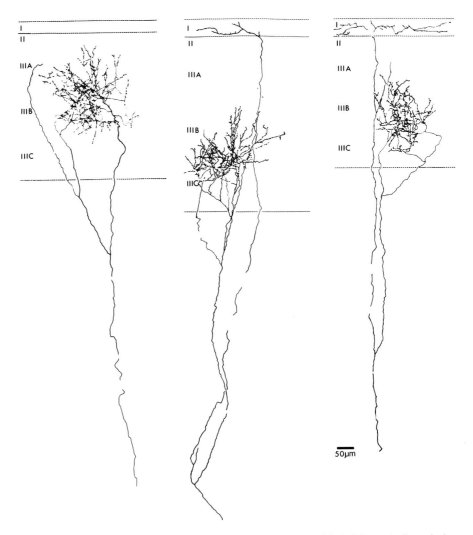

Figure 9. Reconstructions of three different varieties of PHA-L-labeled K geniculocortical axon arbors. The majority of K geniculostriate axons terminated in layer IIIB, and had collateral branches that arborized in layer I. Modified from Lachica and Casagrande (1992) with permission.

and interblobs, respectively, and each projects to separate zones within V2 (Rockland and Pandya, 1979; Tigges *et al.*, 1981; Livingstone and Hubel, 1984b, 1987; Krubitzer and Kaas, 1990b; Van Essen *et al.*, 1990; Rockland, 1992; see Lachica *et al.*, 1993, for review). There are also several major pathways to subcortical areas and these arise from infragranular layers V and VI, which project to several distinct zones in the thalamus, midbrain, and pons (see Kaas and Huerta, 1988).

4.2. Basic Primate Plan

The intrinsic connections between the input and output pathways within primate V1 are exceedingly complex, as illustrated in the many elegant studies of Lund and colleagues in macaque monkeys (e.g., Lund, 1988, 1990; see Lund, this volume). Since the details of projections of individual classes of cells have been reviewed in detail recently (Lund, 1990; Henry, 1991), they will not be considered here. There are three major features of internal connections that have been consistently observed across primate species. First, most layers in V1 send and receive heavy vertical projections from several other layers as well as inputs from outside V1. Thus, the direction of flow of information is not strictly serial. For example, in macaque monkeys layer IV receives input from layers V and VI as well as from the LGN; layer IV, in turn, sends projections back to layer VI (itself an LGN target) and also to layers IIIB and IIIC, which themselves get input from layers V and VI (Blasdel *et al.*, 1985; Fitzpatrick *et al.*, 1985; Lund, 1987, 1990; Lachica *et al.*, 1992, 1993). In spite of these complexities, a consistent suborder can be discerned in specific circuits, particularly those that relate to the output pathways. Layer IIIC, which, as mentioned above, sends the largest projection to extrastriate area MT, receives its major projection directly from M-recipient layer IVα (see Fig. 10 and Fitzpatrick *et al.*, 1985; Lachica *et al.*, 1993). In contrast, layer IIIA, which provides the major output to area V2, does not appear to get any direct input from M- and P-recipient IVα and IVβ; instead, signals from both M- and P-recipient divisions of layer IV appear to be initially relayed to IIIB, IIIC, and other layers before reaching IIIA (see Fig. 11 and 12 and Fitzpatrick *et al.*, 1985; Lachica *et al.*, 1992, 1993). This arrangement suggests that cells in IIIB may function as a set of interneurons specifically for the construction of new output signals in IIIA (Fitzpatrick *et al.*, 1985; Lachica *et al.*, 1992, 1993). Since layers IIIC and IIIA form the initial substrates for information entering the two proposed major processing streams for analysis of object location (or "where") and object identification (or "what"), respectively (Ungerleider and Mishkin, 1982), the differences in intrinsic wiring of cells in IIIC and IIIA may offer clues concerning the initial coding of information for these basic visual functions. The lower layers also show some sublaminar specialization in all primates. Layer V exhibits a distinct thin subdivision at its upper border termed VA by Lund and colleagues (Lund, 1987). Layer VA appears to be particularly well developed in simian primates and has been described in macaque monkeys as a set of interneurons (Lund, 1987). This zone, unlike the remainder of layer V, shows connections with IVα (Lund, 1987; Lachica *et al.*, 1993) and also exhibits distinct connections with the remaining layers (Lund, 1987; Lachica *et al.*, 1993). Our data suggest that, in simian primates, the cells in the upper and lower halves of layer V show differences in connections with the CO blob and

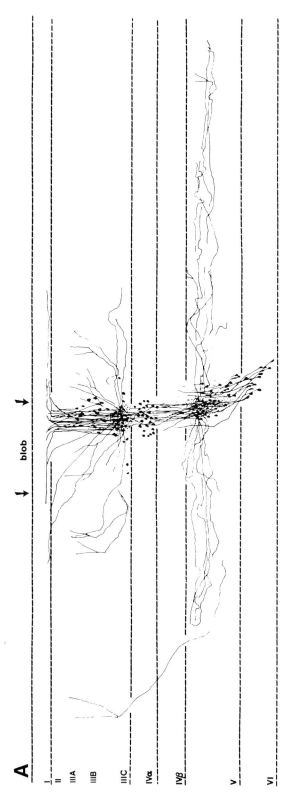

Figure 10. Camera lucida reconstructions showing labeled cells and axons following injections into blob (A) and nonblob (B) layer IIIC in galago. Note labeled cells are absent from layer IVβ following all injections into IIIC. Roman that numerals indicate layers according to a modification of the nomenclature of Hässler, Reproduced from Lachica *et al.* (1993) with permission. See text for details.

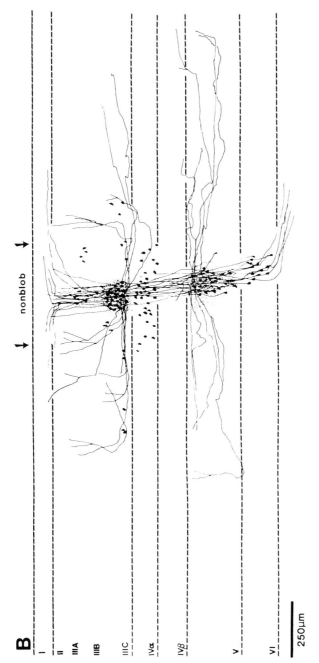

Figure 10. (*Continued*)

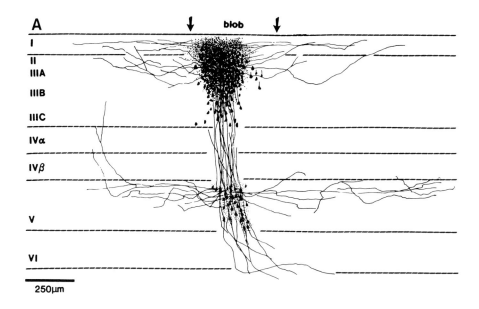

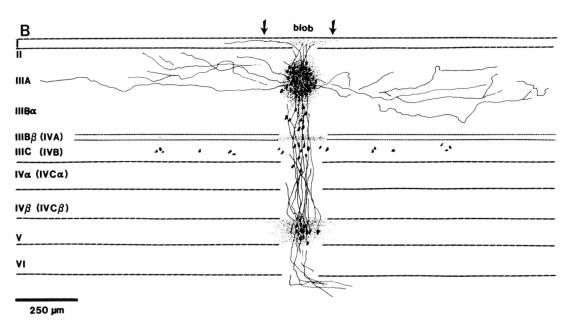

Figure 11. Camera lucida reconstructions showing labeled cells and axons following injections into blob IIIA in a galago (A) and a squirrel monkey (B). Arrows indicate the edges of a blob. Roman numerals indicate layers according to a mod- ification of the nomenclature of Hässler with Brodmann nomenclature in parentheses. Reproduced from Lachica *et al.* (1993) with permission.

interblob compartments and with the substrata of layer III (Casagrande *et al.*, 1992; Lachica *et al.*, 1992, 1993). Examples are shown in Figs. 12 and 15. Layer VI shows a clear division in all primates examined such that the lower division sends and receives input from the M LGN layers and its target layer IVα, while the upper division relates in a similar manner to the P pathway (Lund *et al.*, 1975, 1988; Blasdel and Lund, 1983; Swadlow, 1983; Florence and Casagrande, 1987). Each division of layer VI shows somewhat different patterns of vertical connections with the supragranular layers and CO compartments (Blasdel *et al.*, 1985; Lund *et al.*, 1988; Lachica *et al.*, 1992, 1993).

A third related feature found consistently in primate V1 is the existence of different intrinsic connections of modules identified by CO staining. Within layer III the CO blob and interblob compartments exhibit differences in patterns of vertical connectivity with other layers or sublayers. For example, within IIIB the CO blobs receive indirect input via layer IV from *both M and P* LGN pathways as well as a direct input from the K LGN axons (see Figs. 12–14). Thus, the cells in the CO blobs in layer IIIB are in a position to integrate signals from *all* LGN relay cell classes. In contrast, cells in the IIIB interblobs appear to receive a more restricted input, mainly via layer IV from one relay cell class (either M or P depending on the primate species), and there is no direct input from the K pathway (unless on dendrites extending into layer 1). Examples of patterns of retrogradely labeled cells in different cortical layers following injections into interblobs are shown in Figs. 12 and 15. This pattern of connections does not mean that interblobs are solely influenced by one LGN channel. In fact, physiological studies in which input from either the M or P pathway was blocked pharmacologically at the level of the LGN suggest that interblob cells do get input from both M and P pathways in macaque monkeys (Nealey and Maunsell, 1991; Maunsell *et al.*, 1992). What our anatomical studies suggest is that interblob cells must get this input via a different route than do blob cells, either via cortical layers

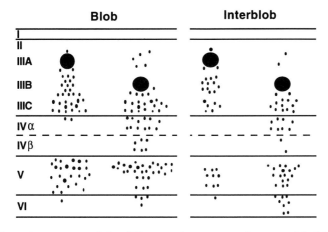

Figure 12. Schematic summary of the differences in patterns of retrogradely labeled cells seen following injections of tracers (biocytin or HRP) into layer IIIA or IIIB blobs and interblobs in an owl monkey. Note that IIIB blobs get input from both M- and P-recipient IVα and IVβ, respectively. Layer IIIB interblobs get input primarily from IVα in this nocturnal primate. Roman numerals indicate layers according to a modification of the nomenclature of Hässler. From Casagrande *et al.* (1993).

below IV or via interconnections with cells in more superficial layers (Fitzpatrick *et al.*, 1985; Blasdel *et al.*, 1985; Casagrande *et al.*, 1992; Lachica *et al.*, 1992, 1993). Note that cells in layers IIIA and IIIC that lie directly above or below the centers of the IIIB CO-rich and -poor compartments also connect with distinct divisions of layers V and VI (Lachica *et al.*, 1992, 1993). In other words, CO blob and interblob differences in intrinsic connections appear to extend beyond the center of blobs and interblobs in layer IIIB to include cell groups within the same vertical cortical domain (see Fig. 11; also compare the patterns shown in Figs. 14 and 15).

These modular differences in wiring reinforce the view that compartments defined by relative CO staining provide substrates for creating separate output pathways. This proposal is further supported by physiological data showing that, as reported for the diurnal simians (macaque monkeys and squirrel monkeys), CO blobs and interblobs can be distinguished in nocturnal prosimians (galagos) (Livingstone and Hubel, 1984a, 1987; Silverman *et al.*, 1989; Hubel and Livingstone, 1990; Born and Tootell, 1991a,b; DeBruyn *et al.*, 1993).

Finally, V1 of primates is characterized not only by elaborate vertical connections, but also by periodic tangential connections that connect cell groups within layer III and within layer V (Rockland and Lund, 1983; Livingstone and Hubel, 1984b; Ts'o *et al.*, 1986; Cusick and Kaas, 1988b; Ts'o and Gilbert, 1988; Gilbert and Wiesel, 1989). These tangential connections are not diffuse, but

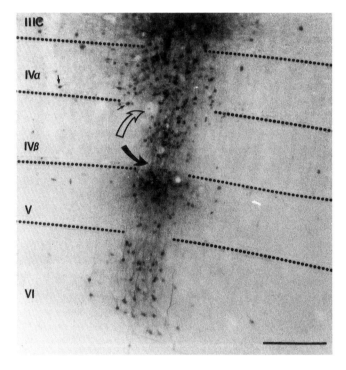

Figure 13. A photomicrograph illustrating the pattern of retrogradely labeled cells following a biocytin injection and into a blob in layer IIIBα of a squirrel monkey. Note that layers IVα, IVβ, and V can be defined solely on the basis of the distribution of labeled cells. Labeled cells are much more widely distributed in layer IVα, some lying well away from the projection column (small arrow). The top (open arrow) and bottom (closed arrow) borders of the narrow column of labeled cells in IVβ are easily defined. Bar = 100 μm. Reproduced from Lachica *et al.* (1993) with permission.

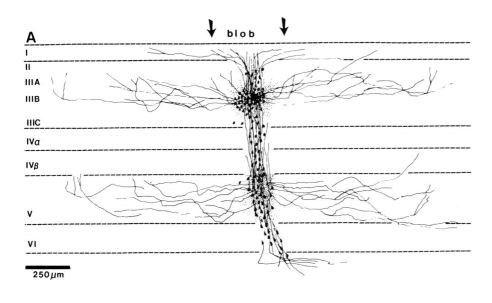

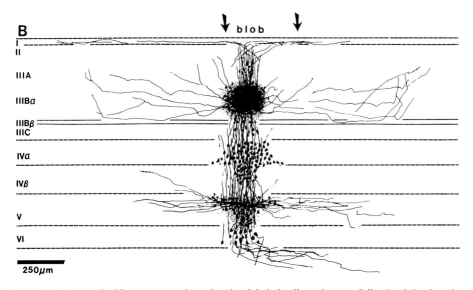

Figure 14. Camera lucida reconstructions showing labeled cells and axons following injections into CO blobs in IIIB in a galago (A) and IIIBα in a squirrel monkey (B). Roman numerals indicate layers according to a modification of the nomenclature of Hässler. Reproduced from Lachica *et al.* (1993) with permission. See text for details.

form highly organized lattices of periodic connections principally, but not exclusively (see Malach *et al.,* 1992), between cells that occupy anatomically and physiologically similar compartments (e.g., connections between CO blobs in III; Rockland and Lund, 1983; Livingstone and Hubel, 1984b; Ts'o *et al.,* 1986; Cusick and Kaas, 1988b; LeVay, 1988). At present, it is still unclear what purpose these lateral interconnections serve. It has been suggested that they may form substrates for feature-linking (Gilbert *et al.,* 1991). Regardless, the presence of periodic tangential interconnections is a feature that has not been described at earlier stages of processing (i.e., the retina or LGN), but appears to be ubiquitous in V1 and other visual cortical areas to which V1 projects.

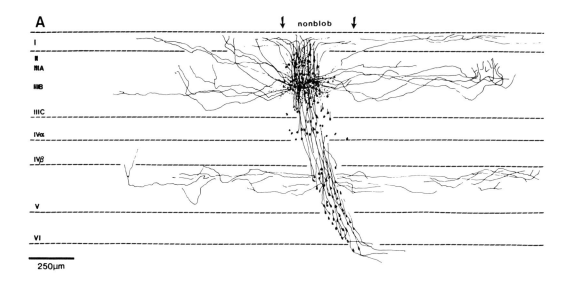

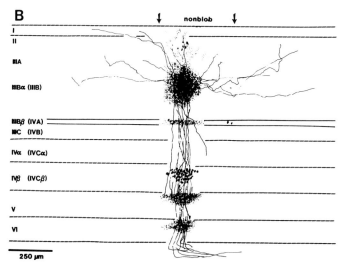

Figure 15. Camera lucida reconstructions showing labeled cells and axons following injections into nonblob IIIB in a galago (A) and nonblob IIIBα in a squirrel monkey (B). Roman numerals indicate layers according to a modification of the nomenclature of Hässler. Reproduced from Lachica *et al.* (1993) with permission. See text for details.

4.3. Nonprimate Pattern

Details of intrinsic cortical connections have only been described for area 17 or V1 of a few mammals. It is not our purpose here to provide an extensive review of detailed similarities and differences. Rather, in this section we examine some of the main features of patterns that other mammals share with primates as well as key points of difference. We restrict our review here primarily to cats and tree shrews.

All mammals appear to share several basic features of V1 organization with primates. Thus, as in primates, inputs from the LGN terminate principally in layer IV (see earlier). In addition, LGN inputs to layers VI and I as well as a portion of layer III have been widely described (e.g., Hendrickson *et al.*, 1978; Diamond *et al.*, 1985). However, there are clear species differences in the detailed organization of LGN inputs to cortex. Two examples make this point.

In cats, LGN cells have been classified in X, Y, and W cell types. However, as described earlier, X and Y cells project in an overlapping fashion to layer IV of area 17 (Humphrey *et al.*, 1985a). W cells appear to project to separate strata in cats, but unlike their counterparts in primates, these cells in cats project both to the upper half of layer V (Gilbert and Wiesel, 1981) and to layer I. In addition, a large number of LGN cells in cats send projections to visual areas outside of area 17 (see Sherman and Spear, 1982). The same appears to be true of other carnivores (e.g., ferrets and mink). Thus, unlike in primates, area 18 in carnivores gets a substantial projection from Y and W LGN cells, and W LGN cells also send axons to a number of visual areas beyond area 18 that are themselves targets of area 17 and/or 18 (e.g., Humphrey *et al.*, 1985b).

In tree shrews, as in primates, area 17 appears to be the major or almost exclusive target of LGN cell axons (Harting *et al.*, 1973; Casagrande and Harting, 1975; Hubel, 1975; Conley *et al.*, 1984; Usrey *et al.*, 1992). However, the organization of LGN layers and their projections to area 17 differ for tree shrews and primates. In tree shrews, X and Y cells have been described in the LGN (Sherman *et al.*, 1975). However, a unique feature of the LGN of tree shrews is the segregation of neurons that respond to the onset and offset of light in the centers of their receptive fields (ON-center versus OFF-center cells) (see Conway and Schiller, 1983; Holdefer and Norton, 1986). The tree shrew LGN has pairs of layers, one innervated by each eye, with layers for ON-center cells (layers 1 and 2) and layers for OFF-center cells (layers 4 and 5) as well as two contralaterally innervated layers (3 and 6) with physiologically distinct W-like cells. Projections of tree shrew LGN cells to cortex maintain the segregation of ON, OFF, and W-like cells. Unlike primates, the upper and lower divisions of layer IV receive separate projections from the ON- and OFF-center LGN layers, respectively, while the W-like layers each show a unique pattern of projections to a combination of layers including III and I (Conley *et al.*, 1984; Usrey *et al.*, 1992). Tree shrew geniculocortical projections to cortex also show a species-unique horizontally stratified pattern of ocular inputs such that ipsilateral inputs end in restricted bands at the top and bottom of layer IV and contralateral inputs extend throughout layer IV (Casagrande and Harting, 1975; Hubel, 1975; Conley *et al.*, 1984; Usrey *et al.*, 1992).

As with LGN inputs, there are features of area 17 intrinsic organization that appear to be common to primates and other mammals as well as features that are not found in primates. As in primates, area 17 of other mammals is organized such that the supragranular layers provide major projections to extrastriate cortical areas and the infragranular layers send major projections out of cortex to sites in the thalamus, midbrain, and hindbrain (e.g., Henry, 1991). The latter projections appear to consistently involve a major pathway from cells in layer V to the superior colliculus and a major reciprocal pathway form layer VI to the LGN (see Swadlow, 1983, for review). Within the cortex itself, several intrinsic patterns have been reported in both primates and other mammals. Thus, layers IV and VI appear to connect; both are also direct targets of the main layers of the LGN. Also, layers V and III have heavy interconnections (see Henry, 1991). Finally, tangential periodic connections appear to be a characteristic feature of V1 in at least some primate and most nonprimate mammals (Rockland *et al.*, 1982; Sesma *et al.*, 1984; Burkhalter and Charles, 1990).

Since one function of area 17 is to combine information from parallel LGN inputs to construct appropriate output channels, the differences in intrinsic circuitry appear to relate to differences in the organization of output pathways. For example, in tree shrews the projections from layer IV to layer III have three separate tiers of innervation within layer III (see Muly and Fitzpatrick, 1992). However, this organization is very different from that just described for primates. In tree shrews, the projections of layer IV to layer III reflect the segregation of ocular dominance and the segregation of ON- and OFF-center cells within layer IV. Muly and Fitzpatrick (1992) have postulated that the differential projections from layer IV to layer III may help to combine ON- and OFF-center pathways in a way that preserves ocular bias. The only similarity to primate organization is that in both tree shrews and primates, stratification in layer IV is reflected in differential patterns of projections within layer III. It is unclear whether the same rule applies to cats where projections from different cell types largely overlap in layer IV. However, it does appear that in cats, cells in the lower portion of layer III project heavily to area 18, while output cells that lie in the more superficial portion of layer III send projections to separate extrastriate visual areas (Symonds and Rosenquist, 1984). The latter observation suggests that functionally relevant stratification may commonly exist in layer III and that its organization may reflect specializations both in the input and in the output pathways.

4.4. Primate Variations

As might be expected from the diversity and behavioral adaptations in primates, some species differences in the intrinsic connections of V1 have been reported. Although only a few studies have addressed the issue of primate species differences in intrinsic circuitry, three observations stand out. First, comparisons of organization of V1 in prosimian galagos (*Galago*) with that of the New and Old World simians, squirrel monkeys, owl monkeys, and macaque monkeys (*Saimiri, Aotus, and Macaca*), suggest that V1 in galagos is less differentiated and compartmentalized. In Nissl-stained sections, laminar boundaries appear less distinct in galagos than in the three simian primates (see Fig. 1). Cells giving rise to the major output pathways in galagos also appear slightly less precisely confined to subdivisions in layer III than do output cells in simians. In addition, sublaminar distinctions of intrinsic connections of layers V and VI with substrata and CO compartments of layer III appear quite distinct in macaque monkeys and squirrel monkeys even though these connections appear only as gradients in galagos (Blasdel *et al.*, 1985; Lachica and Casagrande, 1992, 1993). These species differences may also reflect nocturnal–diurnal adaptations since the intrinsic connections of subdivisions of layers V and VI in the nocturnal simian owl monkeys are also less sharp than those of their diurnal New World squirrel monkey cousins (Casagrande *et al.*, 1992; see also Fig. 12).

Nocturnal–diurnal specializations appear to correlate with a second difference in intrinsic V1 connectivity in primates. In the diurnal macaque monkeys and squirrel monkeys, cells in the CO-poor interblob zones receive connections indirectly via IVβ from the LGN P pathway, whereas in nocturnal owl monkeys and galagos these zones receive connections indirectly mainly via IVα from the M LGN pathway (see Figs. 12 and 15 and Lachica *et al.*, 1992, 1993). Note that differences in the relative strength of input and output pathways have also been found to correlate with nocturnal–diurnal niche specialization. Thus, the ratio of P to M LGN cells in areas of the LGN devoted to central vision is lower for nocturnal primates [e.g., 40 to 1 for macaque monkey and 4 to 1 for galago (Connolly and Van Essen, 1984; Florence and Casagrande, 1987)]. Differences may simply reflect acuity differences between nocturnal and diurnal species (Langston *et al.*, 1986). However, species differences are also seen in the proportion of tissue devoted to homologous extrastriate areas (see next section). Relevant to the present argument is the fact that nocturnal primates have proportionately larger amounts of tissue devoted to the middle temporal visual area (MT) than do diurnal primates (Krubitzer and Kaas, 1990b). Since area MT receives M-dominated input via layer IVα, it may be that M-pathway signals provide information that is particularly important to vision in dim illumination (see also discussion of this point in Lachica *et al.*, 1993). Alternatively, it may be more appropriate to conclude that the P pathways are expanded and more important in diurnal primates.

A third species difference in V1 intrinsic connectivity relates to differences in cortical laminar specialization. In the diurnal catarrhine simians (e.g., macaque monkeys and other Old World monkeys), as well as some diurnal platyrrhine simians (e.g., squirrel monkeys), a thin subdivision of layer IIIB, IIIBβ, receives a specialized input from the P layers of the LGN. As discussed earlier, IIIBβ is absent in prosimian primates. It is weakly evident in some species even when LGN input is lacking, as is the case in the owl monkey (see Fig. 1). In any case, in species that possess a distinct IIIBβ that receive P LGN input (squirrel monkeys and macaque monkeys), it is clear that this sublayer has intrinsic connection patterns with other cortical layers that are distinct from the connections of the remaining sublayers of layer III (IIIC, IIIBα, and IIIA) (see Lund and Yoshioka, 1991; Peters and Sethares, 1991b). At present, it is unclear why some simians possess a well-developed layer IIIBβ, while others do not. No differences in the LGN laminar location of P cells that innervate IIIBβ versus IVβ have been identified. Physiological studies suggest that cells in IIIBβ of macaque monkeys reflect P inputs from the LGN as do cells in IVβ, yet again no distinctions between these two layers have been reported that would help to identify why the P LGN cells split their input to V1 to two layers in some primate species and not others.

5. Ipsilateral Cortical Connections

In all primates examined, V1 projects to several extrastriate areas that are similar in location and which presumably constitute homologous visual areas across primate taxa (see Fig. 16). However, this conclusion is not straightforward, since the terminologies used by different investigators vary and schemes for subdividing extrastriate cortex in primates vary across species. The major cortical target of V1 is the second visual area, V2 or area 18, a visual area present in most or all mammals. A second target is the middle temporal visual area, MT, also known as the superior temporal area or V5. A third target is the dorsomedial area, DM, also known as dorsal V3 and V3a. A fourth target is the

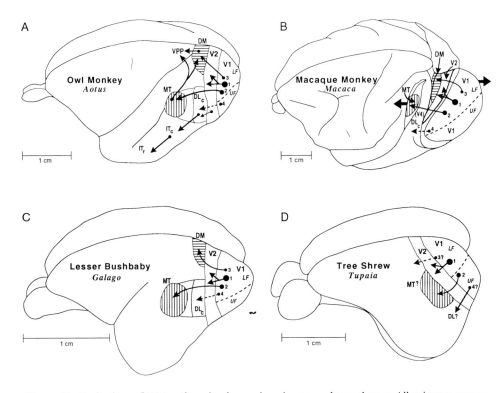

Figure 16. Projections of V1 to other visual areas in primates and tree shrews. All primates appear to have projections to the second visual area, V2, the middle temporal visual area, MT, and the dorsomedial visual area, DM (solid arrows 1–3 with dot size and arrow thickness reflecting the proportions of the projections). In B the large filled arrows indicate where sulci are opened to reveal areas which are normally buried from view within sulci (e.g., MT). The lateral part of V1 representing central vision also projects to the caudal division of the dorsolateral complex, DL_c (dashed arrow 4). Tree shrews (D) are close relatives of primates and they have some of the primate patterns of V1 projections. As in all mammals, V1 projects directly to V2 of tree shrews (projection 1). Other projections (2) are to a lateral field that adjoins V2, but may be the primitive location of the precursor of MT (see Kaas and Preuss, 1993). Other projections are to cortex that may correspond to DM (3) and DL (4) in primates. In V1 of A–D, a dashed line divides the representation of the lower field (LF) from the upper field (UF) in V1. In A (owl monkey), additional projections from MT to DM and ventral posterior parietal cortex (VPP) are shown to illustrate parts of the dorsal stream (Ungerleider and Mishkin, 1982) of processing, while projections from DL_c to caudal (IT_c) and then to rostral (IT_r) divisions of inferior temporal cortex reflect parts of the ventral stream (Weller and Kaas, 1987). These streams appear to exist in all primates.

caudal division of the dorsolateral complex or DL$_c$, also known as V4. Other targets have been inconsistently reported and may be species-variable, individually variable, or artifactually-labeled false positives.

5.1. Connections with V2

In all primates studied, the majority of the cortical projections of V1, perhaps 80% or more, are to the second visual area, V2. The architectonic equivalent of V2 is often considered to be area 18 of Brodmann (1909), but since Brodmann distinguished area 18 as a field that varied in extent across species, from closely corresponding to V2 in a New World monkey to being nearly twice as large in an Old World monkey, the identification of area 18 with V2 can be misleading. Here we define area 18, in contrast to the historical area 18, as the field that is coextensive with a systematic, second-order representation of the contralateral visual hemifield or V2 (Allman and Kaas, 1974).

A complication that may have hindered earlier attempts at an architectonic description of V2 is that the field is not structurally homogeneous. In many mammals, V2 appears to be modularly organized so that the density of myelination and connections with V1 and with the other hemisphere are unevenly distributed (e.g., Kaas *et al.*, 1979; Sesma *et al.*, 1984; Cusick and Kaas, 1986b). However, this modular specialization seems more pronounced in simian primates. The architectonic manifestations of the modular organization of V2 are most apparent in brain sections from cortex that have been stained for CO or myelin (Fig. 17; Krubitzer and Kaas, 1989). In such preparations, V2 can be identified by a series of dense CO bands separated by light CO bands, the CO interbands. The bands are roughly perpendicular to the V1/V2 boundary and they span the width of V2. When V2 is characterized by these bands, the area can be seen as a long, narrow belt along approximately 90% of the outer border of V1 and occupying a surface area less than V1 (Krubitzer and Kaas, 1990b). The bands shorten and the field narrows in the middle of the belt in the portion representing central vision, and toward the ends of the belt, representing the extremes of peripheral vision of the upper and lower quadrants (see Allman and Kaas, 1974; Gattass *et al.*, 1981). Although it is difficult to obtain brain sections where all the bands are evident, it appears that owl monkeys have roughly 30–40 CO-dense bands and the same number of interbands (e.g., Kaas and Morel, 1993) while macaque monkeys have about 56 CO-dense bands (Van Essen *et al.*, 1990). The interbands are further distinguished by dense myelination, so that CO and myelin procedures reveal patterns of dense staining (Fig. 17; Krubitzer and Kaas, 1990a).

The CO-dense bands have been subdivided into two alternating types. One set of bands projects to MT (e.g., Weller *et al.*, 1984; DeYoe and Van Essen, 1985; Ship and Zeki, 1985) and another set of bands projects to caudal DL (DL$_c$) or caudal V4 (DeYoe and Van Essen, 1985; Shipp and Zeki, 1985; Cusick and Kaas, 1988a). In macaque monkeys, the set of bands projecting to MT have been called "thick" bands, and the set projecting to DL$_c$ (V4$_c$) "thin" bands (Livingstone and Hubel, 1982), but the widths of the bands do not seem to reliably indicate the type of projection in macaque monkeys (e.g., Van Essen *et al.*, 1990). In macaque monkeys and apparently humans (Hockfield *et al.*, 1983), the monoclonal antibody Cat-301 preferentially labels bands projecting to MT (Hockfield *et al.*,

1983; DeYoe *et al.*, 1990). In some New World monkeys, the thinner bands appear to project to MT, whereas in other New World simians the bands are not notably different in width (see Fig. 17 and Krubitzer and Kaas, 1990b). Thus, we define the bands projecting to MT as MT bands or dorsal stream bands, because they send information to MT and subsequently into the parietal lobe, and the bands projecting to DL_c-V4_c as ventral stream bands because information from DL_c-V4 is sent into a hierarchy of visual areas in the temporal lobe (Ungerleider and Mishkin, 1982). We avoid the terms M bands and P bands (see DeYoe and Van Essen, 1985) to refer to these zones in V2 since the terms M and P imply that LGN input pathways can be equated with specific V1 output pathways or their functions. As discussed previously, cells in the output streams leaving V1 have very different properties from the M, P, or K input cells of the LGN, and this difference undoubtedly reflects the complex intrinsic circuitry of V1. The CO-dense bands have an internal structure, so that at high magnifications they may

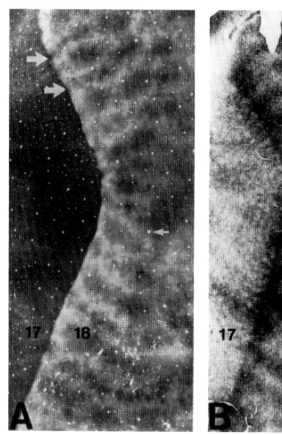

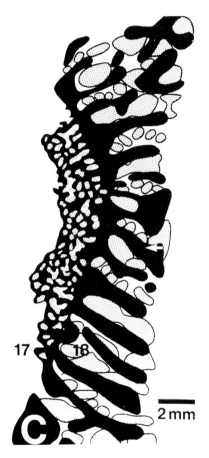

Figure 17. Pattern of cytochrome oxidase (CO)- and myelin-dense regions in areas 17 and 18 of squirrel monkeys. (A) A section stained for CO cut at the level of layer IV parallel to the surface of flattened cortex. Layer IV stains densely in area 17, while alternating CO-dense and -light bands characterize area 18. Large arrows mark two CO-dense bands. (B) A slightly more superficial section stained for myelin. In area 17, denser myelination surrounds and outlines CO blobs of adjacent sections. The myelin-dense bands in area 18 correspond to the CO-light or interbands of panel A. (C) A schematic with the CO-dark (shaded) and myelin-dark (black) regions superimposed. A small arrow in A–C marks one of the blood vessels used to align brain sections. From Krubitzer and Kaas (1989) with permission.

appear as rows of CO-dense patches that nearly fuse (Carroll and Wong-Riley, 1984; Cusick and Kaas, 1988a; Tootell and Hamilton, 1989; Krubitzer and Kaas, 1990b). Finally, CO-dense stripes in V2 appear to be absent or poorly developed in prosimians (Condo and Casagrande, 1990; Krubitzer and Kaas, 1990b; Preuss *et al.*, 1993), although bandlike patterns of connections indicate that V2 is functionally organized in a similar manner in both simian and prosimian primates.

The outputs of V1 are related very clearly to the banding pattern in V2. Small injections of tracers into the blob regions of V1 label neurons and terminals in the ventral stream bands of V2 while injections in the interblob regions label the pale interbands (Livingstone and Hubel, 1984a). Output neurons to V2 originate in layer IIIA of V1 (Rockland and Pandya, 1979; Lund *et al.*, 1981; Tigges *et al.*, 1981; Rockland, 1992) and from a few neurons in layers V and VI, at least in macaques (Kennedy and Bullier, 1985). Small injections in V2 indicate that neurons in these locations project to the ventral stream bands and interbands (Livingstone and Hubel, 1987). In contrast, injections in the dorsal stream bands of V2 label neurons in IIIC of V1 (Livingstone and Hubel, 1987). As a result of these patterns of connections, randomly placed injections in V2 can label different modules in V1 depending on the bands affected by the injections (Cusick and Kaas, 1988a). Thus, sections parallel to the brain surface and through layer III of V1 can reveal zones of labeled neurons and fine processes in CO blobs or interblob regions in a pattern that reflects the involvement of ventral stream bands or interbands (Fig. 18).

Because of the modular nature of V2, the terminal patterns of the projections of V1 to V2 can be quite patchy (see Weller *et al.*, 1979; Wong-Riley, 1979; Lin *et al.*, 1982; Cusick and Kaas, 1988b), reflecting both the matching of modular output zones and target zones, and the patchy nature of the bands. In addition, the fact that V2 is a "second order" representation, split along the horizontal meridian (see Allman and Kaas, 1974), means that injections placed along the representation of the horizontal meridian in V1 label separate locations in the dorsal and ventral wings of V2.

5.2. Connections with MT

The other major target of V1 is the middle temporal visual area (MT), a small oval of densely myelinated cortex in the upper temporal lobe that contains a systematic representation of the contralateral visual hemifield (Allman and Kaas, 1971). Unlike V2, MT is a visual area that has only been definitely identified in primates. Yet, all mammals have projection targets in cortex rostral or lateral to V1, and one of these targets may represent a less specialized homologue of MT (see Kaas, 1993; Kaas and Preuss, 1993). Cortex in the region now identified as MT has been known to receive inputs from V1 (Kuypers *et al.*, 1965; Myers, 1965) since well before MT was identified as a visual area by its other characteristics. Subsequently, evidence from a range of primate species, including humans (Clark and Miklossy, 1990; Kaas, 1992), has indicated that MT is part of the basic primate plan of cortical organization, and likely exists in all primates (see Kaas and Krubitzer, 1992; Preuss *et al.*, 1993).

In simians, the projections to MT originate from IIIC neurons and from a scattering of different types, depending on species, in layer V along the layer VI border and variably in layer VI (e.g., Lund *et al.*, 1975; Tigges *et al.*, 1981; Shipp and Zeki, 1985). In prosimian galagos, the neurons projecting to MT are more widely distributed, such that although the majority are in IIIC, some are also located in upper sublayers of layer III (Diamond *et al.*, 1985). Thus, IIIC, as an output layer to MT, seems to be more differentiated and specialized in simians than in prosimians. Nevertheless, cells in the inner part of layer III are the major source of MT projections from V1 in all primates. The projections to MT are topographic, in that retinotopic locations in the two representations are interconnected (e.g., Symonds and Kaas, 1978; Spatz, 1979; Lin *et al.*, 1982; Weller and Kaas, 1983). However, the projections from any location in V1 to MT are patchy (e.g., Montero, 1980), suggesting the existence of separate classes of processing modules in MT (Fig. 19; see Kaas, 1986). The projections of V1 to MT terminate in layer IV, where they constitute the major source of activation (Kaas and Krubitzer, 1992; Maunsell *et al.*, 1992), but not the only one, since in macaque monkeys there is evidence that many neurons in MT remain responsive to visual stimuli after the inactivation of V1 (Rodman *et al.*, 1990; see Bullier, this volume). As a major component of the dorsal stream of visual processing for visual attention and spatial aspects of vision (Ungerleider and Mishkin, 1982; Goodale and

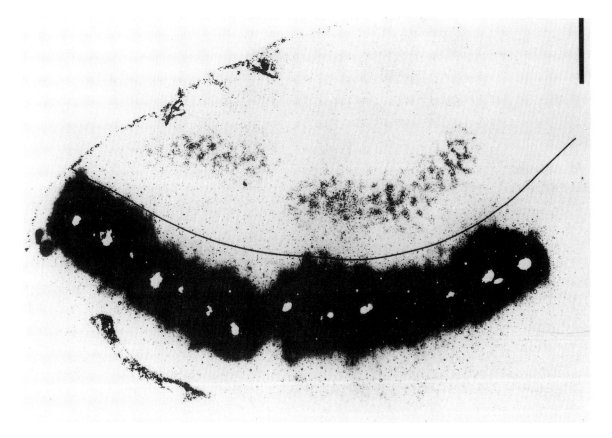

Figure 18. A row of injection sites in area 18 of a squirrel monkey and labeling in area 17. The effective injection sites (around the holes) are smaller than the dense band of label in area 18 so that bands and interbands were randomly involved. As a result, label in area 17 was sometimes concentrated in interblob surrounds in upper and lower 17 and sometimes in blobs (middle 17). Cortex was flattened before sectioning, and processed for HRP. Bar = 2 mm. From Cusick and Kaas (1988) with permission.

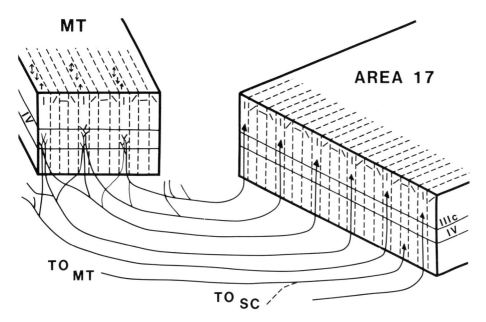

Figure 19. Proposed patterns of divergence and convergence of projections from area 17 and MT. Neurons of specific orientation and direction of movement selectivity in area 17 project to reveal groups of neurons with similar selectivity in MT. Likewise, each group of neurons in MT receives from several matched groups of neurons in area 17. These features of connectivity would account for the anatomical patterns of connections actually observed (e.g., Krubitzer and Kaas, 1990). From Kaas (1986) with permission.

Milner, 1992), the outputs of MT are directed to posterior parietal cortex and to DM (see below), which relays directly to posterior parietal cortex (see Krubitzer and Kaas, 1990a).

5.3. Connections with DM (Dorsal V3)

A third target of V1, that appears to exist in all primates, is located in the dorsomedial cortex just rostral to V2 (Krubitzer and Kaas, 1993). In owl monkeys this cortex has been called the dorsomedial visual area (DM) (Allman and Kaas, 1975), and it contains a systematic, but split or second-order, representation of the contralateral visual hemifield. Injections of tracers in V1 (Fig. 20) label reciprocal connections with area DM (Lin *et al.*, 1982; Krubitzer and Kaas, 1990a, 1993). V1 projections to DM terminate in layer IV and thus may constitute the feedforward or major visual drive for cells in this area, although it is unclear whether the less direct inputs from MT and V2, which are also dense, actually provide the main visual drive for DM cells. The projections from V1 to DM originate from a scattering of neurons that are located primarily within CO blobs. These projections probably constitute less than 5% of the total output of area V1. Feedback projections from DM to V1 originate mainly in layer V neurons, with a few originating from cells in layer III; these projections terminate in supragranular layers of V1.

DM likely exists in all primates since V1 sends axons to cortex in the location of DM in all investigated primates. However, DM has been established as a visual area by additional morphological and physiological criteria only in owl monkeys. In Old World macaque monkeys, cortex in the region of DM has long been

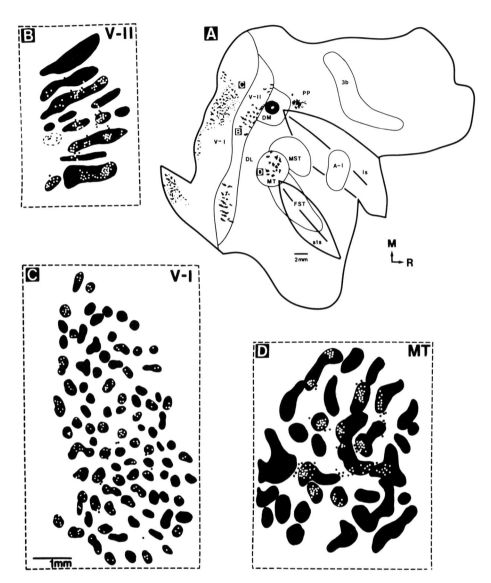

Figure 20. Connections of VI, VII, and MT with the dorsomedial area (DM) in owl monkeys. (A) The connection pattern on cortex that has been separated from the brain, flattened, and cut parallel to the surface. VI, VII, DM, the middle temporal visual area (MT), the fundal superior temporal area (FST), the middle superior temporal area (MST), auditory cortex (A1), and somatosensory cortex (area 3b) were determined architectonically. The dark circle indicates the injection of WGA-HRP in DM, and black dots indicate labeled neurons and terminations in other areas. (B) Projections from VII originated from neurons (dots) that were largely in the CO-dense bands (black). (C) Projections from VI were from neurons largely in CO-dense blobs. (D) Projections from MT were from neurons largely in CO-dense bands. From Krubitzer and Kaas (1990) with permission.

considered to be part of a visual beltlike region, V3, bordering V2 (e.g., Myers, 1965; Cragg, 1969; Zeki, 1969), but the concept of a single beltlike V3 with direct inputs from V1 is inconsistent with the evidence given that the dorsal and ventral parts of V3 have different characteristics (Burkhalter *et al.*, 1986; Van Essen *et al.*, 1986). In particular, ventral "V3" lacks a projection from V1. However, the region defined as dorsal "V3" in macaque monkeys differs from DM in that DM constitutes a complete representation of the contralateral hemifield, whereas dorsal V3 represents only the lower visual quadrant. Nevertheless, cortex adjoining dorsal V3, termed V3a by Zeki (Zeki, 1980a), also receives inputs from V1, is densely myelinated like DM (and like dorsal V3), and represents the upper visual quadrant (Gattass *et al.*, 1988). Thus, dorsal V3 and V3a of macaque monkeys, together, may be homologous to DM in owl monkeys (see Krubitzer and Kaas, 1993). Projections to a densely myelinated area in the expected location of DM have been found in other Old World monkeys (talapoins), New World squirrel monkeys and marmosets, and prosimian galagos (Cusick and Kaas, 1988b; Weller *et al.*, 1991; Krubitzer and Kaas, 1993). It seems unlikely that the dorsomedial target of V1 projections represents different visual areas in different primates. Rather, it seems more parsimonious to conclude that the projections are to DM, a single area common to all primate groups.

DM appears to be linked by its inputs from MT and its projections to posterior parietal cortex (Fig. 20), to the dorsal stream of visual processing concerned with visual attention and localization (Ungerleider and Mishkin, 1982). Yet, direct inputs from the CO blobs of V1, as well as both the dorsal and ventral stream bands in V2 indicate that DM has inputs associated with the ventral as well as the dorsal stream. The ventral stream is critical for object recognition (Ungerleider and Mishkin, 1982). Thus, DM is a visual area with access to both streams of information processing, and it may play a central role in providing ventral stream information to posterior parietal cortex.

5.4. Connections with Dorsolateral Cortex (DL or V4)

V1 also projects to a cortical region that lies between V2 and MT called the dorsal lateral area (DL) in owl monkeys (Allman and Kaas, 1974) and in most other primates (see Steele *et al.*, 1991, for review), and V4 in macaque monkeys (Zeki, 1971). The main output of $DL–V4_c$ is to the inferior temporal cortex (see Steele *et al.*, 1991, for review). It has been difficult to define the exact borders of this region, and uncertainties remain, but it now appears that the region contains three visual areas, each with a crude representation of the visual field. In brief these areas are: (1) a caudal division, DL_c, which is more densely myelinated, expresses more CO activity (Cusick and Kaas, 1988b; Steele *et al.*, 1991; Preuss *et al.*, 1993), and receives the majority of forward projections of V2 (Cusick and Kaas, 1988b); (2) a more rostral division, DL_r, which has connections that associate it with the dorsal rather than the ventral stream of processing (Cusick and Kaas, 1988b; Steele *et al.*, 1991; Weller *et al.*, 1991); and (3) a narrow ringlike area around most of MT that has been called the MT crescent or MT_c (Kaas and Morel, 1993). The region termed "V4t" of macaque monkeys (Ungerleider and Desimone, 1986) appears to correspond to MT_c of owl monkeys. Evidence for these three visual areas, or subdivisions thereof, within the DL or V4 complex largely depends on research in New World monkeys (e.g., Steele *et al.*, 1991), but there is evidence for these areas also in Old World monkeys (see Perkel *et al.,* 1986); in prosimians (e.g., Preuss *et al.,* 1993) there is evidence for DL_c and DL_r but not yet for MT_c. While the major input to DL_c is from V2, a moderate input is from V1 (e.g., Steele *et al.* 1991; for review: Zeki, 1978; Lin *et al.,* 1982; Cusick and Kaas, 1988a; Krubitzer and Kaas, 1993). This layer VI input originates mainly from the part of V1 that represents the central few degrees of vision, but since receptive fields for DL neurons are large, this input terminates over much of the middle portion of DL_c. The input to DL_c is from a scattering of supragranular cells throughout V1, and the major feedforward output of DL_c is to caudal inferior temporal cortex, although DL_r and MT also receive minor inputs.

The connections of DL_r have been less fully described (see Cusick and Kaas, 1988b; Steele *et al.,* 1991). The inputs from V2 are sparse, in contrast to DL_c, and the inputs from V1 are very weak, if present (however, see Preuss *et al.,* 1993). Little is known about the connections of MT_c (see Kaas and Morel, 1993). Cortex in the region of MT_c, however, has been described as having connections with V1 in *Cebus* (Sousa *et al.,* 1991) and macaque monkeys (Perkel *et al.,* 1986).

5.5. Connections with Other Fields

In addition to the connections of V1 described above, other connections have been reported occasionally (Perkel *et al.,* 1986; Sousa *et al.,* 1991; Rockland, this volume). Extremely sparse and variable connections between V1 and posterior parietal cortex as well as cortex medial to DM have been described (Perkel *et al.,* 1986; Sousa *et al.,* 1991). These areas include area M (Allman and Kaas, 1976), also referred to as PO (Colby *et al.,* 1988), the inferior temporal cortex, and the medial superior temporal area (MST) just rostral to MT. To the extent that such connections exist, they would seem to play an extremely minor role in visual processing.

5.6. Feedback Connections

All of the major targets of V1 also project back to V1, and thus these connections are reciprocal. As first stressed by Wong-Riley (1979), Rockland and Pandya (1979), and Tigges *et al.* (1981), and subsequently by others (e.g., Maunsell and Van Essen, 1983; Burkhalter and Bernardo, 1989), the feedback connections to V1 have different laminar origins and terminations than feedforward projections from V1. Feedforward projections originate largely in layer III of V1 and terminate largely in layer IV of V2, MT, DM, and DL_c, while in these same areas neurons in layers V, VI, and to a lesser extent layer III project back to supragranular and infragranular layers of V1 (see Rockland, this volume). Feedback from both areas V2 and MT terminates in the supragranular layers. However, feedback from these areas to the infragranular layers differs such that V2 projects to layer V and also layer IIIC of area V1; MT projects to layer VI. There are area-specific differences in the laminar patterns of feedback as well as feedforward connections. In addition, feedback connections have the general feature of being less topographically specific than feedforward connections. Most notably, the feedback connections from MT to V1 are broadly distributed in a manner that suggests that the feedback from orientation and direction of movement columns in MT include mismatched as well as matched columns in V1 (Shipp and Zeki, 1985; Krubitzer and Kaas, 1989). Also, the feedback connections from MT to V1 are denser in the interblob regions of layer III, although cells in both blobs and interblobs appear to get input from MT. Thus, feedback from MT as well as other targets of V1 has the potential to influence the neurons that provide their own inputs from V1 as well as outputs to other areas.

6. Callosal Connections of V1

The general impression of many investigators is that V1 of primates is completely devoid of callosal connections. This incorrect impression stems, in part, from the results of early studies of V1 in macaque monkeys, using less sensitive degeneration methods, which failed to find evidence for callosal connections (e.g., Myers, 1965), and, in part, from subsequent reports where only the outer margin of V1 was found to have callosal connections (e.g., Kennedy *et al.*, 1985). With more information the picture has changed. The use of more sensitive axonal-transport tracing procedures, coupled with studies of a greater range of species, have led to the general conclusion that area 17 in all mammals investigated, including primates, has callosal connections. The extent and magnitude of these connections are quite variable across species. Nevertheless, a consistent feature of this pathway is that connections are always concentrated along the V1–V2 border (see Cusick and Kaas, 1986b, for review).

In most mammals, area 17 has rather extensive callosal connections. Even in opossums, which lack a corpus callosum, most of area 17 contributes to strong interhemispheric pathways which travel through an anterior commissure (Cusick and Kaas, 1986b). Tree shrews, generally regarded as close relatives of primates, demonstrate the typical pattern with the majority of callosally projecting neurons densely packed within 0.5–1.0 mm of the 17/18 border, but with some cells also projecting callosally from more peripheral regions of area 17 (Cusick *et al.*, 1984). The callosally projecting neurons in most mammals originate in layer IIIC and in layer V, and terminations are concentrated in the same layers. Weyland and Swadlow (1980) first demonstrated that the callosal connections of V1 in prosimians (galagos) can be substantial when they showed that labeled cells extend several millimeters into V1 after HRP injections in the other hemisphere. Subsequent studies have shown that the callosal pattern in primates is quite variable. Several aspects of the species differences in callosal connection patterns of V1 are apparent in Fig. 21 (Cusick *et al.*, 1984; Cusick and Kaas, 1986b). First, as described by Weyland and Swadlow (1980), callosal connections of V1 in galagos do extend inward with decreasing density several millimeters from the border. Moreover, away from the border, the connections have a patchy pattern that matches the pattern of CO blobs. The projecting neurons and terminations are concentrated in layer IIIC and in layer V. This pattern further reinforces the view that CO blob and interblob compartments are not limited to layer IIIB but may involve all cortical layers within a vertical column.

In contrast to galagos, dense callosal connections extend only a short distance into V1 of New World owl monkeys, and hardly at all into V1 of macaque monkeys. Again, the callosally projecting cells are located in both layer IIIC and layer V, although the majority are in layer IIIC. Terminations of callosal axons are found in the same layers, but terminations seem to be more widely distributed than projecting cells, suggesting that individual callosal projection neurons have large axon arbors (see also Newsome and Allman, 1980; Kennedy *et al.*, 1986;

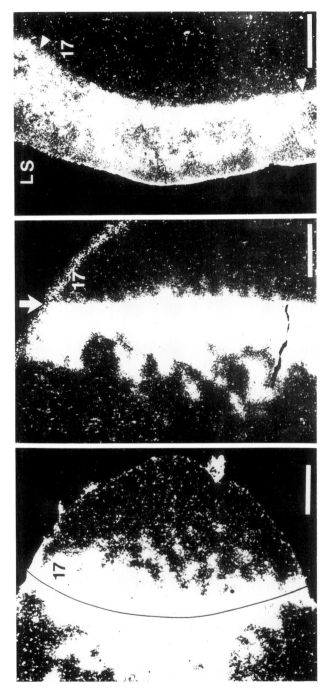

Figure 21. Callosal connections of area 17 in primates. The dark-field photomicrographs are of cortex cut parallel to the surface after injections of HRP in the opposite hemisphere. The 17/18 border is marked in each section by a line or arrow. The label extends from the 17/18 border into area 17 for several millimeters in prosimian galagos (left), only slightly in owl monkeys (middle), and hardly at all in macaque monkeys (right). Bars = 1 mm. LS, lunate sulcus. From Cusick and Kaas (1986a) with permission.

Gould *et al.*, 1987; Kennedy and Dehay, 1988). In owl monkeys, patches of connections are periodically spaced along the V1 border. Diurnal New World monkeys such as squirrel monkeys (Gould *et al.*, 1987) and marmosets (Cusick *et al.*, 1984) appear to have a more restricted pattern than nocturnal New World owl monkeys, so that they more closely resemble diurnal Old World monkeys.

Callosal connections in V1 of primates may subserve some of the basic functions of intrinsic connections for parts of V1 near the border where the horizontal spread of intrinsic connections would be truncated. Nocturnal primates with lower visual acuity and larger receptive fields in V1 apparently have more widespread intrinsic connections in V1 (see Cusick and Kaas, 1988b), and more widespread callosal connections would complement the widespread intrinsic system. Thus, the patterns of callosal connections of V1 in primates closely reflect the patterns of intrinsic connections, and thus they are likely to mediate similar functions.

7. Subcortical Connections

Area 17 projects to a variety of subcortical targets. In all mammals examined, these projections have been described as arising from cells in layers V and VI. As a rule, projections from layer V target a variety of zones in the thalamus, midbrain, and pons, with a strong projection to the superior colliculus being a consistent feature. In contrast, the subcortical targets of cells in layer VI are principally limited to the thalamus and always include a feedback pathway to the LGN.

In primates, cells in layer V of area V1 project to a number of targets, the main ones being the pulvinar, superior colliculus, pretectum, and pons (see Kaas and Huerta, 1988, for review). In galagos and macaque monkeys, similar pyramidal cells in layer V appear to send projections to both the inferior pulvinar and the superficial layers of the superior colliculus (Trojanowski and Jacobson, 1976; Raczkowski and Diamond, 1980), although double-label studies have not been performed to prove this point. However, double-label studies have shown clearly that the large Meynert cells at the lower border of layer V project to both cortical area MT and the superior colliculus (Fries *et al.* 1985). Thus, area 17 pyramidal cells in layer V can innervate multiple targets via collaterals (see also O'Leary and Stanfield, 1985).

Projections to the superior colliculus are retinotopically organized. The corticocollicular projection, however, terminates within the deepest zone of the superficial gray layer and within the upper portion of the stratum opticum, only partially overlapping the projection from the retina (e.g., Huerta and Harting, 1984). In macaque monkeys, it has been argued that the V1 pathway to the colliculus relates most strongly to the M layers of the LGN since silencing the M layers of the macaque LGN or cooling V1 also inactivates cells in the deep layers (i.e., below the stratum opticum; Schiller *et al.*, 1974). Because V1 does not project directly to the deep collicular layers, it is not entirely clear how the information from the M LGN layers and V1 influence the deep collicular neurons. Moreover, in macaque monkeys, galagos, and squirrel monkeys, it is clear that colliculogeniculate projections avoid the M layers and end in the interlaminar zones and the K layers and their equivalents (see Lachica and Casagrande, 1993). Thus, the superior colliculus does not appear to directly modulate M layer activity that is relayed to V1.

Projections from V1 to the pulvinar are also complex and appear to involve zones that get input from cells in the superior colliculus that, themselves, are innervated by projections from V1. At least two divisions of the pulvinar complex have reciprocal connections with V1. The pulvinar is commonly divided into inferior, lateral, medial, and anterior divisions or "nuclei" although each division may actually include several nuclei (see Kaas and Huerta, 1988). Both the central nucleus of the inferior pulvinar and the lateral nucleus of the lateral pulvinar project to V1 and receive topographically organized projections from V1 (Benevento and Rezak, 1976; Ogren and Hendrickson, 1977; Symonds and Kaas, 1978; Carey *et al.*, 1979; Graham *et al.*, 1979; Lin and Kaas, 1979; Rezak and Benevento, 1979; Raczkowski and Diamond, 1980; Graham, 1982; Ungerleider *et al.*, 1983; Dick *et al.*, 1991). Two points are noteworthy regarding inputs to the primate pulvinar. First, it has been demonstrated for macaque monkeys that in regions where projections from V1 and the colliculus overlap (e.g., the inferior pulvinar) in the pulvinar, the main drive for cells in the pulvinar is from the cortex, not the colliculus (Bender, 1983); cells in the inferior pulvinar do not respond to visual stimulation in the absence of input from V1. However, this dependence on V1 may not be the case for other mammals, such as tree shrews, that are able to see well without striate cortex, but not without the temporal lobe target of the pulvinar (Snyder and Diamond, 1968). Second, in macaque monkeys the pulvinar also sends patchy input back to layers I and II of V1 (Ogren and Hendrickson, 1977; Rezak and Benevento, 1979). The pulvinar also projects back to V1 in other primates, but it is not clear if the patterns are identical (see Kaas and Huerta, 1988, for review).

In addition to projections to the colliculus and pulvinar, cells in layer V of primate V1 show projections to several pretectal nuclei (Graham *et al.*, 1979; Hoffman *et al.*, 1991), and to a set of specific zones in the pons. In macaque monkeys, projections from V1 to the pons terminate in several eye-movement-related nuclei (Glickstein *et al.*, 1980).

As mentioned earlier, the major subcortical projection of the cells in layer VI of area V1 is to the LGN. In macaque monkeys and galagos, it has been shown that separate tiers of layer VI project to specific classes of LGN cells from which they also receive a minor direct LGN projection. Preliminary data in galagos indicate that the projections back to the LGN from cortex may be even more complex with some axons innervating pairs of LGN layers as well as the reticular nucleus, and others innervating a single layer, or several interlaminar zones (Lachica *et al.*, 1987, and unpublished). Patterns of projections from layer VI to the LGN in other mammals suggest that considerable variability exists across species (e.g., Swadlow, 1983). Thus, in ferrets single axons have been shown to innervate functionally distinct components of the LGN (e.g., the C layers, A layers, and the interlaminar zones) as well as the perigeniculate nucleus (Claps and Casagrande, 1990). In other species such as tree shrews, V1 projections to the LGN appear to mainly concentrate in the interlaminar zones (Brunso-Bechtold *et al.*, 1983).

Layer VI also sends a projection to the thalamic reticular formation (e.g., Symonds and Kaas, 1978). The latter appears to be part of a loop involving the LGN. Thus, the "visual" portion of the thalamic reticular formation projects heavily to and receives from the LGN. This loop forms a portion of the inhibitory circuitry that regulates the flow of information through LGN relay cells (see

Casagrande and Norton, 1991). Projections from V1 to the LGN and reticular nucleus provide a means by which the cortex can regulate its own input by engaging this inhibitory circuitry or by projecting directly back to cells that send information to layer IV.

8. Conclusions

In humans and other primates, V1 appears to be critical for object recognition and conscious perception (Weiskrantz, 1986; Bullier, this volume). In contrast, some nonprimate species, such as tree shrews, retain good object recognition and spatial localization despite the complete removal of area 17 and subsequent complete degeneration of the LGN (Snyder and Diamond, 1968). The relative importance of V1 to primate vision is also reflected by the size of this area, which occupies nearly 20% of neocortex in both prosimians and New and Old World simians (Felleman and Van Essen, 1991; Krubitzer and Kaas, 1990b). In primates, V1 also represents the main link between visual signals from the eye, via the LGN, to all other visual cortical areas. Thus, primate V1 represents the first staging area for the reorganization of perceptually relevant visual information to be distributed and utilized for perception in subsequent target visual areas. There are across-species similarities in the way parallel inputs and outputs of V1 are organized and relate to intrinsic circuitry in primates. The intrinsic circuitry allows new output pathways to be created from the parallel inputs to V1 in order to support further analysis by higher visual areas located in the dorsal and ventral streams (see Martin, 1992; Merigan and Maunsell, 1993; DeBruyn *et al.*, 1993). In this review we focused on the anatomical framework for the transformation and distribution of signals in V1 of primates, highlighting those features that are common across primate taxa. Several conclusions can be drawn from such comparisons. Some of the connections described earlier are also summarized in Fig. 22.

1. The widespread use of Brodmann's nomenclature for cortical layers, defined in Nissl stain, in V1 of primates generates confusion and error, and this terminology should be replaced with one compatible with both traditional architectonic and current experimental observations. In brief, the evidence indicates that Brodmann included two sublayers of layer III (IIIC and IIIBβ) in layer IV (IVB and IVA) of primates. Thus, comparisons of layers and sublayers of area V1 with layers in other areas in the same species are invalid because the homologies have been evaluated incorrectly. Comparisons across species for area 17 also are invalid because the direct homologies are in error. We use a modified version of Hässler's nomenclature. This system retains the distinctions for the unique primate specializations of layer IV (IVα and IVβ) and also defines the layer III specialization of some simian primates (e.g., IIIBβ).

2. An architectonic feature of area V1 that appears to be universal or nearly so among primates is the distinct periodic pattern of elevated metabolic activity that is revealed by staining for CO. The CO dark (blob) and light (interblob) regions mark zones of distinct vertical connectivity within V1 and connectivity between V1 and other areas. Thus, blobs and interblobs

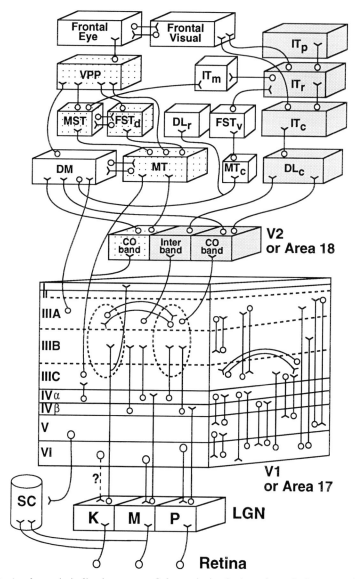

Figure 22. A schematic indicating some of the main intrinsic and extrinsic connections of V1 in primates as described in the text. No effort was made to define the strength of connections, or to indicate true axon collaterals or species-unique features. The major input to V1 is from the lateral geniculate nucleus (LGN) which arrives via three pathways, the koniocellular (K), magnocellular (M), and parvocellular (P) pathways. The retina also projects to other targets, one of which, the superior colliculus (SC), is shown. Within V1, cell layers are heavily interconnected, not only by some of the axonal pathways shown but also via dendritic arbors (not shown). The main ipsilateral connections to extrastriate cortex exit from layers IIIA and IIIC. Within IIIA, the cells within cytochrome oxidase (CO)-rich blobs, indicated by dotted ovals, and CO-poor interblobs send information to different target cells within bands in V2. These three output pathways project into dorsal (light stipple) and ventral (dense stipple) streams. Visual areas within each stream may represent functional clusters; however, visual areas within these streams are also heavily interconnected. Many of these areas also send feedback connections to VI (not shown). Areas are as follows: DL$_c$, caudal dorsolateral (V4$_c$); DL$_r$, rostral dorsolateral; DM, dorsomedial (V3$_d$); FST$_d$, dorsal subdivision of superior temporal; FST$_v$, ventral subdivision of superior temporal; MT, middle temporal; MT$_c$, crescent of middle temporal (V4$_t$); MST, medial superior temporal; IT, inferior temporal (c, caudal; r, rostral; p, posterior; m, medial); VPP, ventral posterior parietal.

mark functionally distinct subdivisions of V1 that are basically similar across primate taxa. While it has been common to relate blobs to a subsystem subserving color vision, blobs are equally prominent and occupy proportionately as much, or more, of V1 in nocturnal primates with limited color vision (e.g., galagos and owl monkeys) as in diurnal primates with well-developed color vision (e.g., squirrel and macaque monkeys). Thus, blobs are likely to participate in functions other than color vision, and they may help mediate a broad range of functions.

3. In some primates, although not all, ocular-specific inputs (via the LGN) are highly segregated into diverging and merging bands that gradually change to a dot and surround pattern in the portion of area V1 devoted to peripheral vision. All catarrhine (Old World) simian primates have ocular dominance bands, but they are well segregated in only the larger platyrrhine (New World) simian primates. Other platyrrhine primates have little or no tendency to exhibit ocular dominance bands, and bands, defined anatomically, are only weakly expressed in prosimian galagos. There are no known functional correlates of bands, although roles in functions such as stereoscopic vision have been postulated. For reasons that are as yet unclear, the ocular dominance bands also appear to have an organizing influence on CO blobs because blobs, or vice versa, are centered on bands.

4. In all primates, the LGN sends major inputs to area V1 over three anatomically and physiologically distinct parallel pathways via the neurons located in magnocellular, parvocellular, and interlaminar or koniocellular LGN layers (M, P, and K). These three LGN pathways terminate, by and large, separately in layers IVα (M), IVβ (P), and the CO blobs of layer III and in layer I (K), respectively. Differences in the physiological characteristics and response properties of neurons in the M and P pathways have been used to support arguments that the M pathway is especially important in providing inputs used in motion perception, and the P pathway in detailed object and color perception. However, given the overlap in the spatial and temporal resolution of P and M LGN cells and the anatomical substrates that provide for mixing of information (see below), it seems unlikely that both of these pathways contribute directly to higher-order perceptual attributes (see Casagrande and Norton, 1991; Merigan and Maunsell, 1993, for discussion). At present, it is not clear what contribution the K pathway makes to the integration of visual processes in V1. In primates the physiological properties of K cells have only been examined in galagos, where studies indicate that these cells are physiologically heterogeneous and resemble W cells in cats. A strong projection of W or W-like retinal ganglion cells to the superior colliculus has suggested to some that this pathway is more primitive and may be associated with general orientation in space and "ambient vision" (Stone, 1983). The strong collicular input to this class of cells in primates and their highly specific output to the CO blobs also hint at a role for this pathway in local modulation of activity (perhaps physiological priming for shifts of local attention) in V1 of primates.

5. In all primates studied, there are major vertical intrinsic connections in V1. Since all layers send and receive vertical interconnections from several other layers, the direction of flow of information cannot be defined anatomically into strict serial steps. Major input from the LGN enters V1

within layer IV, which itself is influenced by projections from the deeper layers. From this point information can travel to both the superficial and deep layers. Within the superficial layers, layer III consists of at least three main sublayers; IIIA, which sends projections to V2, IIIC, which sends projections to MT and the other hemisphere, and IIIB, which appears to act mainly as an interneuronal pool. The subdivisions of layer III have heavy interconnections with other layers, especially layer V. Layer III is also divided tangentially into the CO blob and interblob zones which themselves show differences in vertical connectivity and project to different ventral stream bands of V2. Although there are differences in the details of connections among species, CO blobs in layer IIIB in all species appear to receive not only a direct projection from K LGN cells, but also indirect inputs, via projections from layers IVα and IVβ, from both LGN M and P cells. The interblobs appear to get more restricted inputs but, at least in some species (e.g., owl monkeys), they also get indirect input from both M-recipient IVα and P-recipient IVβ. Thus, there is a mixing and integration of LGN streams in these cortical modules. Layer IIIC receives a heavy projection from M-recipient layer IVα, in addition to projections from both deep and superficial layers. Whereas layer IIIC cells may be dominated more by input from LGN M cells via layer IVα, inputs from infra- and supragranular layers would be subject to P and K influences. The deeper cortical layers V and VI also show sublaminar differences in connections. Within layer V, the most superficial strip, layer VA, may act as a set of interneurons, while layer VB sends major projections to subcortical targets (e.g., pulvinar and superior colliculus). The upper subdivision of layer VI appears to project mainly to P LGN layers and it receives a minor projection from them as well. The lower division of layer VI sends a projection mainly to M LGN layers and it also receives a minor projection from them. Layer VI also has connections with virtually all other cortical layers including layer IV.

6. In all primates studied, lateral or horizontal intrinsic connections within VI are pronounced in layers III and V, and they are extensive for CO blob modules. These connections unite groups of neurons of matched response properties, and they provide a potential substrate for more global processing not found within subcortical levels of the visual system.

7. In all primates studied, three major output pathways distribute information from V1 to other areas of cortex.

   a. Direction-selective neurons in layer IIIC and in layer VB provide output to MT in all investigated primates. MT is a major station in the dorsal stream of visual processing directed to posterior parietal cortex that is important in localizing objects, visual tracking, other spatial aspects of vision, and visual attention.

   b. Layer IIIC cells also provide inputs to the dorsal stream bands of V2. These dorsal stream bands relay to MT, DM, and ventral subdivision of superior temporal (FST$_v$), visual areas associated with the dorsal stream. Layer IIIA cells in the CO blob modules project to the ventral stream bands, which then relay largely to DL$_c$ and then to inferior temporal cortex. Layer IIIA cells in the interblob modules project to the interbands of V2, which then also relay to DL$_c$.

c. Layer IIIA cells, largely associated with the blob modules, project to DM, thereby providing an input usually associated with the ventral stream to a dorsal stream area. Other inputs to DM are from both sets of CO-dense bands in V2. DM further relays to posterior parietal cortex.

8. The callosal connections of V1 appear to subserve the functions of the intrinsic connections for portions of V1 near the border. Thus, they are most dense along the margin of the area where they span all layers and resemble the dense vertical intrinsic connections. Other connections extend with decreasing density up to several millimeters away from the border, especially in layer III and more notably in the CO blob modules, both subdivisions of V1 that have the most extensive horizontal intrinsic connections. These more widespread callosal connections are species variable, being very limited in diurnal simians and most prevalent in nocturnal prosimians (i.e., galagos), where V1 is also characterized by widespread intrinsic connections.

9. Subcortical connections follow the general mammalian pattern with layer VI cells projecting to the LGN and claustrum, and layer V cells projecting to nuclei of the pulvinar and the superficial layers of the superior colliculus. These connections allow feedback to modulate the relay of visual information to cortex. Thus, area 17 projections to the LGN can directly modulate the major inputs to area 17 and projections to the pulvinar and claustrum activate feedback that also modulates area 17 neurons as well as neurons in other visual fields. More indirect feedback loops may involve the projections to the striatum, superior colliculus, and brain-stem structures, especially the visual pons.

In summary, the organization of V1 inputs, intrinsic connections, and output targets is remarkably similar across primate species. Some of the features, but by no means all, are present in nonprimate mammals. The similarities among primates are somewhat surprising, given that the primate order is one of the most varied in body features, and varies enormously in body size and brain size from the mouse lemur to the gorilla and large-brained human. The reassuring implication of this basic similarity is that much of visual processing and distribution of visual functions are similar across primates, and generalizations based on findings limited to a few species are likely to apply to other species including humans.

ACKNOWLEDGMENTS. This work was supported by NIH grants EY01778 (V. A. C.) and EY02686 (J. H. K.) and core grants HD15052 and EY08126. We thank Kathy Rockland, John Allison, and Roz Weller for helpful comments on the manuscript, Barbara Westendorf for help with the figures, Thea Williams for word processing, and Julia Mavity-Hudson for help on all phases of preparation of the manuscript.

9. References

Allman, J. M., and Kaas, J. H., 1971, A representation of the visual field in the caudal third of the middle temporal gyrus of the owl monkey (*Aotus trivirgatus*), *Brain Res.* **31**:85–105.
Allman, J. M., and Kaas, J. H., 1974, The organization of the second visual area (VII) in the owl monkey: A second order transformation of the visual field, *Brain Res.* **76**:247–265.

Allman, J. M., and Kaas, J. H., 1975, The dorsomedial cortical visual area: A third tier area in the occipital lobe of the owl monkey (*Aotus trivirgatus*), *Brain Res.* **100:**473–487.

Allman, J. M., and Kaas, J. H., 1976, Representation of the visual field on the medial wall of occipital–parietal cortex in the owl monkey, *Science* **191:**572–576.

Anderson, P. A., Olavarria, J., and Van Sluyters, R. C., 1988, The overall pattern of ocular dominance bands in cat visual cortex, *J. Neurosci.* **8:**2183–2200.

Bender, D. B., 1983, Visual activation of neurons in primate pulvinar depends on cortex but not colliculus, *Brain Res.* **279:**258–261.

Benevento, L. A., and Rezak, M., 1976, The cortical projections of the inferior pulvinar and adjacent lateral pulvinar in the rhesus monkey (*Macaca mulatta*), *Brain Res.,* **108:**1–24.

Benevento, L. A., and Standage, G. P., 1982, Demonstration of lack of dorsal lateral geniculate nucleus input to extrastriate areas MT and visual 2 in the macaque monkey, *Brain Res.* **252:**161–166.

Benevento, L. A., and Yoshido, K., 1981, The afferent and efferent organization of the lateral geniculoprestriate pathways in the macaque monkey, *J. Comp. Neurol.* **203:**455–474.

Billings-Gagliardi, S., Chan-Palay, V., and Palay, S. L., 1974, A review of lamination in area 17 of the visual cortex of *Macaca mulatta*, *J. Neurocytol.* **3:**619–629.

Blasdel, G. G., and Lund, J. S., 1983, Termination of afferent axons in macaque striate cortex, *J. Neurosci.* **3:**1389–1413.

Blasdel, G. G., Lund, J. S., and Fitzpatrick, P., 1985, Intrinsic connections of macaque striate cortex: Axonal projections of cells outside lamina 4C, *J. Neurosci.* **5:**3350–3369.

Born, R. T., and Tootell, R. B., 1991a, Single-unit and 2-deoxyglucose studies of side inhibition in macaque striate cortex, *Proc. Natl. Acad. Sci. USA* **88:**7071–7075.

Born, R. T., and Tootell, R. B., 1991b, Spatial frequency turning of single units in macaque supra-granular striate cortex, *Proc. Natl. Acad. Sci. USA* **88:**7066–7070.

Braak, H., 1984, Architectonics as seen by lipofuscin stains, in: *Cerebral Cortex,* Volume 1 (A. Peters and E. G. Jones, eds.), Plenum Press, New York, pp. 59–104.

Brodmann, K., 1909, *Vergleichende Lokalisationlehre der Grosshirnrinde in ihren Prinzipien dargestellt auf Grund des Zellenbaues,* J. A. Barth, Leipzig.

Brunso-Bechtold, J. K., Florence, S. L., and Casagrande, V. A., 1983, The role of retinogeniculate afferents in the development of connections between visual cortex and the dorsal lateral geniculate nucleus, *Dev. Brain Res.* **10:**33–39.

Bullier, J., and Kennedy, H., 1983, Projection of the lateral geniculate nucleus onto cortical area V2 in the macaque monkey, *Exp. Brain Res.* **53:**168–172.

Burkhalter, A., and Bernardo, K. L., 1989, Organization of corticocortical connections in human visual cortex, *Proc. Natl. Acad. Sci. USA* **89:**1071–1075.

Burkhalter, A., and Charles, V., 1990, Organization of local axon collaterals of efferent projection neurons in rat visual cortex, *J. Comp. Neurol.* **302:**920–934.

Burkhalter, A., Felleman, D. J., Newsome, W. T., and Van Essen, D. C., 1986, Anatomical and physiological asymmetries related to visual area V3 and VP in macaque extrastriate cortex, *Vision Res.* **26:**63–80.

Carey, R. G., Fitzpatrick, D., and Diamond, I. T., 1979, Layer I of striate cortex of *Tupaia glis* and *Galago senegalensis:* Projections from thalamus and claustrum revealed by retrograde transport of horseradish peroxidase, *J. Comp. Neurol.* **186:**393–438.

Carroll, E. W., and Wong-Riley, M. T. T., 1984, Quantitative light and electron microscopic analysis of cytochrome oxidase-rich zones in the striate cortex of the squirrel monkey, *J. Comp. Neurol.* **222:**1–17.

Casagrande, V. A., and DeBruyn, E. J., 1982, The galago visual system: Aspects of normal organization and developmental plasticity, in: *The Lesser Bushbaby (Galago) as an Animal Model: Selected Topics* (D. E. Haines, ed.), CRC Press, Boca Raton, Fla., pp. 137–168.

Casagrande, V. A., and Harting, J. K., 1975, Transneuronal transport of tritiated fucose and proline in the visual pathways of tree shrew (*Tupaia glis*), *Brain Res.* **96:**367–372.

Casagrande, V. A., and Norton, J. T., 1991, Lateral geniculate nucleus: A review of its physiology and function, in: *Electrophysiology of Vision* (A. G. Leventhal, ed.), Macmillan Press, London, pp. 41–84.

Casagrande, V. A., Mavity-Hudson, J. A., and Taylor, J. G., 1992, Intrinsic connections of owl monkey cortex: Difference between cytochrome oxidase (CO) blobs and interblobs, *Soc. Neurosci. Abstr.* **18:**389.

Chaudhuri, A., Dyck, A., Matsubara, J. A., and Cynader, M. S., 1992, Ocular dominance columns in monkey striate cortex revealed by activity-dependent expression of zif268, *Soc. Neurosci. Abstr.* **18:**209.

Claps, A., and Casagrande, V. A., 1990, The distribution and morphology of corticogeniculate axons in ferrets, *Brain Res.* **530:**126–129.

Clark, W. E. L. G., 1925, The visual cortex of primates, *J. Anat. (London)* **59:**350–357.

Clark, S., and Miklossy, J., 1990, Occipital cortex in man: Organization of callosal connections, related myelo- and cytoarchitecture, and putative boundaries of functional visual areas, *J. Comp. Neurol.* **298:**188–214.

Clarke, P. G. H., and Whitteridge, D., 1976, The cortical visual areas of sheep, *J. Physiol. (London)* **256:**497–508.

Clarke, P. G., Donaldson, I. M., and Whitteridge, D., 1976, Binocular visual mechanisms in cortical areas I and II of the sheep, *J. Physiol. (London)* **256:**509–526.

Colby, C. L., Gattass, R., Olson, C. R., and Gross, C. G., 1988, Topographical organization cortical afferents to extrastriate area PO in the macaque: A dual tracer study, *J. Comp. Neurol.* **269:**392–413.

Colonnier, M., and Sas, E., 1978, An anterograde degeneration study of the tangential spread of axons in cortical areas 17 and 18 of the squirrel monkey (*Saimiri sciureus*), *J. Comp. Neurol.* **179:**245–262.

Condo, G. J., and Casagrande, V. A., 1990, Organization of cytochrome oxidase staining in the visual cortex of nocturnal primates (*Galago crassicaudatus* and *Galago senegalensis*): I. Adult patterns, *J. Comp. Neurol.* **293:**632–645.

Conley, M., Fitzpatrick, D., and Diamond, I. T., 1984, The laminar organization of the lateral geniculate body and the striate cortex in the tree shrew (*Tupaia glis*), *J. Neurosci.* **4:**171–198.

Connolly, M., and Van Essen, D., 1984, The representation of the visual field in parvicellular and magnocellular layers of the lateral geniculate nucleus in the macaque monkey, *J. Comp. Neurol.* **226:**544–564.

Conway, J. L., and Schiller, P. H., 1983, Laminar organization of the tree shrew lateral geniculate nucleus, *J. Neurophysiol.* **50:**1330–1342.

Cowey, A., and Stoerig, P., 1989. Projection patterns of surviving neurons in the dorsal lateral geniculate nucleus following discrete lesions of striate cortex: Implications for residual vision, *Exp. Brain Res.* **75:**631–638.

Cragg, B. G., 1969, The topography of the afferent projections in the circumstriate visual cortex of the monkey studied by the Nauta method, *Vision Res.* **9:**733–747.

Cresho, H. S., Rasco, L. M., Rose, G. H., and Condo, G. J., 1992, Blob-like pattern of cytochrome oxidase staining in ferret visual cortex, *Soc. Neurosci. Abstr.* **18:**298.

Cusick, C. G., and Kaas, J. H., 1986a, Interhemispheric connections of cortical, sensory and motor maps in primates, in: *Two Hemispheres—One Brain* (M. P. F. Lepore and H. H. Jasper, eds.), Liss, New York, pp. 83–102.

Cusick, C. G., and Kaas, J. H., 1986b, Interhemispheric connections of cortical connections of visual cortex of owl monkeys (*Aotus trivirgatus*), marmosets (*Callithrix jacchus*), and galagos (*Galago crassicaudatus*), *J. Comp. Neurol.* **230:**311–336.

Cusick, C. G., and Kaas, J. H., 1988a, Cortical connections of area 18 and dorsolateral visual cortex in squirrel monkey, *Visual Neurosci.* **1:**211–237.

Cusick, C. G., and Kaas, J. H., 1988b, Surface view patterns of intrinsic and extrinsic cortical connections of area 17 in a prosimian primate, *Brain Res.* **458:**383–388.

Cusick, C. G., Gould, H. J., III, and Kaas, J. H., 1984, Interhemispheric connections of visual cortex of owl monkeys (*Aotus trivirgatus*), marmosets (*Callithrix jacchus*), and galagos (*Galago crassicaudatus*), *J. Comp. Neurol.* **230:**311–336.

DeBruyn, E. J., and Casagrande, V. A., 1981, Demonstration of ocular dominance columns in a New World primate by means of monocular deprivation, *Brain Res.* **207:**453–458.

DeBruyn, E. J., Casagrande, V. A., Beck, P. D., and Bonds, A. B., 1993, Visual resolution and sensitivity of single cells in the primary visual cortex (V1) of a nocturnal primate (bush baby), *J. Neurophysiol.* **69:**3–18.

Desimone, R., Wessinger, M., Thomas, L., and Schneider, W., 1990, Attentional control of visual perception: Cortical and subcortical mechanisms, *Cold Spring Harbor Symp. Quant. Biol.* **55:**963–971.

DeYoe, E. A., and Van Essen, D. C., 1985, Segregation of efferent connections and receptive-field properties in visual area V2 of the macaque, *Nature* **317:**58–61.

DeYoe, E. A., Hockfield, S., Garren, H., and Van Essen, D. C., 1990, Antibody labeling of functional subdivisions in visual cortex: Cat-301 immunoreactivity in striate and extrastriate cortex of the macaque monkey, *Visual Neurosci.* **5:**67–81.

Diamond, I. T., Conley, M., Itoh, K., and Fitzpatrick, D., 1985, Laminar organization of geniculocortical projections in *Galago senegalensis* and *Aotus trivirgatus, J. Comp. Neurol.* **242**:584–610.

Dick, A., Kaske, A., and Creutzfeldt, O. D., 1991, Topographical and topological organization of the thalamocortical projection to the striate and prestriate cortex in the marmoset (*Callithrix jacchus*), *Exp. Brain Res.* **84**:233–253.

Dräger, U. C., 1974, Autoradiography of tritiated proline and fucose transported transneuronally from the eye to the visual cortex in pigmented and albino mice, *Brain Res.* **82**:284–292.

Felleman, D. J., and Van Essen, D. C., 1991, Distributed hierarchical processing in the primate cerebral cortex, *Cereb. Cortex* **1**:1–47.

Ferster, D., and LeVay, S., 1978, The axonal arborizations of lateral geniculate neurons in the striate cortex of the cat, *J. Comp. Neurol.* **182**:923–944.

Fitzpatrick, D., Itoh, K., and Diamond, I. T., 1983, The laminar organization of the lateral geniculate body and the striate cortex in the squirrel monkey (*Saimiri sciureus*), *J. Neurosci.* **3**:673–702.

Fitzpatrick, D., Lund, J. S., and Blasdel, G. G., 1985, Intrinsic connections of macaque striate cortex. Afferent and efferent connections of lamina 4C, *J. Neurosci.* **5**:3329–3349.

Florence, S. L., and Casagrande, V. A., 1987, The organization of individual afferent axons in layer IV of striate cortex of a primate (*Galago senegalensis*), *J. Neurosci.* **7**:3850–3868.

Florence, S. L., and Casagrande, V. A., 1990, The development of geniculocortical axon arbors in a primate, *Visual Neurosci.* **5**:291–311.

Florence, S. L., and Kaas, J. H., 1992. Ocular dominance columns in area 17 of Old World macaque and talapoin monkeys: Complete reconstructions and quantitative analyses, *Visual Neurosci.* **8**:449–462.

Florence, S. L., Sesma, M. A., and Casagrande, V. A., 1983, Morphology of geniculo-striate afferents in a prosimian primate, *Brain Res.* **270**:127–130.

Florence, S. L., Conley, M., and Casagrande, V. A., 1986, Ocular dominance columns and retinal projections in New World spider monkeys (*Ateles ater*), *J. Comp. Neurol.* **243**:234–248.

Freund, T. F., Martin, K. A., Soltesz, I., Somogyi, P., and Whitteridge, D., 1989, Arborization pattern and postsynaptic targets of physiologically identified thalamocortical afferents in striate cortex of the macaque monkey, *J. Comp. Neurol.* **289**:315–336.

Fries, W., Keizer, K., and Kuypers, H. G. J. M., 1985, Large layer VI cells in macaque striate cortex (Meynert cells) project to both superior colliculus and prestriate area V5, *Exp. Brain Res.* **58**:613–616.

Fulton, J. F., 1937, A note on Francesco Gennari and the early history of cytoarchitectural studies of the cerebral cortex, *Bull. Inst. Hist. Med.* **5**:895–913.

Garey, L. J., 1971, A light and electron microscopic study of the visual cortex of the cat and monkey, *Proc. R. Soc. London Ser. B* **179**:21–40.

Gattass, R., Gross, C. G., and Sandell, J. H., 1981, Visual topography of V2 in the macaque, *J. Comp. Neurol.* **201**:519–539.

Gattass, R., Sousa, A. P., and Gross, C. G., 1988, Visuotopic organization and extent of V3 and V4 of the macaque, *J. Neurosci.* **8**:1831–1845.

Gilbert, C. D., and Wiesel, T. N., 1981, Laminar specialization and intracortical connections in cat primary cortex, in: *Organization of the Cerebral Cortex* (F. O. Schmitt, F. G. Worden, G. Adelman, and S. G., Dennis, eds.), MIT Press, Cambridge, Mass., pp. 163–191.

Gilbert, C. D., and Wiesel, T. N., 1989, Columnar specificity of intrinsic horizontal and corticocortical connections in cat visual cortex, *J. Neurosci.* **9**:2432–2442.

Gilbert, C. D., Hirsch, J. A., and Wiesel, T. N., 1991, Lateral interactions in visual cortex, *Cold Spring Harbor Symp. Quant. Biol.* **55**:663–677.

Glendenning, K. K., Kofron, E. A., and Diamond, I. T., 1976, Laminar organization of projections of the lateral geniculate nucleus to the striate cortex in *Galago, Brain Res.* **105**:538–546.

Glickstein, M., Cohen, J. L., Dixon, B., Gibson, A., Hollins, M., Labossiere, E., and Robinson, F., 1980, Corticopontine visual projections in macaque monkey, *J. Comp. Neurol.* **190**:521–541.

Goodale, M. A., and Milner, A. D., 1992, Separate visual pathways for perception and action, *Trends Neurosci.* **15**:20–25.

Gould, H. U., Weber, J. T., and Rieck, R. W., 1987, Interhemispheric connections in the visual cortex of the squirrel monkey (*Saimiri sciureus*), *J. Comp. Neurol.* **256**:14–28.

Graham, J., 1982, Some topographical connections of the striate cortex with subcortical structures in *Macaca fascicularis, Exp. Brain Res.* **47**:1–14.

Graham, J., Lin, C., and Kaas, J. H., 1979, Subcortical projections of six visual cortical areas in the owl monkey (*Aotus trivirgatus*), *J. Comp. Neurol.* **187**:557–580.

Harting, J. K., Diamond, T. T., and Hall, W. C., 1973, Anterograde degeneration study of the cortical projections of the lateral geniculate and pulvinar nuclei in the tree shrew (*Tupaia glis*), *J. Comp. Neurol.* **150**:393–440.

Harting, J. K., Casagrande, V. A., and Weber, J. T., 1978, The projection of the primate superior colliculus upon the dorsal lateral geniculate nucleus: Autoradiographic demonstration of interlaminar distribution of tectogeniculate axons, *Brain Res.* **150**:593–599.

Hässler, R., 1967, Comparative anatomy of the central visual systems in day- and night-active primates, in: *Evolution of the Forebrain* (R. Hässler and S. Stephen, eds.), Thieme, Stuttgart, pp. 419–434.

Hendrickson, A. E., Wilson, J. R., and Ogren, M. P., 1978, The neuroanatomical organization of pathways between the dorsal lateral geniculate nucleus and visual cortex in Old World and New World primates, *J. Comp. Neurol.* **182**:123–136.

Henry, G. H., 1991, Afferent inputs, receptive field properties and morphological cell types in different layers, in: *Vision and Visual Dysfunction*, Volume 4 (A. G. Leventhal, ed.), Macmillan Press, London, pp. 223–240.

Hess, D. T., and Edwards, M. A., 1987, Anatomical demonstration of ocular segregation in the retinogeniculocortical pathway of the New World capuchin monkey (*Cebus apella*), *J. Comp. Neurol.* **264**:409–420.

Hitchcock, P. F., and Hickey, T. L., 1980, Ocular dominance columns: Evidence for their presence in humans, *Brain Res.* **182**:176–179.

Hockfield, S., McKay, R. D. G., Hendry, S. H. C., and Jones, E. G., 1983, A surface antigen that identifies ocular dominance columns in the visual cortex and laminar features of the lateral geniculate nucleus, *Cold Spring Harbor Symp. Quant. Biol.* **48**:877–889.

Hoffman, K.-P., Distler, C., and Erickson, R., 1991. Functional projections from striate cortex and superior temporal sulcus to the nucleus of the optic tract (NOT) and the dorsal terminal nucleus of the accessory optic (DTN) of macaque monkeys, *J. Comp. Neurol.* **313**:707–724.

Holdefer, R. N., and Norton, T. T., 1986, Laminar organization of receptive-field properties in the lateral geniculate nucleus of the tree shrew, *Soc. Neurosci. Abstr.* **12**:8.

Horton, J. C., 1984, Cytochrome oxidase patches: A new cytoarchitectonic feature of monkey visual cortex, *Philos. Trans. R. Soc. London Ser. B* **304**:199–253.

Horton, J. C., and Hedley-Whyte, E. T., 1984, Mapping of cytochrome oxidase patches and ocular dominance columns in human visual cortex, *Philos. Trans. R. Soc. London Ser. B* **304**:255–272.

Horton, J. C., and Hubel, D. H., 1981, Regular patchy distribution of cytochrome oxidase staining in primary visual cortex of macaque monkey, *Nature* **292**:762–764.

Hubel, D. H., 1975, An autoradiographic study of the retino-cortical projections in the tree shrew (*Tupaia glis*), *Brain Res.* **96**:41–50.

Hubel, D. H., and Livingstone, M. S., 1990, Color and contrast sensitivity in the lateral geniculate body and primary visual cortex of the macaque monkey, *J. Neurosci.* **10**:2223–2237.

Hubel, D. H., and Wiesel, T. N., 1978, Distribution of inputs from the two eyes to striate cortex of squirrel monkeys (abstract), *Soc. Neurosci.* **4**:632.

Hubel, D. H., and Wiesel, T. N., 1968, Receptive fields and functional architecture of monkey striate cortex, *J. Physiol. (London)* **195**:215–243.

Hubel, D. H., and Wiesel, T. N., 1972, Laminar and columnar distribution of geniculocortical fibers in macaque monkey, *J. Comp. Neurol.* **146**:421–450.

Huerta, M. F., and Harting, J. K., 1984, The mammalian superior colliculus: Studies of its morphology and connections, in: *Comparative Neurology of the Optic Tectum* (H. Vanegas, ed.), Plenum Press, New York, pp. 687–773.

Humphrey, A. L., and Hendrickson, A. E., 1983, Background and stimulus-induced patterns of high metabolic activity in the visual cortex (area 17) of the squirrel and macaque monkey, *J. Neurosci.* **3**:345–358.

Humphrey, A. L., Sur, M., Ulrich, D. J., and Sherman, S. M., 1985a, Termination patterns of individual X- and Y-cell axons in the visual cortex of the cat: Projections to area 18, to the 17/18 border region, and to both areas 17 and 18, *J. Comp. Neurol.* **233**:190–212.

Humphrey, A. L., Sur, M., Ulrich, D. J., and Sherman, S. M., 1985b, Projection patterns of individual X- and Y-cell axons from the lateral geniculate nucleus to cortical area 17 in the cat, *J. Comp. Neurol.* **233**:159–189.

Irvin, G. E., Norton, T. T., Sesma, M. A., and Casagrande, V. A., 1986, W-like response properties of interlaminar zone cells in the lateral geniculate nucleus of a primate (*Galago crassicaudatus*), *Brain Res.* **363**:254–274.

Kaas, J. H., 1986, The structural basis for information processing in the primate visual system, in: *Visual Neuroscience* (J. D. Pettigrew and J. Sanderson, eds.), Cambridge University Press, London, pp. 315–340.

Kaas, J. H., 1987, The organization of neocortex in mammals: Implications for theories of brain function, *Annu. Rev. Psychol.* **38**:129–151.

Kaas, J. H., 1988, Development of cortical sensory maps, in: *Neurobiology of Neocortex* (P. Rakic and W. Singer, eds.), Wiley, New York, pp. 101–113.

Kaas, J. H., 1992, Do humans see what monkeys see? *Trends Neurosci.* **15**:1–3.

Kaas, J. H., 1993, The organization of visual cortex in primates: Problems, conclusions, and the use of comparative studies in understanding the human brain, in: *The Functional Organization of the Human Visual Cortex* (B. Gulyas, D. Ottoson, and P. E. Roland, eds.), Pergamon Press, Oxford, pp. 1–11.

Kaas, J. H., and Huerta, M. F., 1988, Subcortical visual system of primates, in: *Comparative Primate Biology*, Volume 4 (H. P. Steklis, ed.), Liss, New York, pp. 327–391.

Kaas, J. H., and Krubitzer, L. A., 1992, Area 17 lesions deactivate area MT in owl monkeys, *Visual Neurosci.* **9**:399–407.

Kaas, J. H., and Morel, A., 1993, Connections of visual areas of the upper temporal lobe of owl monkeys: The MT crescent and dorsal and ventral subdivisions of FST, *J. Neurosci.* **13**:534–546.

Kaas, J. H., and Preuss, T. M., 1993, Archonton affinities as reflected in the visual system, in: *Mammal Phylogeny* (F. Szulay, M. Novacek, and M. McKenna, eds.), Springer-Verlag, New York, pp. 115–128.

Kaas, J. H., Lin, C.-S., and Casagrande, V. A., 1976, The relay of ipsilateral and contralateral retinal input from the lateral geniculate nucleus to striate cortex in the owl monkey: A transneuronal transport study, *Brain Res.* **106**:371–378.

Kaas, J. H., Nelson, R. J., Sur, M., Lin, C.-S., and Merzenich, M. M., 1979, Multiple representations of the body within the primary somatosensory cortex of primates, *Science* **204**:521–523.

Kageyama, G. H., and Wong-Riley, M., 1986, Laminar and cellular localization of cytochrome oxidase in the cat striate cortex, *J. Comp. Neurol.* **245**:137–159.

Kennedy, H., and Bullier, J. C., 1985, A double-labelling investigation of the afferent connectivity to cortical areas V1 and V2 of the macaque monkey, *J. Neurosci.* **5**:2815–2830.

Kennedy, H., and Dehay, C., 1988, Functional implications of the anatomical organization of the callosal projections of visual areas V1 and V2 in the macaque monkey, *Behav. Brain Res.* **29**:225–236.

Kennedy, H., Bullier, J., and Dehay, C., 1985, Cytochrome oxidase activity in the striate cortex and lateral geniculate nucleus of the newborn and adult macaque monkey, *Exp. Brain Res.* **61**:204–209.

Kennedy, H., Dehay, C., and Bullier, J., 1986, Organization of the callosal connections of visual areas V1 and V2 in the macaque monkey, *J. Comp. Neurol.* **247**:398–415.

Krubitzer, L. A., and Kaas, J. H., 1989, Cortical integration of parallel pathways in the visual system of primates, *Brain Res.* **478**:161–165.

Krubitzer, L. A., and Kaas, J. H., 1990a, Convergence of processing channels in the extrastriate cortex of monkeys, *Visual Neurosci.* **5**:609–613.

Krubitzer, L. A., and Kaas, J. H., 1990b, Cortical connections of MT in four species of primates: Areal, modular, and retinotopic patterns, *Visual Neurosci.* **5**:165–204.

Krubitzer, L. A., and Kaas, J. H., 1993, The dorsomedial visual area (DM) of owl monkeys: Connections, myeloarchitecture, and homologies in other primates, *J. Comp. Neurol.*, **334**:497–528.

Kuypers, H. G., Szwarcbart, M. K., Mishkin, M., and Rosvold, H. E., 1965, Occipitotemporal corticocortical connections in the rhesus monkey, *Exp. Neurol.* **11**:245–262.

Lachica, E. A., and Casagrande, V. A., 1992, Direct W-like geniculate projections to the cytochrome oxidase (CO) blobs in primate visual cortex: Axon morphology, *J. Comp. Neurol.* **319**:141–159.

Lachica, E. A., and Casagrande, V. S., 1993, The morphology of collicular and retinal axons ending on small relay (W-like) cells of the primate lateral geniculate nucleus, *Visual Neurosci.*, in press.

Lachica, E. A., Hutchins, J. B., and Casagrande, V. A., 1987, Morphology of corticogeniculate axon arbors in a primate, *Soc. Neurosci. Abstr.* **13**:1434.

Lachica, E. A., Beck, P., and Casagrande, V. A., 1992, Parallel pathways in macaque monkey striate cortex: Anatomically defined columns in layer III, *Proc. Natl. Acad. Sci. USA* **89**:3566–3570.

Lachica, E. A., Beck, P. D., and Casagrande, V. A., 1993, Intrinsic connections of layer III of striate cortex in squirrel monkey and bush baby: Correlations with patterns of cytochrome oxidase, *J. Comp. Neurol.*, in press.

Langston, A. L., Casagrande, V. A., and Fox, R., 1986, Spatial resolution of the galago, *Vision Res.* **26:**791–796.

LeVay, S., 1988, The patchy intrinsic projections of visual cortex, *Prog. Brain Res.* **75:**147–161.

LeVay, S., and Gilbert, C. D., 1976, Laminar patterns of geniculocortical projections in the cat, *Brain Res.* **113:**1–19.

LeVay, S., Connolly, M., Houde, J., and Van Essen, D. C., 1985, The complete pattern of ocular dominance stripes in the striate cortex and visual field of the macaque monkey, *J. Neurosci.* **5:**486–501.

LeVay, S., McConnell, S. K., and Luskin, M. G., 1987, Functional organization of primary visual cortex in the mink (*Mustela vison*) and a comparison with the cat, *J. Comp. Neurol.* **257:**422–441.

Lin, C.-S., and Kaas, J. H., 1979, The inferior pulvinar complex in owl monkeys: Architectonic subdivisions and patterns of input from the superior colliculus and subdivisions of visual cortex, *J. Comp. Neurol.* **187:**655–678.

Lin, C.-S., Weller, R. E., and Kaas, J. H., 1982, Cortical connections of striate cortex in the owl monkey, *J. Comp. Neurol.* **211:**165–176.

Livingstone, M. S., and Hubel, D. H., 1982, Thalamic inputs to cytochrome oxidase-rich regions in monkey visual cortex, *Proc. Natl. Acad. Sci. USA* **79:**6098–6101.

Livingstone, M. S., and Hubel, D. H., 1984a, Anatomy and physiology of a color system in the primate visual cortex, *J. Neurosci.* **4:**309–356.

Livingstone, M. S., and Hubel, D. H., 1984b, Specificity of intrinsic connections in primate primary visual cortex, *J. Neurosci.* **4:**2830–2835.

Livingstone, M. S., and Hubel, D. H., 1987, Connections between layer 4B of area 17 and thick cytochrome oxidase stripes of area 18 in the squirrel monkey, *J. Neurosci.* **7:**3371–3377.

Livingstone, M., and Hubel, D., 1988, Segregation of form, color, movement, and depth: Anatomy, physiology, and perception, *Science* **240:**740–749.

Löwel, S., and Singer, W., 1987, The pattern of ocular dominance columns in flat-mounts of the cat visual cortex, *Exp. Brain Res.* **68:**661–666.

Lund, J. S., 1987, Local circuit neurons of macaque monkey striate cortex: I. Neurons of laminae 4C and 5A, *J. Comp. Neurol.* **257:**60–92.

Lund, J. S., 1988, Anatomical organization of macaque monkey striate visual cortex, *Annu. Rev. Neurosci.* **11:**253–288.

Lund, J. S., 1990, Excitatory and inhibitory circuitry and laminar mapping strategies in the primary visual cortex of the monkey, in: *Signal and Sense: Local and Global Order in Perceptual Maps* (G. M. Edelman, W. E. Gall, and W. M. Cowan, eds), Wiley, New York, pp. 51–66.

Lund, J. S., and Yoshioka, T., 1991, Local circuit neurons of macaque monkey striate cortex: III. Neurons of laminae 4B, 4A, and 3B, *J. Comp. Neurol.* **311:**234–258.

Lund, J. S., Lund, R. D., Hendrickson, A. E., Bunt, A. H., and Fuchs, A. F., 1975, The origin of afferent pathways from the primary cortex, area 17, of the macaque monkey as shown by retrograde transport of horseradish peroxidase, *J. Comp. Neurol.* **164:**287–304.

Lund, J. S., Henry, G. H., MacQueen, C. L., and Harvey, A. R., 1979, Anatomical organization of the primary visual cortex (area 17) of the cat. A comparison with area 17 of the macaque monkey, *J. Comp. Neurol.* **184:**599–618.

Lund, J. S., Hendrickson, A. E., Ogren, M. P., and Tobin, E. A., 1981, Anatomical organization of primate visual cortex area VII, *J. Comp. Neurol.* **202:**19–45.

Lund, J. S., Hawken, M. J., and Parker, A. J., 1988, Local circuit neurons of macaque monkey striate cortex: II. Neurons of laminae 5B and 5, *J. Comp. Neurol.* **276:**1–29.

McGuiness, E., McDonald, C., Sereno, M., and Allman, J. M., 1986, Primates without blobs: The distribution of cytochrome oxidase activity in *Tarsius, Hapalemur,* and *Cheirogaleus, Soc. Neurosci. Abstr.* **12:**130.

Malach, P., Amin, Y., Bartfeld, E., and Grinvald, A., 1992, Biocytin injections guided by optical imaging reveal relationships between functional architecture and intrinsic connections in monkey visual cortex, *Soc. Neurosci. Abstr.* **18:**389.

Malpeli, J. G., Schiller, P. H., and Colby, C. L., 1981, Response properties of single cells in monkey striate cortex during reversible inactivation of individual lateral geniculate laminae, *J. Neurophysiol.* **46:**1102–1119.

Martin, K. A. C., 1992, Parallel pathways converge: Recent results raise doubts about the popular view that different aspects of vision, such as form, colour and motion, are processed through separate, parallel pathways in the brain, *Vis. Cortex* **2**:555–557.

Maunsell, J. H. R., and Van Essen, D. C., 1983, The connections of the middle temporal visual area (MT) and their relationship to a cortical hierarchy in the macaque monkey, *J. Neurosci.* **3**:2563–2586.

Maunsell, J. H. R., and Van Essen, D. C., 1987, Topographic organization of the middle temporal visual area in the macaque monkey: Representational biases and the relationship to callosal connections and myeloarchitectonic boundaries, *J. Comp. Neurol.* **266**:535–555.

Maunsell, J. H. R., Nealey, T. A., and DePriest, D. D., 1990, Magnocellular and parvocellular contributions to responses in the middle temporal visual area (MT) of the macaque monkey, *J. Neurosci.* **10**:3323–3334.

Maunsell, J. H. R., Nealey, T. A., and Ferrera, V. P., 1992, Magnocellular and parvocellular contributions to neuronal responses in monkey visual cortex, *Invest. Ophthalmol. Vis. Sci. Suppl.* **33**:901.

Merigan, W. H., and Maunsell, J. H. R., 1993, How parallel are the primate visual pathways? *Annu. Rev. Neurosci.* **16**:369–402.

Miklossy, J., 1992, Thalamocortical connections and rostral visual areas in man, in: *The Functional Organization of the Human Visual Cortex* (B. Gulyas, D. Ohoson, and P. E. Rowland, eds.), Pergamon Press, Oxford, pp. 123–136.

Montero, V. M., 1980, Patterns of connections from the striate cortex to cortical visual areas in superior temporal sulcus of macaque and middle temporal gyrus of owl monkey, *J. Comp. Neurol.* **189**:45–59.

Muly, E. C., and Fitzpatrick, D., 1992, The morphological basis for binocular and ON/OFF convergence in tree shrew striate cortex, *J. Neurosci.* **12**:1319–1334.

Murphy, K. M., Van Sluyters, R. C., and Jones, D. G., 1991, The organization of cytochrome-oxidase blobs in cat visual cortex, *Soc. Neurosci. Abstr.* **18**:1088.

Myers, R. E., 1965, The neocortical commissures and interhemispheric transmission of information, in: *Functions of the Corpus Callosum* (E. G. Ettinger, ed.), Little, Brown, Boston, pp. 1–17, 133–193.

Nealey, T. A., and Maunsell, J. H. R., 1991, Magnocellular contributions to the superficial layers, *Suppl. Invest. Opthal. Vis. Sci.* **32**:1117.

Newsome, W. T., and Allman, J. M., 1980, Interhemispheric connections of visual cortex in the owl monkey, *Aotus trivirgatus*, and the bushbaby *Galago senegalensis, J. Comp. Neurol.* **194**:209–233.

Norton, T. T., and Casagrande, V. A., 1982, Laminar organization of receptive-field properties in lateral geniculate nucleus of bush baby (*Galago crassicaudatus*), *J. Neurophysiol.* **47**:715–741.

Norton, T. T., Casagrande, V. A., Irvin, G. E., Sesma, M. A., and Petry, H. M., 1988, Contrast-sensitivity functions of W-, X-, and Y-like relay cells in the lateral geniculate nucleus of bush baby, *Galago crassicaudatus, J. Neurophysiol.* **59**:1639–1656.

Ogren, M. P., and Hendrickson, A. E., 1977, The distribution of pulvinar terminals in visual areas 17 and 18 of the monkey, *Brain Res.* **137**:343–350.

O'Leary, D. D., and Stanfield, B. B., 1985, Occipital cortical neurons with transient pyramidal tract axons extend and maintain collaterals to subcortical but not intracortical targets, *Brain Res.* **336**:326–333.

Perkel, D. J., Bullier, J., and Kennedy, H., 1986, Topography of the afferent connectivity of area 17 in the macaque monkey: A double-labelling study, *J. Comp. Neurol.* **253**:374–402.

Peters, A., and Sethares, C., 1991a, Layer IVA of rhesus monkey primary visual cortex, *Cereb. Cortex* **1**:495–462.

Peters, A., and Sethares, C., 1991b, Organization of pyramidal neurons in area 17 of monkey visual cortex, *J. Comp. Neurol.* **306**:1–23.

Pettigrew, J. D., Ramachandran, V. S., and Bravo, H., 1984, Some neural connections subserving binocular vision in ungulates, *Brain Behav. Evol.* **24**:65–93.

Pospical, M. W., Florence, S. L., and Kaas, J. H., 1994, The postnatal development of geniculocortical axon arbors in owl monkeys, manuscript in preparation.

Preuss, T. M., Beck, P. D., and Kaas, J. H., 1993, Areal, modular, and connectional organization of visual cortex in a prosimian primate, the slow loris, *Nycticebus coucang, Brain Behav. Evol.,* **42**:237–251.

Purves, D., and LaMantia, A., 1990, Number of "blobs" in the primary visual cortex of neonatal and adult monkeys, *Proc. Natl. Acad. Sci. USA* **87**:5764–5767.

Raczkowski, D., and Diamond, I. T., 1980, Cortical connections of the pulvinar nucleus in *Galago, J. Comp. Neurol.* **193**:1–40.

Raczkowski, D., and Fitzpatrick, D., 1990, Terminal arbors of individual, physiologically identified geniculocortical axons in the tree shrew's striate cortex, *J. Comp. Neurol.* **302**:500–514.

Redies, C., Diksic, M., and Riml, H., 1990, Functional organization in the ferret visual cortex: A double-labeled 2-deoxyglucose study, *J. Neurosci.* **10**:2791–2803.

Rezak, M., and Benevento, L. A., 1979, A comparison of the organization of the projections of the dorsal lateral geniculate nucleus, the inferior pulvinar and adjacent lateral pulvinar to primary visual cortex (area 17) in the macaque monkey, *Brain Res.* **167**:19–40.

Ribak, C., and Peters, A., 1975, An autoradiographic study of the projections from the lateral geniculate body of the rat, *Brain Res.* **92**:341–368.

Rockland, K. S., 1992, Laminar distribution of neurons projecting from area V1 to V2 in macaque and squirrel monkeys. *Cereb. Cortex* **2**:38–47.

Rockland, K. S., and Lund, J. S., 1983, Intrinsic laminar lattice connections in primate visual cortex, *J. Comp. Neurol.* **216**:306–318.

Rockland, K. S., and Pandya, D. K., 1979, Laminar origins and terminations of cortical connections of the occipital lobe in the rhesus monkey, *Brain Res.* **179**:3–20.

Rockland, K. S., Lund, J. S., and Humphrey, A. L., 1982, Anatomical banding of intrinsic connections in striate cortex of tree shrews (*Tupaia glis*), *J. Comp. Neurol.* **209**:41–58.

Rodman, H. R., Gross, C. G., and Albright, T. D., 1990, Afferent basis of visual response properties in area MT of the macaque. II. Effects of superior colliculus removal, *J. Neurosci.* **10**:1154–1164.

Rosa, M. G. P., Gattass, R., and Fiorani, M., Jr., 1988, Complete pattern of ocular dominance stripes in V1 of a New World monkey, *Cebus apella*, *Exp. Brain Res.* **72**:645–648.

Rosa, M. G. P., Gattass, R., and Soares, J. G. M., 1991, A quantitative analysis of cytochrome oxidase-rich patches in the primary visual cortex of cebus monkeys: Topographic distribution and effects of late monocular enucleation, *Exp. Brain Res.* **84**:195–209.

Rowe, M. H., Benevento, L. A., and Rezak, M., 1978, Some observations on the patterns of segregated geniculate inputs to the visual cortex in New World primates: An autoradiographic study, *Brain Res.* **159**:371–378.

Sanderson, K. J., Haight, J. R., and Dearson, L. J., 1980, Transneuronal transport of tritiated fucose and proline in the visual pathways of the brushtailed possum, *Trichosurus vulpecula*, *Neurosci. Lett.* **20**:243–248.

Schiller, P. H., Stryker, M. P., Cynader, M., and Berman, N., 1974, Response characteristics of single cells in the monkey superior colliculus following ablation or cooling of the visual cortex, *J. Neurophysiol.* **37**:181–194.

Sesma, M. A., Casagrande, V. A., and Kaas, J. H., 1984, Cortical connections of area 17 in tree shrews, *J. Comp. Neurol.* **230**:337–351.

Shaw, C., and Cynader, M., 1986, Laminar distribution of receptors in monkey (*Macaca fascicularis*) geniculostriate system, *J. Comp. Neurol.* **248**:301–312.

Sherk, H., and LeVay, S., 1983, Contribution of the cortico-claustral loop to receptive field properties in area 17 of the cat, *J. Neurosci.* **3**:2121–2127.

Sherman, S. M., and Spear, P. D., 1982, Organization of visual pathways in normal and visually deprived cats, *Physiol. Rev.* **62**:738–855.

Sherman, S. M., Norton, T. T., and Casagrande, V. A., 1975, X- and Y-cells in the dorsal lateral geniculate nucleus of the tree shrew (*Tupaia glis*), *Brain Res.* **93**:152–157.

Shipp, S., and Zeki, S., 1985, Segregation of pathways leading from area V2 to areas V4 and V5 of macaque monkey visual cortex, *Nature* **315**:322–324.

Silverman, M. S., Grosof, D. H., De Valois, R. L., and Elfar, S. D., 1989, Spatial-frequency organization in primate striate cortex, *Proc. Natl. Acad. Sci. USA* **86**:711–715.

Snyder, M., and Diamond, I. T., 1968, The organization and function of the visual cortex in the tree shrew, *Brain Behav. Evol.* **1**:244–288.

Sousa, A. P., Pinon, M. C., Gattass, R., and Rosa, M. G., 1991, Topographic organization of cortical input to striate cortex in the cebus monkey: A fluorescent tracer study, *J. Comp. Neurol.* **308**:665–682.

Spatz, W. B., 1979, The retino-geniculo-cortical pathway in *Callithrix*. II. The geniculo-cortical projection, *Exp. Brain Res.* **36**:401–410.

Spatz, W. B., 1989, Loss of ocular dominance columns with maturity in the monkey, *Callithrix jacchus*, *Brain Res.* **488**:376–380.

Steele, G. E., Weller, R. E., and Cusick, C. G., 1991, Cortical connections of the caudal subdivision of the dorsolateral area (V4) in monkeys, *J. Comp. Neurol.* **306**:495–520.

Stone, J., 1983, Parallel processing, in: *The Visual System: The Classification of Retinal Ganglion Cells and Its Impact on the Neurobiology of Vision*, Plenum Press, New York.

Swadlow, H. D., 1983, Efferent systems of primary visual cortex: A review of structure and function, *Brain Res. Rev.* **6**:1–24.

Symonds, L. L., and Kaas, J. H., 1978, Connections of striate cortex in the prosimian (*Galago senegalensis*), *J. Comp. Neurol.* **181**:477–512.

Symonds, L. L., and Rosenquist, A. C., 1984, Corticocortical connections among visual areas in the cat, *J. Comp. Neurol.* **229**:39–47.

Tigges, J., and Tigges, M., 1979, Ocular dominance columns in the striate cortex of chimpanzee (*Pan troglodytes*), *Brain Res.* **166**:386–390.

Tigges, J., and Tigges, M., 1985, Subcortical sources of direct projections to visual cortex, in: *Cerebral Cortex*, Volume 3 (A. Peters and E. G. Jones, eds.), Plenum Press, New York, pp. 351–378.

Tigges, J., Tigges, M., and Perachio, A. A., 1977, Complementary laminar termination of afferents to area 17 originating in area 18 and in the lateral geniculate nucleus in squirrel monkey, *J. Comp. Neurol.* **176**:87–100.

Tigges, J., Tigges, M., Anschel, S., Cross, N. A., Letbetter, W. D., and McBride, R. L., 1981, Area and laminar distribution of neurons interconnecting the central visual cortical areas 17, 18, 19, and MT in squirrel monkey (Saimiri), *J. Comp. Neurol.* **560**:539.

Tigges, M., Hendrickson, A. E., and Tigges, J., 1984, Anatomical consequences of long-term monocular eyelid closure on lateral geniculate nucleus and striate cortex in squirrel monkey, *J. Comp. Neurol.* **227**:1–13.

Tootell, R. B. H., and Hamilton, S. L., 1989, Functional anatomy of the second visual area (V2) in the macaque, *J. Neurosci.* **9**:2620–2644.

Tootell, R. B. H., Silverman, M. S., Switkes, E., and De Valois, R. L., 1982, Deoxyglucose analysis of retinotopic organization in primate striate cortex, *Science* **218**:902–904.

Tootell, R. B. H., Hamilton, S. L., and Silverman, M. S., 1985, Topography of cytochrome oxidase activity in owl monkey cortex, *J. Neurosci.* **5**:2786–2800.

Trojanowski, J. Q., and Jacobson, S., 1976, Areal and laminar distribution of some pulvinar cortical efferents in rhesus monkey, *J. Comp. Neurol.* **169**:371–392.

Ts'o, D. Y., and Gilbert, C. C., 1988, The organization of chromatic and spatial interactions in the primate striate cortex, *J. Neurosci.* **8**:1712–1727.

Ts'o, D. Y., Gilbert, C. D., and Wiesel, T. N., 1986, Relationships between horizontal interactions and functional architecture in cat striate cortex as revealed by cross-correlation analysis, *J. Neurosci.* **6**:1160–1170.

Ungerleider, L. G., and Desimone, R., 1986, Projections to the superior temporal sulcus from the central and peripheral field representations of V1 and V2, *J. Comp. Neurol.* **248**:147–163.

Ungerleider, L. G., and Mishkin, M., 1982, Two cortical visual systems, in: *Analysis of Visual Behavior* (M. A. Goodale and R. J. W. Mansfield, eds.), MIT Press, Cambridge, Mass., pp. 549–586.

Ungerleider, L. G., Galkin, T. W., and Mishkin, M., 1983, Visuotopic organization of projections from striate cortex to inferior and lateral pulvinar in rhesus monkey, *J. Comp. Neurol.* **217**:137–157.

Usrey, W. M., Muly, E. C., and Fitzpatrick, D., 1992, Lateral geniculate projections to the superficial layers of visual cortex in the tree shrew, *J. Comp. Neurol.* **319**:159–171.

Van Essen, D. C., Newsome, W. T., and Maunsell, J. H. R., 1984, The visual-field representation in striate cortex of the macaque monkey: Asymmetries, anisotropies, and individual variability, *Vision Res.* **24**:429–448.

Van Essen, D. C., Newsome, W. T., Maunsell, J. H. R., and Bixby, J. L., 1986, The projections from striate cortex (V1) to areas V2 and V3 in the macaque monkey: Asymmetries, areal boundaries, and patchy connections, *J. Comp. Neurol.* **244**:451–480.

Van Essen, D. C., Felleman, D. J., DeYoe, E. A., Olavarria, J., and Knierim, J., 1990, Modular and hierarchical organization of extrastriate visual cortex in the macaque monkey, *Cold Spring Harbor Symp. Quant. Biol.* **55**:679–696.

Weber, J. T., Casagrande, V. A., and Harting, J. K., 1977, Transneuronal transport of ^{3}H proline within the visual system of the grey squirrel, *Brain Res.* **129**:346–352.

Weber, J. T., Huerta, M. F., Kaas, J. H., and Harting, J. K., 1983, The projections of the lateral geniculate nucleus of the squirrel monkey: Studies of the interlaminar zones and the S layers, *J. Comp. Neurol.* **213**:135–145.

Weiskrantz, L., 1986, *Blindsight*, Oxford University Press (Clarendon), London.

Weller, R. E., and Kaas, J. H., 1982, The organization of the visual system in *Galago:* Comparisons with monkeys, in: *The Lesser Bush Baby* (Galago) *as an Animal Model: Selected Topics* (D. E. Haines, ed.), CRC Press, Boca Raton, Fla., pp. 107–135.

Weller, R. E., and Kaas, J. H., 1983, Retinotopic patterns of connections of 17 with visual areas V-II and MT in macaque monkeys, *J. Comp. Neurol.* **220**:253–279.

Weller, R. E., and Kaas, J. H., 1987, Subdivisions and connections of inferior temporal cortex in owl monkeys, *J. Comp. Neurol.* **256**:137–172.

Weller, R. E., Kaas, J. H., and Wetzel, A. B., 1979, Evidence for the loss of Y-cells of the retina after long-term ablation of visual cortex in monkeys, *Brain Res.* **160**:134–138.

Weller, R. E., Wall, J. T., and Kaas, J. H., 1984, Cortical connections of the middle temporal visual area (MT) and the superior temporal cortex in owl monkeys, *J. Comp. Neurol.* **228**:81–104.

Weller, R. E., Steele, G. E., and Cusick, C. G., 1991, Cortical connections of dorsal cortex rostral to V II in squirrel monkeys, *J. Comp. Neurol.* **306**:521–537.

Weyland, T. G., and Swadlow, H. A., 1980, Interhemispheric striate projections in the prosimian primate, *Galago senegalensis, Brain Behav. Evol.* **17**:473–477.

Wiesel, T. N., Hubel, D. H., and Lam, D. M. K., 1974, Autoradiographic demonstration of ocular dominance columns in the monkey striate cortex by means of transneuronal transport, *Brain Res.* **79**:273–279.

Wong-Riley, M., 1979, Changes in the visual system of monocularly sutured or enucleated cats demonstrable with cytochrome oxidase histochemistry, *Brain Res.* **171**:11–28.

Wong-Riley, M. T. T., Hevner, R. F., Cutlan, R., Earnest, M., Egan, R., Frost, J., and Nguyen, T., 1993, Cytochrome oxidase in the human visual cortex: Distribution in the developing and the adult brain, *Visual Neurosci.* **10**:41–58.

Zeki, S. M., 1969, Representation of central visual fields in prestriate cortex of monkey, *Brain Res.* **19**:63–75.

Zeki, S. M., 1971, Cortical projections from two prestriate areas in the monkey, *Brain Res.* **34**:19–35.

Zeki, S. M., 1978, The cortical projections of foveal striate cortex in the rhesus monkey, *J. Physiol. (London)* **277**:227–244.

Zeki, S. M., 1980a, A direct projection from area VI to area V3a of the rhesus monkey visual cortex, *Proc. R. Soc. London* **270**:499–506.

Zeki, S. M., 1980b, The representation of color in the cerebral cortex, *Nature* **284**:412–418.

The Organization of Feedback Connections from Area V2 (18) to V1 (17)

KATHLEEN S. ROCKLAND

1. Introduction

One of the major afferent inputs to primate visual cortex is the geniculocortical system. These connections terminate mostly, although not exclusively, in subcomponents of layer 4. Because of the importance of geniculocortical connections, much of the anatomical work on the organization of area V1 (or area 17) has focused on how this input is transformed by intra- and especially interlaminar cortical processes. It has repeatedly been observed, however, that geniculocortical terminations in the monkey constitute only about 30% of the total number of synapses even in layer 4 (Peters, 1987, in Volume 6 of this series). The remaining inputs derive from a variety of intrinsic and extrinsic sources.

Among the nongeniculate extrinsic inputs to area V1 are connections from certain prestriate areas; in particular, areas V2, V3, and MT. Connections from area V2 to V1 were observed in 1965 by Kuypers *et al.*, who visualized them by staining degenerating fibers after lesions in area V2. The existence of these connections was subsequently confirmed in the squirrel monkey by several investigators, first by degeneration (Tigges *et al.*, 1973) and then by autoradiographic methods (Tigges *et al.*, 1977; Wong-Riley, 1978). These studies also confirmed the observation of Kuypers *et al.* that connections from V2 to V1 terminate

KATHLEEN S. ROCKLAND • Department of Neurology, The University of Iowa College of Medicine, Iowa City, Iowa 52242-1053.
Cerebral Cortex, Volume 10, edited by Alan Peters and Kathleen S. Rockland. Plenum Press, New York, 1994.

mainly in layer 1, not in layer 4. Because of their distinctive termination in layer 1, it seemed appropriate to consider these connections as a subsystem distinct from those which targeted layer 4. They were called "backgoing" or feedback connections (also: caudally directed or countercurrent), to distinguish them from "feedforward" cortical connections. These latter originated from area V1 and terminated primarily in layer 4 of area V2.

In 1977 feedback connections were demonstrated, in the marmoset, from another prestriate area, MT, to area 17 (Spatz, 1977). This study, by using both anterograde and retrograde tracers, was able to describe a laminar complementarity of neurons of origin, as well as terminations in the two systems. That is, the projection from area V1 to MT originated from neurons in layer 4B (also called layer 3C) and from large Meynert neurons in layer 6, and it terminated in layers 3 and 4. The feedback projections from area MT to V1 originated from neurons in layers 2/3 and 5/6 and terminated in layers 1, 4B, and 6.

The laminar complementarity of feedback and feedforward connections was demonstrated between several visual cortical areas of the macaque monkey in 1979 by Rockland and Pandya. Paired injections of horseradish peroxidase and ^{3}H-labeled amino acids revealed reciprocal connections between areas V1 and V2, V2 and V4, and V4 and TE. Feedforward connections from areas V1, V2, or V4 originated from neurons mainly in layer 3 and terminated mainly in layer 4. Feedback connections from areas TE, V4, or V2, however, originated from neurons in layer 6 and (secondarily) in layer 3A. They terminated in layer 1 and, to a lesser extent, layers 5 and 6. This laminar distribution was confirmed and extended to other visual areas by several investigators (Tigges *et al.*, 1981; Weller and Kaas, 1983; Ungerleider and Desimone, 1986). In 1983, Maunsell and Van Essen suggested that these connections, viewed in a pairwise fashion, provided a convenient marker for functionally ranking cortical areas, in conformity with a hierarchical organization viewed as deriving from area V1 (see also, Felleman and Van Essen, 1991).

In this chapter, we will focus on several anatomical aspects of feedback connections from area V2 to V1. We will begin by summarizing data from reconstructions of single axons anterogradely labeled by injections of *Phaseolus vulgaris* leucoagglutinin (PHA-L) in area V2. Against this background, we will attempt to highlight some of the particular characteristics of feedback connections, by comparing them with certain other connections and with the overall functional architecture of area V1. We will then briefly discuss morphological features of layer 1, which may be of relevance to understanding the function of feedback connections, and conclude with a discussion of recent experiments and ideas about their possible significance.

2. Axon Configuration

The early autoradiographic descriptions of feedback terminations clearly demonstrated their preferential targeting of layer 1 and avoidance of layer 4. This technique, however, among other limitations, could not easily distinguish between actual terminations and axons en passage. Thus, it was difficult to interpret label at lower density levels or in the deeper cortical layers, because axons destined for layer 1 necessarily travel through the underlying cortex. The introduction of more sensitive anterograde tracers, such as PHA-L and biocytin, made feasible a reexamination of the exact morphology of feedback connections. Because these tracers produce a Golgi-like image of individual axons, it becomes possible to investigate specific morphometric parameters; for example, whether single axons arborize in more than one layer, whether terminations occur in clusters, how large an area is innervated by a single axon, axonal caliber, and numbers and distributions of terminal swellings.

By reconstructing individual axons through sequential sections, several characteristic features of the feedback projection system have been identified. The following description is based on 28 reconstructions previously reported (Rockland and Virga, 1989), and an additional 6 reconstructions prepared more recently for this chapter (Figs. 1–8). Feedback axons will be divided into two groups; namely, those with restricted or those with widespread terminal arbors. This subdivision is convenient for purposes of description. Such morphological differences may be indicative of genuine heterogeneity, but more correlative features will need to be established before any actual groupings can be suggested (see Section 3 for further discussion of this point).

2.1. Axons with Restricted Arbors

Of the 34 axons in our sample, 4 had a single arbor localized in layers 1–3A. The size of a terminal cluster varies from 150 μm (Fig. 1) to 500 μm (Rockland and Virga, 1989, Fig. 9). Two of these axons have no collaterals in other layers; the other two have collaterals in layer 5, but only in the form of delicate, very short sprigs (Rockland and Virga, 1989).

2.2. Axons with Widespread Arbors in Layer 1

The majority of axons in our sample have a widespread distribution in layer 1, with occasional incursion into layer 2. Most axons travel horizontally for 0.75–2.0 mm in layer 1. A minority (15%) extend over even greater distances (3 axons: 2.5–2.7 mm; 1 axon: 3.5 mm; 1 axon: 4.3 mm).

The significance of this widespread distribution depends largely on the arrangement of terminal boutons. Three patterns of bouton distribution can be distinguished. In one pattern, boutons are primarily concentrated in spatially separate clusters along the main axon trunk. The size and spacing of the clusters are variable. Commonly, however, clusters measure 150–200 μm in diameter, and tend to be separated by 450–500 μm (e.g., Fig. 3 in Rockland and Virga, 1989). In a second pattern, boutons are primarily arrayed in a linear fashion, along one to three branches, more or less parallel to the pia and (in the case of multiple branches) parallel to each other. Bouton-studded branches range from 0.30 to 1.25 mm in length (Figs. 3 and 5). This linear configuration does not occur in layer 3A, but rather seems to be preferentially limited to layers 1 and 2. In a third pattern, both terminal clusters and linear branches occur along the same axon (Figs. 6 and 8; and Fig. 2 in Rockland and Virga, 1989).

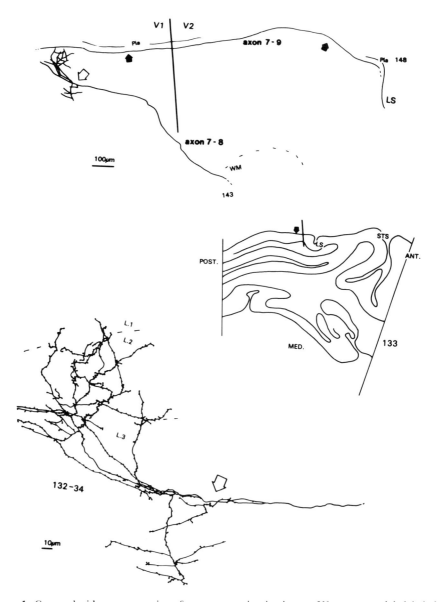

Figure 1. Camera lucida reconstruction of an axon terminating in area V1, anterogradely labeled by a PHA-L injection in area V2. Low-magnification reconstruction (through 12 sections, each 50 μm thick) shows the main axon traveling through the white matter, from the injection site (section 150), posterior and ventral toward area V1. The axon continues, somewhat obliquely, in the gray matter, and then terminates in a circumscribed arbor, about 150 μm in diameter, concentrated in layers 1–3A. The arbor has been redrawn separately at higher magnification below, from the point indicated by the hollow arrow.

Solid arrows in the lower-magnification reconstruction illustrate the trajectory of a second axon (7–9, Fig. 6). The horizontal section outline provides general orientation and indicates approximate position of the terminal arbor (arrow) relative to the border between areas V1 and V2 (slash mark). Smaller section numbers in this case denote more ventral levels.

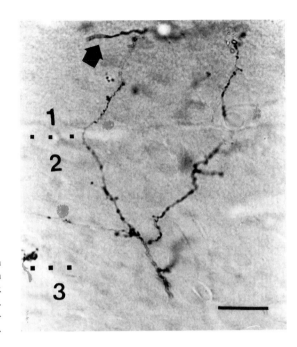

Figure 2. Photomicrograph through the terminal arbor of the axon shown in Fig. 1. A segment of the axon trunk from a second axon (illustrated in Fig. 6) can be seen in layer 1, and is indicated by the solid arrow. Bar = 20 μm.

2.3. Collaterals in Other Layers

In addition to their main arborization in layer 1, widespread axons may have collaterals in other layers. The most common layer to receive collaterals is layer 5. Collaterals to layer 5 occur in about 80% of these 30 axons. Collaterals in this layer appear to have a linear rather than clustered morphology, are always much shorter (≤ 0.75 mm) than the main branches in layer 1, and sometimes appear only as very short and faintly labeled sprigs. Infragranular collaterals are not necessarily in register with the branches in layer 1, and in fact are usually offset from these by about 0.5 mm (Figs. 3 and 8; and Fig. 2 in Rockland and Virga, 1989). Most commonly, infragranular collaterals will occur off the main axon trunk as this ascends from layer 6 toward the pia; but in one case (Figs. 3 and 4), we found a collateral exiting off the main axon in the white matter into layer 6.

The possibility of white matter branches can only be addressed by laboriously extensive reconstructions through the white matter. This factor, as well as the possibility of technically related failures of transport, emphasizes the need for caution in drawing conclusions about the number and extent of collaterals in layer 5. However, given the consistency of our results, we think it reasonable to conclude that collaterals in layer 5 do not always occur but that, when they do, they are always shorter than overlying branches.

Five axons in our sample so far have terminations in layer 3. These are concentrated in the upper portion of layer 3 (3A), are in the form of delimited clusters, and occur in combination with more numerous terminations in layers 1 and 2. One axon has been found (Figs. 6 and 7) with three separate clusters in layer 3A. We have no examples of axons with terminations in layers 4A, 4B, or 4C, and only one of terminations in layer 6 (Fig. 3).

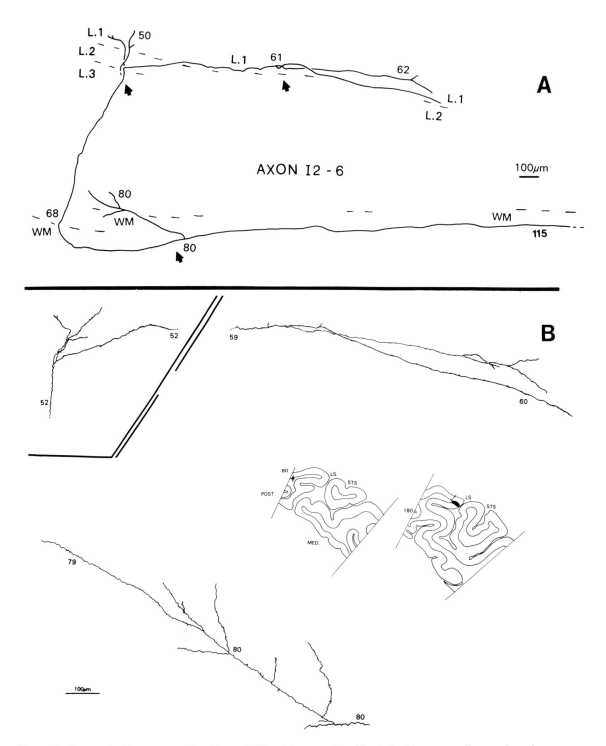

Figure 3. Camera lucida reconstruction (through 65 serial sections, each 50 μm thick) of an axon terminating in area V1, anterogradely labeled by a PHA-L injection in area V2. Horizontal section outlines (in B) indicate injection site (section 180) and approximate location of the axon (arrow) at one dorsal level (in this case, smaller numbers denote more dorsal sections). As shown in A, the axon trunk travels posterior and dorsal from the injection, in a position parallel to and just shallow to layer 6. One cluster of terminations exits from the main axon in the white matter to layer 6. The axon

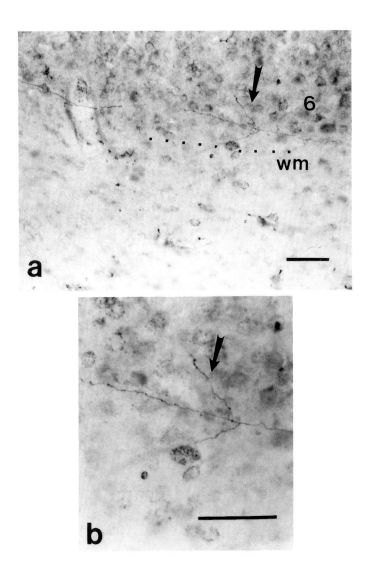

Figure 4. Intermediate (a) and higher magnification (b) photomicrographs, from the axon shown in Fig. 3, of terminal collaterals in layer 6 of area V1. Arrows point to corresponding features in a and b. Bars = 50 μm.

2.4. Bouton Size and Density

Based on an analysis of ten axons, the total number of swellings (assumed to be terminal specializations) per axon ranged from 146 to 644 boutons. Bouton density was assessed for 30 segments of 100-μm length in three axons. This ranged from a maximum of 15 boutons/100 μm to a minimum of 3 boutons/ 100 μm.

Figure 3. (*Continued*) continues slightly more posterior and dorsal, then enters the gray matter, ascends toward the pia, and turns back anteriorly, traveling within layer 1. (B) Details of terminal collaterals in layers 1 and 6 are redrawn at higher magnification. (See also photomicrographs, Fig. 4.)

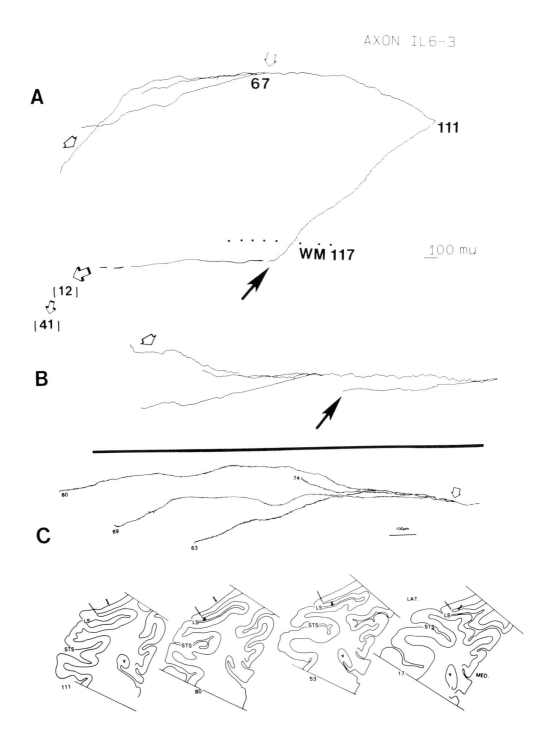

AXON IL6-3

A

67

111

· · · · · · · ·

WM 117

[12]

[41]

100 mu

B

C

80

74

69

63

100µm

LS

STS

111

LS

STS

80

LS

STS

53

LAT.

LS

STS

17

MED.

Terminal specializations appear both as *en passant* swellings or beads and spinelike profiles. The proportion of beads to spines is variable, but there are usually at least twice as many beads as spines. Beads are of small to moderate size, typically 1.5 μm in their long axis. Spines vary in configuration and size, but are often 4.0 μm long and occasionally measure up to 7.5 μm long.

2.5. Trajectory

The axons illustrated in Figs. 3 and 5 show several standard features of this connectional system. From the injection site in lateral area V2, in the lunate sulcus, the main axon travels in the white matter below the occipital operculum. Frequently, the axon will travel at a more or less constant subgriseal depth, posterior and ventral, about 300 μm below layer 6 and obliquely parallel to the pia. The axon typically enters the gray matter at a 90° angle and maintains a basically perpendicular course toward the pia. Usually in layer 1, but occasionally in layer 2 or even 3, the axon will abruptly turn, again almost at 90°, and then continue its course within layer 1. This will often be in the anterior direction, so that the axon doubles back on itself, pointing toward the injection site in V2.

Several deviations from this pattern can occur. For instance, the axon shown in Fig. 8 travels in a short trajectory across the white matter core, ascends in an obliquely ventral path to layer 1, and then, within the gray matter, loops back on itself, traveling dorsal and anterior. Also, especially if the injection site is very near the border between areas V1 and V2, axons can travel exclusively through the gray matter. There is even one instance (Figs. 1 and 6) where an axon apparently exits from the injection site in layer 1 and then maintains its further course entirely within this layer.

3. Neurons of Origin of Feedback Connections

The neurons giving rise to feedback connections from area V2 to V1 have been demonstrated, from injections in area V1, by retrogradely transported HRP (Rockland and Pandya, 1979; Lund *et al.*, 1981; Tigges *et al.*, 1981; Weller

---

Figure 5. Computer-assisted reconstruction of an axon in area V1, anterogradely labeled by a PHA-L injection in area V2. As shown in A, as the axon is followed back toward the injection, it first travels in the white matter in a trajectory parallel to layer 6. The trajectory is complicated, however, in that the axon loops dorsal to the injection site (around section 53), to section 12. At this point, taking a dorsoventral course, it goes back toward the injection site (the axon was followed back ventrally to section 41).

At section 117, the axon enters the gray matter, ascends to layer 1 (section 111), turns anterior and dorsal (to section 67), and then breaks up into four terminal branches. These continue further anteriorly, three in a ventral and one in a dorsal direction. In B, part of the axon is shown rotated, as if looking down on the terminal branches from the pia (solid arrows indicate corresponding points at the base of the axon trunk as it enters layer 6; small hollow arrows point to distal portion of longest branch). In C, the terminal branches have been redrawn manually (from section 67: large hollow arrows point to corresponding regions in A and C) to illustrate details of bouton distribution. The horizontal section outlines depict approximate position of this axon at different levels (arrows). Slash mark denotes the border between areas V1 and V2 on the lateral surface. v, ventricle.

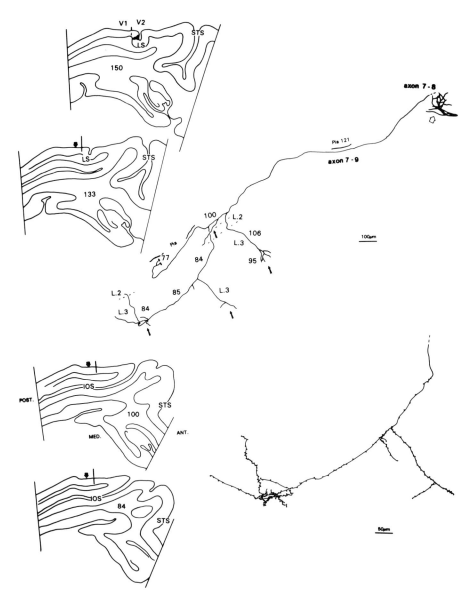

Figure 6. Camera lucida reconstruction of an axon anterogradely labeled by a PHA-L injection in area V2, which terminates in a widespread fashion in area V1. The axon travels from the vicinity of the injection (section 150 at left) through layer 1 along a posterior and ventral trajectory (the proximal portion of the trajectory is shown in Fig. 1). Four terminal foci (arrows) occur at sections 100–101, 95–96, 84–85, and 83–85. The first is a linear branch confined to layer 1 (see Fig. 7c), the other three are more or less well-developed terminal clusters extending well into layer 3A or even 3B (see Fig. 7). The arbors at sections 95–96 and 84–85 are in register in the anteroposterior plane, but offset 0.5 mm dorsoventrally (the apparent displacement in the reconstruction reflects some distortion from flattening of the depth dimension). The fourth arbor (sections 83–85) is accurately shown, as 0.5 mm displaced posteriorly. The ventralmost distal segment is redrawn at higher magnification. Beyond section 77, the axon could not be followed, either because it ended or because it faded out. For orientation, a segment of axon 7–8 from Fig. 1 (hollow arrow) has been included in the reconstruction. The horizontal section outlines at left indicate approximate location of this axon at several dorsoventral levels (smaller numbers are ventral in this case).

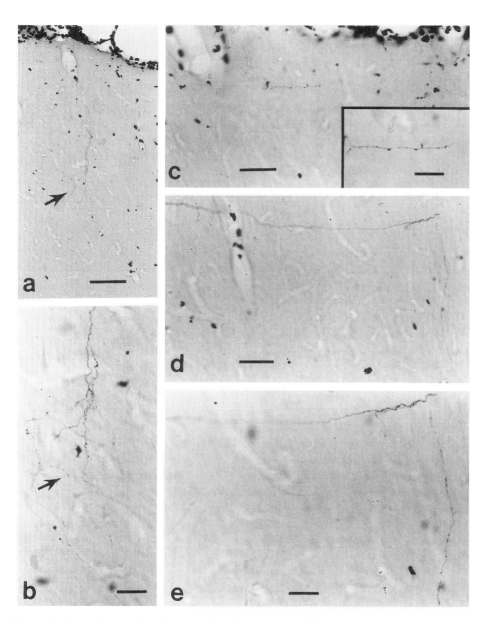

Figure 7. Photomicrographs of terminations from the axon illustrated in Fig. 6. (a, and higher magnification, b) Terminal cluster (sections 95–96) in layer 3A; (c) linear collateral (sections 100–101) in layer 1 (insert = higher magnification); (d, and higher magnification, e): portion of terminal cluster (sections 83–85) in layer 3A. Second cluster (off the print to the left) is shown in a camera lucida reconstruction in Fig. 6. Bars: a = 100 μm; b = 20 μm; c = 50 μm (insert = 20 μm); d = 50 μm; e = 20 μm.

and Kaas, 1983), fluorescent dyes (Kennedy and Bullier, 1985), and PHA-L (Rockland and Virga, 1989). Retrograde filling by PHA-L is successful for only a small percentage of this population; but when it occurs, these neurons can be visualized in Golgi-like detail (Fig. 9).

Feedback neurons consist of at least two distinct subpopulations, which differ in their laminar distribution. One, smaller group is comprised mostly of medium-sized pyramidal cells (15×20 μm soma size), whose somata are located superficially in layer 3. These neurons have numerous basal dendrites, which

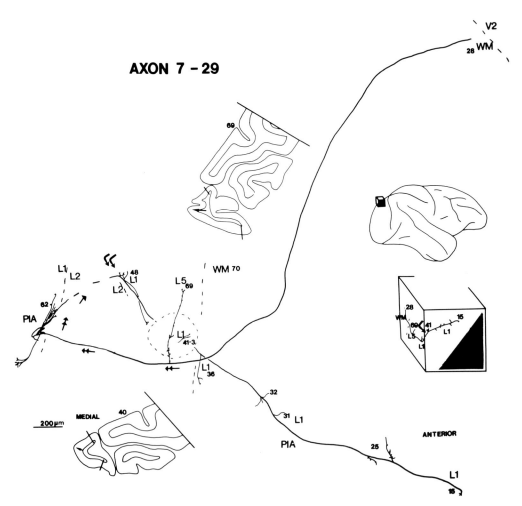

Figure 8. Camera lucida reconstruction of an axon (through 55 sections, each 50 μm) terminating in area V1 and anterogradely labeled by a PHA-L injection in area V2. The axon trunk was followed from the white matter below the injection in area V2 (section 28), across the white matter core to where it enters area V1 (section 70; arrow in section outline). In the gray matter, the axon emits a short collateral (pointing posterior) in layer 5, and continues toward the pia. On reaching layer 1, the axon emits a circumscribed termi- nal arbor, and at first travels posteriorly and dorsally. At section 48 (double curved arrows), the axon emits a second set of terminals, and then loops anteriorly. The axon contin- ues for another 2.6 mm, emitting another six, very fine ter- minal sprigs. (Total trajectory = 4.3 mm.) The brain dia- gram with cube summarizes this trajectory. (The blackened face of the small cube corresponds to the pia surface, and to the partially darkened face of the larger cube, which en- frames a schematic version of axon 7–29.)

typically fill a circular area of 200–250 μm diameter. Their apical dendrites vary in configuration, partly as a function of the laminar depth of the cell body. Neurons with a soma located near the border of layers 2 and 3 usually have modified apical dendrites which fork proximally, near the soma, and have abundant side branches. Neurons with a soma deeper in layer 3 tend to have a longer apical dendrite, with fewer side branches (Rockland and Virga, 1989). The distal apical tuft often extends over 200 μm laterally within layer 1. In particularly good examples of dendritic filling (from PHA-L or biocytin), portions of the apical tuft can be seen to bend conspicuously parallel to the pia surface (Fig. 9). The axon, when this could be followed from the soma in layer 3, was seen to give off collaterals in layers 3 and 5 (Rockland and Virga, 1989).

The larger group of feedback neurons is located in layer 6. The composition of this group is hard to determine exactly because of the complex nature of layer 6 itself. Layer 6 is often subdivided into an outer zone (6A) of pyramidal and pyramidal-like neurons and an inner, polymorphic zone (6B), which includes pyramidal, fusiform, and horizontally oriented neurons (reviewed in Braak, 1982; Tömböl, 1984). In primate area V2, subdivisions of this layer are not particularly apparent; and HRP preparations tend to show a uniform band of

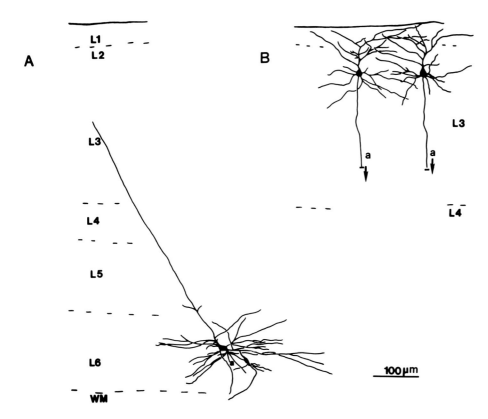

Figure 9. Camera lucida reconstructions of neurons in area V2 retrogradely labeled by a PHA-L injection in area V1. (A) Neuron in layer 6. Note thin apical dendrite, which apparently ends in layer 3. (B) Two pyramidal neurons in layer 3A. Note relatively wide apical tufts, and several dendritic segments aligned parallel to the pia. Dendritic spines were only faintly labeled and have not been drawn. a, axon.

filled neurons, five to seven cells deep (150–200 μm) throughout the thickness of layer 6 and possibly encroaching on the deeper portion of layer 5 (Fig. 10). Labeled profiles can be identified as small ($< 10 \times 10$ μm) and medium-sized (15×20 μm) pyramidal cells, along with some inverted pyramids. Other cell types may be labeled, but cannot be clearly identified with this technique.

Despite the greater number of neurons contributing to this pathway from layer 6 than from layer 3A, the deeper neurons are harder to label retrogradely with PHA-L or biocytin. Ten neurons were adequately well filled for quantitative analysis. Of these, seven had a basal dendritic spread in layer 6 of 200–250 μm (for three neurons, the spread was 350 μm). All of the ten had a slender apical dendrite, which appeared to end in layer 3 (Fig. 9). Their general morphology is consistent with published Golgi drawings of pyramidal neurons in layer 6 of area V2 (Lund *et al.,* 1981).

In order to gauge the relative number of neurons contributing to feedback connections, counts were made of neurons in layer 6 of area V2 labeled by an HRP injection in area V1. Neurons were counted in four separate fields through the inferior occipital sulcus in one case. Each field was subdivided into three standard rectangles, 2.8×10^4 μm^2 in area. Within these areas, all HRP-labeled neurons were counted, and unlabeled cells, counterstained with neutral red, were counted if a nucleolus could be ascertained. This procedure may have resulted in slight undercounting of the unlabeled neurons, but ensured against inclusion of possible glia. By these criteria, 41–68% of the total number of neurons were HRP-labeled. For comparison, similar counts were done of neurons in layer 3 of area V2, which had been retrogradely labeled from an HRP injection in area V4. These "feedforward" neurons in area V2 appear to make up a higher percentage of the total neuron population in their home layer (59–84%).

In summary, neurons giving rise to feedback connections from area V2 to V1 consist of at least two subpopulations with distinct laminar distributions. Since there are no data from intracellular filling of individual identified feedback neurons, it is not known how these neuronal subpopulations may correlate with particular axon morphology. Perhaps the relatively less common axons, those with "restricted" terminations, may originate from the smaller group of neurons in layer 3A, and the more numerous axons, those with widespread terminations, may originate from the more numerous group of neurons in layer 6. Against this interpretation, however, one axon in our sample (Fig. 6) exits from the injection site in layer 1 and travels exclusively within layer 1. It seems somewhat more likely that this axon would originate from a neuron in layer 3A rather than ascend to layer 1 from a deep neuron in layer 6. It may be significant that, although this axon has a widespread, not restricted configuration, it has three distinct terminal clusters. Perhaps the significant distinction is between axons with circumscribed terminal clusters, either single or multiple, and those with boutons in a predominantly linear array. One might hypothesize that the former, smaller group arises from neurons in layer 3, and the latter, from neurons in layer 6.

4. Other Feedback Connections to Area V1

Connections from prestriate area MT to V1 have been demonstrated in the macaque monkey by bulk injections into area MT of ^{3}H-labeled amino acids

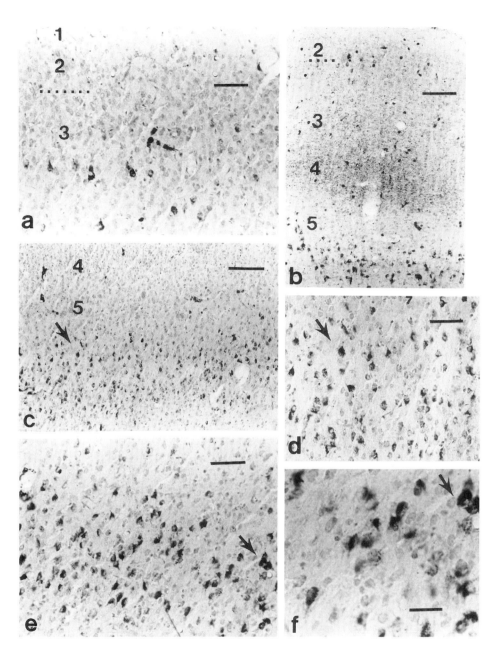

Figure 10. Photomicrographs of neurons in area V2 (of the inferior occipital sulcus), which have been retrogradely labeled by an HRP injection in area V1. As shown in b, most neurons are located in layer 6, with only a few in upper layer 3 (layer 4 has been labeled by anterogradely transported HRP). (a) Higher magnification of neurons labeled in layer 3. (c and e, and higher magnification, d and f, respectively) Two fields of labeled neurons in layer 6. There is some apparent heterogeneity among the labeled neurons; for example, inverted pyramids can be seen in d and f. Arrows in c and d, and in e and f demarcate equivalent features. Bars: a, d, e = 50 μm; b, c = 100 μm; f = 25 μm.

(Maunsell and Van Essen, 1983; Ungerleider and Desimone, 1986) or WGA-HRP (Krubitzer and Kaas, 1989, 1990; Shipp and Zeki, 1989). These connections originate from neurons in layer 3A and, more abundantly, from neurons in layer 6 of area MT; they terminate in layers 6, 4B, and, except for the central 10° of visual field, layer 1 of area V1. Since these results are based on bulk injections, it is still uncertain how far individual axons travel in a given layer, how dense the terminal labeling is in each layer, whether a single axon typically branches in more than one layer, or how much of the label in layer 6 is associated with terminals or axons.

Other connections of a feedback nature have been described to area V1 from areas V4, TEO, and TE. There have been occasional reports over the last 10 years, after fluorescent dye injections in area V1, of labeled neurons in several prestriate areas beyond V2 and MT, and of a few neurons also in the parahippocampal gyrus (Doty, 1983; Kennedy and Bullier, 1985; Perkel *et al.*, 1986; Sousa *et al.*, 1991). Recently, we have confirmed and extended these observations in a series of retrograde and anterograde experiments (Douglas and Rockland, 1992; Rockland *et al.*, 1994; Rockland and Van Hoesen, 1984). Like the feedback projections from area V2, projections from areas V4 and TEO originate from neurons in layers 3A and 6. Those from area TE are concentrated in layer 6. From area V4, terminations target primarily layer 1 and, to a lesser extent, layer 6. From areas TEO and TE, they appear directed more exclusively in layer 1. Tangential trajectories of 4–6 mm in layer 1 are common for axons in all three of these systems. This is a greater span than that seen for feedback axons originating in area V2, which were usually < 2.5 mm long in layer 1.

Connections from area V4 to V1 can be considered part of a reciprocal pathway, as area V1 is reported to project to area V4, at least in its region of foveal representation (Zeki, 1980; Van Essen *et al.*, 1986; Nakamura *et al.*, 1991). Areas TEO and TE, however, are not known to receive connections from area V1. Thus, "feedback" connections from these areas to V1 may be nonreciprocal.

5. Comparison with Geniculocortical Connections

Geniculocortical connections have been described at the level of individual axons by several groups (in macaque: Blasdel and Lund, 1983; Freund *et al.*, 1989; in the galago: Florence and Casagrande, 1987). Geniculocortical afferents are comprised of at least two subpopulations, which terminate in different layers, have arbors of different sizes, and originate from different morphological and functional subdivisions of the LGN. Three points of comparison can be made with the feedback pathway from V2 to V1. First, geniculocortical axons have circumscribed terminal arbors, whereas feedback axons generally have more divergent terminations. In specific, arbors of parvocellular axons are about 0.2 mm in diameter. Those of magnocellular axons cover a larger territory, about 0.3 × 1.2 mm, but in two to three patchlike arbors. This range is more delimited than that of most (although not all) feedback axons. It is worth noting, however, that one geniculocortical axon recovered from layer 1 had a more divergent spread—four separate patches covering "at least 2 mm" (Blasdel and Lund, 1983). Some degree of divergent spread, in other words, may reflect maps of different grain in different layers (Blasdel and Fitzpatrick, 1984).

Second, both geniculocortical and feedback axons from area V2 inconstantly have collaterals in layers outside their main layer of termination (respectively, layer 4 and 1). Some geniculocortical axons can have collaterals in layer 6; feedback axons, usually in layer 5. In both systems, the secondary collaterals are shorter than the principal arborization in the granular or supragranular layers.

Third, geniculocortical axons have more terminal boutons per arbor. Data for two parvocellular axons report counts of 1520 and 1380 boutons (Freund *et al.*, 1989), and for two magnocellular axons, counts of ≥ 3200 (Freund *et al.*, 1989) and 3085 (Blasdel and Lund, 1983). In terms of density, figures of 13.5 and 11 boutons per 50 μm, respectively, have been reported in the galago (Florence and Casagrande, 1987). These are larger numbers than total bouton counts for our feedback axons (≤ 644 boutons) or than their average density distribution (≤ 7 boutons per 50 μm).

More boutons and less divergence, as characterize the geniculocortical system, seem appropriate to a driving, as opposed to "modulatory" pathway. The interpretation of these morphological differences, however, is complicated by various uncertainties, such as how to evaluate synaptic efficacy (see Section 8). Further work on these and other afferents will be required to assess actual morphological–functional relationships.

6. Relationship of Feedback Connections to Compartmental Organization

As reviewed by other contributors to this volume, primate area V1 has an exquisitely ordered modularity. This modularity is presently understood as organized around thalamically based ocular dominance columns in layer 4C and cytochrome oxidase (CO) "blobs" or patches in layer 3. The relationship of various intrinsic and extrinsic connectional systems to this modularity is still an active area of investigation, but it is already clear that this relationship can be complex (see, e.g., Casagrande and Kaas, and Lund *et al.*, this volume).

The relationship of feedback connections to CO compartments (for the latter, see Wong-Riley, this volume) has been investigated by Livingstone and Hubel (1983, 1984; 1987). These investigators made injections of WGA-HRP into area V2 of macaque and squirrel monkeys, and compared the resultant pattern of label in area V1 with the pattern of CO activity. They found that injections confined to thin CO stripes in area V2 preferentially produce label in CO-dense patches in area V1 whereas injections confined to the CO-poor matrix in area V2 result in labeled patches corresponding to the CO-poor matrix in V1. Label in these experiments consisted of retrogradely filled "feedforward" neurons in area V1 and diffuse granular labeling interspersed between the cells, but by and large coincident with them. This granular label may result from anterograde transport from the injection site and, if so, would suggest a compartmental specificity of the feedback connections from area V2 to V1. The interpretation, however, may be complicated by potential technical issues. For example, microinjections provide good localization, but often at the cost of incomplete transport, especially in the anterograde direction. Also, as these investigators note, it is often difficult to determine how much of the "anterograde" transport of WGA-HRP may be attributable to collateral, rather than true anterograde transport.

Similar experiments have been carried out to analyze how feedback connections from area MT to V1 might relate to the CO compartments (Krubitzer and Kaas, 1989, 1990; Shipp and Zeki, 1989). Injections of WGA-HRP in area MT resulted in label in area V1, in layers 1, 4B, and 6. In layer 4B, cellular and granular label was patchy, but the patches could not easily be related to the CO pattern. In layers 1 and 6, it was more continuous and extended over a wider area than the neuronal labeling. This laminar pattern and the continuous nature of the label in layer 1 was also reported by Ungerleider and Desimone (1986), after injections of ³H-labeled amino acids in area MT.

In our material, the need to preserve uninterrupted serial sections for axon reconstruction was not compatible with histochemical counterstains for CO. Thus, it was not possible to demonstrate directly the relationship of terminal arbors to the CO-defined modules. Several inferences can nevertheless be made from the axon reconstructions alone.

First, since linearly configured terminal branches commonly extend over distances of 0.5–1.25 mm, it is clear that a single axon potentially contacts neurons in several CO compartments. Further work will be necessary to determine whether this distance is extended within a single ocular dominance slab or whether, in contrast, it may cross over two or more different dominance domains.

Second, collaterals in layer 5 are typically offset from the main arbor. The distance of offset is variable and hard to measure, but tends to be around 0.5 mm. Again, this interval suggests that a single axon can, in the supra- and infragranular layers, innervate neurons belonging to different CO compartments and, possibly, even different ocular dominance domains.

A third observation, that axons can have delimited terminal clusters, either single or multiple, is harder to interpret without direct counterstains to demonstrate the modular structure of area V1. The most conservative interpretation might be that clusters, which are commonly 200–500 μm in diameter and are separated by 0.5 mm, terminate in similar type CO compartments. The larger clusters, however, are larger than a single CO patch. There is, moreover, clearly potential for various combinations. In the axon shown in Fig. 6, for example, three terminal clusters are spatially separated in the dorsoventral plane by 250 and 400 μm, as if targeting three CO patches with a given ocular dominance row. There is a fourth cluster, however, where the spacing extends in the orthogonal plane (anteroposterior instead of dorsoventral). One can only speculate whether this cluster lies within an adjacent column or whether, as is also possible, its offset merely reflects the curvature of a single dominance slab.

In summary, the long trajectories and linear terminal configuration of most feedback axons would seem to imply that they contact neurons associated with several different CO "blobs" or patches. One likely possibility, given the morphological heterogeneity of these axons, is that some axons cross CO boundaries, while others have terminations that are more specific.

7. Composition of Layer 1

Since feedback connections to area V1 preferentially terminate in layer 1, in understanding these connections, some consideration must be given to the overall organization of this layer. As has been repeatedly remarked, neither the physiology nor even the morphology of this layer is particularly well understood, largely because of its vulnerability to experimental procedures (see Vogt, 1991, in Volume 9 of this series). In terms of morphology, layer 1 has been described as consisting of six basic elements (Marin-Padilla, 1984, p. 448): three types of fibers (specific afferents, axonal terminals of Martinotti neurons in layer 6, and terminals from afferent systems of the underlying layers), two types of neurons (Cajal–Retzius cells and smaller interneurons), and an extensive receptive surface formed by apical dendritic tufts of underlying pyramidal neurons. To these six, it may be useful to add a seventh element, namely, the abundant glial population of this layer, which, among other roles, contributes to the specialized pial–glial membrane. Each of these components will be briefly discussed below.

7.1. Fiber Systems

In addition to cortical feedback connections, layer 1 of the primary visual cortex in primates receives major connections from the lateral geniculate nucleus (Blasdel and Lund, 1983), the inferior and lateral pulvinar (Ogren and Hendrickson, 1977), and the lateral basal nucleus of the amygdala (Amaral and Price, 1984; Iwai and Yukie, 1987). These connections target layer 1 preferentially but not necessarily exclusively. Layer 1 also receives dense cholinergic afferents, probably originating mainly from the basal nucleus of Meynert (Hedreen *et al.*, 1984), and some serotonergic and noradrenergic fibers, although in the macaque these monoaminergic terminations are denser in layer 4 (Kosofsky *et al.*, 1984; Foote and Morrison, 1987). Connections from various subcortical nuclei to area V1 have been previously described from retrograde experiments (Tigges and Tigges, 1985); and some of these terminate at least in part in layer 1 (Avendaño *et al.*, 1990). In addition, there is a dense plexus of somatostatin-positive fibers, of intrinsic origin, concentrated in the outer half of layer 1 (de Lima and Morrison, 1989). This is probably the same plexus that can be visualized by neuropeptide Y (Kuljis and Rakic, 1989a) or NADPH diaphorase (Fig. 11; Sandell, 1986). The neuropil in layer 1 is immunoreactive for calbindin, but not for parvalbumin (van Brederode *et al.*, 1990).

The significance of this particular convergence of afferents within layer 1 is hard to evaluate. In part, considerably more data will be necessary concerning the precise morphology of individual axons and their postsynaptic targets. For example, do some or all of these systems have, like the feedback connections, widespread arborizations within layer 1, or can they have more localized arbors? Blasdel and Lund (1983, Fig. 17) illustrate one axon from the LGN, filled intra-axonally with HRP, which probably covers over 2.0 mm across layer 1, but in four separate, well-localized patches. Most available data do not provide comparable detail, because they are based on retrograde transport from injections in area V1 or else, in the case of anterograde tracers, only on aggregate populational labeling by WGA-HRP or ^{3}H-labeled amino acids. Monoaminergic fibers are well delineated by immunohistochemical techniques, but are too dense for serial reconstructions to be a practical option.

Another problem is that this particular combination of afferents is characteristic only of primate area V1 and is subject to a high degree of interspecies and interareal variability. Areas V2 or V4, for example, are not reported to receive input to layer 1 from the LGN (Lund *et al.*, 1981). But area V4 and other

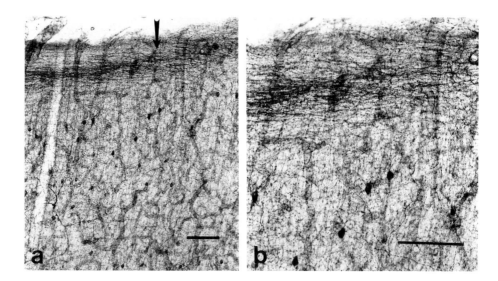

Figure 11. Photomicrographs of fiber plexus, labeled by NADPH diaphorase, in layer 1 of area V1. Plexus often occupies layer 1B but (arrow) can also shift at points to fill the thickness of layer 1 (b is a higher magnification of a). Bars = 100 μm.

associational cortices do receive callosal, cortical, and probably other afferents whose distribution can include layer 1. Those systems which do seem constant to most cortical areas (i.e., from the amygdala, locus coeruleus, and raphe) terminate with differential density in different areas (Foote and Morrison, 1987).

Finally, although it is tempting to speculate about possible connectional interactions, in the sense of patterns of excitation and inhibition, transmitter effects in fact have a well-known diversity and plasticity of action. Somatostatin, for example, can diminish or enhance the inhibitory action of GABA. Somatostatin alone has an inhibitory effect on pyramidal neurons, but it can be facilitatory in the presence of ACh (reviewed in de Lima and Morrison, 1989). Thus, pending further data on the specific synaptology and transmitter effects in layer 1, very little can be said about the interactions of different afferent systems, or about the particular contribution made by feedback connections to the functional architecture of this layer.

7.2. Preferential Orientation of Fibers?

One of the most distinctive features of layer 1 is the relatively uniform orientation of its fibers, which run predominantly tangential to the surface of the brain. This attribute is strikingly obvious in myelin stains (Fig. 12). The tangential orientation is also obvious in histological preparations for somatostatin or serotonin (both showing mainly unmyelinated fibers), and in our own PHA-L-labeled material (Fig. 13).

Our reconstructions of feedback axons show that most of these fibers, in their tangential course within layer 1, have a characteristic anisotropic configuration. That is, they tend to take a direct, near-perpendicular path from the white matter toward layer 1, and then, on reaching layer 1 or its vicinity, make an abrupt L-shaped turn (Figs. 3, 5, and 8). This L-shaped geometry can easily be

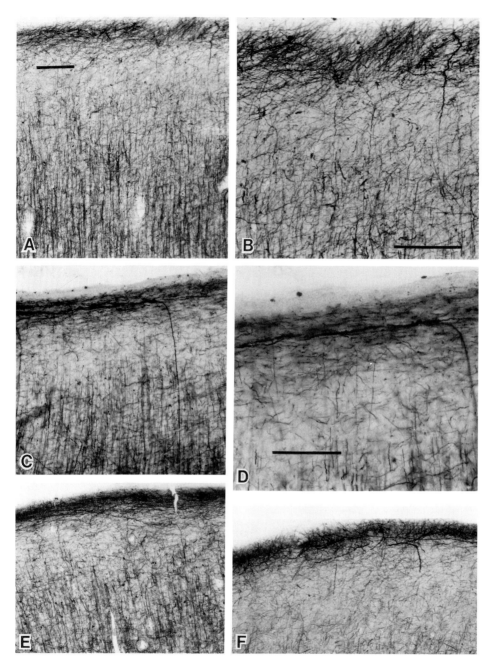

Figure 12. Photomicrographs of horizontal sections, through area V1 or V2 (C, D, F) stained for myelinated fibers by the Gallyas procedure. Note myelinated plexus, concentrated in layer 1. Examples can easily be found of fibers which ascend straight to layer 1 and then take an elbow-shaped turn. (A, and higher magnification, B) Fiber in area V1, turning in an anterior direction. (C, and higher magnification, D) Fiber in area V2, posterior bank of the lunate sulcus, which "points" posteriorly. (E) Fiber in area V1, turning posteriorly. (F) Fiber in area V2 on the medial surface, pointing posteriorly. Bars: A = 100 μm (C, E, and F are the same magnification as A); B, D = 100 μm.

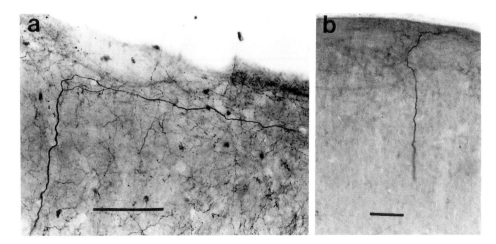

Figure 13. Photomicrographs of two axons, anterogradely labeled by PHA-L injections in area V2, which demonstrate the typical elbow-shaped turn of these axons as they enter layer 1 of area V1. Bars = 100 μm.

seen in random, myelin-stained sections (Fig. 12; see also Ramón y Cajal, Volume II, 1911, Fig. 379), both in primate area V1 and in other cortical areas (unpublished observations). In one earlier report, which used silver impregnations of tangentially sectioned rabbit cortex, the authors describe an orderly orientation of fibers over the surface of the hemisphere, in the anteromedial to posterolateral direction (Fleischhauer and Laube, 1977). This bias, however, has not been confirmed in later work. Furthermore, examples can easily be found in our PHA-L-labeled material where, even within the same tissue section, axons make their "turns" in opposite directions. Particularly long axons (> 4.0 mm) can even change direction several times during their trajectory; for example, one axon, anterogradely labeled by an injection in area TE, started by turning dorsally within layer 1 of area V1, then changed to an anteroposterior course for about 0.5 mm, and then reverted to a dorsal trajectory for another 4.0 mm (Douglas and Rockland, 1992).

7.3. Cellular Elements

The persistence of Cajal–Retzius cells in the adult brain is controversial, because of the difficulty of selectively staining these neurons and because of the sparse and possibly irregular distribution of their somata (see Marin-Padilla, 1984, 1990; Huntley and Jones, 1990). When they are described, these neurons are noted as having a particularly widespread axonal arbor, which may even cross architectonic boundaries.

The small interneurons of layer 1 have been demonstrated by classical Golgi techniques (Marin-Padilla, 1984); by immunolabeling for neuropeptide Y (Kuljis and Rakic, 1989b), for somatostatin (de Lima and Morrison, 1989), and for GABA (Fitzpatrick *et al.*, 1987; Hendry *et al.*, 1987); by NADPH diaphorase (Sandell, 1986); and, in a single instance in the cat, by intracellular filling with HRP (Martin *et al.*, 1989). These neurons are morphologically heterogeneous, as judged by the size and shape of their soma and dendritic field. Dendrites can extend 200 μm or more in one direction from the cell body, and can either be confined to layer 1 or extend into layers 2 and 3A (Fig. 14).

According to Fitzpatrick *et al.* (1987), 77–81% of the neurons in layer 1 of the macaque are GABAergic. Some proportion of these must also contain somatostatin and/or NPY. Few or no cells which are positive for calcium-binding proteins have been described in layer 1 (Bluemcke *et al.*, 1990; DeFelipe *et al.*, 1990; van Brederode *et al.*, 1990).

Neurons in layer 1 which potentially receive input from feedback axons are predominantly GABAergic. These are relatively few in number, but there is a dense band of GABAergic neurons in layer 2 (Fitzpatrick *et al.*, 1987; Hendry *et al.*, 1987), the dendrites of which undoubtedly extend into layer 1. Many of the interneurons in layer 2 are also immunopositive for calbindin (DeFelipe *et al.*, 1990; Van Brederode *et al.*, 1990). Thus, there is significant opportunity for feedback axons to synapse with inhibitory interneurons.

7.4. Apical Dendritic Tufts

Many, but not all, pyramidal neurons have apical dendrites which ascend to layer 1, where they break up into an elaborate apical tuft near the border of layers 1 and 2. The configuration and extent of this apical tuft have been analyzed by Golgi preparations (Lund, 1973; Feldman, 1984, in Volume 1 of this series; Valverde, 1985, in Volume 3; Marín-Padilla, 1992) and by retrograde filling with HRP or biocytin (Rockland, 1992). As can be seen from published figures, the apical tuft is predominantly cone-shaped. This implies that tangentially oriented axons in layer 1 make "crossing," and thus relatively sparse, synaptic contacts with the apical tufts (Szentágothai, 1978). There are, however, at least some instances where the apical tuft is oriented mainly parallel to the pia surface. This configuration is commonly depicted for neurons with their soma in layer 5 (Martin and Whitteridge, 1984). At least some subpopulations of neurons in the supragranular layers also can have a portion of their apical tufts aligned parallel to the pia (Figs. 15–21).

Some feedforward neurons which project from area V1 to V2 have distal dendrites oriented parallel to the pia. These would thus be parallel to the tangentially running fibers in layer 1, and potentially in "close," rather than "crossing" contact with these fibers. Figure 17 shows one such neuron, retrogradely labeled in Golgi-like detail by a biocytin injection in area V2. Two segments of the apical tuft (≥ 70 μm long) are oriented tangential to the pia. The tangential segment is thin (≤ 0.25 μm, versus 1.5 μm for the distal apical shaft). On this attenuated segment, spines tend to be stubby or may even appear as "en passant" thickenings along the dendritic membrane. They are relatively low in density (average: 1 spine per 4 μm). Since feedback axons average < 15 boutons per 100 μm, a given dendritic segment (with about 25 spines per 100 μm) may receive multiple, but clearly not exclusive input from a single axon, even when this is in close parallel apposition.

The same biocytin injections in area V2 result in sporadic labeling of neurons in layers 3A and 6 of areas V3 and V4; that is, of neurons that send their axons to layer 1 of area V2. The population of feedback-projecting neurons in layer 3A of areas V3 and V4 consistently has apical tufts with tangentially oriented segments (Figs. 18–21). These segments measure 150–200 μm long, with

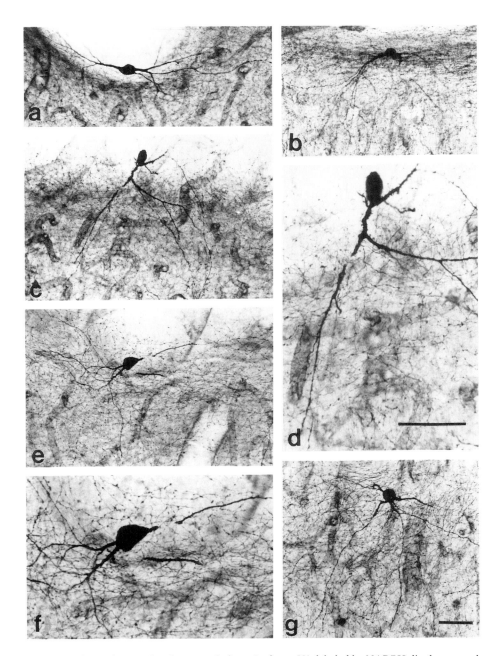

Figure 14. Photomicrographs of neurons in layer 1 of area V1, labeled by NADPH diaphorase. a, b, c (and higher magnification, d), e (and higher magnification, f), and g illustrate five examples of these neurons. Dendrites can be horizontally oriented and narrowly confined within layer 1 (a, b), or they can dip conspicuously into layer 2 in a semicircular formation (g). Bars = 50 μm (a, b, c, e, and g are same magnification; d and f are same magnification).

a total span for the apical tuft of 300–500 μm (this can be greater than the diameter of the basal arbor, which typically measures 225–250 μm). As for the feedforward-projecting neurons in area V1, these tangential dendritic segments are relatively attenuated (< 0.25 μm thick) and have a low density of rather bulbous spines or membrane thickenings (1 per 3–4 μm). Tangentially oriented dendrites also characterize feedback-projecting neurons in layer 3A of area V2, which have been labeled by PHA-L retrogradely transported from an injection in area V1 (Fig. 9).

These results suggest that many or most feedback-projecting neurons in layer 3A have a portion of their distal apical dendrites oriented parallel to the pia, as do at least some feedforward-projecting neurons in area V1. Further

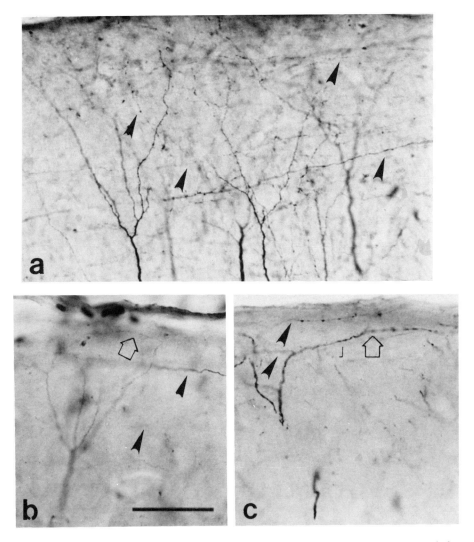

Figure 15. Photomicrographs showing the relationship between apical dendrites of neurons in layer 3 of area V1, and axons (arrowheads) from area V2. Both elements are labeled, retrogradely or anterogradely, by PHA-L injections in area V2. Note segments of dendrites oriented parallel to the pia (hollow arrows in b and c). Bar = 50 μm (a, b, and c are same magnification).

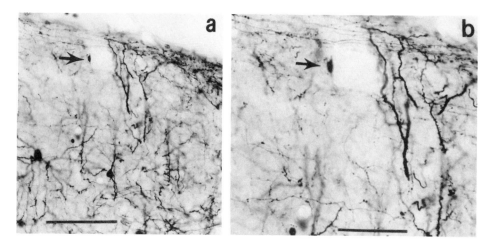

Figure 16. Photomicrographs of apical dendrites from neurons in layer 3 of area V1, retrogradely labeled by PHA-L injections in area V2, along with some anterogradely labeled axons from V2. Apical dendrites are by and large conical in shape, but there are also portions parallel to the pia. b is higher magnification of a (arrows point to equivalent features). Bars: a = 100 μm; b = 50 μm.

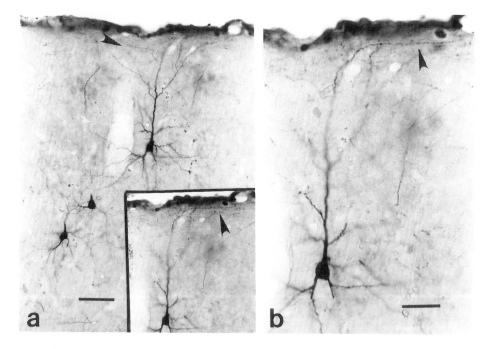

Figure 17. Photomicrograph of a pyramidal neuron in layer 3A of area V1, retrogradely labeled from a biocytin injection in area V2. (a and insert) Low-magnification views, in two focal planes, of soma, basal dendrites, and apical dendrite of this neuron. (b) Higher magnification. Arrowheads point to dendritic segments which are aligned parallel to the pia. Bars: a = 50 μm; b = 25 μm.

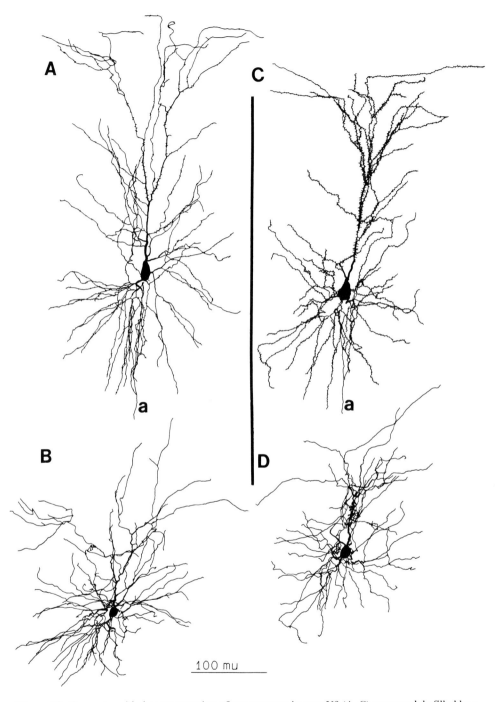

Figure 18. Computer-aided reconstruction of two neurons in area V3 (A, C) retrogradely filled by a biocytin injection in area V2 (see photomicrographs in Figs. 19 and 20). Note dendritic segments aligned parallel to the pia. Also note wide extent of apical tuft relative to the basal dendrites, especially in C. In B and D, neurons have been rotated to allow viewing down onto the apical tuft, as if slantwise from the pia surface. a, axon.

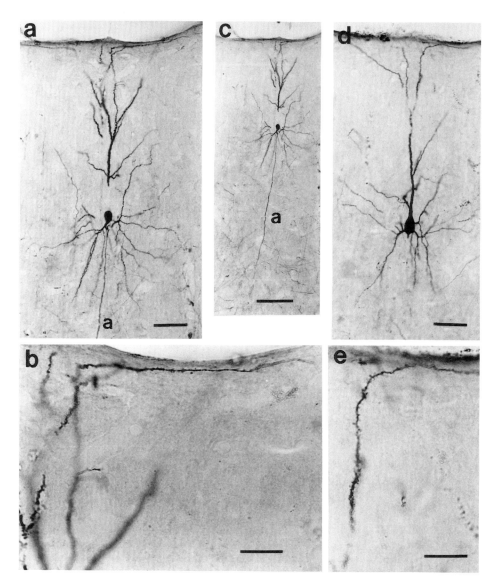

Figure 19. Photomicrograph of a pyramidal neuron in layer 3A of area V3, retrogradely filled by a biocytin injection in area V2. (a and d) Sequential sections showing portions of the cell body, basal dendrites, apical dendrite, and axon (a). (c is lower magnification of a.) (b and e) Higher magnification of apical dendrite from, respectively, a and d. Note dendritic segments aligned parallel to the pia. Spine density is somewhat underrepresented. (See reconstruction in Fig. 18C.) Bars: a, d = 50 μm; c = 100 μm; b, e = 20 μm.

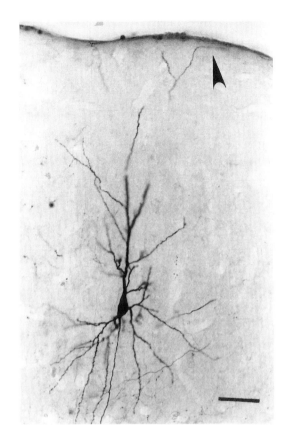

Figure 20. Photomicrograph of a second pyramidal neuron in layer 3A of area V3, retrogradely filled by a biocytin injection in area V2. Arrowhead points to dendritic segment oriented parallel to the pia. (See reconstruction in Fig. 18A.) Bar = 50 μm.

work is required to determine whether the tangentially oriented dendrites are common and whether they make up a constant proportion of the apical tuft. These dendritic specializations might be a significant way to enhance contact between feedback and other axons in layer 1, and the apical tufts of pyramidal cells (see further discussion, Section 8). Contacts between apical tufts and feedback axons are likely to be excitatory in nature (Fig. 22), as the latter originate from pyramidal cells. This cell type is usually associated with asymmetric, excitatory terminations.

7.5. Glia

Layer 1, although relatively poor in neuronal cell bodies, has a high density of astrocytes, whose outer endfeet form a continuous subpial layer. In general, glia are known to participate in the regulation of neural activity by their role in potassium homeostasis (Somjen, 1987; Walz, 1989; Müller, 1992). They are also known to possess receptors for a variety of neurotransmitter substances, including norepinephrine, and may well be involved in interactions among other receptors, second messengers, and additional signal substances. The specific impact of these interactions remains to be elucidated, but is probably complex, given the highly idiosyncratic microenvironment of layer 1.

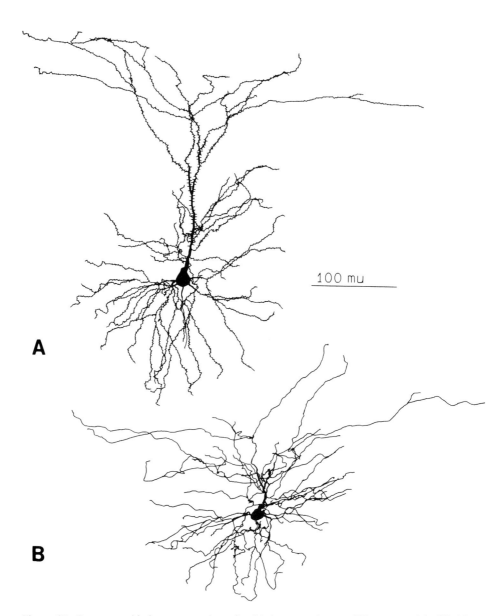

Figure 21. Computer-aided reconstruction of a third neuron in area V3, retrogradely filled by a biocytin injection in area V2. Note dendritic segments aligned parallel to the pia, and the relatively wide extent of the apical dendritic tuft. In B, the neuron has been rotated to allow viewing down onto the apical tuft, as if through the pia surface. (Spine density is somewhat underrepresented.)

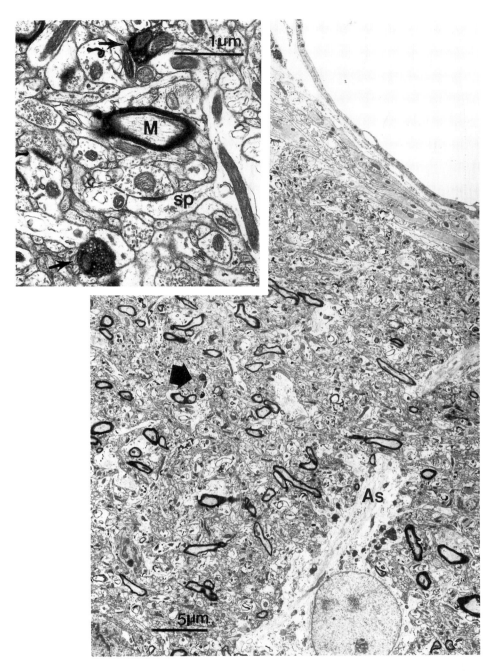

Figure 22. Low-power electron micrograph of layer 1 of area V1 (*Macaca mulatta*). Pia is at the top. There are only a few cell bodies overall in this layer, but there are abundant dendritic profiles which belong to the apical tufts of underlying pyramidal cells (As, astrocyte). The neuropil contains myelinated axons and is dominated by small, unmyelinated axons that run in bundles. Higher magnification insert (from arrow) shows myelinated axons (M), apical dendrites, and synapses, some of which are onto spines (sp). Two dark profiles (arrows) are terminal boutons that have been biocytin-labeled by an injection in area V2.

Investigations of the functional correlates of feedback connections are still in the early stages, for several reasons. Terminations are relatively inaccessible to physiological experiments, concentrated as they are in the fragile superficial layers. Nor can the originating neurons be as easily localized or characterized as those giving rise to, for example, the geniculocortical pathway. Functional analyses are, in fact, still largely directed at determining whether feedback connections have a driving or only a "modulatory" role.

Several factors seem to suggest a modulatory role; for example, the entire set of inputs to layer 1 are often thought to be nonspecific, in contrast to those to layer 4. In area V1, geniculocortical afferents to layer 4 are clearly modality specific. Layer 4, however, also receives some of the same "nonspecific" inputs that project to layer 1 and, in the case of monoaminergic and cholinergic systems, terminations are even denser in layer 4 than in layer 1 (Hedreen *et al.*, 1984; Foote and Morrison, 1987). Moreover, connections to layer 1 from the LGN and the inferior and lateral pulvinar are probably modality specific. Without additional information, it is difficult to argue whether feedback connections are more similar in their action to "nonspecific" afferents from the amygdala and thalamic intralaminar nuclei, or to the presumably specific pulvino- and geniculocortical afferents to layer 1.

The functional significance of feedback connections has been probed to some degree by several experimental approaches and computational models. In one approach, area V2 is either cooled (Sandell and Schiller, 1982; Mignard and Malpeli, 1991) or electrically stimulated (Bullier *et al.*, 1988), and the effects are monitored by recording from neurons in area V1. These experiments can be difficult to interpret because of such factors as anesthesia effects, the nature of the presented stimulus, and the difficulty of isolating out complex underlying connectional interactions. Results from these experiments do, however, consistently show changes in the response properties of many neurons in area V1. In the cat, Mignard and Malpeli (1991) report that projections from area V2 to V1 are sufficient to generate visual responses even in the absence of thalamocortical responses in layers 4 and 6.

In a different experimental paradigm, involving an *in vitro* slice preparation, Cauller and Connors (1992) addressed the synaptic efficacy of inputs to distal dendritic locations. Work on pyramidal cells in the hippocampus implies that dendrites can be electrotonically compact and that distal synaptic inputs are not necessarily less effective than ones located more proximal to the soma (see Cauller and Connors, 1992, for recent review). This seemingly goes against the general assumption that distal inputs, comparatively removed from the axon hillock "trigger" zone, contribute only a small value to the total synaptic weight of any given neuron.

In one part of their experiments, Cauller and Connors recorded from an *in vitro* slice of rat somatosensory cortex. Horizontal fibers in layer 1 were isolated in one part of the slice by making a perpendicular cut from layer 2 into the white matter. Layer 1 was stimulated on one side of the cut, and the response mediated by horizontal fibers passing to the other side (presumed to be of cortical origin) was recorded extra- and intracellularly. In the second part of their experiments, these investigators examined the computational properties of distal inputs by simulating the passive electrotonic structure of a physiologically and morphometrically identified pyramidal cell in layer 5. By comparing a simulated somatic EPSP to experimentally observed intracellular responses produced by distal synaptic activity, they concluded that distal inputs are probably amplified by active conductance along the apical dendrite. These experiments have their own problems of interpretation, relating to the identity of the fibers stimulated and the possibility of slice artifacts, but they do provide evidence for the potential efficacy of distally located synapses. This work will undoubtedly be extended, especially on its computational side, as more biophysically relevant data become available on dendritic and axonal elements of layer 1.

There is, thus, some experimental evidence in support of a driving, rather than merely modulatory role for feedback connections. The probable importance of these connections is also supported by their widespread occurrence. That is, feedback connections, as defined by their laminar distribution, occur in the auditory (Pandya and Sanides, 1973; Galaburda and Pandya, 1983) and somatosensory (Friedman, 1983) cortical systems, and have been reported in many species, including man (Burkhalter and Bernardo, 1989).

Aside from the difficulties already mentioned, another factor complicating our understanding of feedback connections is that the modes of interaction of various cortical systems are not well defined. Important basic findings concerning cortical organization, such as temporal interactions and oscillations (Eckhorn *et al.*, 1988; Gray and Singer, 1989; Engel *et al.*, 1992), are still being gathered, and several are of a controversial nature. Feedback connections, by virtue of their divergent pattern, might be an anatomical substrate for temporal interactions, especially those described between areas. They might also be associated with what have been designated transmodal reentrant pathways. These have been proposed (Finkel and Edelman, 1989; Tononi *et al.*, 1992) as a means of synthesizing signals from multiple, functionally segregated cortical areas. In a somewhat different context, the termination of the connections in layer 1 is compatible with a role in learning and memory (see, e.g., Vogt, 1991, in Volume 9 of this series).

The bidirectional relationship and laminar complementarity of feedforward and feedback connections has been linked to some kind of matching process between incoming stimuli and previous experience (e.g., Ullman, 1991; Mumford, 1992, and other papers reviewed therein). Because the actual microcircuitry, even as incompletely understood, is bound to be elaborate, it is hard to assign any specific neuronal substrate for these processes. In particular, as described earlier in this chapter, feedforward-projecting pyramidal neurons in layer 3 of area V1 may well receive *direct* feedback projections onto their apical dendrites. Any "match" with incoming stimuli is likely to be only indirect. That is, these neurons receive incoming geniculocortical stimuli, but only via interlaminar relays from layer 4 and probably mainly onto their basal and proximal dendrites (Fig. 23). Similarly, feedback-projecting neurons in area V2, in layers 3A and 6, receive little or no *direct* feedforward connections from area V1. Thus, it is possible that single neurons compare feedforward and feedback inputs terminating on different portions of their dendritic tree, but such comparisons would be between inputs of different degrees of "directness." Such input "mismatch" must have temporal implications that will need to be further explored.

The importance of feedback connections is also indicated by experimental neuropsychological results. As described and reviewed by Damasio (1989, 1990), these results pose several questions about the traditional concept of unidirectional "information flow," originating from the primary sensory cortices. Detailed and integrated representations of reality, which typically survive damage to anterior cortical regions, are unlikely to depend exclusively or primarily on stepwise, progressive refinements of signal extraction. Rather, Damasio suggests, processes of perception and recall may entail synchronous reactivation of neural assemblies previously activated by real-life perceptual–motor interactions between the organism and the environment. Feedback connections from anterior levels are a plausible anatomical substrate for achieving temporally synchronous activation of neural assemblies at early cortical levels.

9. Summary and Conclusions

This chapter has reviewed the detailed morphology of feedback connections from area V2 to V1, as this is demonstrated by serial section reconstruction of individual axons, and has attempted to relate this to current discussions of their role in cortical function.

This work has emphasized several salient feedback features of axons. The laminar termination of these connections is mainly in layer 1. Some terminations can occur in layers 2, 3A, and 5 or, infrequently, 6. This distribution had been shown with earlier anterograde tracers, but reconstruction of axons labeled by PHA-L has further clarified important details of individual axon configura-

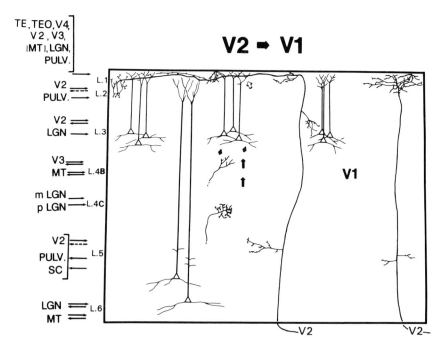

Figure 23. Schematic diagram summarizing the characteristics of feedback axons from V2 to V1. Two examples are shown of restricted (at right) and widespread feedback axons from area V2. These probably terminate at least in part onto apical dendrites (hollow arrow) of underlying pyramidal cells. For comparison, two geniculocortical axons are indicated (adapted from Blasdel and Lund, 1983) in layers 4Cβ and 3B. This input may reach basal and proximal dendrite (solid arrows) of pyramidal neurons in overlying layer 3A. Modified from Rockland and Virga (1989).

tion. A single axon can have collaterals in several layers, but there is no evidence so far that different axons target different layers.

The widespread, divergent arborization of feedback axons, especially in layer 1, implies that one axon may contact large numbers of neurons and, in particular, neurons in different CO compartments. About 15% of widespread axons have both linear terminations in layer 1 as well as one to three terminal clusters extending into layer 3A. Terminal clusters in layer 3A can be about the same size as CO modules, but are usually larger. Widespread linear terminations do not occur in layer 3A; but in layer 5, by contrast, terminations always seem organized in a more or less linear geometry. Thus, the terminal geometry of feedback connections may observe different rules, depending on the layer and the particular portion of dendritic surface contacted.

Reconstructions demonstrate that feedback axons are morphologically heterogeneous. This heterogeneity is consistent with the known laminar and morphological diversity of the parent neurons in area V2, although precise correlations of individual neurons and axons are not available with current techniques. Heterogeneity may correlate with functionally distinct categories, as it does in the geniculocortical pathway, but it may also denote some kind of structural–functional gradient.

One aspect of feedback connections that has attracted considerable attention is their relationship with feedforward connections. This has been customarily described in terms of laminar complementarity and reciprocity of connections between area V1 and areas V2, V4, and MT. Recent results, however, indicate that there are also feedback connections from areas TEO and TE to V1, which do not "reciprocate" any identified feedforward pathways from V1. There are abundant examples of other connectional systems which are not reciprocal: corticocollicular, monoaminergic, and some amygdalofugal connections are not reciprocal. For that matter callosal connections are not strictly reciprocal, in that they are not directed just to the mirror-image contralateral territory, but rather have more diffuse heterotopic, as well as homotopic components. The finding of nonreciprocal, as well as reciprocal feedback connections further underlines the importance of these systems, and suggests that, rather than view the cortical visual areas as a series of pairwise related structures within a strict hierarchy, it will be important to consider additional and richer interaction patterns.

In conclusion, the data suggest that geniculocortical afferents provide driving input for area V1 in primates, but that, at a minimum, normal visual function essentially depends on interactions with feedback and other connections.

ACKNOWLEDGMENTS. I would like to thank Kelly Douglas for her skilled assistance in the preparation of the materials and photographs for this chapter; Ann Reedy for her patient care in manuscript preparation; and Drs. A. Damasio, D. Pollen, and G. Van Hoesen for their helpful discussions and critical reading of the manuscript. The work described in the author's laboratory was supported by NIH grants EYO7058 and NS19632 and by the Roy J. Carver Charitable Trust.

10. References

Amaral, D. G., and Price, J. L., 1984, Amygdalo-cortical projections in the monkey (*Macaca fascicularis*), *J. Comp. Neurol.* **230**:465–496.

Avendaño, C., Stepniewska, I., Rausell, E., and Reinoso-Suárez, F., 1990, Segregation and hetero-geneity of thalamic cell populations projecting to superficial layers of posterior parietal cortex: A retrograde tracer study in cat and monkey, *Neuroscience* **3**:547–559.

Blasdel, G. G., and Fitzpatrick, D., 1984, Physiological organization of layer 4 in macaque striate cortex, *J. Neurosci.* **4**:880–895.

Blasdel, G. G., and Lund, J. S., 1983, Termination of afferent axons in macaque striate cortex, *J. Neurosci.* **3**:1389–1413.

Bluemcke, I., Hof, P. R., Morrison, J. H., and Celio, M. R., 1990, Comparison in the distribution of parvalbumin in the visual cortex of Old World monkeys and humans, *J. Comp. Neurol.* **301**:417–432.

Braak, E., 1982, *On the Structure of the Human Striate Area*, Springer-Verlag, Berlin.

Bullier, J., McCourt, M. E., and Henry, G. H., 1988, Physiological studies on the feedback connection to the striate cortex from cortical areas 18 and 19 of the cat, *Exp. Brain Res.* **70**:90–98.

Burkhalter, A., and Bernardo, K. L., 1989, Organization of corticocortical connections in human visual cortex, *Proc. Natl. Acad. Sci. USA* **86**:1071–1075.

Cauller, L. J., and Connors, B. W., 1992, Functions of very distal dendrites: Experimental and computational studies of layer I synapses on neocortical pyramidal cells, in: *Single Neuron Computation* (T. McKenna, J. Davis, and S. F. Zornetzer, eds.), Academic Press, New York, pp. 199–230.

Damasio, A. R., 1989, The brain binds entities and events by multiregional activation from conver-gence zones, *Neural Computation* **1**:123–132.

Damasio, A. R., 1990, Synchronous activation in multiple cortical regions: A mechanism for recall, *Semin. Neurosci.* **2**:287–297.

DeFelipe, J., Hendry, S. H. C., Hashikawa, T., Molinari, M., and Jones, E. G., 1990, A microcolumnar structure of monkey cerebral cortex revealed by immunocytochemical studies of double bou-quet cell axons, *Neuroscience* **3**:655–673.

de Lima, A. D., and Morrison, J. H., 1989, Ultrastructural analysis of somatostatin-immunoreactive neurons and synapses in the temporal and occipital cortex of the macaque monkey, *J. Comp. Neurol.* **283**:212–227.

Doty, R. W., 1983, Nongeniculate afferents to striate cortex in macaques, *J. Comp. Neurol.* **218**:159–173.

Douglas, K. L., and Rockland, K. S., 1992, Extensive visual feedback connections from ventral inferotemporal cortex, *Soc. Neurosci. Abstr.* **18**:390.

Eckhorn, R., Bauer, R., Jordan, W., Brosch, M., Kruse, W., Munk, M., and Reitboeck, H. J., 1988, Coherent oscillations: A mechanism of feature linking in the visual cortex? *Biol. Cybern.* **60**:121–130.

Engel, A. K., König, P., Kreiter, A. K., Schillen, T. B., and Singer, W., 1992, Temporal coding in the visual cortex: New vistas on integration in the nervous system, *Trends Neurosci.* **15**:218–225.

Feldman, M. L., 1984, Morphology of the neocortical pyramidal neuron, in: *Cerebral Cortex*, Volume 1 (A. Peters and E. G. Jones, eds.), Plenum Press, New York, pp. 123–200.

Felleman, D. J., and Van Essen, D. C., 1991, Distributed hierarchical processing in the primate cerebral cortex, *Cereb Cortex* **1**:1–47.

Finkel, L. H., and Edelman, G. M., 1989, Integration of distributed cortical systems by reentry: A computer simulation of interactive functionally segregated visual areas, *J. Neurosci.* **9**:3188–3208.

Fitzpatrick, D., Lund, J. S., Schmechel, D. E., and Towles, A. C., 1987, Distribution of GABAergic neurons and axon terminals in the macaque striate cortex, *J. Comp. Neurol.* **264**:73–91.

Fleischhauer, K., and Laube, A., 1977, A pattern formed by preferential orientation of tangential fibres in layer I of the rabbit's cerebral cortex, *Anat. Embryol.* **151**:233–240.

Florence, S. L., and Casagrande, V. A., 1987, Organization of individual afferent axons in layer IV of striate cortex in a primate, *J. Neurosci.* **7**:3850–3868.

Foote, S. L., and Morrison, J. H., 1987, Extrathalamic modulation of neocortical function, *Annu. Rev. Neurosci.* **10**:67–95.

Freund, T. F., Martin, K. A. C., Soltesz, I., Somogyi, P., and Whitteridge, D., 1989, Arborisation pattern and postsynaptic targets of physiologically identified thalamocortical afferents in striate cortex of the macaque monkey, *J. Comp. Neurol.* **289**:315–336.

Friedman, D. P., 1983, Laminar patterns of terminations of cortico-cortical afferents in the so-matosensory system, *Brain Res.* **273**:147–151.

Galaburda, A. M., and Pandya, D. N., 1983, The intrinsic architectonic and connectional organiza-tion of the superior temporal region of the rhesus monkey, *J. Comp. Neurol.* **221**:169–184.

Gray, C. M., and Singer, W., 1989, Stimulus-specific neuronal oscillations in orientation columns of cat visual cortex, *Proc. Natl. Acad. Sci. USA* **86:**1698–1702.

Hedreen, J. C., Uhl, G. R., Bacon, S. J., Fambrough, D. M., and Price, D. L., 1984, Acetylcholinesterase-immunoreactive axonal network in monkey visual cortex, *J. Comp. Neurol.* **226:**246–254.

Hendry, S. H. C., Schwark, H. D., Jones, E. G., and Yan, J., 1987, Numbers and proportions of GABA-immunoreactive neurons in different areas of monkey cerebral cortex, *J. Neurosci.* **7:**1503–1519.

Huntley, G. W., and Jones, E. G., 1990, Cajal–Retzius neurons in developing monkey neocortex show immunoreactivity for calcium binding proteins, *J. Neurocytol.* **19:**200–212.

Iwai, E., and Yukie, M., 1987, Amygdalofugal and amygdalopetal connections with modality-specific visual cortical areas in macaques (*Macaca fuscata, M. mulatta,* and *M. fascicularis*), *J. Comp. Neurol.* **261:**362–387.

Kennedy, H., and Bullier, J., 1985, A double-labelling investigation of the afferent connectivity to cortical areas V1 and V2 of the macaque monkey, *J. Neurosci.* **5:**2815–2830.

Kosofsky, B. E., Molliver, M. E., Morrison, J. H., and Foote, S. L., 1984, The serotonin and nor-epinephrine innervation of primary visual cortex in the cynomolgus monkey (*Macaca fascicularis*), *J. Comp. Neurol.* **230:**168–178.

Krubitzer, L. A., and Kaas, J. H., 1989, Cortical integration of parallel pathways in the visual system of primates, *Brain Res.* **478:**161–165.

Krubitzer, L. A., and Kaas, J. H., 1990, Cortical connections of MT in four species of primates: Areal, modular, and retinotopic patterns, *Visual Neurosci.* **5:**165–204.

Kuljis, R. O., and Rakic, P., 1989a, Distribution of neuropeptide Y-containing perikarya and axons in various neocortical areas in the macaque monkey, *J. Comp. Neurol.* **280:**383–392.

Kuljis, R. O., and Rakic, P., 1989b, Multiple types of neuropeptide Y-containing neurons in primate neocortex, *J. Comp. Neurol.* **280:**393–409.

Kuypers, H. G. J. M., Szwarcbart, M. K., Mishkin, M., and Roswold, H. E., 1965, Occipitotemporal corticocortical connections in the rhesus monkey, *Exp. Neurol.* **11:**245–262.

Livingstone, M., and Hubel, D. H., 1983, Specificity of corticocortical connections in monkey visual system, *Nature* **304:**531–534.

Livingstone, M., and Hubel, D. H., 1984, Anatomy and physiology of a color system in the primate visual cortex, *J. Neurosci.* **4:**309–356.

Livingstone, M., and Hubel, D. H., 1987, Connections between layer 4B of area 17 and the thick cytochrome-oxidase stripes of area 18 in the squirrel monkey, *J. Neurosci.* **7:**3371–3377.

Lund, J. S., 1973, Organization of neurons in the visual cortex, area 17, of the monkey (*Macaca mulatta*), *J. Comp. Neurol.* **147:**455–496.

Lund, J. S., Hendrickson, A. E., Ogren, M. P., and Tobin, E. A., 1981, Anatomical organization of primate visual cortex area VII, *J. Comp. Neurol.* **202:**19–45.

Marín-Padilla, M., 1984, Neurons of layer I. A developmental analysis, in: *Cerebral Cortex,* Volume 1 (A. Peters and E. G. Jones, eds.), Plenum Press, New York, pp. 447–478.

Marín-Padilla, M., 1990, Three-dimensional structural organization of layer I of the human cerebral cortex: A Golgi study, *J. Comp. Neurol.* **299:**89–105.

Marín-Padilla, M., 1992, Ontogenesis of the pyramidal cell of the mammalian neocortex and developmental cytoarchitectonics: A unifying theory, *J. Comp. Neurol.* **321:**223–240.

Martin, K. A. C., and Whitteridge, D., 1984, Form, function and intracortical projections of spiny neurones in the striate visual cortex of the cat, *J. Physiol. (London)* **353:**463–504.

Martin, K. A. C., Friedlander, M. J., and Alones, V., 1989, Physiological, morphological, and cytochemical characteristics of a layer 1 neuron in cat striate cortex, *J. Comp. Neurol.* **282:**404–414.

Maunsell, J. H. R., and Van Essen, D. C., 1983, The connections of the middle temporal visual area (MT) and their relationship to a cortical hierarchy in the macaque monkey, *J. Neurosci.* **3:**2563–2586.

Mignard, M., and Malpeli, J. G., 1991, Paths of information flow through visual cortex, *Science* **251:**1249–1251.

Müller, C. M., 1992, A role for glial cells in activity-dependent central nervous system plasticity? Review and hypothesis, *Int. Rev. Neurobiol.* **34:**215–281.

Mumford, D., 1992, On the computational architecture of the neocortex. II. The role of corticocortical loops, *Biol. Cybern.* **66:**241–251.

Nakamura, H., Gattass, R., Desimone, R., and Ungerleider, L. G., 1991, Comparison of inputs from areas V1 and V2 to areas V4 and TEO in macaques, *Soc. Neurosci. Abstr.* **17:**845.

Ogren, M. P., and Hendrickson, A. E., 1977, The distribution of pulvinar terminals in visual areas 17 and 18 of the monkey, *Brain Res.* **137**:343–350.

Pandya, D. N., and Sanides, F., 1973, Architectonic parcellation of the temporal operculum in rhesus monkey and its projection pattern, *Z. Anat. Entwicklungsgesch.* **139**:127–161.

Perkel, D. J., Bullier, J., and Kennedy, H., 1986, Topography of the afferent connectivity of area 17 in the macaque monkey: A double-labelling study, *J. Comp. Neurol.* **253**:374–402.

Peters, A., 1987, Number of neurons and synapses in primary visual cortex, in: *Cerebral Cortex,* Volume 6 (E. G. Jones and A. Peters, eds.), Plenum Press, New York, pp. 267–294.

Ramón y Cajal, S., 1911, *Histologie du Système Nerveux de l'Homme et des Vertébrés,* Maloine, Paris.

Rockland, K. S., 1992, Laminar distribution of neurons projecting from area V1 to V2 in macaque and squirrel monkeys, *Cereb. Cortex* **2**:38–47.

Rockland, K. S., and Pandya, D. N., 1979, Laminar origins and terminations of cortical connections of the occipital lobe in the rhesus monkey, *Brain Res.* **179**:3–20.

Rockland, K. S., and Van Hoesen, G. W., 1994, Direct temporal-occipital feedback connections to striate cortex (V1) in the macaque monkey, *Cereb Cortex,* in press.

Rockland, K. S., and Virga, A., 1989, Terminal arbors of individual "feedback" axons projecting from area V2 to V1 in the macaque monkey: A study using immunohistochemistry of anterogradely transported *Phaseolus vulgaris*-leucoagglutinin, *J. Comp. Neurol.* **285**:54–72.

Rockland, K. S., Saleem, K. S., and Tanaka, K., 1994, Divergent feedback connections from areas V4 and TEO in the macaque, *Vis. Neurosci.,* in press.

Sandell, J. H., 1986, NADPH diaphorase histochemistry in the macaque striate cortex, *J. Comp. Neurol.* **251**:388–397.

Sandell, J. H., and Schiller, P. H., 1982, Effect of cooling area 18 on striate cortex cells in the squirrel monkey, *J. Neurophysiol.* **48**:38–48.

Shipp, S., and Zeki, S., 1989, The organization of connections between areas V5 and V1 in macaque monkey visual cortex, *Eur. J. Neurosci.* **1**:309–332.

Somjen, G. G., 1987, Functions of glial cells in the cerebral cortex, in: *Cerebral Cortex,* Volume 6 (E. G. Jones and A. Peters, eds.), Plenum Press, New York, pp. 1–40.

Sousa, A. P. B., Piñon, M. C. G. P., Gattass, R., and Rosa, M. G. P., 1991, Topographic organization of cortical input to striate cortex in the *Cebus* monkey: A fluorescent tracer study, *J. Comp. Neurol.* **308**:665–682.

Spatz, W. B., 1977, Topographically organized reciprocal connections between areas 17 and MT (visual area of superior temporal sulcus) in the marmoset Callithrix jacchus, *Exp. Brain Res.* **27**:559–572.

Szentágothai, J., 1978, The neuron network of the cerebral cortex: A functional interpretation, *Proc. R. Soc. London Ser. B* **201**:219–248.

Tigges, J., and Tigges, M., 1985, Subcortical sources of direct projections to visual cortex, in *Cerebral Cortex,* Volume 3 (A. Peters and E. G. Jones, eds.), Plenum Press, New York, pp. 351–378.

Tigges, J., Spatz, W. B., and Tigges, M., 1973, Reciprocal point-to-point connections between parastriate and striate cortex in the squirrel monkey (*Saimiri*), *J. Comp. Neurol.* **148**:481–490.

Tigges, J., Tigges, M., and Perachio, A. A., 1977, Complementary laminar terminations of afferents to area 17 originating in area 18 and in the lateral geniculate nucleus in squirrel monkey, *J. Comp. Neurol.* **176**:87–100.

Tigges, J., Tigges, M., Anschel, S., Cross, N. A., Letbetter, W. D., and McBride, R. L., 1981, Areal and laminar distribution of neurons interconnecting the central visual cortical areas 17, 18, 19, and MT in the squirrel monkey (*Saimiri*), *J. Comp. Neurol.* **202**:539–560.

Tömböl, T., 1984, Layer VI cells, in: *Cerebral Cortex,* Volume I (A. Peters and E. G. Jones, eds.), Plenum Press, New York, pp. 479–519.

Tononi, G., Sporns, O., and Edelman, G. M., 1992, Re-entry and the problem of integrating multiple cortical areas: Simulation of dynamic integration in the visual system, *Cereb. Cortex* **2**:310–335.

Ullman, S., 1991, Sequence-seeking and counter-streams: A model for information processing in the cortex, *AI Memo* 1311, MIT.

Ungerleider, L. G., and Desimone, R., 1986, Cortical connections of visual area MT in the macaque, *J. Comp. Neurol.* **248**:190–222.

Valverde, F., 1985, The organizing principles of the primary visual cortex in the monkey, in: *Cerebral Cortex,* Volume 3 (A. Peters and E. G. Jones, eds.), Plenum Press, New York, pp. 207–257.

van Brederode, J. F. M., Mulligan, K. A., and Hendrickson, A. E., 1990, Calcium-binding proteins as markers for subpopulations of GABAergic neurons in monkey striate cortex, *J. Comp. Neurol.* **298**:1–22.

Van Essen, D. C., Newsome, W. T., Maunsell, J. H. R., and Bixby, J. L., 1986, The projections from striate cortex (V1) to areas V2 and V3 in the macaque monkey: Asymmetries, areal boundaries, and patchy connections, *J. Comp. Neurol.* **224:**451–480.

Vogt, B. A., 1991, The role of layer I in cortical function, in: *Cerebral Cortex,* Volume 9 (A. Peters, ed.), Plenum Press, New York, pp. 49–80.

Walz, W., 1989, Role of glial cells in the regulation of the brain ion microenvironment, *Prog. Neurobiol.* **33:**309–333.

Weller, R. E., and Kaas, J. H., 1983, Retinotopic patterns of connections of area 17 with visual areas V-II and MT in macaque monkeys, *J. Comp. Neurol.* **220:**253–279.

Wong-Riley, M., 1978, Reciprocal connections between striate and prestriate cortex in the squirrel monkey as demonstrated by combined peroxidase histochemistry and autoradiography, *Brain Res.* **147:**159–164.

Zeki, S. M., 1980, A direct projection from area V1 to area V3a of rhesus monkey visual cortex, *Proc. R. Soc. London Ser. B* **207:**499–506.

The Role of Area 17 in the Transfer of Information to Extrastriate Visual Cortex

JEAN BULLIER, PASCAL GIRARD, and
PAUL-ANTOINE SALIN

1. Introduction

1.1. Effects of Lesions of Area 17 on Visual Behavior

The ideas concerning the role of area 17 in the transfer of visual information to the rest of the cerebral cortex have for a long time been influenced by the results of behavioral studies of primates following cortical lesions. Since the last century it has been known that lesions of area 17 lead to blindness in humans (for a review see Weiskrantz, 1986, and Rizzo, this volume). This critical role of area 17 in vision was used in the beginning of the 20th century by Inouye in Japan and Holmes in Great Britain to map the representation of the visual field in area 17 of humans by delimiting the extents of scotomata resulting from focal lesions in area 17 of wounded soldiers. In 1942, the results of an extensive study by Klüver of monkeys with cortical lesions seemed to leave little doubt that, for this species as well, area 17 is necessary for any kind of vision beyond a simple discrimination between light and dark. Rudimentary sensitivity to light had also been noted to

JEAN BULLIER, PASCAL GIRARD, and PAUL-ANTOINE SALIN • Cerveau et Vision, INSERM Unité 371, 69500 Bron/Lyon, France.
Cerebral Cortex, Volume 10, edited by Alan Peters and Kathleen S. Rockland. Plenum Press, New York, 1994.

persist in humans with lesions of area 17 by Holmes (1918) and Riddoch (1917). Interestingly, this last author noted a weak residual sensitivity to moving targets, but no perception of stationary objects.

Such devastating effects of lesions of area 17 were in keeping with the concept of a primary area inherited from the studies of Flechsig, a concept which was much in vogue in those days. According to Flechsig (1896), for each sensory modality, the information from the periphery is related in the thalamus and reaches a unique cortical area called the primary area. Thus, the blindness resulting from lesions of area 17 was thought to result from this bottleneck role of area 17 in distributing the information to the rest of the cerebral cortex.

This simple model of organization was first questioned in 1942 when Talbot found that visually evoked potentials could still be recorded in area 18 of the cat after a lesion of area 17. This finding, controversial at first, was subsequently replicated by a number of investigators (Doty, 1958; Berkley et al., 1967) before it was demonstrated that all neurons in area 18 of the cat remain visually responsive when area 17 is lesioned or inactivated (Dreher and Cottee, 1975; Donaldson and Nash, 1975; Sherk, 1978; Casanova et al., 1992). Anatomical studies using the degeneration technique provided a possible explanation for the residual activity in area 18 by showing that the LGN sends a direct projection to area 18 in the cat (Garey, 1965; Glickstein et al., 1967).

At the same time, others were conducting experiments on the behaviors of cats after restricted lesions of the cerebral cortex. It became clear that cats with area 17 lesions show very little impairment of their visuomotor behavior and their capacity for shape recognition (Sprague, 1966; Doty, 1971; Sprague et al., 1977). Experiments done in the years 1960–1970 on the visual behavior of mammals with cortical lesions demonstrated that all nonprimate species studied show extensive residual visual capacities after lesion of area 17 (hamster: Schneider, 1967; hedgehog: Hall and Diamond, 1968; tupaia: Killackey et al., 1971; rabbit: Murphy and Chow, 1974, rat: Hughes, 1977).

Although less readily accepted than in nonprimate species, the results of behavioral experiments conducted on destriated monkeys showed that the bottleneck model of information processing in primate visual cortex was no longer tenable. As early as 1963, Weiskrantz demonstrated that macaque monkeys with area 17 lesions are still capable of visual behavior (Weiskrantz, 1963; Cowey and Weiskrantz, 1963). This was followed by the description of the visual behavior of the now-famous macaque monkey "Helen" which, despite a nearly complete lesion of area 17, showed a remarkable ability to explore its environment and manipulate objects without being able to recognize them on the basis of visual inspection (Humphrey and Weiskrantz, 1967; Humphrey, 1974). These results were greeted at first with skepticism and the residual visual capacities were usually attributed to remaining islands of striate cortex spared by the lesion. This interpretation was contradicted by the results of the Pasiks who showed that, after a total bilateral lesion of area 17, monkeys are still able to discriminate between targets on the basis of contrast, orientation, form, and color (Pasik et al., 1969; Pasik and Pasik, 1971; Schilder et al., 1972). At the same time, it was demonstrated that humans with an area 17 lesion exhibit unconscious residual vision usually referred to as blindsight (see Rizzo, this volume).

During the same period as the behavioral studies conducted on nonprimate species with lesions of area 17, projections of the lateral geniculate nucleus (LGN) to extrastriate cortex were demonstrated in many nonprimate species using recently developed neuroanatomical techniques such as retrograde transport of horseradish peroxidase or anterograde transport of radioactive amino acids. Results of such studies in the cat showed that neurons in the A layers project exclusively to areas 17 and 18 and that the C layers project to practically all the known visual areas in this animal (Rosenquist *et al.*, 1974; Gilbert and Kelly, 1975; Maciewicz, 1975; Holländer and Vanegas, 1977; Raczkowski and Rosenquist, 1980). In nonprimate species other than the cat, it was found that the LGN sends projections not only to area 17 but also to cortical areas surrounding it (rat: Hughes, 1977; Coleman and Clerici, 1980; hamster: Dürsteler *et al.*, 1979; goat: Pettigrew *et al.*, 1984; sheep: Karamanlidis *et al.*, 1979; Pettigrew *et al.*, 1984; rabbit: Höllander and Hälbig, 1980; Towns *et al.*, 1982).

In the 1970s, the last support that remained for the bottleneck schema of primate organization of thalamocortical connections was the specific projections of the LGN to area 17. This last support was removed by the results of anatomical studies using sensitive retrograde tracers injected in extrastriate cortex of the macaque monkey. A number of reports demonstrated direct projections of LGN neurons to a region of the prelunate gyrus belonging to area V4 (Benevento and Yoshida, 1981; Fries, 1981; Yukie and Iwaï, 1981), to the posterior bank of the superior temporal sulcus where area MT is located (Fries, 1981), and to area V2 (Bullier and Kennedy, 1983).

The extrastriate projections of the LGN in macaque monkeys, however, are not necessarily homologous to those demonstrated in other species. For example, in the cat, the LGN projection to area 18 involves neurons in laminae A and A1 which do not receive afferents from the superior colliculus (Torrealba *et al.*, 1981) but are directly innervated by retina fibers. Also, a number of neurons projecting to area 18 send an axonal branch to area 17 (Geisert, 1980; Bullier *et al.*, 1984; Birnbacher and Albus, 1987). This is clearly different from the situation in the macaque monkey since extrastriate LGN projections arise mostly from the interlaminar and S layers (Yukie and Iwaï, 1981; Benevento and Yoshida, 1981; Bullier and Kennedy 1983). These are the recipients of superior colliculus afferents (Partlow *et al.*, 1977; Harting *et al.*, 1980). Figure 1 provides an example of the labeling observed in the LGN after simultaneous injections of two different tracers in V1 and V2 of the macaque monkey. Note that most neurons labeled by the injection in V2 are outside of the main laminae of the LGN which contain cells labeled by the V1 injection. Note also in Fig. 1 that very few LGN neurons send bifurcating axons to area 17 and extrastriate cortical areas. Similarly, after injections in areas V4 and V1, different populations of labeled cells were also observed in the LGN (Lysakowski *et al.*, 1988).

Extrastriate projections from the interlaminar zones and the S layers of the LGN in the macaque monkey appear to be homologous to those of the C layers of the cat. In both cases, neurons in these regions of the LGN receive afferents from the superior colliculus, project widely to many extrastriate areas via bifurcating axons (Bullier and Kennedy, 1983), and are located in different LGN laminae from those projecting to area 17.

The organization of the primate visual system therefore appears homologous to that in other species in possessing a pathway from the superior colliculus to extrastriate cortex through the tecto-recipient layers of the LGN. On the other hand, there appears to be no projection system in primates which would be the homologue of the direct geniculate projection to area 18 in the cat.

In the cat, after lesions or inactivation of area 17, residual activity is observed in area 18 (Dreher and Cottee, 1975; Sherk, 1978; Casanova *et al.*, 1992) and in areas of the lateral suprasylvian sulcus (Spear and Baumann, 1979). This is also true, to a lesser extent, in area 21a (Michalski *et al.*, 1992), and it is likely that most visual areas retain some activity. Thus, the residual visual capacities observed in this species after area 17 lesions probably reflect neural activity in many cortical areas. Given the differences presented above between primate and nonprimate species concerning the anatomy of the geniculostriate projections and the severity of the effects of area 17 lesions, one may question to what extent the situation is the same in primates. In other words, among the visual areas of the macaque monkey, presented in Fig. 2, which ones are active during inactivation or lesions of area 17 and can therefore be considered as the possible structural basis for residual vision in destriated primates? It is the purpose of this review to provide answers to this question. Because of the similarity between the organizations of the visual systems of monkeys and humans, these questions have obvious implications concerning the possible structural basis for blindsight in humans.

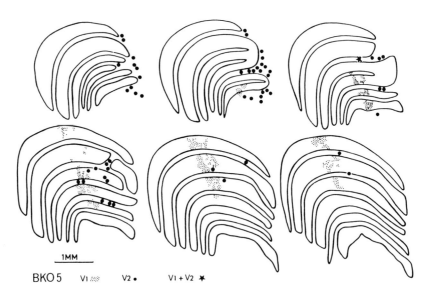

Figure 1. Neurons labeled in the LGN of a macaque monkey which received an injection of a retrograde tracer in V1 and another in area V2. Note the different location of neurons projecting to V1 and those projecting to V2. Note also the small number of neurons sending a bifurcating connection to V1 and V2 (single neuron represented by the star). Reprinted from Bullier and Kennedy (1983) with permission.

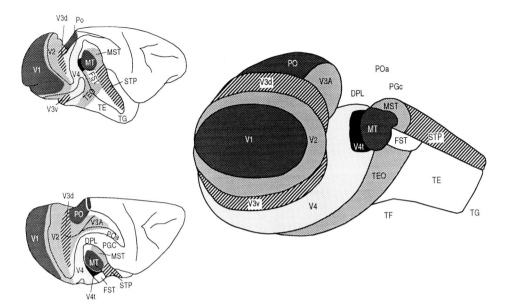

Figure 2. Lateral view and upper views of macaque visual cortex to show the different cortical areas. On the right is a schematic representation of the visual cortical areas after flattening the cortex.

2. Experimental Evidence for Activity in Extrastriate Cortex following Removal of Area 17 Input in Primates

2.1. Methodological Considerations

The evidence concerning the persistence of visual responses in extrastriate cortex without input from V1 is based on the results of two types of experiments: lesions or reversible inactivation of V1. These two techniques do not address the same aspects of the question. Immediately after a lesion, there is usually a traumatic reaction caused by bleeding or edema which can lead to a temporary loss of response in surrounding areas and, at least in the case of thermal lesions, major changes in the firing pattern of neighboring neurons (Eysel and Schmidt-Kastner, 1991). During postoperation recovery, compression related to edema and bleeding is resorbed and activity can be recovered. On a longer time scale, neural responses can be modified in connected structures through the phenomenon of diaschisis. This refers to a temporary functional inactivation of certain cortical or subcortical structures which are connected to the lesioned areas. For example, diaschisis could explain the temporary loss of visual responses in the pulvinar following a lesion of V1 in macaque monkeys (Bender, 1983). After 3 weeks, visual responses return as do visuomotor abilities (Mohler and Wurtz, 1977). The neural explanation for diaschisis as originally proposed by von Monakow (1914) remains, however, controversial (West *et al.*, 1976; Miller, 1984).

Another aspect of the functional reaction to cortical lesioning is that of adult plasticity. This has recently attracted much attention, following its initial demonstration in somatosensory cortex (for a review see Kaas *et al.*, 1983). For example, after a restricted lesion of the retina, the deafferented region of V1 is initially silent but regains visual responses with displaced receptive fields following several months of recovery (Kaas *et al.*, 1990; Gilbert and Wiesel, 1992). All the recent evidence points out that deafferented cortical neurons usually regain some functional role, possibly through a mechanism of changes in synaptic strength of those inputs which played only a modulatory role in the normal cortex.

The lesion method demonstrates the maximum extent of the capacity for recuperation by the nervous system following destruction of a given neural structure, but it does not necessarily reveal the functional contribution of this structure to normal functioning. In the case of restricted retinal lesions, it is clear that the appearance of displaced receptive fields in the deafferented cortical region does not correspond to a normal function of the system but constitutes a reaction of plasticity in compensation to the lesion.

Studies which use reversible inactivation of a given structure do not suffer from these limitations: during the short period of the inactivation of the afferent pathway, there is no possibility for long-term plasticity reactions and therefore the residual response is more likely to reflect the presence of a normal functional input. Chino *et al.* (1992) and Gilbert and Wiesel (1992) have observed the occurrence of plasticity only a few minutes after retinal lesions, a time interval that corresponds to a cooling session. However, during cooling, we never saw variations of receptive field sizes or locations as observed during rapid plastic reorganization following retinal lesions. This observation implies that the reversible inactivation method does not trigger similar reorganization mechanisms as lesions. In order to interpret the residual activity in extrastriate areas following removal of input from area V1, it is therefore important to be able to refer to different studies including those done with inactivation as well as lesions of area V1.

2.2. Residual Visual Activity in Inferotemporal Cortex and Area STP

The first attempt at determining the effect of area V1 lesions on the visual responses of neurons in cortical regions beyond V1 was made by Rocha-Miranda *et al.* (1975). After a complete bilateral V1 lesion, no visual response was observed in a cortical region corresponding to the lateral part of area TE on the inferotemporal gyrus (Fig. 2). After a unilateral V1 lesion, recordings in areas TEO and TE revealed that the visual response disappeared in the deafferented contralateral half of the receptive field. In a more recent study, Bruce *et al.* (1986) showed that most cells located in the ventral bank of the superior temporal sulcus also failed to respond in the contralateral part of their receptive fields following a unilateral V1 lesion. They found a few (4 out of 48) responsive neurons but it is unclear whether these belong to TE/TEO or to FST (Fig. 2).

The demonstration that visual activity in areas TEO and TE depend on input from V1 relayed through extrastriate cortex was in keeping with the bottleneck concept of organization of the flow of information in the visual cortex. It was therefore a surprise when researchers from the same group demonstrated for the first time that visual activity can survive in extrastriate cortex following a lesion of area V1. This discovery was made by Bruce *et al.* (1986) who recorded in the anterior two-thirds of the upper bank and fundus of the superior temporal sulcus, an area in which they found polysensory neurons and which was consequently termed STP (superior temporal polysensory area; Bruce *et al.*, 1981). In the normal animal, 92% of the STP neurons have bilateral receptive fields. If STP neurons were entirely dependent on the V1 input for their visual responses, following a unilateral V1 lesion one would expect that all the receptive fields would be entirely confined to the nondeafferented visual hemifield. On the contrary, Bruce and his colleagues found that half of the neurons recorded in STP remained responsive to stimulation in the deafferented part of their receptive fields. This suggests that, in contrast to the situation in TEO and TE, where all the neurons were inactivated by the removal of their input from V1, half the neurons in STP receive significant visual input from a source other than V1. Figure 3 provides an example of recordings in deafferented STP: after a lesion of the left area 17, responses to a flashing light could be recorded not only in the left visual hemifield (corresponding to the intact area 17), but also in the right visual hemifield which corresponds to the lesioned area 17. By comparing the visual responses elicited in the deafferented and nondeafferented parts of the RF, these authors concluded that removal of striate cortex input leads to a decrease in the strength of the response but not to its elimination. By the same method, they also studied the effect of V1 lesions on direction selectivity, which is a common feature in STP neurons (Bruce *et al.*, 1981), and found that it was mostly abolished. Area STP has been subdivided into TAa, TPO, and PGa on cytoarchitectural grounds (Seltzer and Pandya, 1978; Baylis *et al.*, 1987). These areas differ in the functional characteristics of their neurons (Baylis *et al.*, 1987) and in the pattern of corticocortical connectivity (Morel and Bullier, 1990) and it would therefore be interesting to know specifically which of these areas contain visually responsive neurons in the absence of V1.

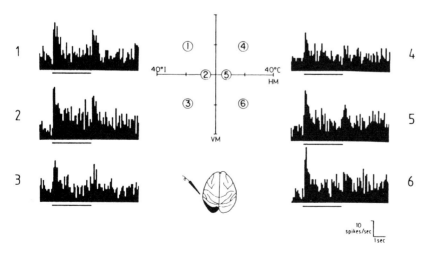

Figure 3. Response of a neuron located in STP to a flashing stimulus (light ON corresponds to the bar below the PST histogram) in a monkey with area 17 removed on the left side. Notice the clear response in the locations (4, 5, and 6) corresponding to the deafferented part of the receptive field. Reprinted from Bruce *et al.* (1986) with permission.

2.3. Residual Activity in Area MT

The presence of residual visual activity in STP following a V1 lesion raised the question of what other areas might reserve visual responses after striate deafferentation. This led Rodman *et al.* (1989) to study the responsiveness of cells in area MT. They found that, even after a total bilateral lesion of V1, about 60% of the neurons were visually responsive 5 to 6 weeks after the operation. The average level of the response was severely decreased with respect to that in normal MT, but the direction selectivity which is characteristic of MT neurons did not differ significantly in strength and tuning width from that recorded in normal animals.

Because of the long period of recovery following the removal of V1, it was unclear whether the visual responses of neurons in deafferented MT were the result of a plastic reorganization of the cortex by reinforcement of an initially modulating input or whether they reflected the presence of a functional visual drive independent from V1. To test for this, Rodman and her collaborators studied the activity of a small number of MT neurons when V1 was inactivated by cooling. They found that six cells out of seven thus tested still responded to visual stimulation, suggesting that most cells in area MT are normally activated by visual inputs other than V1. This was confirmed by Girard *et al.* (1992) who found that 80% of the neurons in area MT remain visually responsive when the corresponding region of area V1 is silenced by cooling. Figure 4 presents two examples of MT neurons recorded during inactivation of a large region of area V1 located below a cooling plate represented by the stippled disk on the lateral view of the brain. Photic stimulation was confined to the region of the visual field represented in the inactivated region of V1, as identified by electrophysiological mapping (this region is called the cooling perimeter). Blocking of the region of V1 below the cooling plate results in a strong decrease of the response in one unit (13.5 on the left) whereas the visual response of the other (10.4 on the right) is hardly affected. The histogram at the upper right illustrates the distribution of the blocking index for MT neurons during cooling of V1 (blocking index = 0 for no change of the response, = 1 for total inactivation). It is clear that many units give a strong response (blocking index below 0.6) in the absence of input from V1.

Since the same cell is studied before, during, and after the inactivation of V1, it is possible by this method to determine the specific contribution of the input from V1 to the receptive field properties of neurons in MT. We studied the effects of blocking the input from V1 on the direction selectivity of MT neurons. Figure 5A illustrates the angular changes in optimal direction for neurons in MT during V1 inactivation with respect to control (direction 0 corresponds to no change). As shown in Fig. 5A, the optimal direction was maintained or changed by only a small amount in most neurons. When the selectivity of the response to movement in the optimal direction versus its opposite was quantified by the direction index DI, it was found that most units showed only a small decrease in their selectivity (Fig. 5B). This decrease in selectivity is usually caused by a specific weakening of the response in the optimal direction. This suggests that the contribution of the V1 input to direction selectivity in MT neurons is mostly achieved by the excitatory convergence of inputs from neurons with the same optimal direction.

These results therefore demonstrate that the remaining direction selectivity in MT neurons that was observed by Rodman and her collaborators after V1

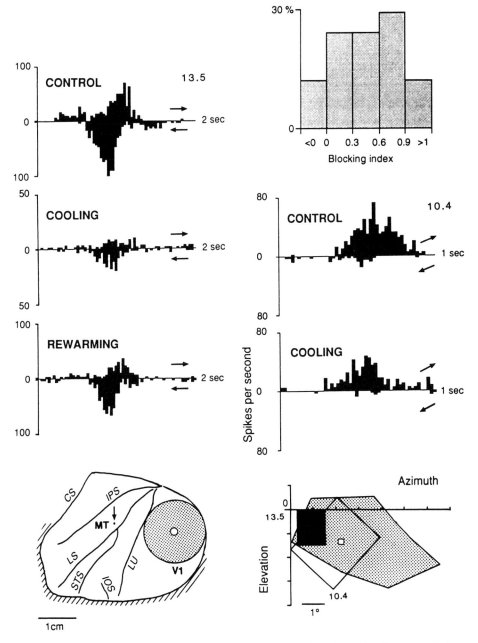

Figure 4. Visual response of two neurons in area MT during cooling of V1. The diagram at the lower left shows the cooling plate (stippled disk) applied to the surface of area V1. On the right, in gray, is represented the extent of visual field represented in the inactivated portion of V1, a region called the cooling perimeter. The small white square illustrates the aggregate receptive field of neurons recorded in V1 in the center of the cooling plate. Activity of these neurons was monitored during the cooling in order to ensure proper inactivation of V1. Two sets of responses are shown for MT neurons No. 13.5 and 10.4. The receptive fields of these neurons are shown on the lower right (black for 13.5, open rectangle for 10.4). In both cases, visual stimulation was confined to the part of the receptive field contained within the cooling perimeter. Responses of neurons are shown as PST histograms. On the left, response of neuron 13.5 whose activity is strongly diminished by the inactivation of V1 (blocking index 0.9); on the right, neuron 10.4 whose activity is only little affected by the cooling of V1 (blocking index 0.3). Upper right: frequency histogram of the blocking index for 17 MT neurons (0 for no effect, 1 for complete blocking of the response). Redrawn with permission from Girard *et al.* (1992).

lesions are not the sole result of cortical reorganization following the lesion. The respective contribution of the different afferents to area MT in the maintenance of direction selectivity during inactivation of V1 is examined in Section 3.

In a recent publication, Kaas and Krubitzer (1992) report that, in the owl monkey, acute lesions of area 17 lead to a complete disappearance of visual responses in the retinotopically corresponding region of MT. It is unclear at the moment whether this discrepancy between the results obtained in the owl monkey and in the macaque monkey is related to species or technical differences. We recorded in macaque area MT after a large acute lesion in area 17 and found visually responsive units several hours after the lesion (Girard *et al.*, 1992). Therefore, the discrepancy does not appear to result from the use of reversible inactivation of V1 in one study and an acute lesion in the other. It may be that anesthesia is a critical factor when examining residual responses. This is suggested by the fact that we could abolish residual responses in MT neurons during cold inactivation of V1 by increasing slightly the concentration of halothane in the gas mixture (Girard *et al.*, 1992).

2.4. Visual Activity in Other Cortical Areas

The presence of residual visual responses in MT after lesion or inactivation of V1 is not surprising since MT provides an indirect input to area STP through areas MST and FST (Boussaoud *et al.*, 1990) and, as described above, visual activity persists in STP after striate deafferentation. In a similar way, the absence of visual activity in areas TE and TEO after V1 lesions suggested that area V4, which provides the major input to this region (Desimone *et al.*, 1980; Morel and Bullier, 1990; Baizer *et al.*, 1991), would be silenced by inactivation of area V1. This prediction was tested with methods similar to those used for the investigation conducted in area MT (Girard *et al.*, 1991a). Figure 6 illustrates the results obtained in V4 by showing the response of two neurons during inactivation of

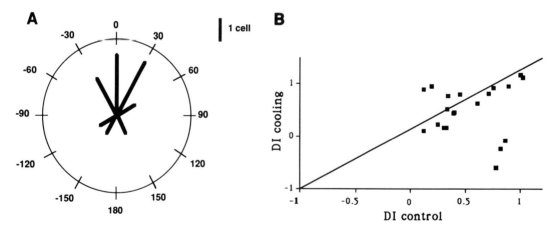

Figure 5. (A) Change in optimal direction of movement for MT cells when V1 is blocked by the cold. Direction 0 corresponds to no change. It appears that most neurons show only a slight change in optimal direction. (B) Effect of blocking V1 on the direction index DI (DI is 0 for non-direction-selective units, 1 for total direction selectivity). For most units, DI is only diminished by a small amount (small decrease in direction selectivity). Redrawn with permission from Girard *et al.* (1992).

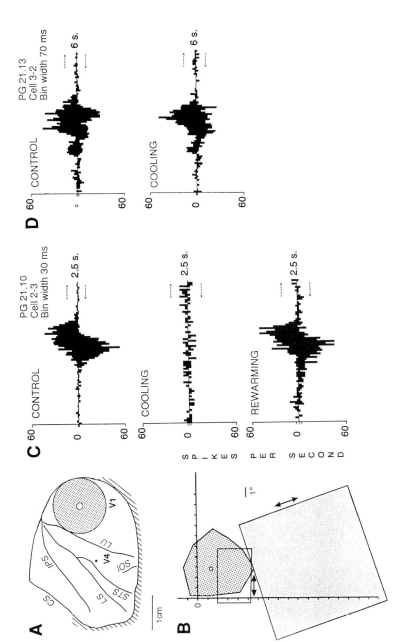

Figure 6. Examples of neurons recorded in area V4 during cold-blocking of V1. The presentation here is similar to Fig. 4, with the cooling plate and perimeter represented by the stippled surfaces, and the two receptive fields by the two rectangles. The unit on the left (C) had its receptive field included in the cooling perimeter. When photic stimulation was confined to the cooling perimeter, visual responses disappeared during the cooling of V1. Note the slight increase in spontaneous activity. The unit on the right (D) had a receptive field outside the cooling perimeter. No significant effect was observed when V1 was blocked by the cold. Redrawn with permission from Girard *et al.* (1991a).

V1. The visual response of the neuron whose receptive field is inside the cooling perimeter (Fig. 6C) is completely abolished by the inactivation of V1, while the spontaneous activity was hardly changed. On the other hand, the neuron with the receptive field outside the cooling perimeter (Fig. 6D) shows only a marginal modification of its response to visual stimulation during cooling of V1. The retinotopic specificity of the results provides a control for possible indirect effects of cooling on the neural activity outside V1, such as spreading depression or disruption of the cerebral circulation or of the oxygen uptake. This specificity also shows that, at least in terms of transfer of visual drive, the corticocortical connections from V1 to V4 are visuotopically organized (for a different result in the cat see Salin *et al.*, 1992).

The vast majority (91%) of the sites tested in V4 were completely inactivated by the blockage of V1. Except for one unit which gave a clear response, all other cells gave a response which was too weak or inconsistent to plot the receptive field. This is in contrast to cells that remained active in MT, for which it was easy to plot the receptive field and measure the response selectivity during the blockage of V1. It appears therefore that neurons in area V4, as well as in areas TEO and TE, are dependent on the presence of an active input from V1.

In the chain of information transfer from V1 to extrastriate cortex, area V2 is likely to play an important role. It is heavily interconnected wth V1 (Kennedy and Bullier, 1985; Perkel *et al.*, 1986) and distributes information to many cortical areas which are not directly innervated by V1 (Felleman and Van Essen, 1991). It was therefore logical to test whether some neurons in V2 remain active in the absence of input from V1. The question is particularly interesting since V2 neurons projecting to V4 are segregated from those projecting to MT, and the zones of projections to these two areas can be identified on sections stained for cytochrome oxidase activity (De Yoe and Van Essen, 1985; Shipp and Zeki, 1985, 1989). The proximity between V1 and V2 precludes the use of lesions to investigate this question since such a lesion of V1 is likely to undercut most other input to V2. It was therefore necessary to use a method of reversible inactivation. An early report by Schiller and Malpeli (1977) concluded that all visual activity is abolished in V2 by blocking V1. This was confirmed by the study of Girard and Bullier (1989) who took special care to minimize the direct effects of the cold on the responses of neurons in V2 by recording at a sufficient distance in the depth of the lunate sulcus. In this later study, it was also shown that all types of neurons are inactivated by cold-blocking V1, including those located in the thick cytochrome oxidase bands which are projecting to MT. Thus, it does not appear that the residual visual input to MT comes from V2. The absence of visual response in area V2 of the macaque monkey during inactivation of area V1 stands in contrast with the presence of residual activity in area 18 of the cat (Sherk, 1978; Casanova *et al.*, 1992), which is sometimes believed to be homologous to area V2. This finding is consistent with the differences in organization of the retinogeniculate pathway between primates and nonprimates (see Section 1.2).

We also studied the visual responses of neurons in areas V3 and V3a during cold-blocking of V1 (Girard *et al.*, 1991b). Practically all neurons were inactivated in area V3, whereas approximately 30% of the neurons in V3a remained active when V1 was blocked. The contrast between recordings in areas V3 and V3a was particularly striking when neurons in both areas were sampled in the same penetration. Visually responsive neurons were encountered in area V3a located in the anterior bank of the lunate sulcus, whereas when the microelectrode crossed the sulcus to enter V3, all visual response disappeared, as illustrated in Fig. 7.

Figure 8 presents a summary diagram of the areas which remain active in the absence of V1 input. Areas in dark gray correspond to active areas, areas in

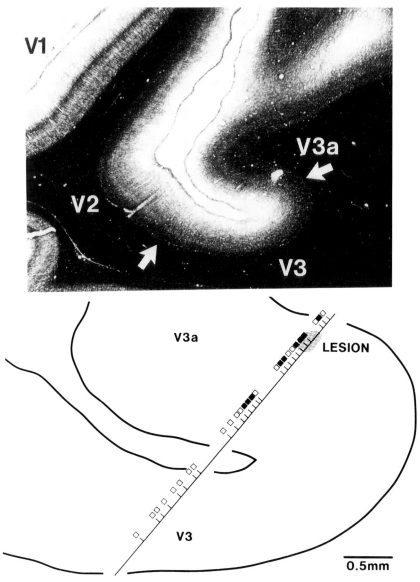

Figure 7. Illustration of a penetration through the anterior bank and the fundus of the lunate sulcus. The upper panel presents a myelin-stained parasagittal section of macaque monkey cortex showing areas V1, V2, V3, and V3a and the lower panel illustrates the cortical regions penetrated by the microelectrode. The penetration directed from upper right toward lower left first traverses area V3a and, after crossing the sulcus, it enters area V3. In V3a, numerous neurons are found that still respond to visual stimulation during cold-blocking of the retinotopically corresponding region of V1 (dark squares). In V3 all neurons are silenced by the blocking of V1 (open squares). Reprinted from Girard *et al.* (1991b).

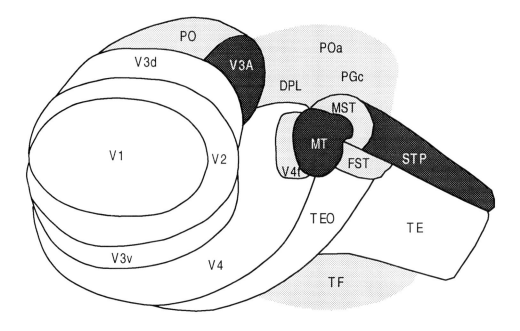

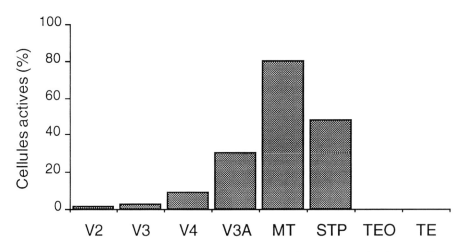

Figure 8. Summary of the visual areas containing visually responsive neurons in the absence of input from V1. Areas in dark gray still contain substantial proportions of visually responsive neurons; areas in white contain no or very few poorly responsive neurons; areas in light gray have not been tested. Lower panel: proportions of active cells in different areas during inactivation or lesions of area 17. (V2 N = 209, Girard and Bullier, 1989; V3 N = 37, Girard *et al.*, 1991a; V3a N = 76, Girard *et al.*, 1991a; V4 N = 85, Girard *et al.*, 1991b; MT N = 57, Girard *et al.*, 1992; STP N = 216, Bruce *et al.*, 1986; TEO/TE N = 106, Rocha-Miranda *et al.*, 1975.)

white represent areas which contain no or very few responsive neurons and for which receptive field plotting is impossible. Areas in light gray have not been tested. The lower panel presents the proportions of active neurons in each of the areas which have been tested so far. In brief, areas V2, V3, V4, and the inferotemporal cortex are silenced by the removal of input from V1. Several areas in the more medial section of cortex and especially in the superior temporal sulcus remain active. Since areas MST and FST are heavily interconnected with areas MT and STP (Ungerleider and Desimone, 1986; Boussaoud *et al.*, 1990), it is highly likely that they also remain visually active. In addition, areas POa and PGc of the parietal cortex probably remain visually active as well, since they are also strongly interconnected with MT, MST and STP (Morel and Bullier, 1990; Baizer *et al.*, 1991; Andersen *et al.*, 1990a; Cavada and Goldman-Rakic, 1989). Given the special functional properties of area PO, which contains a large region devoted to the representation of the periphery of the visual field, it would be interesting to know whether neurons in this area remain visually responsive when V1 is inactivated or lesioned.

3. Pathways Responsible for the Residual Responses

A dense network of corticocortical connections has been traced between the visual cortical areas of macaque monkeys (e.g., Felleman and Van Essen, 1991). Subcortical structures have also been shown to project to each node of this network, but they are often omitted from the general schematic organization of the visual system, as if they did not participate in the transfer of cortical information. Given that residual responses can be observed in extrastriate cortical areas, the question arises as to which of these inputs might serve as relays for the visual information.

3.1. Subcortical Pathways

There are four main subcortical routes which enable visual signals to bypass area V1: First is the well-known retinotectopulvinar pathway that projects massively and directly to most visual areas. Another pathway reaches extrastriate cortex from the superior colliculus via the interlaminar and S layers of the LGN. The third route from the retina bypasses the superior colliculus and directly reaches the pulvinar, and thence to visual cortical areas. Lastly, retinal information may be conveyed directly through the interlaminar and S layers of the LGN to extrastriate cortex.

Only a few studies have tested which of these pathways may be responsible for residual visual information. Some researchers have tried to eliminate the residual activity observed in extrastriate cortex after a V1 lesion by lesioning specific subcortical territories that were suspected to act as relays. Thus, both Rodman *et al.* (1990) and Bruce *et al.* (1986) have observed the disappearance of residual visual responses, respectively in MT and STP, after a lesion of the superior colliculus. Moreover, such combined V1–superior colliculus lesions prevent monkeys from making a saccade toward a visual target appearing within the scotoma (Mohler and Wurtz, 1977; Solomon *et al.*, 1981). The results of these

studies therefore point to the necessity of an intact colliculus for the maintenance of residual visual abilities.

From the early studies of Diamond and his collaborators (Diamond and
Hall, 1969; Diamond 1976), it has been hypothesized that visual information
reaches extrastriate cortex through a tectopulvinar route. However, if the role of
the superior colliculus is clearly established in residual vision, it is less well known
whether the pulvinar constitutes the next relay and, if it does, which part of the
pulvinar plays this role. Standage and Benevento (1983) in the macaque monkey
and Lin and Kaas (1979) in the owl monkey report that the region of the pulvinar that projects to MT does not correspond to the tecto-recipient zone. However, by comparing directly the figures of retrograde labeling in the pulvinar
after an injection in MT (Standage and Benevento, 1983) and anterograde
labeling in the pulvinar after a tectum injection (Benevento and Standage,
1983), one can see several overlap regions which could act as relays. Clearly some
direct test of this question needs to be made using a combination of anterograde
and retrograde labels.

The pulvinar nucleus in primates is a very large nucleus in which at least
three main divisions can be made, the lateral, the inferior, and the medial pulvinar (Baleydier and Morel, 1992). Visual responses have been recorded in the
lateral and inferior pulvinar after a V1 lesion (Burman et al., 1982; Bender,
1983) and these nuclei could therefore serve as relays for residual visual information. However, these nuclei project widely to extrastriate cortex and it is
therefore not clear why some of these areas remain active but not others in the
absence of a V1 input. In this context, the recent results of Baleydier and Morel
(1992) are interesting since they demonstrate that the projections to the inferotemporal cortex arise from different pulvinar regions than those connected to
the parietal cortex. It is possible that a similar segregation exists between the
pulvinar regions projecting to different extrastriate areas and that visual activity
can only be recorded in pulvinar regions connected to areas V3a, MT, and STP
and not in those projecting to V2, V3, V4, TEO, and TE. Clearly more work is
needed to clarify these questions.

Several studies have shown that the retina projects directly onto the pulvinar
(Mizuno et al., 1982; Itaya and Van Hoesen, 1983; Nakagawa and Tanaka, 1984)
and that such connections can terminate on neurons projecting to the prestriate
cortex (Mizuno et al., 1983). Very recently, it has been shown that these retinal
afferents come primarily from a particular subset of ganglion cells, namely Pβ
and Pγ ganglion cells (Stoerig et al., 1991). Since these authors previously showed
that a subset of ganglion cells do not degenerate after long-term V1 lesions
(Stoerig and Cowey, 1989), it is possible that the direct retinopulvinar pathway
also participates in the transfer of residual visual information to extrastriate
cortex after a lesion or inactivation of V1.

It is therefore likely that the superior colliculus and the pulvinar act as relays
for visual information subserving residual vision. We showed above that area MT
contains a large proportion of neurons which remain active and that these
neurons retain their direction selectivity in the absence of a V1 input. One can
wonder whether the direction selectivity of deafferented MT neurons is simply a
consequence of their subcortical inputs. Concerning the superior colliculus, the
region of interest is the superficial layers that project to the pulvinar and remain
visually responsive following a V1 lesion (Schiller et al., 1974; Marrocco and Li,
1977). Early data from Goldberg and Wurtz (1972) and Cynader and Berman

(1972) show a proportion of 5 to 10% direction-selective or direction-biased cells in the superficial layers of the superior colliculus. On the other hand, Moors and Vendrick (1979) could not confirm the presence of direction-selective cells in the superior colliculus. As for the pulvinar, in normal monkeys, visual properties of neurons in this nucleus are similar to those of neurons in extrastriate cortex (Bender, 1982; Petersen *et al.*, 1985). Whether the direction selectivity of residual visual responses in area MT neurons could be the consequence of their pulvinar input is still an open question. Burman *et al.* (1982) found that, after occipital lesions, direction selectivity is not entirely lost for cells in the lateral pulvinar, whereas this property seems to disappear in the inferior pulvinar (Bender, 1983). Since the lateral pulvinar projects onto area MT (Standage and Benevento, 1983), this input could be partly responsible for the maintenance of direction selectivity in area MT. Another possibility is that direct selectivity is rebuilt or reinforced in MT itself from nonselective projections via intrinsic connections.

Lesions or inactivation of the LGN suggest that this structure might also play an important role in the transfer of residual visual information. Maunsell *et al.* (1990) report a few cases of simultaneous inactivation of parvo- and magnocellular layers of the LGN with lidocaine or magnesium chloride that led to a complete disappearance of visual responses in area MT. A second group (Schiller *et al.*, 1990) has studied the behavioral effects of combined lesions of parvo- and magnocellular layers of the LGN. They came to the conclusion that monkeys were completely unable to discriminate or detect visual stimuli and that their behavior did not show a similar recovery as observed in the case of a V1 lesion (Mohler and Wurtz, 1977). However, it is not clear whether the animals studied by Schiller and his collaborators were retrained after the LGN lesions using high-flux stimuli as recommended by Mohler and Wurtz for observing recovery.

The studies reviewed above suggest that, in addition to the superior colliculus, the LGN is a necessary relay for residual vision. Unlike the situation in the cat LGN in which retinal terminals appear to terminate on relay neurons of the A and A1 laminae which can then transfer excitatory projections to area 18 (see Section 1.2), in the macaque monkey, a direct retinogeniculocortical pathway is unlikely to be able to transfer visual information in destriated monkeys. Indeed, a recent study demonstrates that, after a lesion of area 17 in the monkey, retinal axons terminate predominantly on GABAergic interneurons and not on relay neurons projecting to extrastriate cortex (Kisvárday *et al.*, 1991). Therefore, a direct relay of retinal input to extrastriate cortex through the LGN does not appear to constitute a functional pathway for residual vision. On the other hand, the other route, namely from the retina to the superior colliculus and the interlaminar and S layers of the LGN to extrastriate cortex, may play a crucial role in residual vision. This hypothesis would be consistent with the observation that the superior colliculus and the LGN both constitute essential relays for transferring residual visual information.

3.2. Cortical Pathways

It is not necessary that all visual areas which remain visually active during inactivation of V1 receive information directly from subcortical centers. For example, residual visual activity could be carried by feedback pathways from higher-order areas such as STP to extrastriate cortical areas. The fact that the receptive fields were not enlarged during V1 inactivation or after lesions of area 17 (Rodman *et al.,* 1989) argues against such a possibility. Also, if feedback connections are important in residual activity, one would expect to find visually active neurons in the thick cytochrome oxidase bands of V2 which are reciprocally connected to MT (Shipp and Zeki, 1989). The silence of units recorded in thick cytochrome oxidase bands in V2 during inactivation of V1 (Girard and Bullier, 1989) suggests that the feedback connections cannot alone transmit the residual activity present in area MT.

Experimental results also allow us to reject the corpus callosum as being crucial for the maintenance of residual activity. First, all the residual responses in V3a during V1 inactivation have been recorded in callosotomized monkeys (Girard *et al.,* 1991b) and this was also the case for two animals in the recordings from MT (Girard *et al.,* 1992). It is significant, too, that most neurons were inactive in areas V2 and V4, despite an intact contralateral cortex and corpus callosum (Girard and Bullier, 1989; Girard *et al.,* 1991a,b). Furthermore, persistent visual responses have been recorded in area MT of animals with bilateral ablations of area 17 (Rodman *et al.,* 1989).

If it is unlikely that residual activity can be mainly driven by inputs from contralateral cortex, such inputs could have a modulatory action on neurons that remain visually responsive during V1 inactivation. We observed that, during inactivation of V1, residual responses in area MT were weaker for neurons recorded in callosotomized monkeys (Girard *et al.,* 1992). A comparable observation was made by Rodman *et al.* (1989) who noted that MT neurons with receptive fields close to the vertical meridian showed strong responses in the case of unilateral lesions of area 17, but that such strong responses disappeared in the case of a bilateral lesion.

4. Functional Aspects of Residual Activity in Extrastriate Cortex

Results presented in Section 2 lead to a map of the cortical areas that remain active in the absence of a V1 input (Fig. 8) and which could constitute the substrate of residual visual abilities in the absence of V1. It is interesting therefore to examine to what extent the functional properties of neurons located in those areas are consistent with the characteristics of residual visual abilities. First, it is important to consider our results with respect to which stream of cortical visual processing these areas belong to. Although they are less segregated than originally thought (Ungerleider and Mishkin, 1982), it has become customary to distinguish two major streams of processing in extrastriate cortical areas. The dorsal occipitoparietal stream involves areas V3a, MT, MST, FST, POa and areas of the parietal cortex. The ventral occipitotemporal stream involves areas V3, V4, TEO, and TE (for recent discussion of these issues see Morel and Bullier, 1990; Baizer *et al.,* 1991). The question of residual activity specifically delimited to the areas of the dorsal occipitoparietal pathway is crucial since the functional characteristics of neurons in this group of areas are related to the type of vision that is considered as more resistant to V1 lesions. On the other hand, residual responses in the ventral occipito-inferotemporal pathway should be associated with form and color since these are the main attributes of this pathway.

4.1. The Occipitoparietal Pathway Remains Active

That the occipitoparietal pathway remains visually active as a coherent network during inactivation or after lesions of V1 is suggested by the results of anatomical studies. Area STP, which remains active when V1 is lesioned, receives a strong contingent of afferents from area MST (Boussaoud *et al.*, 1990). MST, in turn, receives projections from areas V3a and MT (Boussaoud *et al.*, 1990) which, as reviewed above, remain visually responsive. In comparison, areas V3, V4 and the central part of V2, which are silent during inactivation of V1, have little or no projection to area MST.

MT, MST, and STP are more densely connected with parietal than with inferotemporal cortex (Morel and Bullier, 1990; Baizer *et al.*, 1991). Area V3a projects onto both parietal and inferotemporal cortex, but the latter projection is relatively weak and is not observed in all studies (Ungerleider and Desimone, 1986; Cavada and Goldman-Rakic, 1989; Morel and Bullier, 1990; Boussaoud *et al.*, 1990; Andersen *et al.*, 1990a; Baizer *et al.*, 1991). On the other hand, areas which are inactive in the absence of V1 input (areas V2, V3, V4, TEO, and TE) send weak or no connections to parietal cortex except in the case of those parts of V2 and V3 representing the peripheral part of the visual field (Baizer *et al.*, 1991; Morel and Bullier, 1990).*

An important point arises from the above considerations. Areas which do not entirely depend on V1 for their visual activity belong to a pathway routing through MT, V3a, MST, and STP that reaches parietal cortex (some authors consider the pathway from MT to STP through MST and FST as being distinct from the occipitoparietal and occipito-inferotemporal pathways). It remains to be tested whether, in the case of inactivation or lesions of V1, visual responses can be observed in parietal cortex. From the results of anatomical studies, one can predict that residual activity will be recorded in areas like MST, FST, VIP, or POa (also called LIP), which receive strong connections from areas MT and/or V3a (Boussaoud *et al.*, 1990; Morel and Bullier, 1990; Andersen *et al.*, 1990a).

Consideration of the functional characteristics of residual vision in monkeys and humans is in keeping with the participation of specific dorsoparietal areas. First, there are indications that the presence of extrastriate cortex is necessary for the preservation of certain abilities in the case of a V1 lesion. For example, Perenin (1989) shows that, in humans, the ability to discriminate the direction of motion of a large stimulus is eliminated when an entire cortical hemisphere is removed whereas this task is still possible after a lesion restricted to area 17. This observation rejects the possibility that residual direction selectivity could be

*One limitation of our studies of inactivation of V1 comes from the fact that only the regions representing central visual field were examined. It is possible that neurons in V2, V3, and V4 representing peripheral visual field may remain visually responsive during inactivation or lesion of V1. Such a possibility will be difficult to test, however, because the representation of the peripheral visual field is located in the depth of the calcarine sulcus in V1. This makes it difficult to block this region without also blocking the surrounding white matter. It is important to mention in this context that blindsight or residual vision have been demonstrated in the central part of the visual field (Weiskrantz, 1986; Blythe *et al.*, 1987).

processed only subcortically. On the basis of their known receptive field properties in monkey, areas of the dorsoparietal pathway appear to be well suited to perform this kind of processing.

Other authors have shown that primates retain some perception of motion in the absence of area V1. Keating (1980) has shown that monkeys with a V1 lesion invading extrastriate areas, but leaving cortex of the superior temporal sulcus partially spared, are still able to discriminate the angular speed of moving targets. In normal animals, an elaborate mechanism for the processing of motion information is found at the level of area MST where neurons are selective to rotation or expansion of visual stimuli. Tanaka *et al.,* (1989) and Orban *et al.* (1992) argue that such receptive field properties arise from the proper combination of the direction-selective receptive fields of MT neurons converging on neurons of MST. As argued above, it is likely that neurons in MST remain visually active when V1 is inactivated or lesioned. In that case, most MST neurons would probably retain their direction selectivity as do their inputs from MT.

Residual direction-selective activity in areas homologous to MT and MST could explain the ability of patients with V1 lesions to discriminate the direction of large stimuli (Perenin, 1991). Residual activity in these two areas may also be the reason why, in a study by Pizzamiglio *et al.* (1984), patients with occipital lesions could, in their blind fields, achieve tasks that required large rotating stimuli. These are effective stimuli for neurons in areas MT and MST of the monkey (Saito *et al.,* 1986).

Visuomotor behavioral aspects of residual vision (e.g., Mohler and Wurtz, 1977; Pöppel *et al.,* 1973) fit well with the fact that V3a and MT are still active during V1 cooling. Visual guidance of saccades is possible if retinal coordinates of a target are transformed into a signal specifying eye position in a coordinate system centered on head position (Robinson, 1975). Gaze-sensitive neurons could be a basis of such a coordinate system. Indeed, Galletti and Battaglini (1989) have shown that 50% of neurons in V3a have visual responses modulated by gaze position. Since 30% of the neurons in area V3a remain active during inactivation of V1, it is likely that this population includes some gaze-sensitive neurons. The pulvinar could relay such information in the absence of V1 since Robinson *et al.* (1990) have shown that cells in this nucleus possess this property. From V3a, gaze position information could reach parietal cortex since V3a projects onto area LIP which contains gaze-sensitive neurons (Andersen, *et al.,* 1990b).

In case of a V1 lesion, one behavioral consequence of residual direction selectivity might be a residual sensibility to optic flow (Gibson, 1950). Optic flow allows one to evaluate distance and time to collision during locomotion. An example of residual perception of optic flow is given in the study by Humphrey (1974), in which the destriated monkey Helen managed to avoid obstacles placed in her scotoma. Such a residual function could take place in MT, since Albright (1989) has shown that cells in the peripheral part of MT have direction preferences fitting well with those needed for optic flow processing.

4.2. The Occipito–Inferotemporal Pathway Is Strongly Dependent on V1 Input

The results presented in Section 2 suggest that the pathway to the inferotemporal cortex is massively dependent on its input from area V1. Indeed, most inactivated areas shown in Fig. 8 (V3, V4, TEO, and TE) belong to the ventral pathway leading to the inferior temporal cortex. Since a lot of evidence suggests that this pathway is specialized in the treatment of form and color information, it is somewhat surprising that these two attributes are present in residual vision.

Studies of residual vision in humans and monkeys (Schilder *et al.*, 1972; Dineen and Keating, 1981; Weiskrantz, 1987) show capacities for orientation discrimination. This may be related to residual activity in area V3a, since neurons in this area have a marked selectivity to orientation and spatial frequency of visual stimuli (Zeki, 1978; Gaska *et al.*, 1988; Galletti and Battaglini, 1989). We found that, during cooling of area V1, many neurons remain responsive in area V3a and that they all keep their orientation selectivity (Girard *et al.*, 1991b).

The presence of residual color selectivity reported in monkeys and humans is more difficult to interpret in terms of the chart of active cortical areas. Although it remains unclear whether the spectral sensitivity of residual vision in monkey corresponds to that of scotopic vision (Schilder *et al.*, 1972; Keating, 1979; Lepore *et al.*, 1975), recent evidence for color opponency and wavelength discrimination has been found in humans with V1 lesions (Stoerig and Cowey, 1991, 1992). Areas V2 ad V4, which contain the largest proportions of color-selective cells (Felleman and Van Essen, 1987), are silent in the absence of input from V1. Residual color vision could be mediated by area MT since it has been shown that some neurons in this area respond to movement of a border between two isoluminant contours of different colors (Saito *et al.*, 1989). Another possibility is that some color-selective units remain active during inactivation of V1 in a region of V1 not explored by our recordings (Girard *et al.*, 1991a). Finally, the wavelength selectivity in V3a is not yet known and neurons that remain visually responsive in that area could be responsible for the color selectivity of residual vision.

In contrast to cortical areas, we know more about the subcortical pathway that could bring a color-selective input directly to extrastriate cortex. Color opponency has been reported for neurons of the pulvinar (Felsten *et al.*, 1983). Since the pulvinar projects widely to extrastriate cortex, color-selective information can influence neurons in the areas that remain active after lesion or inactivation of V1.

5. Functional Role of Pathways Short-Circuiting V1 in Normal Vision

The question of the role of these short-circuit pathways is the subject of interesting debates. For instance, Gross (1991) claims that the subcortical input to extrastriate cortex is forced to play a role only in the case of a striate lesion. The immediate appearance and strength of the residual visual response recorded in MT and V3a during inactivation of V1 suggest, on the contrary, that this input participates in the processing of visual information when V1 is active. Creutzfeldt (1988) argues that, given the large size of the pulvinar in primates, it is paradoxical to negate the role of a tectopulvinar pathway to extrastriate cortex when this pathway has been shown to play an important role in nonprimate species for which the pulvinar is comparatively smaller. It is therefore important to examine whether subcortical input bypassing V1 could play a role in normal vision. We would like to argue that pathways bypassing area V1 do operate in normal vision, that they could serve to rapidly transmit essential visual information, but that activity in these pathways may not reach consciousness.

There are several lines of evidence for fast and/or unconscious visual processing in intact monkeys or human subjects. First, there is the study of Raiguel *et al.* (1989) who showed that the earliest responses to visual stimulation in MT have shorter latencies than those recorded in V1 neurons. Although this may be simply related to undersampling of certain populations of neurons in V1, it is tempting to argue that these fast responses in MT are mediated by pathways bypassing V1, or alternatively that the convergence of visual input from cortical and subcortical structures leads to a more rapid depolarization of neurons in MT, and hence to shorter latencies. The proper way to test this hypothesis would be to record from neurons in MT during inactivation of the superior colliculus. If the subcortical input relayed by the superior colliculus serves to accelerate the activation of MT neurons, one should observe a lengthening of the latency of MT neurons when the superior colliculus is inactivated.

A number of examples of visual behavior needing fast processing of visual information have been reported. In normal human or monkey subjects, the latency distribution of saccades toward a visual target that always appears at the same position is bimodal, peaking at 85 msec (express saccades) and 250 msec (regular saccades) (Boch *et al.*, 1984; Fischer and Ramsperger, 1986). The express saccades are abolished by lesions of the superior colliculus (Schiller *et al.*, 1987) and it is likely that they are sustained by the tectopulvinar pathway. This may explain why, following a lesion of V4 in macaque monkeys, the express saccades but not the regular ones remain (Weber and Fischer, 1990). This is probably because a lesion of V4 silences the ventral occipitotemporal pathway involved in the regular saccades. It is also interesting to note that express saccades occur only after extensive training and if the target luminance is high enough and its position constant (Boch *et al.*, 1984; Fischer *et al.*, 1984; Fischer and Ramsperger, 1986). Such conditions are reminiscent of those necessary for recovery following lesions of striate cortex in monkeys (Mohler and Wurtz, 1977) or for improving blindsight in humans (Zihl, 1980; Zihl and Werth, 1984).

Although we usually think of vision as a conscious process because of the importance of image recognition in conscious behavior, there are many examples pointing out the presence of unconscious processing of visual stimuli (Kihlstrom, 1987). A form of unconscious vision has been revealed in normal human subjects by masking experiments. Marcel (1988) briefly presented a word in the central visual field followed by another stimulus to mask the afterimage. In these conditions, the first stimulus cannot be consciously perceived since the subject is unable to report either the presence or the meaning of the word. However, it can be shown that the meaning of this word has been processed since it can bias the interpretation of subsequently presented ambiguous words. Two aspects of this study are relevant to our hypothesis: First, the interval between stimulus and mask presentation is short (20 msec) and therefore analysis of the priming stimuli probably implicates fast processing of visual signals. Second, these tasks can also be carried out by subjects with lesions of area 17. Another study by Meeres and Graves (1990) has revealed that human subjects can detect with precision the position in space of rapidly displaced targets that they do not perceive consciously, a behavior reminiscent of blindsight. Yet another example comes from the work of Goodale *et al.* (1986) showing that subjects point with precision to a target that has been displaced imperceptibly during the hand movement, and that they are unaware of the correction.

What could be the biological significance of being able to deal rapidly and unconsciously with fast visual stimuli? The presence of residual activity in area STP after lesions of area V1 suggests that this area, like V3a and MT, receives visual signals allowing fast and unconscious processing. Area STP is classified within the polymodal areas and some studies (Bruce *et al.*, 1981; Hikosaka *et al.*, 1988) indicate its participation in motion perception and ambient vision (as defined in Trevarthen, 1968). In the awake monkey, STP neurons display selectivity to biological motion, like a walking animal or a looming face (Perrett *et al.*, 1985). This may translate into the possibility that STP is carrying visual signals necessary for interactions with other individuals (of the same or different species). Part of this processing should be done rapidly and, probably as a matter of consequence, unconsciously, in order to produce an efficient behavior, e.g., the rapid avoidance of predators.

Despite species differences, these properties of STP neurons can be compared with those of neurons in the superior colliculus of the rat. Some cells in that structure are selectively activated by looming stimuli (Dean *et al.*, 1989). This selectivity is not abolished by a lesion of area 17. On the basis of these findings, Dean and his collaborators propose that one role of collicular input could be, in the monkey as in the rat, the elaboration of a defensive response to looming stimuli. Unfortunately, this property has not been studied in the monkey colliculus, but we believe that selectivities of neurons in the pathway leading from the superior colliculus to STP are well suited to play a role in that function. We would then propose that, in normal primates, a subcortical–extrastriate pathway leading to STP short-circuits V1 in order to process more rapidly (and unconsciously) information leading to appropriate defensive or orienting behavior. The direct access to area STP through a relay in the pulvinar or the LGN could be a way of speeding up the transfer of information by decreasing the number of synapses involved.

6. Conclusion

We mentioned at the beginning of this chapter that there are clear differences between primates and nonprimates concerning the organization of the geniculocortical pathway. The logical consequence of the focused projection of the LGN to area V1 in primates was the prediction that total blindness and total loss of neural activity in extrastriate cortical areas would result from a lesion or inactivation of area V1. As reviewed above, none of these predictions is correct: although the effects are severe, a lesion of area 17 does not produce total blindness and visually active neurons are found in extrastriate cortex deafferented from V1. Furthermore, we argued that visual activity is specifically restricted to areas of the dorsal occipitoparietal pathway, which appear to be involved in ambient or praxic vision, whereas areas of the ventral occipitotemporal pathway, which mediate form vision, lose their ability to respond to visual stimuli when their V1 input is removed.

It is unclear why, in primates, the conscious processing of form vision entirely depends on the transfer of visual information through area V1. This appears to differ from what has been demonstrated in nonprimate species (Section 1). On the other hand, the necessity for bypassing area V1 through direct subcortical inputs to extrastriate cortical areas involved in praxic vision may be common to primate and nonprimate species and may involve a common pathway through the superior colliculus and the LGN or the pulvinar. This conservation of a primitive plan of organization of subcortical pathways would be in keeping with our hypothesis that such pathways are important for the rapid and unconscious processing of visual signals, in particular for the recognition of potentially dangerous creatures or objects. It is clear that pathways bypassing area V1 and subserving fast and unconscious processing of looming stimuli are important in our daily life, as experienced by a few minutes of driving a car in the center of Lyon.

7. References

Albright, T. D., 1989, Centrifugal direction bias in the middle temporal visual area (MT) of the macaque, *Visual Neurosci.* **2:**177–188.

Andersen, R. A., Asanuma, C., Essig, G. K., and Siegel, R. M., 1990a, Corticocortical connexions of anatomically and physiologically defined subdivisions within the inferior parietal lobule, *J. Comp. Neurol.* **296:**65–113.

Andersen, R. A., Bracewell, R. M., Barash, S., Gnadt, J. W., and Fogassi, L., 1990b, Eye position effects on visual, memory, and saccade-related activity in areas LIP and 7a of macaque, *J. Neurosci.* **10:**1176–1196.

Baizer, J. S., Ungerleider, L. G., and Desimone, R., 1991, Organization of visual inputs to the inferior temporal and posterior parietal cortex in macaques, *J. Neurosci.* **11:**168–190.

Baleydier, C., and Morel, A., 1992, Segregated thalamocortical pathways to inferior parietal and inferotemporal cortex in macaque monkey, *Visual Neurosci.* **8:**391–405.

Baylis, G. C., Rolls, E. T., and Leonard, C. M., 1987, Functional subdivisions of the temporal lobe neocortex, *J. Neurosci.* **7:**330–342.

Bender, D. B., 1982, Receptive field properties of neurons in the macaque inferior pulvinar, *J. Neurophysiol.* **48:**1–17.

Bender, D. B., 1983, Visual activation of neurons in the primate pulvinar depends on cortex but not colliculus, *Brain Res.* **279:**258–261.

Benevento, L. A., and Standage, G. P., 1983, The organization of projections of the retinorecipient and nonretinorecipient nuclei of the pretectal complex and layers of the superior colliculus to the lateral pulvinar and medial pulvinar in the macaque monkey, *J. Comp. Neurol.* **217:**307–336.

Benevento, L. A., and Yoshida, K., 1981, The afferent and efferent organization of the lateral geniculo-prestriate pathways in the macaque monkey, *J. Comp. Neurol.* **203:**455–474.

Berkley, M., Wolf, E., and Glickstein, M., 1967, Photic evoked potentials in the cat: Evidence for a direct geniculate input to visual II, *Exp. Neurol.* **19:**188–198.

Birnbacher, D., and Albus, K., 1987, Divergence of single axons in afferent projections to the cat's visual cortical areas 17, 18 and 19: A parametric study, *J. Comp. Neurol.* **261:**543–561.

Blythe, I. M., Bromley, J. M., Kennard, C., and Ruddock, K. H., 1987, Residual vision in patients with retrogeniculate lesions of the visual pathways, *Brain* **110:**887–905.

Boch, R., Fischer, B., and Ramsperger, E., 1984, Express-saccades of the monkey: Reaction times versus intensity, size, duration and eccentricity of their targets, *Exp. Brain Res.* **55:**223–231.

Boussaoud, D., Ungerleider, L. G., and Desimone, R., 1990, Pathways for motion analysis: Cortical connexions of the MST and fundus of the superior temporal sulcus visual areas in the macaque, *J. Comp. Neurol.* **296:**462–495.

Bruce, C. J., Desimone, R., and Gross, C. G., 1981, Visual properties of neurons in a polysensory area in superior temporal sulcus of the macaque, *J. Neurophysiol.* **46:**369–384.

Bruce, C. J., Desimone, R., and Gross, C. G., 1986, Both striate cortex and superior colliculus contribute to visual properties of neurons in superior temporal polysensory area of macaque monkey, *J. Neurophysiol.* **55:**1057–1075.

Bullier, J., and Kennedy, H., 1983, Projection of the lateral geniculate nucleus onto cortical area V2 in the macaque monkey, *Exp. Brain Res.* **53**:168–172.

Bullier, J., Kennedy, H., and Salinger, W., 1984, Bifurcation of subcortical afferents to visual areas 17, 18 and 19 in the cat cortex, *J. Comp. Neurol.* **228**:309–328.

Burman, D., Felsten, G., and Benevento, L. A., 1982, Visual properties of neurons in the lateral pulvinar of normal and occipital lobectomized macaques, *ARVO Abstr.* **22**:237.

Casanova, C., Michaud, Y., Morin, C., McKinley, P. A., and Molotchnikoff, S., 1992, Visual responsiveness and direction selectivity of cells in area 18 during local reversible inactivation of area 17 in cats, *Visual Neurosci.* **9**:581–593.

Cavada, C., and Goldman-Rakic, P. S., 1989, Posterior parietal cortex in rhesus monkey: I. Parcellation of areas based on distinctive limbic and sensory corticocortical connections, *J. Comp. Neurol.* **287**:393–421.

Chino, Y. M., Kaas, J. H., Smith, E. L., III, Langston, A. L., and Cheng, H., 1992, Rapid reorganization of cortical maps in adult cats following restricted deafferentation in retina, *Vision Res.* **32**:789–796.

Coleman, J., and Clerici, W. J., 1980, Extrastriate projections from thalamus to posterior occipital-temporal cortex in rat, *Brain Res.* **194**:205–209.

Cowey, A., and Weiskrantz, L., 1963, A perimetric study of visual field defects in monkeys, *Q. J. Exp. Psychol.* **15**:91–115.

Creutzfeldt, O. D., 1988, Extrageniculo-striate visual mechanisms: Compartmentalization of visual functions, *Prog. Brain Res.* **75**:307–320.

Cynader, M., and Berman, N., 1972, Receptive field organization of monkey superior colliculus, *J. Neurophysiol.* **35**:187–201.

Dean, P., Redgrave, P., and Westby, G. W. M., 1989, Event or emergency? Two response systems in the mammalian superior colliculus, *Trends Neurosci.* **12**:137–147.

Desimone, R., Fleming, J., and Gross, C. G., 1980, Prestriate afferents to inferior temporal cortex: An HRP study, *Brain Res.* **184**:41–55.

DeYoe, E. A., and Van Essen, D. C., 1985, Segregation of efferent connections and receptive field properties in visual area V2 of the macaque, *Nature* **317**:58–61.

DeYoe, E. A., and Sisola, L. C., 1991, Distinct pathways link anatomical subdivisions of V4 with V2 and temporal cortex in the macaque monkey, *Soc. Neurosci. Abstr.* 511.8.

Diamond, I. T., 1976, Organization of the visual cortex: Comparative anatomical and behavioral studies, *Fed. Proc.* **35**:60–67.

Diamond, I. T., and Hall, W. C., 1969, Evolution of neocortex, *Science* **164**:251–262.

Dineen, J., and Keating, E. G., 1981, The primate visual system after bilateral removal of striate cortex (survival of complex pattern vision), *Exp. Brain Res.* **41**:338–345.

Donaldson, I. M. L., and Nash, J. R. G., 1975, The effect of a chronic lesion in cortical area 17 on the visual responses of units in area 18 of the cat, *J. Physiol, (London)* **245**:325–332.

Doty, R., 1958, Potentials evoked in cat cerebral cortex by diffuse and by punctiform photic stimuli, *J. Neurophysiol.* **21**:437–464.

Doty, R. W., 1971, Survival of pattern vision after removal of striate cortex in the adult cat, *J. Comp. Neurol.* **143**:341–369.

Dreher, B., and Cottee, L. J., 1975, Visual receptive-field properties of cells in area 18 of cat's cerebral cortex before and after lesions in area 17, *J. Neurophysiol.* **38**:735–750.

Dürsteler, M. R., Blakemore, C., and Garey, L. J., 1979, Projections of the visual cortex in the golden hamster, *J. Comp. Neurol.* **183**:185–204.

Eysel, U. T., and Schmidt-Kastner, R., 1991, Neuronal dysfunction at the border of focal lesions in cat visual cortex, *Neurosci. Lett.* **131**:45–48.

Felleman, D. J., and Van Essen, D. C., 1987, Receptive field properties of neurons in area V3 of macaque monkey extrastriate cortex, *J. Neurophysiol.* **57**:889–920.

Felleman, D. J., and Van Essen, D. C., 1991, Distributed hierarchical processing in the primate cerebral cortex, *Cereb. Cortex* **1**:1–47.

Felsten, G., Benevento, L. A., and Burman, D., 1983, Opponent-color responses in macaque extrageniculate pathways: The lateral pulvinar, *Brain Res.* **288**:363–367.

Fischer, B., and Ramsperger, E., 1986, Human express-saccades: Effects of daily practice and randomization, *Exp. Brain Res.* **64**:569–578.

Fischer, B., Boch, R., and Ramsperger, E., 1984, Express-saccades of the monkey: Effect of daily training on probability of occurrence and reaction time, *Exp. Brain Res.* **55**:232–242.

Flechsig, P., 1896, *Ueber die Lokalisation der geistigen Vorgänge, insbesondere der Sinnesempfindungen des Menschen,* Veit, Leipzig.

Fries, W., 1981, The projection from the lateral geniculate nucleus to the prestriate cortex of the macaque monkey, *Proc. R. Soc. London Ser. B* **213:**73–80.

Galletti, C., and Battaglini, P. P., 1989, Gaze-dependent visual neurons in area V3a of monkey prestriate cortex, *J. Neurosci.* **9:**1112–1125.

Garey, L. J., 1965, Interrelationships of the visual cortex and superior colliculus in the cat, *Nature* **207:**1410–1411.

Gaska, J. P., Jacobson, L. D., and Pollen, D. A., 1988, Spatial and temporal frequency selectivity of neurons in visual cortical area V3a of the macaque monkey, *Vision Res.* **28:**1179–1191.

Geisert, E. E., 1980, Cortical projections of the lateral geniculate nucleus in the cat, *J. Comp. Neurol.* **190:**793–812.

Gibson, J. J., 1950, *The Perception of the Visual World,* Houghton Mifflin, Boston.

Gilbert, C. D., and Kelly, J. P., 1975, The projections of cells in different layers of the cat's visual cortex, *J. Comp. Neurol.* **163:**81–106.

Gilbert, C. D., and Wiesel, T. N., 1992, Receptive field dynamics in adult primary visual cortex, *Nature* **356:**150–152.

Girard, P., and Bullier, J., 1989, Visual activity in area V2 during reversible inactivation of area 17 in the macaque monkey, *J. Neurophysiol.* **62:**1287–1302.

Girard, P., Salin, P. A., and Bullier, J., 1991a, Visual activity in macaque area V4 depends on area 17 input, *Neuroreport* **2:**81–84.

Girard, P., Salin, P. A., and Bullier, J., 1991b, Visual activity in areas V3A and V3 during reversible inactivation of area V1 in the macaque monkey, *J. Neurophysiol.* **66:**1493–1503.

Girard, P., Salin, P. A., and Bullier, J., 1992, Response selectivity of neurons in area MT of the macaque monkey during reversible inactivation of area V1, *J. Neurophysiol.* **67:**1–10.

Glickstein, M., King, R. A., Miller, J., and Berkley, M., 1967, Cortical projections from the dorsal lateral geniculate nucleus of cats, *J. Comp. Neurol.* **130:**55–76.

Goldberg, M. E., and Wurtz, R. H., 1972, Activity in the superior colliculus in behaving monkey. I. Visual receptive fields of single neurons, *J. Neurophysiol.* **35:**542–559.

Goodale, M. A., Pélisson, D., and Prablanc, C., 1986, Large adjustments in visually guided reaching do not depend on vision of the hand or perception of target displacement, *Nature* **320:**748–750.

Gross, C. G., 1991, Contribution of striate cortex and the superior colliculus to visual function in area MT, the superior temporal polysensory area and inferior temporal cortex, *Neuropsychologia* **29:**497–515.

Hall, W. C., and Diamond, I. T., 1968, Organization and function of the visual cortex in hedgehog. II. An ablation study of pattern discrimination, *Brain Behav. Evol.* **1:**215–243.

Harting, J. K., Huerta, M. F., Frankfurter, H. J., Strominger, N. L., and Royce, G. J., 1980, Ascending pathways from the monkey superior colliculus: An autoradiographic analysis, *J. Comp. Neurol.* **192:**853–882.

Hikosaka, K., Iwai, E., Saito, H. A., and Tanaka, K., 1988, Polysensory properties of neurons in the anterior bank of the caudal superior temporal sulcus of the macaque monkey, *J. Neurophysiol.* **60:**1615–1637.

Holländer, H., and Hälbig, W., 1980, Topography of retinal representation in the rabbit cortex: An experimental study using transneuronal and retrograde technique, *J. Comp. Neurol.* **193:**701–710.

Holländer, H., and Vanegas, H., 1977, The projections from the lateral geniculate nucleus onto the visual cortex in the cat. A quantitative study with horseradish peroxidase, *J. Comp. Neurol.* **173:**519–536.

Holmes, G., 1918, Disturbances of visual orientation, *Br. J. Ophthalmol.* **2:**449–486, 506–516.

Hughes, H. C., 1977, Anatomical and neurobehavioral investigations concerning the thalamo-cortical organization of the rat's visual system, *J. Comp. Neurol.* **175:**311–336.

Humphrey, N. K., 1974, Vision in a monkey without striate cortex: A case study, *Perception* **3:**241–255.

Humphrey, N. K., and Weiskrantz, L., 1967, Vision in monkeys after removal of striate cortex, *Nature* **215:**595–597.

Itaya, S. K., and Van Hoesen, G. W., 1983, Retinal projections to the inferior and medial pulvinar nuclei in the Old World monkey, *Brain Res.* **269:**223–230.

Kaas, J. H., and Krubitzer, L. A., 1992, Area 17 lesions deactivate area MT in owl monkey, *Visual Neurosci.* **9:**399–407.

Kaas, J. H., Merzenich, M. M., and Killackey, H. P., 1983, The reorganization of somatosensory cortex following peripheral nerve damage in adult and developing mammals, *Annu. Rev. Neurosci.* **6**:325–356.

Kaas, J. H., Krubitzer, L. A., Chino, Y. M., Langston, A. L., Polley, E. H., and Blair, N., 1990, Reorganization of retinotopic cortical maps in adult mammals after lesions of the retina, *Science* **248**:229–231.

Karamanlidis, A. N., Saigal, R. P., Giolli, R. A., Mangana, O., and Michaloudi, H., 1979, Visual thalamocortical connections in sheep studied by means of the retrograde transport of horseradish peroxidase, *J. Comp. Neurol.* **187**:245–260.

Keating, E. G., 1979, Rudimentary color vision in the monkey after removal of striate and preoccipital cortex, *Brain Res.* **179**:379–384.

Keating, E. G., 1980, Residual spatial vision in the monkey after removal of striate and preoccipital cortex, *Brain Res.* **187**:271–290.

Kennedy, H., and Bullier, J., 1985, A double-labeling investigation of the afferent connectivity to cortical areas V1 and V2 of the macaque monkey, *J. Neurosci.* **5**:2815–2830.

Kihlstrom, J. F., 1987, The cognitive unconscious, *Science* **237**:1445–1452.

Killackey, H., Snyder, M., and Diamond, I. T., 1971, Function of striate and temporal cortex in the tree shrew, *J. Comp. Physiol. Psychol.* **74**:1–29.

Kisvárday, Z. F., Cowey, A., Stoerig, P., and Somogyi, P., 1991, Direct and indirect retinal input into degenerated dorsal lateral geniculate nucleus after striate cortical removal in monkey: Implications for residual vision, *Exp. Brain Res.* **86**:271–292.

Klüver, H., 1942, Functional significance of the geniculo-striate system, *Biol. Symp.* **7**:253–299.

Lepore, F., Cardu, B., Rasmussen, T., and Malmo, R. B., 1975, Rod and cone sensitivity in destriate monkeys, *Brain Res.* **93**:203–221.

Lin, C. S., and Kaas, J. H., 1979, The inferior pulvinar complex in owl monkeys: Architectonic subdivisions and patterns of input from the superior colliculus and subdivisions of visual cortex, *J. Comp. Neurol.* **187**:655–678.

Lysakowski, A., Standage, G. P., and Benevento, L. A., 1988, An investigation of collateral projections of the dorsal lateral geniculate nucleus and other subcortical structures to cortical areas V1 and V4 in the macaque monkey: A double label retrograde tracer study, *Exp. Brain Res.* **69**:651–661.

Maciewicz, R. J., 1975, Thalamic afferents to areas 17, 18 and 19 of cat traced with horseradish peroxidase, *Brain Res.* **84**:308–312.

Marcel, A. J., 1988, Phenomenal experience and functionalism, in: *Consciousness in Contemporary Science* (A. J. Marcel and E. Bisiach, eds.), Oxford Science Publication, pp. 121-158.

Marrocco, R. T., and Li, R. H., 1977, Monkey superior colliculus: Properties of single cells and their afferent inputs, *J. Neurophysiol.* **40**:844–860.

Maunsell, J. H. R., Nealey, T. A., and DePriest, D. D., 1990, Magnocellular and parvocellular contributions to responses in the middle temporal visual area (MT) of the macaque monkey, *J. Neurosci.* **10**:3323–3334.

Meeres, S. L., and Graves, R. E., 1990, Localization of unseen visual stimuli by humans with normal vision, *Neuropsychologia* **28**:1231–1237.

Michalski, A., Wimborne, B. M., and Henry, G. H., 1993, The effect of reversible cooling of cat's primary visual cortex on the responses of area 21a neurons, *J. Physiol. (London)* **466**:133–156.

Miller, E., 1984, *Recovery and Management of Neuropsychological Impairments*, Wiley, New York.

Mizuno, N., Itoh, K., Uchida, K., Uemura-Sumi, M., and Matsushima, R., 1982, A retino-pulvinar projection in the macaque monkey as visualized by the use of anterograde transport of horseradish peroxidase, *Neurosci. Lett.* **30**:199–203.

Mizuno, N., Takahashi, O., Itoh, K., and Matsushima, R., 1983, Direct projections to the prestriate cortex from the retino-recipient zone of the inferior pulvinar nucleus in the macaque monkey, *Neurosci. Lett.* **43**:155–160.

Mohler, C. W., and Wurtz, R. H., 1977, Role of striate cortex and superior colliculus in visual guidance of saccadic eye movements in monkeys, *J. Neurophysiol.* **40**:74–94.

Moors, J., and Vendrick, A. J. H., 1979, Responses of single units in the monkey superior colliculus to moving stimuli, *Exp. Brain Res.* **25**:349–369.

Morel, A., and Bullier, J., 1990, Anatomical segregation of two cortical visual pathways in the macaque monkey, *Visual Neurosci.* **4**:555–578.

Murphy, E. H., and Chow, K. L., 1974, Effects of striate and occipital cortical lesions on visual discrimination in the rabbit, *Exp. Neurol.* **42**:78–88.

Nakagawa, S., and Tanaka, S., 1984, Retinal projections to the pulvinar nucleus of the macaque monkey: A re-investigation using autoradiography, *Exp. Brain Res.* **57**:151–157.

Orban, G. A., Lagae, L., Verri, A., Raiguel, S., Xiao, D., Maes, H., and Torre, V., 1992, First-order analysis of optical flow in monkey brain, *Proc. Natl. Acad. Sci. USA* **89:**2595–2599.

Partlow, G. D., Colonnier, M., and Szabo, J., 1977, Thalamic projections of the superior colliculus in the rhesus monkey Macaca mulatta. A light and electron microscopic study, *J. Comp. Neurol.* **171:**285–318.

Pasik, P., Pasik, T., and Schider, P., 1969, Extrageniculostriate vision in the monkey: Discrimination of luminous flux-equated figures, *Exp. Neurol.* **24:**421–437.

Pasik, T., and Pasik, P., 1971, The visual world of monkeys deprived of striate cortex: Effective stimulus parameters and the importance of the accessory optic system, *Vision Res.* **3:**419–435.

Perenin, M. T., 1989, Visual motion processing in perimetrically blind fields, *Eur. J. Neurosci.* Suppl **2:**86.3

Perenin, M. T., 1991, Discrimination of motion direction in perimetrically blind fields, *Neuroreport* **2:**397–400.

Perkel, D. J., Bullier, J., and Kennedy, H., 1986, Topography of the afferent connectivity of area 17 in the macaque monkey: A double-label study, *J. Comp. Neurol.* **253:**374–402.

Perrett, D. I., Smith, P. A. J., Mistlin, A. J., Chitty, A. J., Head, A. S., Potter, D. D., Broennimann, R., Milner, A. D., and Jeeves, M. A., 1985, Visual analysis of body motion by neurones in the temporal cortex of the macaque monkey: A preliminary report, *Behav. Brain Res.* **16:**153–170.

Petersen, S. E., Robinson, D. L., and Keys, W., 1985, Pulvinar nuclei of the behaving rhesus monkey: Visual responses and their modulation, *J. Neurophysiol.* **54:**867–886.

Pettigrew, J. D., Ramachandran, V. S., and Bravo, H., 1984, Some neural connections subserving binocular vision in ungulates, *Brain Behav. Evol.* **24:**65–93.

Pizzamiglio, L., Antonucci, G., and Francia, A., 1984, Response of cortically blind hemifields to a moving visual scene, *Cortex,* **20:**89–99.

Pöppel, E., Held, R., and Frost, D., 1973, Residual visual function after brain wounds involving the central visual pathways in man, *Nature* **243:**295–296.

Raczkowski, D., and Rosenquist, A. C., 1980, Connections of the parvocellular C laminae of the dorsal lateral geniculate nucleus with the visual cortex of the cat, *Brain Res.* **199:**447–451.

Raiguel, S. E., Lagae, L., Gulyas, B., and Orban, G. A., 1989, Response latencies of visual cells in macaque areas V1, V2 and V5, *Brain Res.* **493:**155–159.

Riddoch, G., 1917, Dissociation of visual perceptions due to occipital injuries, with especial reference to appreciation of movement, *Brain* **40:**15–57.

Robinson, D. A., 1975, Oculomotor control signals. Part III. Are saccades retinotopically or spatially organized? in: *Basic Mechanisms of Ocular Motility and Their Clinical Implications* (G. Lennerstrand and P. Bach-Y-Rita, eds.), Pergamon Press, New York, pp. 366–374.

Robinson, D. L., McClurkin, J. W., and Kertzman, C., 1990, Orbital position and eye movement influences on visual responses in the pulvinar nuclei of the behaving macaque, *Exp. Brain Res.* **82:**235–246.

Rocha-Miranda, C. E., Bender, D. B., Gross, C. G., and Mishkin, M., 1975, Visual activation of neurons in inferotemporal cortex depends on striate cortex and forebrain commissures, *J. Neurophysiol.* **38:**475–491.

Rodman, H. R., Gross, C. G., and Albright, T. D., 1989, Afferent basis of visual response properties in area MT of the macaque: I. Effects of striate cortex removal, *J. Neurosci.* **9:**2033–2050.

Rodman, H. R., Gross, C. G., and Albright, T. D., 1990, Afferent basis of visual response properties in area MT of the macaque. II. Effects of superior colliculus removal, *J. Neurosci.* **10:**1154–1164.

Rosenquist, A. C., Edwards, S. B., and Palmer, L. A., 1974, An autoradiographic study of the projections of the dorsal lateral geniculate nucleus and the posterior nucleus in the cat, *Brain Res.* **80:**71–93.

Saito, H. A., Yukie, M., Tanaka, K., Hikosaka, K., Fukuda, Y., and Iwai, E., 1986, Integration of direction signals of image motion in the superior temporal sulcus of the macaque monkey, *J. Neurosci.* **6:**145–157.

Saito, H., Tanaka, K., Isono, H., Yasuda, M., and Mikami, A., 1989, Directionally selective response of cells in the middle temporal area (MT) of the macaque monkey to the movement of equi-luminous opponent color stimuli, *Exp. Brain Res.* **75:**1–14.

Salin, P. A., Girard, P., Kennedy, H., and Bullier, J., 1992, The visuotopic organization of corticocortical connections in the visual system of the cat, *J. Comp. Neurol.* **320:**415–434.

Schilder, P., Pasik, P., and Pasik, T., 1972, Extrageniculostriate vision in the monkey. III. Circle vs triangle and red vs green discrimination? *Exp. Brain Res.* **14:**436–448.

Schiller, P. H., and Malpeli, J. G., 1977, The effect of striate cortex cooling on area 18 cells in the monkey, *Brain Res.* **126**:366–369.

Schiller, P. H., Stryker, M., Cynader, M., and Berman, N., 1974, Response characteristics of single cells in the monkey superior colliculus following ablation or cooling of visual cortex, *J. Neurophysiol.* **37**:181–194.

Schiller, P. H., Sandell, J. H., and Maunsell, J. H. R., 1987, The effect of frontal eye field and superior colliculus lesions on saccadic latencies in the rhesus monkey, *J. Neurophysiol.* **57**:1033–1049.

Schiller, P. H., Logothetis, N. K., and Charles, E. R., 1990. Role of the color-opponent and broadband channels in vision, *Visual Neurosci.* **5**:321–346.

Schneider, G. E., 1967, Contrasting visuomotor functions of tectum and cortex in the golden hamster, *Psychol. Forsch.* **31**:52–62.

Seltzer, B., and Pandya, D. N., 1978, Afferent cortical connections and architectonics of the superior temporal sulcus and surrounding cortex in the rhesus monkey, *Brain Res.* **149**:1–24.

Sherk, H., 1978, Area 18 cell responses in cat during reversible inactivation of area 17, *J. Neurophysiol.* **41**:204–215.

Shipp, S., and Zeki, S. M., 1985, Segregation of pathways leading from area V2 to areas V4 and V5 of macaque monkey visual cortex, *Nature* **315**:322–325.

Shipp, S., and Zeki, S., 1989, The organization of connections between areas V5 and V2 in macaque monkey visual cortex, *Eur. J. Neurosci.* **1**:333–354.

Solomon, S. J., Pasik, T., and Pasik, P., 1981, Extrageniculostriate vision in the monkey. VIII. Critical structures for spatial localization, *Exp. Brain Res.* **44**:259–270.

Spear, P. D., and Baumann, T. P., 1979, Effects of visual cortex removal on receptive field properties of cells in the lateral suprasylvian visual area of the cat, *J. Neurophysiol.* **42**:31–56.

Sprague, J. M., 1966, Interaction of cortex and superior colliculus in mediation of visually guided behavior in the cat, *Science* **153**:1544–1547.

Sprague, J. M., Levy, J., DiBernardino, A., and Berlucchi, G., 1977, Visual cortical areas mediating form discrimination in the cat, *J. Comp. Neurol.* **172**:441–488.

Standage, G. P., and Benevento, L. A., 1983, The organization of connections between the pulvinar and visual area MT in the macaque monkey, *Brain Res.* **262**:288–294.

Stoerig, P., and Cowey, A., 1989, Wavelength sensitivity in blindsight, *Nature* **342**:916–917.

Stoerig, P., and Cowey, A., 1991, Increment-threshold spectral sensitivity in blindsight, *Brain* **114**:1487–1512.

Stoerig, P., and Cowey, A., 1992, Wavelength discrimination in blindsight, *Brain* **115**:425–444.

Stoerig, P., Cowey, A., and Bannister, M., 1991, Retinal ganglion cells that project to the pulvinar nucleus in macaque monkeys, *Soc. Neurosci. Abstr.* 282.11

Talbot, S. A., 1942, A lateral localization in the cat's visual cortex. *Fed. Proc.* **1**:84.

Tanaka, K., Fukuda, Y., and Saito, H. A., 1989, Underlying mechanisms of the response specificity of expansion/contraction and rotation cells in the dorsal part of the medial superior temporal area of the macaque monkey, *J. Neurophysiol* **62**:642–656.

Torrealba, F., Partlow, G. D., and Guillery, R. W., 1981, Organization of the projection from the superior colliculus to the dorsal lateral geniculate nucleus of the cat, *Neuroscience* **6**:1341–1360.

Towns, L. C., Burton, S. L., Kimberly, C. J., and Fetterman, M. R., 1982, Projections of the dorsal lateral geniculate and lateral posterior nuclei to visual cortex in the rabbit, *J. Comp. Neurol.* **210**:87–98.

Trevarthen, C. B., 1968, Two mechanisms of vision in primates, *Psychol. Forsch.* **31**:299–337.

Ungerleider, L. G., and Desimone, R., 1986, Cortical connections of visual area MT in the macaque, *J. Comp. Neurol.* **248**:190–222.

Ungerleider, L. G., and Mishkin, M., 1982, Two cortical visual systems, in: *Analysis of Visual Behavior* (D. J. Ingle, M. A. Goodale, and R. J. W. Mansfield, eds.), MIT Press, Cambridge, Mass., pp. 549–586.

Van Essen, D. C., and Zeki, S., 1978, The topographic organization of rhesus monkey prestriate cortex, *J. Physiol. (London)* **277**:193–226.

von Monakow, C., 1914, *Die Lokalisationim im Grosshirn und der Abbau Funktion durch Kortikale Herde,* J. F. Bergmann, Wiesbaden.

Weber, H., and Fischer, B., 1990, Effect of a local ibotenic acid lesion in the visual association area on the prelunate gyrus (area V4) on saccadic reaction times in trained rhesus monkeys, *Exp. Brain Res.* **81**:134–139.

Weiskrantz, L., 1963, Contour discrimination in a young monkey with striate cortex ablation, *Neuropsychologia* **1:**145–164.

Weiskrantz, L., 1986, *Blindsight: A Case Study and Implications*, Oxford Psychology Series, Volume 12, Oxford University Press, London.

Weiskrantz, L., 1987, Residual vision in a scotoma (a follow-up study of "form" discrimination), *Brain* **110:**93–105.

West, J. R., Deadwyler, S. A., Cotman, C. W., and Lynch, G. S., 1976, An experimental test of diaschisis, *Behav. Biol.* **22:**419–425.

Yukie, M., and Iwaï, E., 1981, Direct projection from the dorsal lateral geniculate nucleus to the prestriate cortex in macaque monkeys, *J. Comp. Neurol.* **201:**81–97.

Zeki, S. M., 1978, The third visual complex of rhesus monkey prestriate cortex: The third visual complex of rhesus monkey prestriate cortex, *J. Physiol (London)* **277:**245–272.

Zihl, J., 1980, "Blindsight": Improvement of visually guided eye movements by systematic practice in patients with cerebral blindness, *Neuropsychologia* **18:**71–77.

Zihl, J., and Werth, R., 1984, Contributions to the study of blindsight. II. The role of specific practice for saccadic localization in patients with postgeniculate visual field defects, *Neuropsychologia* **22:**13–22.

8

What Does *in Vivo* Optical Imaging Tell Us about the Primary Visual Cortex in Primates?

RON D. FROSTIG

1. Introduction

In this chapter I will describe findings related to the functional organization of the primary visual cortex of primates, obtained by using the unique advantages of optical imaging techniques. Two optical imaging methods will be described: optical imaging using voltage-sensitive dyes and optical imaging using intrinsic signals. One method, optical imaging using voltage-sensitive dyes, excels in monitoring the temporal aspects of the functional organization of the cortex, while the other, imaging of intrinsic signals, excels in monitoring the spatial aspects. Thus, the two methods complement each other by enabling high-resolution visualization of the spatial and temporal aspects of the functional organization. This chapter will describe the imaging methods, their applications, and recent findings that have advanced our understanding of the organization of the primary visual cortex.

RON D. FROSTIG • Department of Psychobiology, University of California, Irvine, California 92717.

Cerebral Cortex, Volume 10, edited by Alan Peters and Kathleen S. Rockland. Plenum Press, New York, 1994.

2. Why Optical Imaging?

Detailed microelectrode investigation of the primary visual cortex pioneered by Hubel and Wiesel revealed that cells in the primary visual cortex are organized in groups. The first principle of grouping was for cells with similar receptive field position. Next, cells are grouped according to response properties. Most neurons in the primary visual cortex were characterized by two response properties: ocular dominance and orientation selectivity (Hubel and Wiesel, 1962, 1968, 1972, 1974a,b, 1977). Neurons responding to a certain orientation of the stimulus or to a certain eye were thought to be stacked vertically, spanning the full cortical depth, creating "columns" or parallel slabs. An "icecube" model was suggested (Hubel and Wiesel, 1977) describing the three-dimensional relationships between orientation and ocular-dominance columns. Based on cytochrome oxidase staining (Wong-Riley, 1979, this volume), a third functional system was found in the primary visual cortex (Horton and Hubel, 1981). This system, known as the blob system, contained cells that responded primarily to one eye and to color (Livingstone and Hubel, 1984a). This led to a revised ice-cube model, describing the three-dimensional relationships between blobs, ocular-dominance, and orientation columns (Livingstone and Hubel, 1984a). These models were based on findings obtained with microelectrodes and confirmed by the use of the 2-deoxyglucose (2DG) method (Sokoloff, 1977) in the visual cortex (Hubel *et al.*, 1978; Schoppmann and Stryker, 1981). However, the direct visualization of spatial relationships between the different systems in the visual cortex was hampered by the limited success in extrapolating from findings obtained by microelectrode sampling and by the fact that the 2DG technique could be applied only for one visual stimulus at a time.

Although the importance of *in vivo* visualization of functional organization has been recognized, techniques for high-resolution spatial and/or temporal visualization have been scarce. Existing *in vivo* methods, including positron emission tomography (PET), magnetic resonance imaging (MRI), thermal imaging, electroencephalography, and magnetoencephalography, suffer from either limited spatial or temporal resolution needed for visualization of cortical organization.

High-resolution visualization, in both temporal and spatial dimensions, of the functional organization of the visual cortex has been significantly advanced by *in vivo* optical imaging methods. Research interests dictated the choice of imaging method. When visualization of the functional organization was considered ("where are the active modules?"), and thus no requirements for high temporal resolution, intrinsic signal imaging with its high spatial resolution—about 50 μm—was the technique of choice. On the other hand, when the flow of information in the cortex was emphasized ("where and how is information flowing between modules?"), temporal resolution was essential and thus imaging based on voltage-sensitive dyes (1-msec temporal resolution) was the technique of choice. Finally, the two imaging techniques could be combined within the same experiment as originally demonstrated by Grinvald *et al.* (1986) for the primary somatosensory cortex of the rat.

For more details on findings beyond the primate visual cortex and on technical aspects of both imaging methods described here, see recent reviews by Grinvald (1985), Cohen and Lesher (1986), Blasdel (1989a,b), Lieke *et al.* (1989), Frostig *et al.* (1991), and Grinvald *et al.* (1991b).

Optical imaging of intrinsic signals is based on the finding that when the living cortex is illuminated with light, neurally active areas in the cortex reflect less light than nonactive ones. Thus, in the illuminated cortex, activity is transformed to changes in reflectance, and the more a certain area in the cortex is active the less light it reflects (Grinvald *et al.*, 1986). Such changes in reflectance are very small, about 0.1% of the light that was reflected back from the illuminated cortex. The changes in reflectance from active areas were found to be relatively slow—typical rise time to peak reflectance was about 1 to 2 sec (Grinvald *et al.*, 1986; Frostig *et al.*, 1990)—and thus were unsuitable to follow the millisecond time resolution that was possible with voltage-sensitive dyes. Such activity-related changes in reflectance were termed intrinsic signals. Although such signals are weak and relatively slow, they provided *in vivo* high-resolution (about 50 μm) images of the functional organization of large areas of the cerebral cortex in rats (Grinvald *et al.*, 1986; Masino *et al.*, 1992), cats (Grinvald *et al.*, 1986; Bonhoeffer and Grinvald, 1991), monkeys (Ts'o *et al.*, 1990; Frostig *et al.*, 1990; Grinvald *et al.*, 1991a), and humans (Haglund *et al.*, 1992), unmatched by any other existing *in vivo* technique. Since only illumination of the brain is involved, intrinsic signal imaging is relatively noninvasive and thus has several unique advantages. Such advantages include the ability to obtain high-resolution images for an indefinite period of time during an experiment and the ability to repeatedly obtain images from the same animal at different points along the animal's life span. Steps have been taken to improve the noninvasiveness of the technique by demonstrating the ability of the technique to obtain images through the dura and thinned skulls of cats (Frostig *et al.*, 1990).

The first experiments to use the intrinsic signal imaging concentrated on obtaining images of known functional organizations and thus establishing the validity of the technique (Grinvald *et al.*, 1986; Ts'o *et al.*, 1990; Frostig *et al.*, 1990). Since the validity of the technique was established, new findings about the functional organization of the primary cortex have accumulated and they will be described below.

3.1. Intrinsic Signal Recording and Imaging

A system setup for intrinsic signal imaging is shown in Fig. 1. In a typical experiment, performed on an anesthetized and paralyzed monkey, the skull and dura above the visual cortex were removed and a special stainless-steel "optical chamber" was mounted around the opening. This optical chamber had to accomplish several tasks: allow the illumination from the light guides to reach all parts of the exposed cortex through a transparent glass mounted at the ceiling of the chamber; allow microelectrodes to reach the surface of the cortex through a specialized rubber portion of the glass; minimize cortical movements relating to circulatory and respiratory pulsation by sealing the chamber and filling it with oil; and allow dyes or drugs to reach the cortex via specialized side tubings. Since the intrinsic signals were small—about 0.1% of the reflected light—the cortical illumination had to be stabilized to ensure that its fluctuations would be much smaller than the intrinsic signals.

Since the amplitude and temporal characteristics of the intrinsic signals depend on the wavelength of the illumination (described below), a narrow band filter was used to obtain the proper wavelength of illumination. Three different types of light-sensitive devices have been used to record intrinsic signals: diode arrays, slow-scan CCD cameras, and video cameras. The advantage of a diode array (using an array of 10×10 photodiodes) was its fast time resolution in the submillisecond range (Cohen and Lesher, 1986; Grinvald *et al.*, 1988). Such a temporal response was ideal for monitoring voltage-sensitive dye responses and was instrumental in the characterization of the temporal properties of intrinsic-signal responses (Grinvald *et al.*, 1986; Frostig *et al.*, 1990). The temporal characterization of the intrinsic signals revealed their relatively slow time course and thus prompted the use of a slow-scan CCD camera that offered a better spatial resolution than the diode array (Ts'o *et al.*, 1990). Video cameras have also been used for cortical imaging using voltage-sensitive dyes (Blasdel and Salama, 1986; Blasdel, 1992a,b). The results of using the video for dye-based imaging will be described in the intrinsic signal imaging section of this chapter rather than in the voltage-sensitive imaging section. Such organization is more appropriate since the static images obtained with the video were identical to images obtained with intrinsic signals and since the researchers did not use the dynamic time-domain characteristics that voltage-sensitive dyes excel in.

A major advantage of both CCD and video cameras compared with a diode array was that the latter could obtain only temporal traces of the intrinsic signals from different parts of the cortex (Fig. 2), while both CCD and video obtained *images* that enabled a crisp visualization of activity patterns in the imaged area.

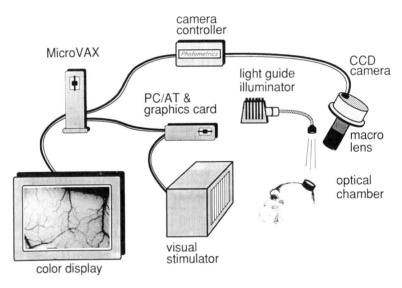

Figure 1. Setup for intrinsic-signal imaging. A sealed optical chamber with a transparent glass window was implanted over the exposed visual cortex. The cortex was illuminated with light that could be filtered to any desired spectral range. The visual cortex was stimulated with gratings moving on a video screen. The gratings were moving in different orientations that the monkey saw through one or the other eye by using computer-controlled shutters (not shown). A slow-scan CCD camera was used to obtain high-resolution images from the exposed cortex. The images were digitized and sent to a computer memory where they were analyzed. The resulting images of activity patterns were displayed on a video display monitor. From Ts'o *et al.* (1988).

Light-sensitive devices (CCD, video, or photodiode array) were hooked to a computer that controlled the visual stimulation and data collection. The same computer or other computers connected to it analyzed the data and produced images of activity patterns.

In most experiments involving anesthetized and paralyzed animals, the data acquisition was synchronized to both the heart and the respiration to reduce biological sources of noise. Noise reduction technique, routinely performed in dye-based imaging (Orbach *et al.*, 1985; Lieke *et al.*, 1988), was also used for intrinsic signal imaging (Grinvald *et al.*, 1986; Blasdel and Salama, 1986; Ts'o *et al.*, 1990; Frostig *et al.*, 1990). However, as eventually shown by Grinvald *et al.* (1991a), synchronization was not essential for intrinsic signal imaging. This finding enabled the application of intrinsic signal imaging to awake monkey (Grinvald *et al.*, 1991a) and human patients (Haglund *et al.*, 1992).

3.2. Origins of Intrinsic Signals

Studies using high-resolution images of the visual cortex in cat and monkey demonstrated that intrinsic signals had several sources (Frostig *et al.*, 1990). Using different narrow-band cortical illuminations, a detailed comparison was performed on: (1) images of a known functional system such as ocular-dominance columns in the primary visual cortex of a monkey (Fig. 3); (2) temporal characteristics (e.g., delays, amplitudes, rise times) of intrinsic signals. The study revealed three major sources for the intrinsic signals: changes in blood volume; changes in oxygen saturation of hemoglobin; and changes in light scattering. All of these changes were driven by neural activity in the cortical tissue. Changes in blood volume that related to local capillary recruitment and dilation of venules appeared as an increase in hemoglobin absorption and were dominant when the cortex was illuminated in the blue–green part of the spectrum. Changes in the oxygen saturation of hemoglobin related to local metabolic demands—and thus to the ratio of hemoglobin to oxyhemoglobin—were dominant when the cortex was illuminated in the red part of the spectrum. Changes in light scattering related to neuronal activation (expansion and contraction of activated neurons, ion movements, neurotransmitter release; see review by Cohen, 1973) were dominant when the cortex was illuminated in the near-infrared part of the spectrum. Although different mechanisms were dominant in different parts of the spectrum, images of functional organization looked identical when obtained under different wavelength illumination. An example is shown in Fig. 3, where images of the ocular-dominance column system obtained from the green part of the spectrum are compared with the near-infrared.

These studies (Frostig *et al.*, 1990) have also demonstrated the following: (1) although the largest amplitude intrinsic signals were detected with green illumination, they were less useful for localization than signals obtained with red illumination; (2) though signals in the near-infrared part of the spectrum had the smallest amplitude, they had advantages over visible light, such as the ability to sample deeper below the cortical surface, the ability to better penetrate the dura and thinned skull and thus make the technique more noninvasive; (3) blood-vessel artifacts were reduced by moving from shorter to longer wavelength illumination [Fig. 3; see also Blasdel (1992a,b) for detailed information on blood-vessel artifacts].

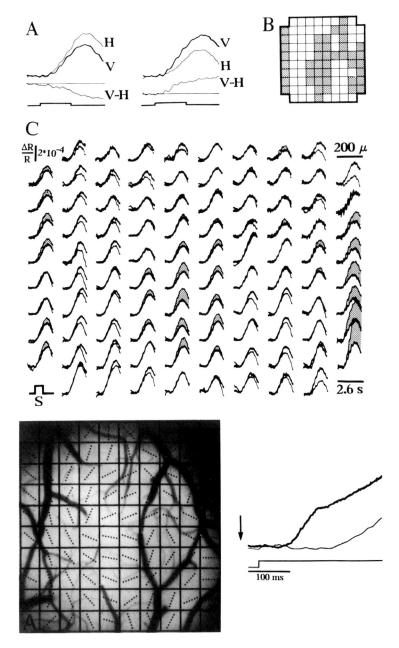

Figure 2. Relationship between intrinsic signals and functional architecture of visual cortex. (A) Intrinsic signals recorded with a single photodiode over a cortical patch where neurons preferentially responded to horizontal moving grating (left) as opposed to another cortical patch where neurons preferentially responded to vertical moving grating (right). The x axis is time and the y axis is amplitude (see part C). Note that in both areas there was a response to both stimuli, but the response to the preferred orientation of each patch [horizontal (H), thin line; vertical (V), thick line] had the highest amplitude. By subtracting the traces for each response (V − H), one is left with the part of the intrinsic signal that related to localization of horizontal versus vertical responses on the cortex. Stimulus duration 1 sec. (B) Spatial distribution of areas in the cortex preferring the horizontal stimulus (shaded) as seen by a matrix of 96 photodiodes. (C) Intrinsic-signal imaging from 96 photodiodes showing the spatial distribution of responses to horizontal (thin trace) and vertical (thick trace) stimuli. Areas with stronger response to horizontal stimuli are shaded. Each photodiode

A natural target for the global visualization offered by intrinsic signal imaging was the "columns" or "modules" of functional organization in the visual cortex. The basic strategy was to obtain data about stimulus-related changes in intrinsic signals separately for each stimulus condition. Thus, in a typical experiment, stimuli were delivered to either the left or the right eye by controlling shutters located in front of the eyes. When the animal was viewing the screen, either through the right or left eye, a moving grating stimulus with a specific orientation (e.g., 0, 22.5, 45°) was presented while keeping other variables, such as speed of movement and contrast of grating, constant. Since intrinsic signals were small, averaging of data was needed to obtain a better signal-to-noise ratio. Thus, every experiment consisted of repeated cycles of data sampling for every orientation to each eye, and for control a condition consisting of no stimulus. Best results were typically generated by subtracting or dividing orthogonal conditions, e.g., left versus right eye, horizontal versus vertical, and one oblique (45°) versus the other (135°), and not by subtracting or dividing each condition by the control condition (Blasdel and Salama, 1986; Grinvald *et al.*, 1986; Ts'o *et al.*, 1990; Frostig *et al.*, 1990; Blasdel, 1992a,b).

Figure 2 shows how mapping of the orientation columns was achieved using a photodiode array to record intrinsic signals from the cat visual cortex (area 18). Figure 2C shows a superimposed figure of intrinsic signals obtained by a matrix of 10×10 photodiodes each sampling a cortical patch of about 200×200 μm. For each photodiode, one can see two traces superimposed: a thin line which was the average response to a moving horizontal grating and a thick line which was the average response to a vertical tracing. From Fig. 2C it is clear that: (1) unlike single-unit recording, both stimuli activated the entire imaged area; (2) the difference between horizontal and vertical activation could be detected only in the amplitude of the responses. That is, responses alternated between areas responding stronger to vertically moving gratings and areas responding stronger to horizontally moving gratings (shaded). Fig 2B shows the spatial distribution of the vertical and horizontal (shaded) preferring areas as a two-state coded map.

To highlight the difference between areas preferring horizontal to areas preferring vertical stimuli, the two signals are subtracted as shown in Fig. 2A. Subtraction was needed to cancel the "common mode" of the responses to both vertical and horizontal gratings and thus to highlight the difference, or "mapping signal." The mapping signal was the part of the intrinsic signal that con-

Figure 2. (*Continued*) recorded intrinsic signals from an area of 200×200 μm. The amplitude of the response is shown as the fractional change of the response obtained by dividing the signal size by the total amount of light reflected from the cortex. Bottom left: Superimposed on a picture of the exposed visual cortex are the squares representing the cortical areas "seen" by each photodiode. Dashed bars represent the optimal stimulus orientation of each area. The optimal response for each area was obtained by vectorial addition of responses to four different stimulus orientations. Bottom right: A comparison between temporal characteristics of voltage-sensitive dye signal (thick trace) and an intrinsic signal (thin trace). In this figure, as in the rest of the figures, a decrease in reflectance or fluorescence activity is plotted in the upward direction.

The results shown in this figure were obtained from a cat visual cortex but were identical to results obtained from the monkey visual cortex. Modified from Grinvald *et al.* (1986).

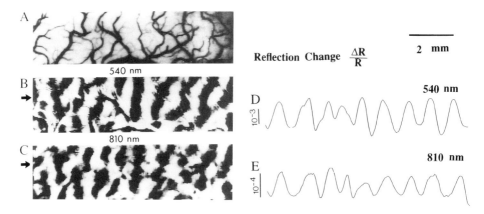

Figure 3. Relationship between illumination wavelengths and functional images of the visual cortex. (A) Image of a 9 × 2.5-mm portion of the primary visual cortex of a macaque taken with green light to highlight the contrast between vasculature and cortical tissue. (B) Image of ocular-dominance columns from the same piece of cortex obtained with 540-nm (green) illumination. (C) Same as B but illumination was filtered to 810 nm (near infrared). (D), (E) Plots of pixel values in a horizontal line crossing the area at the arrow seen in B and C. Note that although the traces are similar, the signal obtained for 540 nm was approximately 10 times larger than the 810-nm signal. Also, stronger blood-vessel artifacts appear in the 540-nm image as opposed to the 810-nm image. Images and traces were obtained from a slow-scan CCD camera. From Frostig *et al.* (1990).

tained the localization-related information (bottom traces of Fig. 2A). Although intrinsic signals were slower than dye-based signals, they already start about 250 msec after stimulation—compared with 80 msec for the dye-based signal—as shown at the lower right of Fig. 2. Since the intrinsic signals developed slowly—1–2 sec to peak—sampling for 1–3 sec improved the signal-to-noise ratio of the images. However, images of functional organization could already be detected 300–500 msec after stimulation using a CCD camera (Frostig, Lieke, Ts'o and Grinvald, unpublished results).

Using analytical procedures described above for data obtained with the slow-scan CCD camera, all of the known systems of functional architecture of the macaque primary visual cortex were imaged. Figure 4 shows the ocular-dominance columns of area 17 where left eye activity was coded as black and right eye activity was coded as white. This image highlights the area 17–area 18 border in the upper third of the image and the orthogonal relationships between the ocular-dominance columns and this border. Figure 5 shows orientation columns activated by an oblique moving grating in area 17. Figure 6 shows the cytochrome oxidase blob system as obtained both optically and anatomically in area 17. In area 18, thin and thick cytochrome oxidase stripes were imaged (Ts'o *et al.*, 1990). Similar images of orientations and ocular dominance were obtained with a sensitive video camera (Blasdel and Salama, 1986; Blasdel, 1992a,b).

One of the advantages of intrinsic-signal imaging was the ability to repeatedly obtain functional maps from the same area of cortex. For example, Fig. 7 shows four different images of the same imaged area of macaque area 17, each demonstrating the location of a different set of orientation columns. By combining vectorially such images, a composite image could be obtained that would show the information from all such pictures together [an example is shown in Fig. 2 (bottom left) and Fig. 8]. Data from every pixel in each image were converted to vectors and added using polar coordinates to create two maps: an angle map that shows where responses to different angles were located on the cortex, and a magnitude map that indicates how strong such responses were at the corresponding areas. This vectorial approach, pioneered by Blasdel and Salama (1986), was also used by Grinvald *et al.* (1986) Ts'o *et al.* (1990), Blasdel (1992b), and Bartfeld and Grinvald (1992).

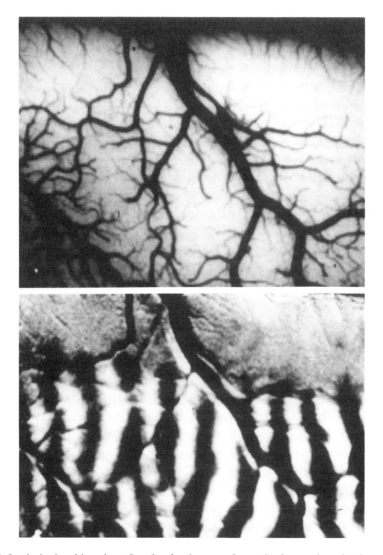

Figure 4. Intrinsic-signal imaging of ocular-dominance columns in the monkey visual cortex. Top: 9 × 6-mm image of the exposed visual cortex highlighting the vasculature of the imaged area. Bottom: Same area showing the ocular-dominance columns in black and white. Note the lack of columnar organization for eye dominance in area 18 (upper third of the image) and the orthogonal relationships between the ocular-dominance columns and the 17–18 border. Images in this figure and the following images were obtained with a slow-scan CCD camera. From Ts'o, Frostig, Lieke, and Grinvald (unpublished).

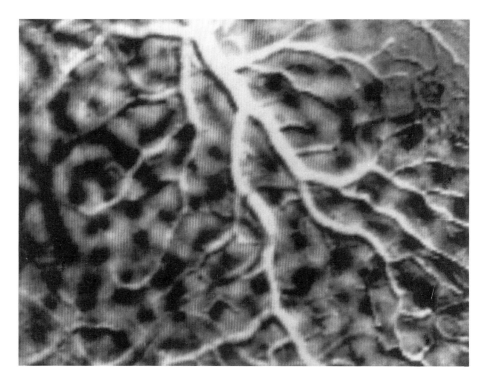

Figure 5. Intrinsic-signal imaging of orientation columns in the monkey visual cortex. The image shows a 9 × 6-mm image of the visual cortex. Black patches and beaded strings signify a preference for oblique-moving gratings. From Frostig (1990).

3.4. The Emergence of a New, Detailed Picture of Area 17 in the Macaque

The ability for repeated high-resolution sampling from the same area of cortex combined with analytical tools developed for adding and comparing images of functional maps have recently added to a better understanding of the functional architecture of area 17 (Blasdel, 1992a,b; Bartfeld and Grinvald, 1992). Not only had optical imaging added details about the functional organization of each system in area 17, it significantly enriched our understanding about the relationships between these functional organizations.

Orientation columns. Earlier work using single-unit recording techniques and 2DG techniques found that groups of cells responding to the same orientation—also termed "iso-orientation columns"—were organized in relatively long parallel slabs (Hubel *et al.,* 1978). Such orientation columns were later found to contain regions that were nonselective for orientation (Horton and Hubel, 1981; Humphrey and Hendrickson, 1983; Livingstone and Hubel, 1984a). When optical imaging methods were applied to imaging only one iso-orientation column, elongated bands were in fact evident, mostly appearing as darker "beaded areas" connected by lighter "interbeaded areas" (Bonhoeffer and Grinvald, 1991) (see Figs. 5 and 7). When several images of different iso-orientation were compared, however, it turned out that areas that looked as if they were part of one iso-orientation system actually belonged to another iso-orientation system. Detailed comparison of the iso-orientation images showed that the columns were not elongated bands with beaded appearance but were in fact comprised of patches. The patches were colocalized with the dark, beaded areas, while interbeaded areas actually corresponded to beaded areas of neighboring orientations (Fig. 7). Thus, to define an iso-orientation patch unambiguously, one needs to compare the functional maps obtained for several orientations (Bonhoeffer and Grinvald, 1991; Blasdel, 1992a).

The iso-orientation patches, representing groups of cells preferring the same orientation, were rarely larger than 0.5 mm (Blasdel, 1992b). A distance of 0.5 mm or less for an iso-orientation column was shorter than expected from single-unit recordings (Hubel and Wiesel, 1974a; Hubel *et al.*, 1978) or 2DG (Hubel *et al.*, 1978). A composite image, representing the data obtained from many single iso-orientation maps (Fig. 8), showed that linear sequences of iso-orientation were rarely longer than 1.0 mm (Blasdel, 1992b).

The reason why linear sequences of iso-orientations were rarely longer than 1.0 mm was related to the existence of "breaks in sequence regularity" originally described by Hubel and Wiesel (1974a). There were two types of breaks, or discontinuities, revealed by imaging: singularities and fractures. Fractures are discontinuities smaller than 90° that extend as lines in one dimension along the cortex, while singularities are discontinuities greater than 90° confined to points on the cortex (Blasdel, 1992b). Singularities wee generated when orientation preferences rotated continuously through 180° around a certain point—the singularity—creating a "pinwheel-like" organization (Fig. 8). Each orientation was represented once around the pinwheel center while the sequence of orientations could rotate around the center in a clockwise or counterclockwise fashion (Fig. 8). Pinwheels were smoothly connected to each other and frequently had an opposite direction of orientation sequence, clockwise and counterclockwise (Bartfeld and Grinvald, 1992). Both Blasdel (1992b) and Bartfeld and Grinvald (1992) found that the magnitude of the response to orientation was weakened near pinwheel centers. This result could reflect either a reduction of orientation tuning of cells in these areas or it could arise from the spatial mixture of many orientations converging at the pinwheel centers. Bartfeld and Grinvald (1992) further investigated this point with single-unit recording at or near the center of pinwheels. They found, in more than 75% of such recordings, that cells located at these regions were sharply tuned. They concluded that the decline in magnitude of response near the centers of pinwheels originated from the mixture of many orientations near the center of a pinwheel. Pinwheels, originally described by Swindale *et al.* (1987) in the cat visual cortex, were recently imaged in the cat visual cortex (areas 17 and 18) (Bonhoeffer and Grinvald, 1991) and thus seem to be a general organization feature for the visual cortex (Swindale, 1992).

Orientation organization in relation to ocular dominance organization. Hubel and Wiesel (1974a) suggested, on the basis of microelectrode recordings, that iso-orientation columns tended to cross the borders of ocular-dominance columns at roughly orthogonal angles. Subsequent studies, however, failed to demonstrate this relationship (Hubel *et al.*, 1978; Schoppmann and Stryker, 1981). Both Bartfeld and Grinvald (1992) and Blasdel (1992b) have confirmed the original suggestion of Hubel and Wiesel (1974a) and showed how such orthogonal relationships were kept over large regions of the visual cortex. Such orthogonal relationships were kept even in regions where the ocular-dominance columns were highly curved (Bartfeld and Grinvald, 1992).

To examine the interaction between ocular dominance and orientation, Blasdel (1992a) compared images of ocular dominance and orientation obtained when both eyes were open versus images obtained from one eye only. According to Hubel and Wiesel's ice-cube model of cortical organization (Hubel and Wiesel, 1977), ocular-dominance columns and orientation columns were hypothesized to interact. The ice-cube model predicted, in particular, that when a certain single orientation was viewed through one eye, active areas in the cortex should be expected only where the ocular dominance for the open eye intersected with the active orientation column. Attenuation of activity was expected in areas that were dominant by the closed eye and other, nonstimulated orientations. In global imaging, one would expect to view patches of activity corresponding to these intersections between the proper ocular dominance and orientation, as

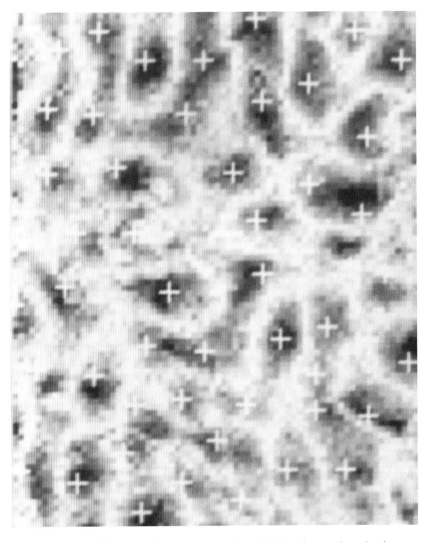

Figure 6. Intrinsic-signal imaging of cytochrome oxidase blobs in the monkey visual cortex. *Left:* Black patches represent cytochrome oxidase blob activity obtained from areas with high monocularity.

seemed to be the case with 2DG (Hubel *et al.,* 1978). Images obtained from the visual cortex for each of several orientations showed, however, only subtle differences when obtained through one or both eyes, underscoring unexpected independence of the ocular-dominance system and the orientation system. Blasdel (1992a) related the difference between the results obtained with 2DG and optical imaging to the constant activity of blobs (Horton and Hubel, 1981; Humphrey and Hendrickson, 1983). Blob activity was added to sensory-evoked 2DG results but not to the optical imaging results, since it was common to all images and thus eliminated by division.

Another modification resulting from the imaging work relates to orientation selectivity and binocularity. According to the ice-cube model, it was hypothesized that orientation selectivity was constant across the cortical surface. Near

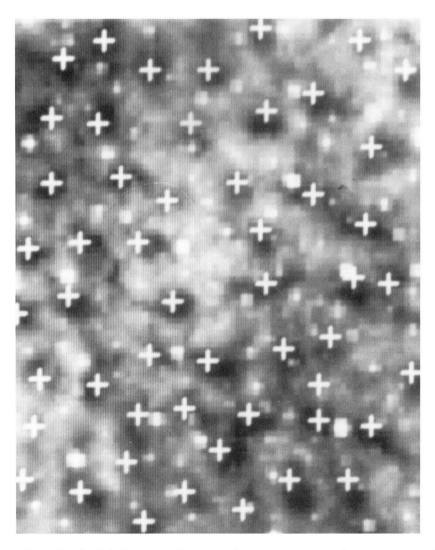

Figure 6. *(Continued). Right:* Same area of cortex, with postmortem staining for cytochrome oxidase. Cross marks were placed on the blobs seen in histology and then transferred to the imaging map to underscore the correspondence between the two maps. Modified from Ts'o *et al.* (1990).

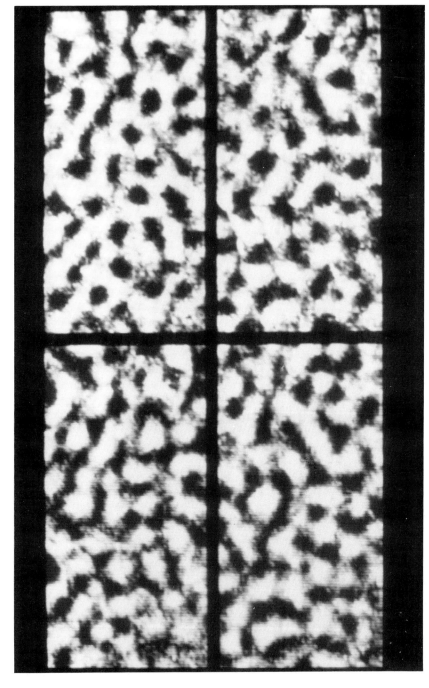

Figure 7. Intrinsic-signal imaging of different orientation columns obtained from the same piece of visual cortex. Four 6 × 3-mm images obtained from the same cortical area, each showing black patches and beaded strings in response to a different orientation of moving gratings. Upper left: vertical; upper right: horizontal; lower left: oblique (45°); lower right: oblique (135°). Modified from Frostig *et al.* (1989).

blobs, however, orientation selectivity always decreased (Horton and Hubel, 1981; Livingstone and Hubel, 1984a; Bartfeld and Grinvald, 1992; Blasdel, 1992b). Since blobs were mostly monocular, a model was proposed where regions that were most selective for orientations were also the most binocular (Blasdel, 1992a). According to this model, orientation selectivity should be stronger at the edges of ocular dominance compared with their centers since blobs, and hence monocularity, were found in the center of ocular-dominance columns (Horton and Hubel, 1981; Blasdel and Salama, 1986; Tootell *et al.*, 1988; Ts'o *et al.*, 1990). These predictions were confirmed by direct measurements of the optical signals for different orientations at the centers and edges of ocular dominance (Blasdel, 1992a).

Cytochrome oxidase blobs. Livingstone and Hubel (1984a) demonstrated that blobs contained a predominance of monocular color-selective cells. Subsequent work by Ts'o and Gilbert (1988) found clustering of red–green versus blue–yellow opponency such that individual blobs seemed to be dedicated to one opponency type or the other. They also found color-sensitive cells located in "bridges" between blobs. These bridges could be found either within or between ocular-dominance columns. The question arose whether bridges connected only blobs of the same color-opponency or blobs of opposite opponency. Using optical imaging, Landisman and Ts'o (1992) confirmed the overlap of color-selective regions with blobs and imaged the bridges between blobs. Using imaging and imaging-guided single-unit recording, they showed that bridges crossing ocular-

Figure 8. Intrinsic-signal imaging of "pinwheels" around orientation centers in the monkey visual cortex. The image is of approximately 1.5 × 1 mm of macaque visual cortex. Different orientation columns were coded with different gray levels. Darker gray levels correspond to areas preferring horizontal orientations, lighter gray levels correspond to areas preferring vertical orientation. Modified from Bartfeld and Grinvald (1992).

dominance boundaries could connect blobs of the same opponency system as well as blobs having different opponency system.

Relationships between singularities, fractures, blobs, orientation, and ocular dominance. Bartfeld and Grinvald (1992), concentrating on pinwheel organization, pointed out that 81% of the pinwheel centers (or singularities) were located along the midline of ocular-dominance columns. The remaining pinwheel centers were located close to the borders of ocular-dominance columns. Only 17% of the pinwheel centers, however, were located in the blobs. Blasdel (1992b), lumping both fractures and singularities, found that the rate of change in orientations per area was 55% higher in the centers of ocular-dominance columns as compared with the edges. Fractures tended to align parallel to the ocular-dominance columns at the columns' center and tended to be orthogonal to the ocular-dominance columns at the edge of the columns. When the relationships between fractures, singularities, and cytochrome oxidase blobs were examined by Blasdel (1992b), he noticed a higher rate of changes in orientation within the blobs but also noted that the relationships were complicated since fractures tended to run between blobs and singularities did not usually lie in the center of blobs.

In conclusion, intrinsic-signal imaging has served to verify predictions such as the relationships between orientation and ocular-dominance columns, it has succeeded in settling discrepancies such as the one involving results obtained with the 2DG method, it has provided a glimpse of the more detailed features of the functional organization of the visual cortex—such as pinwheels, singularities, fractures, blobs' bridges—and their relationships to ocular-dominance columns, orientation columns, and blobs.

3.5. New Functional Divisions of the Primary Visual Cortex

A hypercolumn was described by Hubel and Wiesel (1974b, 1977) as containing one set of all orientations and an ocular-dominance column from each eye. They found that during tangential penetration in the upper layers, approximately 2×2 mm of the visual cortex had to be crossed to record from cells that do not have overlapping receptive fields. This 2×2-mm area roughly contained two hypercolumns and had more than enough machinery to process information from a small region of the visual world. This area of cortical tissue was suggested as a repeating fundamental unit, or module, of the visual cortex.

Searching for a repeating cortical unit in images obtained optically, Bartfeld and Grinvald (1992) observed only one type of repeating unit. This unit was composed of a pinwheel centered on a part of an ocular-dominance column. Such units could process information from all orientations derived from one eye. Two adjacent units, each containing complementary eye dominance, one pinwheel, and a few blobs, seemed like the best candidate for a repeating fundamental processing module. Although many such units were found, they could not define a fundamental unit in a regular, repeating manner as suggested by Hubel and Wiesel. Thus, Bartfeld and Grinvald (1992), unable to detect an appropriate single, repeating module and impressed by the appearance of repeated modules of all subsystems (patches of iso-orientation, blobs, etc.), proposed an alternative model for the functional organization of the visual cortex. According to their model, there was no single, repeating fundamental module, but multiple repeating units of each functional subsystem in the primary visual cortex. They claimed that a single, repeating fundamental unit was actually not required since within an area of 2×2 mm of the visual cortex each of the subsystems was adequately represented.

Based on his images of angles and magnitudes, Blasdel (1992b) emphasized the division of the cortex into small regions of rapid change in orientation that were intermixed between larger regions of slower, linear change in orientation. This arrangement represented a trade-off between linearity and density in the mapping of orientation preferences. When linearity was optimized, orientation preferences were defined with high resolution but low density, since large areas of the cortex were needed to represent all possible orientations. When density was optimized, all orientation preferences were brought close together while sacrificing linearity and resolution. In view of the uniformity of upper layer dendrite sizes (250–300 μm; Lund and Yoshioka, 1991), it seemed that neurons with dendrites located in the linear zones would receive linearly organized binocular information from a small range of selective orientation preferences; presumably this would enable them to process information about edge position and orientation with high resolution. In contrast, in the dense areas, particularly around singularities where information from all orientations was converging, comparison of information about edges from all orientations could be performed, with lower resolution, by dendrites of upper layer neurons located there. Thus, according to Blasdel (1992b), the divisions between slowly changing, linear areas and rapidly changing, high-density areas represented a functional division where each such area processed a different type of visual information. Specifically, he hypothesized that linear areas specialized in detection of contours while regions near or at singularities specialized in detection of surface textures.

3.6. Combining Imaging with Anatomy

One of the main advantages of optical imaging was the high-resolution image of functional activity from a large area of the cortex identified by landmarks such as blood vessels. Thus, an image could be used to precisely direct electrodes for unit recording (Bartfeld and Grinvald, 1992; Roe and Ts'o, 1992) and for tracer injections. The combination of tracer injections guided by high-resolution imaging has a promising potential in revealing relationships between functional organization of the visual cortex and its anatomy. This approach had been demonstrated recently by two groups investigating the primary visual cortex of the macaque monkey. These groups (Malach *et al.*, 1992; Blasdel *et al.*, 1992) have used imaging of orientation columns to guide small injections of the anterograde tracer biocytin into a given iso-orientation column or pinwheel center. Preliminary results indicated that a given orientation could be anatomically connected with identical iso-orientation as well as with nonidentical iso-orientations, and that injections in the center of pinwheels did provide clear, clustered projections to many neighboring columns.

3.7. Imaging of Cortical Organization in the Awake Monkey

Obtaining functional images from an awake monkey offers many advantages, especially for the study of learning, memory, and other higher cognitive functions. Two crucial steps were necessary to succeed in the transition between an anesthetized, paralyzed preparation and an awake preparation. First, it had to be shown that images of the functional organization of the visual cortex, obtained from an anesthetized monkey, were identical whether they were obtained with the noise reduction procedure of triggering the data acquisition to the respiration and ECG or without this triggering. Second, it was necessary to restrict the head position by a solid head holder to eliminate movement noise in the awake monkey (Grinvald *et al.*, 1991a). The temporal characteristics of the intrinsic signals at different wavelengths were similar to those obtained from an anesthetized monkey, indicating that the signal sources were similar (Grinvald *et al.*, 1991a). Images of the ocular-dominance system and the blob system were obtained while the monkey was watching video movies. An example of a high-resolution image of the ocular-dominance system, obtained from an awake monkey, is shown in Fig. 9.

4. Optical Imaging with Voltage-Sensitive Dyes

Optical imaging with voltage-sensitive dyes used the property of the dye to transform changes in membrane potential to either fluorescence or absorption changes. Voltage-sensitive dyes attach to membranes of cells. If these cells are excitable, the dyes transduce the voltage changes into small optical changes (Tasaki *et al.*, 1968; Salzberg *et al.*, 1973; Cohen *et al.*, 1974). From simultaneous intracellular and optical recording, it has been demonstrated that optical changes were identical to the fast intracellular recording (Grinvald *et al.*, 1983; Loew *et al.*, 1985). Dye-based optical changes could reliably monitor potential changes in the submillisecond range, and in the case of a single neuron, or a few neurons in a dish, the interpretation of the optical changes was straightforward. However, when *in vivo* recordings from the brain were obtained, the interpretation of the dye signal became more complex (Grinvald *et al.*, 1984; Orbach *et al.*, 1985). In whole, living brain, however, the optical signal had many sources, including pre- and postsynpatic elements, cell somata, axonal arborizations, and dendritic arborizations. In fact, since the area of a membrane of both dendritic and axonal arborization is approximately 1000-fold larger than the cell soma, dye-based imaging is strongly biased to activity in these neuronal compartments. The bias of dye-based imaging for dendritic and axonal arborization is advantageous as it permits research into these domains that were unapproachable by traditional neurophysiological techniques.

The relationship between optical and electrical activity in the cortex was correlated by simultaneous recording of the dye-based optical signal and the local field potential. The field potential was recorded from an electrode centered at the optical recording area in the monkey visual cortex. There was a clear correspondence between peaks in the optical signal and peaks in the local field potential signal. The similarity between optical recording and local field potential recording indicated that the dye response was directly related to electrical activity. Since optical signals were restricted to the membranes while field potentials may spread in unpredictable fashion, optical signals provided a better spatial resolution than field potentials.

There are several limitations associated with dye-based imaging. One limitation stems from photodynamic damage (Cohen *et al.*, 1974; Ross *et al.*, 1977;

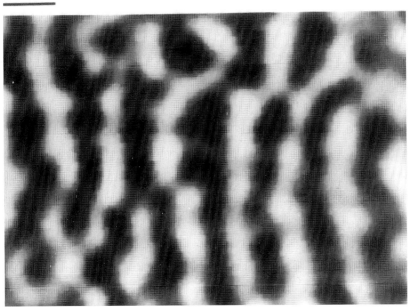

Figure 9. Intrinsic-signal imaging of ocular dominance obtained from an awake monkey. The ocular-dominance image was obtained while the monkey was watching Winnie the Pooh on a VCR screen with either the left or the right eye. Scale bar 1 mm. From Grinvald *et al.* (1991a).

Grinvald *et al.*, 1988). The duration of a dye-based imaging experiment is limited by such damage. Thus, dye-based experiments have to be designed so as to minimize such potential damage and cortical activity has to be closely monitored for signs of deterioration. The other limitation is associated with pharmacological side effects that could alter the activity of the stained area. Recently, fabricated dyes such as RH-795 (Lieke *et al.*, 1988; Grinvald *et al.*, 1993) have shown negligible side effects. Results from dye-based experiments reported below were obtained using this dye.

Each session of data acquisition in a dye-based experiment was kept brief– 200 to 400 msec—for two reasons: (1) to minimize photodynamic damage and bleaching of the dye, and (2) to reduce the chance of mixing the dye-based signal with intrinsic signals, since 250–300 msec after stimulation intrinsic-signal activity starts to mix with the dye-based signal.

4.1. Setup for Dye-Based Imaging

The setup for dye-based imaging is similar to the one described for intrinsic signals as shown in Fig. 1. The optical chamber mounted on the monkey's head is identical to the one described above for intrinsic-signal imaging. Since dye-based imaging excelled in the temporal domain, only a photodiode array was optimal to follow the rapid changes in fluorescence. In the past, dye-based *in vivo* imaging was obtained through a microscope with the photodiode array mounted on the image plane of the microscope (Grinvald *et al.*, 1984; Orbach *et al.*, 1985; Lieke *et al.*, 1988). Although the imaging was performed through the best available microscope objective, conventional microscope objectives were not suitable for imaging from large areas of the monkey visual cortex. A custom-made macroscope, mounted directly on the photodiode array and fabricated from commercially available 35-mm camera lenses, enabled dye-based imaging from large areas of the visual cortex. The macroscope also improved the fluorescence collection and allowed a greater working distance between the optical elements and the cortex (Ratzlaff and Grinvald, 1991). In a typical experiment, a 6 × 6-mm image of the primary visual cortex was projected to a 10 × 10 photodiode array. Information from each photodiode was amplified by a three-stage amplifier, digitized, and sent to the computer. The time course of the evoked optical signals, obtained from the center of the array, was displayed on a graphical terminal. The output of all the elements of the array was converted into a three-dimensional surface plot. A millisecond-by-millisecond surface plot was shown as a movie and was found to be the most effective display technique for the huge data based produced by even a few minutes of a dye-based experiment.

In a dye-based experiment, the noise time-locked to the heartbeat and respiration was at least as large as the sensory-evoked optical signal. To reduce the contribution of such noise, data acquisition was triggered by the ECG (Grinvald *et al.*, 1984; Orbach *et al.*, 1985; Blasdel and Salama, 1986). Furthermore, the respirator was synchronized to the ECG prior to data acquisition. This arrangement enabled the reduction of biological sources of noise by subtracting the optical signals obtained with the visual stimulation from optical signals obtained without stimulation. In some experiments, to ensure a steady heart rate, the heart activity was stabilized with drugs (Lieke *et al.*, 1988; Grinvald *et al.*, 1993).

One of the major advantages of dye-based imaging is the ability to visualize and quantify the topographical distribution of sensory-evoked responses. Previous results obtained from the frog optic tectum and the primary somatosensory cortex of the rat—two systems that are topographically organized with respect to the input—showed two common characteristics: focus and spread of activity. In both systems, small sensory stimuli evoked activity focused at the areas expected from the topographical organization. However, beyond the first few milliseconds of the response, the spatial distribution of the responses continued to spread and covered much larger areas from those expected by single-unit mapping (Grinvald *et al.*, 1984; Orbach *et al.*, 1985).

Sensory-evoked responses in the monkey primary visual cortex showed similar characteristics (Lieke *et al.*, 1988; Grinvald *et al.*, 1993). To measure the lateral spread of retinotopically restricted stimuli, a small visual stimulus (1 × 1°) of a drifting grating was used. The responses of neurons located in the center of the imaged area (6 × 6 mm) were established using microelectrodes. The size of the response area was about 2 × 4 mm, a size expected from neuronal responses at this eccentricity. When dye-based imaging was obtained from the same area, the optical response started and peaked at the center, corresponding to the center of the single-unit responses. By obtaining millisecond-by-millisecond three-dimensional surface plots of optical activity, as shown in Figs. 10 and 11, it became apparent how the optical activity spreads over time to cover large areas of the cortex. From Figs. 10 and 11 it is clear that 350 msec after the onset of the stimulus, optical activity was measured from an area larger than 6 × 6 mm (size of the imaged area). This size was significantly larger than the area expected by single-unit recording. Grinvald *et al.* (1993) discussed possible noncellular mechanisms that might contribute to this optical spread. Such mechanisms include artifacts originating from residual eye movements, artifacts originating from light scattering of the fluorescence, image blur related to out-of-focus layers, and possible contamination of the dye-based signal with slower intrinsic signals. They concluded that such mechanisms could not account for the observed spread. The possibility that a significant contribution to the spread originates from the glial cell syncytium was also rejected. The most likely interpretation was that such spread represents lateral neuronal activity spreading over the visual cortex. Such an interpretation was also consistent with the anatomical findings of the morphology of intrinsic, long-range, horizontal connections in the monkey visual cortex (Fisken *et al.*, 1975; Rockland and Lund, 1983; Livingstone and Hubel, 1984b; McGuire *et al.*, 1991). The observed spread could also originate from the known feedback connections from higher visual areas (see Rockland, this volume; Krubitzer and Kaas, 1990; Felleman and Van Essen, 1991).

A possible reason why this large spread was undetectable with extracellular single-unit recording techniques was related to the bias of dye-based imaging to subthreshold dendritic and axonal activation (Grinvald *et al.*, 1993). As explained above, dye-based imaging is biased to detect activity in domains such as dendritic trees that are typically inaccessible to single-unit recording techniques.

The spread of optical activity was not symmetrical from the peak. The spread showed a clear anisotropy where the long axis was approximately parallel to the V1–V2 border and was approximately double the length of the short axis (optical recordings were performed close to the V1–V2 border at an eccentricity

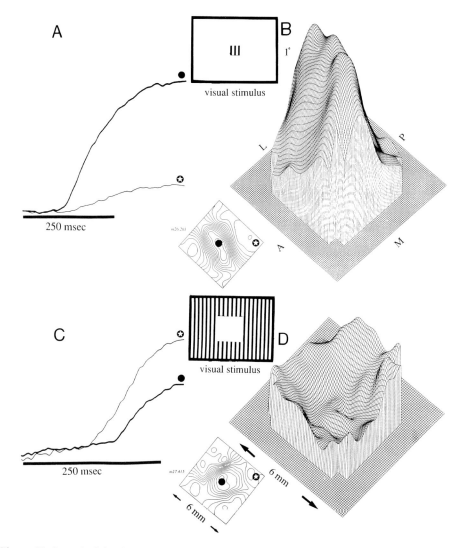

Figure 10. Spread of the dye-based signal beyond retinotopic boundaries. (A) Optical signals from the center of the photodiode array (bold trace) and the periphery of the matrix (thin trace). The stimulation was a small (1 × 1°) moving grating ("center-only" stimulus). Note the differences in temporal characteristics between the two lines. (C) Optical signals obtained from the same piece of cortex for a large moving grating stimulus with a mask on its center ("surround" stimulus). Note that for this stimulus the center response lagged behind the periphery. (B, D) Three-dimensional figures of the spatial pattern of the two signals, each from one frame—duration of frame 1 msec—350 msec after the stimulus onset. Insets display the stimulus (top) and a contour map of the signals (bottom) at the same time. The star and bold circle on the contour map indicate the optical recording sites for the traces shown in the left part of the figure. Size of imaged area 6 × 6 mm. Note how the signal spreads beyond this area. When mapped with single units, only a smaller response area of about 2 × 4 mm was detected at the center of the imaged area. Modified from Grinvald *et al.* (1993).

of 6–8°). Speed of spread was between 0.09 and 0.25 m/sec and was independent of the direction of the spread either from center stimulus to surround areas or from surround stimulus to center areas (Fig. 11). Such a speed was compatible with either nonmyelinted connections or polysynaptic, myelinated connections as well as feedback connection from higher visual areas.

4.3. Surround Inhibition

To evaluate the functional nature of the spread of activity and the possible interactions between simultaneously activated areas, more complex stimuli were used (Lieke *et al.*, 1988; Grinvald *et al.*, 1993). As shown in Fig. 10, each experiment consisted of four interlaced stimulus conditions: center stimulus alone, surround stimulus alone, and simultaneous activation of both center and surround stimuli either with the same orientation or with orthogonal orientations. As shown in Figs. 10 and 11, the center stimulus created a "mountain" of activity with the peak at the center and considerable spread toward the periphery, while the surround stimulus created a "craterlike" pattern of activity with the peak in the periphery of the imaged area and a spread toward the center. The amplitude of the response to the simultaneous activation was attenuated by 34% relative to

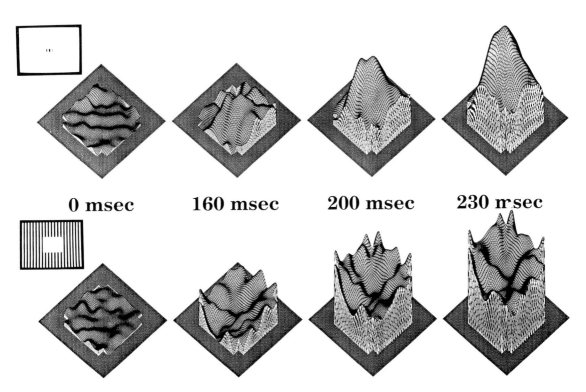

0 msec   **160 msec**   **200 msec**   **230 msec**

Figure 11. Spatiotemporal spread of dye signal. Selected 1 msec movie frames demonstrating the spatiotemporal spread of the dye signal. Four frames are shown for each of the two stimuli at 0, 160, 200, and 230 msec after the onset of the stimulus. Top trace: Spread of signal from the center to the periphery evoked by a "center" stimulus. Bottom trace: Spread of signal from the periphery to the center evoked by a "surround" stimulus. Modified from Grinvald *et al.* (1993).

the one evoked by the center activation alone and by 39% relative to the computed linear combination of center and surround. Surround inhibition persisted even when the gap between the center and surround stimuli was expanded. Even when the surround was 3° away from the center stimulus (6 mm apart on the cortex), about 10% inhibition was still observed. In this experiment, there was no difference in the amount of inhibition by a surround having the same orientation as compared with surround with different orientations.

When the same experiment was performed with a more spatially restricted surround (Fig. 12), one which did not extend along the long axis of the bars in the center stimulus, the results were orientation dependent. The results in Fig. 12 are shown 350 msec after the onset of the stimulus. In this experiment a surround with the same orientation as the center had a stronger inhibitory effect (B) compared with a surround having an orthogonal orientation (A). The histogram in Fig. 12 shows the average amplitude of responses from eight neighboring detectors covering the center region of the activation. Figure 12 demonstrates that when the same orientation was in the center and surround, it produced about twice as much inhibition (23%) than did orthogonal orientations (11%).

These results illuminate several properties of surround inhibition. First, surround inhibition occurs in response to various types of stimuli and could be detected even when the center and the surround stimuli were relatively far apart. Furthermore, since the inhibition was dependent on both the orientation and size of the stimuli, it could be described as context-dependent. Recent studies using single units and 2DG also demonstrated orientation-dependent surround inhibition in the primary visual cortex of the monkey (Van Essen *et al.*, 1989; Born and Tootell, 1991). The orientation-dependency of the surround inhibition points to the cortex as a source for at least part of this inhibition, since orientation-specific responses are first encountered only in the visual cortex. The fact that surround inhibition was different for long and short surround bars might be related to mechanisms of end-stopping that were also first encountered in the primary visual cortex and thus add support for the cortical origin of such inhibition.

In conclusion, dye-based imaging demonstrated an unexpected spatiotemporal extent of both sensory-evoked activity and surround inhibition. This unexpected extent of processing should help to catalyze new insights into cortical mechanisms of sensory processing in the primary visual cortex.

5. Conclusions

This chapter has described the findings in primate primary visual cortex obtained with two *in vivo* optical techniques. It is clear that these methods have a significant impact on our understanding of the functional organization of the primary visual cortex. Such an impact, despite the rich history of research in the visual cortex with many other techniques, has occurred since the optical imaging techniques have invaded an empty niche of research: *in vivo*, high-resolution imaging of large cortical areas. Further growth is expected in the near future in various directions, such as technical improvements in both hardware and software, and simultaneous application of optical imaging techniques in combina-

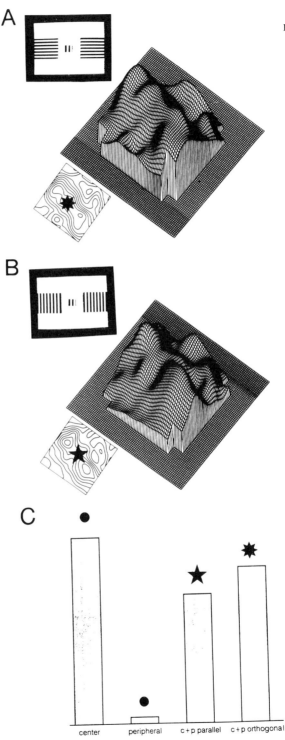

Figure 12. Surround inhibition in primary visual cortex. Selected 1-msec movie frames obtained 350 msec after the onset of the stimulus for each of two different stimuli, both containing center and surround stimuli but the surround orientation varies between the two experiments. When surround orientation matches center orientation (B) there is a greater inhibition of activity evoked by the center stimulus compared with the case when the surround inhibition is orthogonal (A). The histogram (C) demonstrates the same results for central detectors of the array. Modified from Grinvald *et al.* (1993).

tion with neurophysiological, neuropharmacological, and neuroanatomical techniques. These advances will allow expansion to other related fields such as plasticity and development of the visual cortex.

ACKNOWLEDGMENTS. I thank S. Masino and C. Chen for their helpful comments on the manuscript and Dr. A. Grinvald for supplying original figures. Supported by a grant from the Beckman Foundation.

6. References

Bartfeld, E., and Grinvald, A., 1992, Relationships between orientation-preference pinwheels, cytochrome oxidase blobs, and ocular-dominance columns in the primate striate cortex, *Proc. Natl. Acad. Sci. USA* **89**:11905–11909.

Blasdel, G. G., 1989a, Topography of visual function as shown with voltage sensitive dyes, in: *Sensory Systems in the Mammalian Brain* (J. S. Lund, ed.), Oxford University Press, London, pp. 242–268.

Blasdel, G. G., 1989b, Visualization of neuronal activity in monkey striate cortex, *Annu. Rev. Physiol.* **5**:561–581.

Blasdel, G. G., 1992a, Differential imaging of ocular dominance and orientation selectivity in monkey cortex, *J. Neurosci.* **12**:3117–3140.

Blasdel, G. G., 1992b, Orientation selectivity, preference, and continuity in monkey striate cortex, *J. Neurosci.* **12**:3141–3163.

Blasdel, G. G., and Salama, G., 1986, Voltage-sensitive dyes reveal a modular organization in monkey striate cortex, *Nature* **321**:579–585.

Blasdel, G. G., Yoshioka, T., Levitt, J. B., and Lund, J. S., 1992, Correlation between patterns of lateral connectivity and patterns of orientation preference in monkey striate cortex, *Soc. Neurosci. Abstr.* **18**:389.

Bonhoeffer, T., and Grinvald, A., 1991, Iso-orientation domains in cat visual cortex are arranged in pinwheel like patterns, *Nature* **353**:429–431.

Born, R. T., and Tootell, R. B. H., 1991, Single-unit and 2-deoxyglucose studies of side inhibition in macaque striate cortex, *Proc. Natl. Acad. Sci. USA* **88**:7071–7075.

Cohen, L. B., 1973, Changes in neuron structure during action potential propagation and synaptic transmission, *Physiol. Rev.* **53**:373–418.

Cohen, L. B., and Lesher, S., 1986, Optical monitoring of membrane potential: Methods of multisite optical measurement, *Soc. Gen. Physiol. Ser.* **40**:71–99.

Cohen, L. B., Salzberg, B. M., Davila, H. V., Ross, W. N., Landowe, D., Wagonner, A. S., and Wang, C. H., 1974, Changes in axon fluorescence during activity; molecular probes of membrane potential, *J. Membr. Biol.* **19**:1–36.

Felleman, D. J., and Van Essen, D. C., 1991, Distributed hierarchical processing in the primate cerebral cortex, *Cereb. Cortex* **1**(1):1–47.

Fisken, R. A., Garey, C. J., and Powell, T. P. S., 1975, The intrinsic, association and commissural connections of area 17 of the visual cortex, *Trans. R. Soc. London Ser. B* **272**:487–536.

Frostig, R. D., 1990, Optical imaging of functional changes in the visual system of anesthetized adult macaques after brief monocular occlusion, *Soc. Neurosci. Abstr.* **16**:42.

Frostig, R. D., Gilbert, C. D., Ts'o, D. Y., Grinvald, A., and Wiesel, T. N., 1989, Interactions between adjacent active cortical regions in macaque visualized by optical imaging of intrinsic signals, *Soc. Neurosci. Abstr.* **15**:799.

Frostig, R. D., Lieke, E. E., Ts'o, D. Y., and Grinvald, A., 1990, Cortical functional architecture and local coupling between neuronal activity and the microcirculation revealed by in vivo high-resolution optical imaging of intrinsic signals, *Proc. Natl. Acad. Sci. USA* **87**:6082–6086.

Frostig, R. D., Lieke, E. E., Arieli, A., Ts'o, D. Y., Hildesheim, R., and Grinvald, A., 1991, Optical imaging of neuronal activity in the living brain, in: *Neuronal Cooperativity* (J. Kruger, ed.), Springer-Verlag, Berlin, pp. 30–51.

Grinvald, A., 1985, Real-time optical mapping of neuronal activity: From single growth cones to the intact mammalian brain, *Annu. Rev. Neurosci.* **8**:263–305.

Grinvald, A., Fine, A., Farber, I. C., and Hildesheim, R., 1983, Fluorescence monitoring of electrical responses from small neurons and their processes, *Biophys. J.* **42**:195–198.

Grinvald, A., Anglister, L., Freeman, J. A., Hildesheim, R., and Manker, A., 1984, Real time optical imaging of naturally evoked electrical activity in the intact frog brain, *Nature* **308:**848–850.

Grinvald, A., Lieke, E., Frostig, R. D., Gilbert, C. D., and Wiesel, T. N., 1986, Functional architecture of cortex revealed by optical imaging of intrinsic signals, *Nature* **324:**361–364.

Grinvald, A., Frostig, R. D., Lieke, E., and Hildesheim, R., 1988, Optical imaging of neuronal activity, *Physiol. Rev.* **68:**1285–1366.

Grinvald, A., Frostig, R. D., Siegal, R., and Bartfeld, E., 1991, High resolution optical imaging of neuronal activity in awake monkey, *Proc. Natl. Acad. Sci. USA* **88:**11559–11563.

Grinvald, A., Bonhoeffer, T., Malonek, D., Shoham, D., Bartfeld, E., Arieli, A., Hildesheim, R., and Ratzlaff, E., 1991b, Optical imaging of architecture and function in the living brain, in: *Memory: Organization and Locus of Change* (L. Squire, N. M. Weinberger, G. Lynch and J. L. McGaugh, eds.), Oxford University Press, London, pp. 49–85.

Grinvald, A., Lieke, E. E., Frostig, R. D., and Hildesheim, R., 1993, Cortical point images and long range lateral interactions revealed by real-time optical imaging of macaque monkey primary visual cortex, *J. Neurosci.*, in press.

Haglund, M. M., Ojemann, G. A., and Hochman, D. W., 1992, Optical imaging of epileptiform and functional activity in the human cerebral cortex, *Nature* **358:**668–671.

Horton, J. C., and Hubel, D. H., 1981, Regular patchy distribution of cytochrome oxidase staining in primary visual cortex of macaque monkey, *Nature* **358:**668–671.

Hubel, D. H., and Wiesel, T. N., 1962, Receptive fields, binocular interactions and functional architecture in the cat's visual cortex, *J. Physiol. (London)* **160:**106–154.

Hubel, D. H., and Wiesel, T. N., 1968, Receptive fields and functional architecture of monkey striate cortex, *J. Physiol. (London)* **195:**215–243.

Hubel, D. H., and Wiesel, T. N., 1972, Laminar and columnar distribution of geniculo-cortical fibers in the macaque monkey, *J. Comp. Neurol.* **146:**421–450.

Hubel, D. H., and Wiesel, T. N., 1974a, Sequence regularity and geometry of orientation columns in the monkey striate cortex, *J. Comp. Neurol.* **158:**267–293.

Hubel, D. H., and Wiesel, T. N., 1974b, Uniformity of monkey striate cortex: A parallel relationship between field size, scatter, and magnification factor, *J. Comp. Neurol.* **158:**295–306.

Hubel, D. H., and Wiesel, T. N., 1977, Functional architecture of macaque monkey visual cortex, *Proc. R. Soc. London Ser. B* **198:**1–59.

Hubel, D. H., Wiesel, T. N., and Stryker, M. P., 1978, Anatomical demonstration of orientation columns in macaque monkey, *J. Comp. Neurol.* **177:**361–380.

Humphrey, A. L., and Hendrickson, A. E., 1983, Background and stimulus-induced patterns of high metabolic activity in the visual cortex (area 17) of the squirrel and macaque monkey, *J. Neurosci.* **3:**345–358.

Kubitzer, L. A., and Kaas, J. H., 1990, Cortical connections of MT in four species of primates: Areal, modular, and retinotopic patterns, *Visual Neurosci.* **5:**165–204.

Landisman, C. E., and Ts'o, D. Y., 1992, Color processing in the cytochrome oxidase-rich blobs and bridges of macaque striate cortex, *Soc. Neurosci. Abstr.* **18:**592.

Lieke, E. E., Frostig, R. D., Ratzlaff, E. H., and Grinvald, A., 1988, Center/surround inhibitory interaction in macaque V1 revealed by real time optical imaging, *Soc. Neurosci. Abstr.* **14:**1122.

Lieke, E., Frostig, R. D., Arieli, A., Ts'o, D. Y., Hildesheim, R., and Grinvald, A., 1989, Optical imaging of cortical activity: Real-time imaging using extrinsic dye signals and high resolution imaging based on slow intrinsic signals, *Annu. Rev. Physiol.* **51:**543–559.

Livingstone, M. S., and Hubel, D. H., 1984a, Anatomy and physiology of a color system in the primate visual cortex, *J. Neurosci.* **4:**309–356.

Livingstone, M. S., and Hubel, D. H., 1984b, Specificity of intrinsic connections in primate primary visual cortex, *J. Neurosci.* **4:**2830–2835.

Loew, L. M., Cohen, L. B., Salzberg, B. M., Obaid, A. L., and Bezanilla, F., 1985, Charge shift probes of membrane potential. Characterization of aminostyrylpyridinum dyes on the squid giant axon, *Biophys. J.* **47:**71–77.

Lund, J. S., and Yoshioka, T., 1991, Local circuit neurons of macaque monkey striate cortex. III. Neurons of laminae 4B, 4A and 3B, *J. Comp. Neurol.* **311:**234–258.

McGuire, B. A., Gilbert, C. D., Rivlin, P. K., and Wiesel, T. N., 1991, Targets of horizontal connections in macaque primary visual cortex, *J. Comp. Neurol.* **305:**370–392.

Malach, R., Amir, E., Bartfeld, E., and Grinvald, A., 1992, Biocytin injections guided by optical imaging reveal relationships between functional architecture and intrinsic connections in monkey visual cortex, *Soc. Neurosci., Abstr.* **18:**389.

Masino, S. A., Chen, C., Dory, Y., and Frostig, R. D., 1992, Optical imaging of functional organization in the rat somatosensory cortex: Representations and interactions of single vs. multiple whiskers, in: *Fifth Conference on the Biology of Learning and Memory*, p. 57.

Orbach, H. S., Cohen, L. B., and Grinvald, A., 1985, Optical mapping of electrical activity in rat somatosensory and visual cortex, *J. Neurosci.* **5**:1886–1895.

Ratzlaff, E. H., and Grinvald, A., 1991, A tandem-lens epifluorescence macroscope: Hundred-fold brightness advantage for wide field imaging, *J. Neurosci. Methods* **36**:127–137.

Rockland, K. S., and Lund, J. S., 1983, Intrinsic laminar lattice connections in primate visual cortex, *J. Comp. Neurol.* **216**:303–318.

Roe, A. W., and Ts'o, D. Y., 1992, Functional connectivity between V1 and V2 in the primate, *Soc. Neurosci. Abstr.* **18**:11.

Ross, W. N., Salzberg, B. N., Cohen, L. B., Grinvald, A., Davila, H. V., Waggoner, A. S., and Chang, C. H., 1977, Changes in absorption, fluorescence, dichroism and birefringence in stained axons: Optical measurement of membrane potential, *J. Membr. Biol.* **33**:141–183.

Salzberg, B. M., Davila, H. V., and Cohen, L. B., 1973, Optical recording of impulses in individual neurones of an invertebrate central nervous system, *Nature* **246**:508–509.

Schoppmann, A., and Stryker, M. P., 1981, Physiological evidence that the 2-deoxyglucose method reveals orientation in cat visual cortex, *Nature* **293**:574–576.

Sokoloff, L., 1977, Relation between physiological function and energy metabolism in the central nervous system, *J. Neurochem.* **19**:13–26.

Swindale, N., 1992, A model for the coordinated development of columnar systems in primate striate cortex, *Biol. Cyber.* **66**:217–230.

Swindale, N. V., Matsubara, J. A., and Cynader, M. S., 1987, Surface organization of orientation and direction selectivity in cat area 18, *J. Neurosci.* **7**:1414–1427.

Tasaki, I., Watanabe, A., Sandlin, R., and Carnay, L., 1968, Changes in fluorescence, turbidity and birefringence associated with nerve excitation, *Proc. Natl. Acad. Sci. USA* **61**:883–888.

Tootell, R. H., Hamilton, S. L., Silverman, M. S., and Switkes, E., 1988, Functional anatomy of macaque striate cortex. I. Ocular dominance interactions, and baseline conditions, *J. Neurosci.* **8**:1500–1530.

Ts'o, D. Y., and Gilbert, C. D., 1988, The organization of chromatic and spatial interactions in the primate striate cortex, *J. Neurosci.* **8**:1712–1727.

Ts'o, D. Y., Frostig, R. D., Lieke, E. E., and Grinvald, A., 1988, Functional organization of visual area 18 of macaque as revealed by optical imaging of activity-dependent intrinsic signals, *Soc. Neurosci. Abstr.* **14**:898.

Ts'o, D. Y., Frostig, R. D., Lieke, E. E., and Grinvald, A., 1990, Functional organization of visual area 18 of macaque as revealed by high resolution optical imaging, *Science* **249**:417–420.

Van Essen, D. C., DeYoe, E. A., Olavarria, J. F., Knierim, J. J., Fox, J. M., Sagi, D., and Julesz, B., 1989, Neural responses to static and moving texture patterns in visual cortex of the macaque monkey, in: *Neural Mechanisms of Visual Perception* (D. M. K. Lam and C. D. Gilbert, eds.), Portfolio Publishing, Texas, pp. 137–156.

Wong-Riley, M. T., 1979, Changes in the visual system of monocularly sutured or enucleated cats demonstrable with cytochrome oxidase histochemistry, *Brain Res.* **171**:11–28.

9

Computational Studies of the Spatial Architecture of Primate Visual Cortex
Columns, Maps, and Protomaps

ERIC L. SCHWARTZ

1. Introduction

Computational neuroscience is a term that has recently come into widespread use, following a symposium* which was organized in the mid-1980s for the purpose of defining an area of research that had more contact with the biology of the brain than was (and is) customary in areas such as neural networks, but which involved significant mathematical and computational techniques and ideas. Some of the themes discussed in this symposium [later published in book form (Schwartz, 1990)] were that the study of brain form and function necessarily involved a hierarchy of spatial scales, and that computational techniques were associated with neuroscience in two complementary ways:

1. The application of methods of computer graphics, image processing, and numerical analysis is critical to the description of the nervous system.

*SDF Symposium on Computational Neuroscience, Carmel, California, 1985, organized by Eric L. Schwartz.

ERIC L. SCHWARTZ • Department of Cognitive and Neural Systems, Department of Electrical and Computer Systems, College of Engineering, Boston University, and Department of Anatomy and Neurobiology, Boston University School of Medicine, Boston, Massachusetts 02215.
Cerebral Cortex, Volume 10, edited by Alan Peters and Kathleen S. Rockland. Plenum Press, New York, 1994.

2. Understanding of the possible functional utility of neural architectures is likely to feed back to computation via the design of machine vision, robotics, and other applications.

In the present chapter, a review of the spatial architecture of primate V1 will be provided, which illustrates the issue of multiple scales of analysis, the descriptive use of computational method, as well as the use of neural architectures to motivate the construction of machine vision systems. These illustrations will be taken from a range of scale that might be termed *supraneuronal*. Cortex is viewed in terms of spatial patterns in which the scale of individual neurons is not addressed. This is much like an approach to fluid mechanics in which the individual molecules are ignored, in favor of large-scale variations in density and velocity fields. Many of the mathematical concepts which are central to classical fluid mechanics, such as the representation of fluid flows in terms of conformal mapping and patterns of vorticity, appear directly in a continuum level approach to cortical patterns, as will be made clear below. And, lest the reader become uneasy with a "continuum" approach to the nervous system,* it should be pointed out that this type of approach is implicit in many of the important anatomical techniques which have been developed during the past two decades. Thus, 2-deoxyglucose (2-DG), cytochrome oxidase (CO), and optical dye recording are ipso facto pattern level methods of description. What has been lacking in this area is not the application of pattern level description, but rather a careful mathematical analysis of the implications of those pattern level techniques which are the principal methods of characterizing the architecture of the cortex.

The first section of this chapter will review the columnar aspect of V1. There are three major columnar systems that have been described in striate cortex:

1. Ocular dominance "columns," whose clearest structure in the primate occurs promptly at the input in layer IV, and whose pattern is reminiscent of "zebra stripes" and other morphological patterns
2. Orientation columns whose somewhat regular pattern is pierced by a net of the more recently discovered CO puffs
3. Cytochrome oxidase puffs (Hubel and Livingstone, 1985), which consist of neurons with receptive fields that are chromatically tuned but orientationally nonspecific.

The spatial structure of both the ocular dominance and orientation column systems has been the subject of many computational models during the past 15 years. In this chapter, we show that the patterns of both systems, and most of the models which have been advanced to account for them, are underlaid by a simple model based on bandpass-filtered random (noise) patterns (Rojer and Schwartz, 1990a). Moreover, the orientation column–puff system may be viewed as a single system whose spatial pattern is characterized jointly by filtered noise, and the

---

*Most workers trained as experimental neuroscientists appear to have a strong preference for computational models based on reducing the conceptual units of the nervous system to their smallest physical scale, i.e., the level of synapses if not macromolecules. However, in the areas of interest of this chapter, it is not clear that synaptic level approaches have provided much insight into the pattern level phenomena addressed in this paper, and it might be argued that, in principle, they are not likely to do so.

topological properties of smooth orientation maps. The vortex singularities that are familiar from fluid flow, and which are in most simple terms the singularities that must occur when orientation is smoothly mapped to a two-dimensional surface, jointly describe both the orientation and puff subsystems. This analysis suggests that the connection of these topological singularities and noise requires a more careful analysis of recent optical dye experiments (see Schwartz and Rojer, 1992). The observations that have been reported in optical dye studies of cat and monkey orientation systems (Blasdel and Salama, 1986; Blasdel, 1992; Bonhoeffer and Grinvald, 1991) are difficult to distinguish from a form of topological noise artifact associated with the relatively poor spatial resolution of these techniques.

In the next section of this chapter, a review of models of cortical topography will be provided. One-dimensional approaches to modeling cortical magnification factor will be presented, followed by a presentation of two-dimensional models of cortical mapping. The principal modeling tool in use for the past 15 years in this area has been the complex logarithm function, and related numerical conformal mappings. Here, we will review these methods, along with the use of computer brain flattening, to model the topographic aspect of V1 architecture. Preliminary data from a recent 2-DG measurement of V1 topography, using computer brain flattening and numerical conformal mapping, will be provided. These data suggest that a conformal model for V1 provides a good approximation. Finally, recent work will be reviewed which provides a conceptual and computationally constructive means of integrating the multimodal columnar aspect of neocortex with its unimodal topographic aspect (Landau and Schwartz, 1992). In this work, several new geometric data structures are introduced (protomaps and protocolumns), and computer simulations of the joint columnar and ocular dominance column structure of primate V1 are demonstrated. Finally, the application of our understanding of cortical architecture to the problem of building high-performance machine vision systems will be briefly reviewed. Recently, space-variant computer vision architectures based on the structure of the primate retinostriate system have been implemented as machine vision systems, and have the promise of making an important contribution to this area of computer science.

2. Columns: Cortical Column Patterns as Bandpass-Filtered White Noise

2.1. Ocular Dominance Columns

We base our investigation of the properties of cortical columns on the properties of random, or "noise," images. The essential property of the noise image is that the values of neighboring pixels are independent. We create a noise image by adding a random quantity to each pixel of an image of constant gray level. In neuronal terms, we can view this random number pattern to represent an array of neurons for which some receptive field descriptor (e.g., ocularity for ocular dominance columns or preferred orientation for orientation columns) has a random structure, i.e., is not yet assigned by the genetic or developmental processes which are responsible for specifying the pattern.

A noise image formed by addition of a Gaussian random offset to each pixel is illustrated in Fig. 1. In this case, the noise is characterized by the mean μ and standard deviation σ of the Gaussian random offset. Figure 1 shows a two-dimensional white noise pattern. For a two-dimensional signal (e.g., an image), the frequency variable s is two-dimensional. These two dimensions may be interpreted as describing the orientation and (scalar) frequency of a planar cosine wave, in analogy to the argument and length of a complex variable. Formally, if $s = (u, v)$, we can define $\rho = (u^2 + v^2)^{1/2}$ as the scalar frequency or frequency magnitude, and $\theta = \arctan(v/u)$ as the orientation. The inverse Fourier transform of a nonzero value at s is the planar sine grating with $1/\rho$ pixels between wave crests, and with normals to the wave crests pointing in direction θ. For a fixed θ, as ρ increases the wave crests move closer together. Similarly, as θ increases by $\delta\theta$, the grating is rotated by $\delta\theta$. Finally, the position (phase) of the grating is determined by the ratio of the real and imaginary components of s.

The decomposition of two-dimensional frequency signals into orientation and scalar frequency permits a characterization of two-dimensional bandpass filters according to orientation preference. When a filter has no preferred orientation, we say it is isotropic. In this case, the filter has an annular shape in the frequency domain, shown in Fig. 2a. In the space domain we find a kernel which has a familiar center–surround structure as in Fig. 2b. Thus, the isotropic bandpass filter is characterized by two parameters, the center frequency ρ_c and the bandwidth δ, which correspond respectively to the center radius and width of the annulus in the frequency domain. An example of a filtered noise image is shown in Fig. 2c. This is a filter (or kernel) which might be associated with lateral inhibition, i.e., with an isotropic "Mexican hat" or difference of Gaussian (DOG) receptive field.

The effects of parameter modification for the isotropic bandpass filter are shown in Fig. 3. Increasing the scalar frequency parameter ρ has the effect of scaling the image (a sort of reverse zoom). Increasing the bandwidth parameter has little effect, although it does seem to reduce the local directional correlation between nearby columns.

When we turn to oriented (anisotropic) filters, we find that the frequency domain representation consists of "humps" centered on symmetric points in the frequency domain (Fig. 4a). The center point of the humps corresponds to a planar sine wave of particular scalar frequency and orientation. The space-domain kernel of the filter consists of alternating positive and negative regions in the direction perpendicular to the preferred orientation, with the regions elongated parallel to the preferred orientation (Fig. 4b). This is the type of filter or kernel which we might associate with an oriented (e.g., simple cell) receptive field.

Filtered noise assumes a wavelike pattern (Fig. 4c). The anisotropic filter is more complicated than the isotropic filter. Formally we characterize the anisotropic bandpass as a symmetric pair of binormal Gaussian blurs centered on s_0, with principal axes oriented normal and parallel to the θ_0. We thus require four parameters for a complete description. Like the isotropic filter, the scalar frequency ρ_c must be specified. In addition, we need to specify the orientation of the center frequency θ and the width δ and eccentricity ϵ (i.e., the widths in the directions parallel and normal to the orientation).

The effects of variation of these parameters are shown in Fig. 5. Here, we systematically vary the filter parameters and then examine the properties of

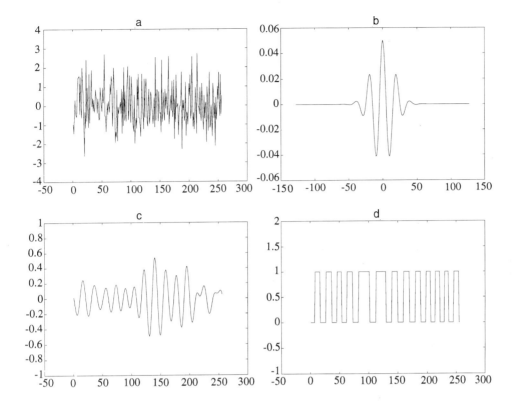

Figure 1. One-dimensional demonstration of the qualitative characteristics of bandpass-filtered noise. (a) Zero mean, unit standard deviation Gaussian noise. (b) The (space domain) kernel of a typical bandpass filter. (c) The convolution of the kernel in (b) with the noise signal in (a). (d) The "columns" which result when the signal in (c) is thresholded. (e) A two-dimensional noise pattern.

thresholded, bandpass-filtered white noise. Like the isotropic filter, scalar frequency provides a scale, determining the column width. The bandwidth in the parallel direction to the filter orientation determines the variation in column width; at low values, the columns have very uniform width. The bandwidth normal to the filter orientation determines the "waviness" of the columns. At high values, the columns are prone to small shifts of orientation, and column boundaries are irregular. Changing the orientation parameter has the effect of rotating the direction of the columns. In other work, we show that the primate ocular dominance column pattern is best accounted for by an anisotropic filter, while the cat ocular dominance column pattern seems best accounted for by an isotropic filter (Rojer and Schwartz, 1990a).

A comparison of anisotropic filtered and thresholded bandpass white noise and the macaque ocular dominance column pattern is shown in Fig. 6. In this figure, we compare a small section of computer-flattened macaque V1. This specimen was obtained from a one-eye enucleated monkey (1 month) whose brain was cut in coronal sections (40 μm), processed for CO, with subsequent computer imaging of the serial sections, three-dimensional reconstruction, and computer flattening of the ocular dominance column pattern (Schwartz *et al.*, 1988). A comparison of the data (and its power spectrum) is shown in Fig. 6(a) and (b), (c) and (d) show the image and power spectrum of bandpass-filtered noise, and (e) and (f) show the thresholded pattern. Both the image (e) and power spectrum (f) show a close resemblance to the data (a) and (b).

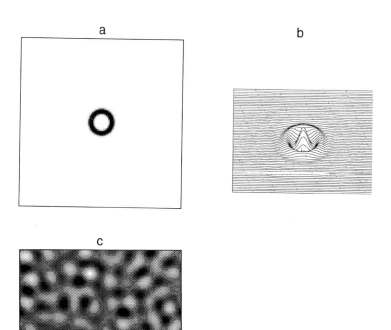

Figure 2. Two-dimensional demonstration of the qualitative characteristics of unoriented (isotropic) bandpass-filtered noise, shown as intensity plots. (a) A typical isotropic bandpass filter in the frequency domain. (b) The (space domain) kernel of the filter shown in (a). (c) The convolution of the kernel in (b) with white noise.

Figure 3. Parametric characteristics of thresholded isotropic (unoriented) bandpass-filtered noise. (a) Variation of the center frequency $\rho_c = \frac{1}{2}(s_1 + s_2)$, constant bandwidth. (b) Midrange threshold of (a). (c) Variation of the bandwidth $\delta = (s_1 - s_2)$, constant center frequency. (d) Midrange threshold of (c).

Figure 7 shows a computer synthesis of the overall pattern of macaque ocular dominance columns compared with the full pattern of macaque ocular dominance columns obtained from the experiment outlined previously. In this synthesis, the local orientation of columns was estimated on a patchwise basis, and then an oriented filter was generated at this angle, using the filter parameters estimated in Fig. 6. The net result produced an image whose overall pattern of ocular dominance columns resembled that of the computer-flattened experimental data.

A variety of theoretical models for column development have been proposed. Although such models differ in their assumptions of primary mechanisms, they share a common perspective in focusing on cellular processes as the basis for columnar structure. These models all postulate a cellular adaptation process which depends on cellular parameters, cellular interactions, and input signal characteristics. The resulting cellular organization is then shown to resemble columnar structure (see Swindale, 1980; Miller *et al.*, 1989).

These models succeed in reducing macroscopic structure to cellular processes, but they suffer from intrinsic limitations. Adaptability at the cellular level itself is poorly understood, forcing modelers to postulate adaptation processes which cannot be verified with existing experimental technique. The complex dynamics of cellular adaptation in subtle interaction with the input signal makes it difficult to see how changes in model parameters affect the macroscopic structure of the columns. Measurements of relevant cellular and signal parameters are rarely available to constrain the model parameters. The consequence is that

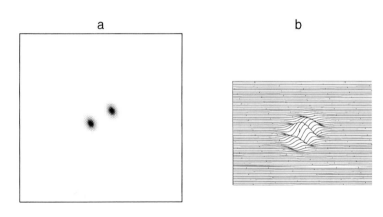

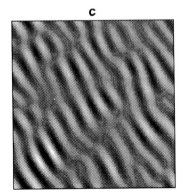

Figure 4. Two-dimensional demonstration of the qualitative characteristics of oriented (anisotropic) bandpass-filtered noise. (a) The frequency domain representation of an anisotropic bandpass filter. (b) The (space domain) kernel of the filter shown in (a). (c) The convolution of the kernel in (b) with a white noise image.

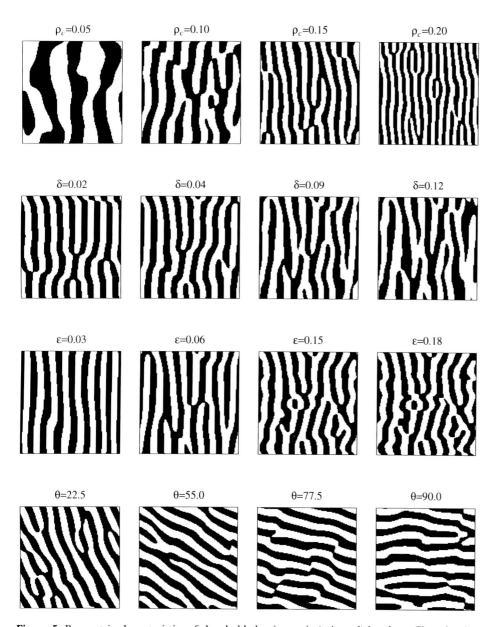

Figure 5. Parametric characteristics of thresholded anisotropic (oriented) bandpass-filtered noise. (a) Variation of center frequency ρ_c. (b) Variation of the bandwidth δ (in the direction of the dominant frequency). (c) Variation of eccentricity ϵ (perpendicular to the direction of the dominant frequency). (d) Variation of the orientation θ.

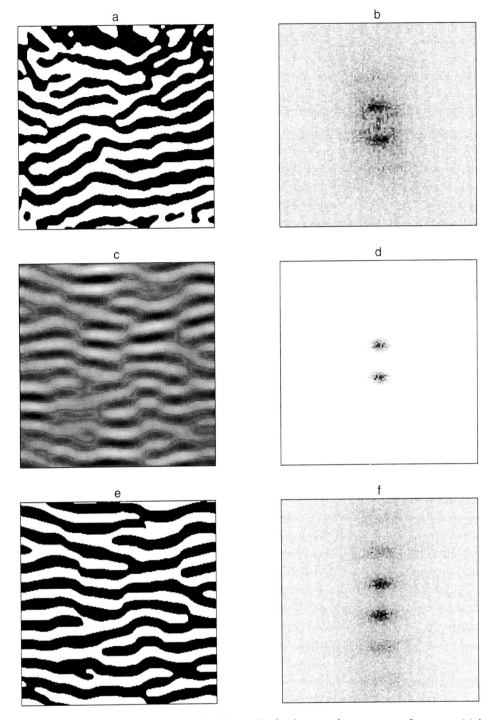

Figure 6. Example of analysis of a patch of the ocular dominance column pattern of macaque. (a) A section of the flattened ocular dominance pattern. (b) The power spectrum of (a). (c) An image synthesized by application of the filter parameters derived from (b) to Gaussian noise. (d) The spectrum of the synthetic image (i.e., convolution of the derived filter with Gaussian noise). (e) Threshold applied to (c). (f) The spectrum of (e)—compare with (b).

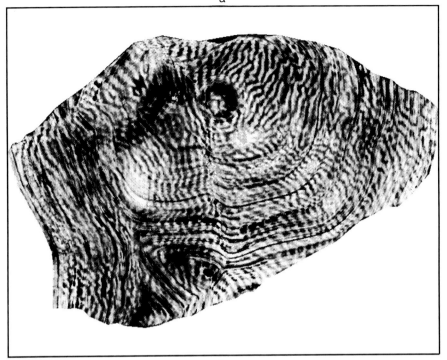

b

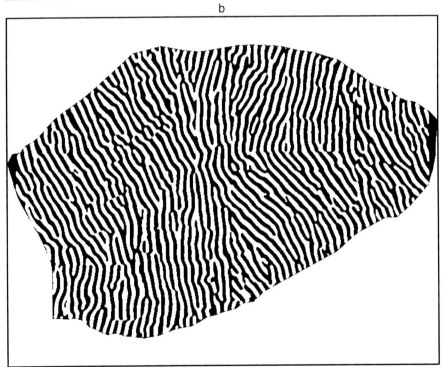

Figure 7. Comparison of actual and synthetic column data. (a) Actual column data from the flattened brain. (b) Synthetic columns generated with parametric filtering and blending.

models are computationally expensive to compute (usually requiring supercomputers to study them) and do not provide concise representations of columnar structure.

Most of the limitations described above are the result of difficulties of reductionist modeling of macroscopic development in the absence of firm knowledge of the underlying microscopic processes. Our requirements for economical *characterization* of columnar structure (independent of the *developmental* processes which engender the structure) have led us to pursue an alternative approach. We ignore the problem of column development and concentrate exclusively on the problem of representation of columnar structure. By neglecting the cellular structure of the brain, treating the cortex as an image, we have been able to apply image processing operations (in particular, bandpass filtering) to a seed image of Gaussian noise to obtain columnar structures which resemble, in fairly close detail, the ocular dominance (and orientation column) patterns of the monkey (as shown here) and the cat (Rojer and Schwartz, 1990a). In particular, reducing the ocular dominance column pattern of the macaque to an anisotropic bandpass filter provides a concise parametric representation of this pattern in terms of four numbers: a frequency, the two principal values of the bandwidth, and an orientation. In the next section, we will show that a similar parametric description may be supplied for the orientation column pattern of V1.

2.2. Orientation Columns and Cytochrome Oxidase "Puffs"

In visual cortex, neurons which respond to similar stimulus orientations cluster together, and these clusters in turn appear to be organized into larger patterns which have been termed hypercolumns (Hubel and Wiesel, 1974). These spatial patterns of functional architecture have been a major area of study, and some controversy, for more than 20 years. Recently, interest in this area has been stimulated by the introduction of new techniques based on optical recording of voltage-sensitive dyes and local blood flow (Blasdel and Salama, 1986; Bonhoeffer and Grinvald, 1991).

According to the original definition (Hubel and Wiesel, 1974), a cortical hypercolumn consists of a set of neurons whose responses represent the full range of stimulus orientations presented to both eyes. Nearby patches of cortex are observed to have a similar orientation response. These patches have been termed *orientation columns*. They are arranged in a progession termed *sequence regularity* which is occasionally interrupted by abrupt changes in orientation sequence. An early model of Hubel and Wiesel (1974) proposed that clusters of neurons of similar orientation tuning were arranged in roughly parallel slabs. More recent experiments have led to variations on this idea, suggesting that the pattern of clustering of orientation-tuned regions is better characterized by "vortices" (Blasdel and Salama, 1986) or, in the case of cat area 18, "pinwheels" (Bonhoeffer and Grinvald, 1991).

The cyclic nature of orientation (i.e., periodicity, identification of 0 and 2π)*
suggests that it is natural to model cortical orientation selectivity as points on the boundary of the unit circle, assigned to (planar) cortical points. This analysis focuses attention on a basic idea from topology, the "nonretraction theorem"

---

*Stimulus orientation takes values from $[0, \pi]$, but $[0, \pi]$, with 0 and π identified, is topologically equivalent to the range $[0, 2\pi]$ with 0 and 2π identified.

(Chinn and Steenrod, 1966), which precludes the existence of a one-to-one, invertible mapping from a planar region to the boundary of the unit circle. It is not possible to make a smooth labeling of orientations to points in a planar region, such that nearby orientations map to nearby points, and such that the entire range of orientations around the unit circle is represented. More formally, it is not possible to retract the boundary of the unit disk to its interior, without creating a singularity. The singularity has the property that all orientations are represented in an infinitesimal region around it. It is the center of a vortexlike region.

The vortex singularity implied by Hubel and Wiesel's hypercolumn model was first pointed out over 15 years ago (Schwartz, 1977a, c). Subsequently, Swindale (1982) provided a nontopological argument for the existence of these patterns, and Winfree (1987) has stated, from a topological viewpoint identical to the present paper, that cortical hypercolumns might be instances of the "phase singularities" that are associated with the cyclic definition of orientation.

The best way to understand the nonretraction theorem is in terms of the "winding number" of the orientation map. If one draws a closed loop over any part of the cortex, each region of cortex under the loop will have some value of orientation associated with the local average "best response" to an oriented stimulus. If one adds up all of the changes in orientation as the loop is traversed one time, then this sum can only be an integral multiple of 2π,[*] if the orientation map is continuous and its inverse is continuous (Chinn and Steenrod, 1966). This quantity is termed the winding number, and takes values of $2n\pi$, $n \in [0, \pm 2, \ldots]$. Because the winding number takes integral values, and the underlying orientation map is presumed continuous, any region of cortex contained within a loop drawn upon it must either have no singularities present (winding number $= 0$), or else an integral number of singularities, whose nature is clarified by imagining the loop to shrink smoothly to zero size. The integral winding number cannot change, and hence there must be some point in the cortex which has, in an arbitrarily small neighborhood, all orientations represented at least once. This is a vortex singularity.[†]

Consider an orientation map on a discretized planar region, i.e., an assignment of an orientation $\theta(x_i, y_i)$ to each discrete point (x_i, y_i) of the region.[‡] We assigned to each point an orientation drawn from a uniform distribution on the interval $[0, 2\pi]$. These random numbers were then locally correlated, or smoothed, by convolution with a Gaussian kernel[§] in order to emulate the "aver-

---

[*] The loop must return to its starting orientation, and, by continuity, must therefore wind around some multiple of 2π.

[†] The cortex of the tree shrew appears to have no vortex singularities, as it consists of long parallel regions of constant orientation, and is thus a contradiction to the generic presence of vortex regions in orientation maps. Thus, the orientation winding number of the tree shrew cortex is everywhere zero: any closed loop drawn upon this cortex accumulates a net orientation change that is zero. There appears to be insufficient "randomness" in the tree shrew orientation map to produce any area of nonzero winding number.

[‡] Consider the points (x_i, y_i) to lie on a regular grid. The orientation map $\theta(x_i, y_i)$ may be considered to be a discrete sampling of a continuous distribution, or even the argument function of a complex analytic function. A function continuous everywhere except at certain singular points whose sample on a discrete grid matched our random orientation distribution could be constructed by (e.g.) spline interpolation.

[§] We used a two-dimensional Gaussian kernel which is circularly symmetric (i.e., isotropic) and hence characterized by one parameter, σ. The kernel has the mathematical form $G(r) = e^{-r^2/2\sigma^2}$, with $r^2 = x^2 + y^2$.

aging" of neuronal orientation response that is provided by multiunit recording, optical recording, and 2-DG studies. The Gaussian kernel introduces a characteristics scale [e.g., the full width at half maximum (FWHM) of the convolution kernel], which is a measure of the size of the region over which the measurement process averages neuronal responses, i.e., of the "resolution" of the measurement.*

Smoothing an array of orientation values must be done with care.† Our method to ensure correct convolution of orientations is to convert them to unit amplitude complex numbers (taking advantage of the equivalence of orientations and points on the unit circle), and separately convolve the real and imaginary parts. We interpret the argument of each complex number as representing the preferred orientation at that position and the magnitude of each complex number as representing the "strength" of the response to the preferred orientation at that position. The magnitude of convolved orientations at each point, together with the average orientation itself, define a vector field, $\Theta(x,y)$. Most experimental studies focus on the orientation component of $\Theta(x,y)$ and not on its amplitude. Figure 8 shows a random orientation map, smoothed by a series of increasingly larger kernels. The original random map resembles white noise, but as the kernel is made larger, pronounced vortex structure emerges. These simulations demonstrate that our usual intuition about the nature of smoothed noise appears to be contradicted in the present case.‡ It is difficult to accept that the robust vortex structure and long range order which emerge with increasing kernel size in Fig. 8 are merely random fluctuations of noise.

To place this figure in a biological context, let us imagine for a moment that the cortex consisted of randomly placed orientation-tuned neurons. In other words, in this (possibly) hypothetical example, there is no spatial structure at all for orientation-tuned cortical neurons. Let us then "observe" this cortex through a "fuzzy" instrument that blends together the orientations of neurons. In mathematical terms, this "fuzzy" observation is modeled by convolution with a Gaussian kernel of increasing size, as shown in Fig. 8.

Now, the effect of the "fuzzy" observation is not simply to blur the original white noise pattern, as might be intuitively imagined. Because of the topological properties of orientation (namely, the identification of the angles 0 and 2π, which gives orientation the topology of a circle), increasing blurring of the pattern actually causes a robust pattern to emerge, which is that of a vortex, as in Fig. 8. In other words, vortex patterns in orientation maps are the signature of blurred random orientation maps!

Now, virtually all of the techniques which have been used to study orientation patterns in cortex view the cortex with a "blur."§ For example, optical dye recording has been estimated to have a resolution of only about 150 μm (Frostig

---

*For a normal distribution, $FWHM = 2.354\sigma$, where σ is the variance of the distribution.

†It is not correct to smooth uniform random variables on $[0, 2\pi]$ with conventional averaging due to phase-wrapping of angles ($0 = 2\pi$). For example, the "average" of α and $2\pi - \alpha$ is 0, not π.

‡One would expect that convolution of a noise image with increasingly large kernels would eventually result in a uniform gray image with no structure, as diffusion over infinite time smears any initial distribution into a uniform one. However, the topological properties of orientation maps prevent this from occurring, and the understanding of this topological property of orientation maps is the key to understanding the nature of the cortical hypercolumn pattern.

§With the exception of single-unit recording.

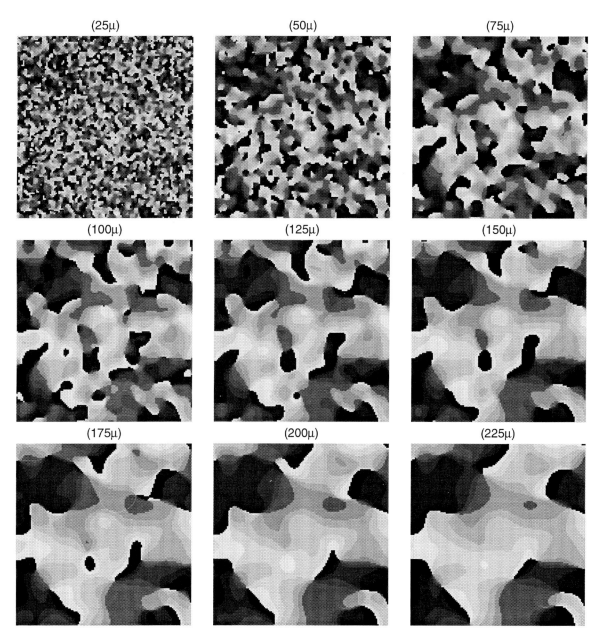

Figure 8. Smoothing random orientation noise with increasing kernel size, in steps of 25 μm, with scale chosen so that the entire frame represents 3.12 × 3.12 mm of "visual cortex." Orientation is represented by increasing gray scale in eight equal steps, between 0 and 2π. Standard deviation of the Gaussian smoothing kernel is indicated on the top of each panel. Initially, the figure resembles the white noise from which it was generated. As the kernel size increases, vortex structure emerges, with small vortices apparently fusing into larger ones, which remain stable over large changes in kernel size. The striking vortex structure that emerges with increasing kernel size is nothing more than a form of smoothed orientation white noise.

et al., 1990).* In other words, optical dye recording is a form of Gaussian convolution with a kernel whose size is in the range of several hundred microns. 2-DG has a similar spatial resolution. Multiunit recording also averages over a range of perhaps 100–200 μm. These observations raise a question about recent reports of strong "vortex" or "pinwheel"-like structure in cat and monkey orientation patterns. Since the vortex pattern is the signature of "noise" imaged with a blurry detector, one can raise the question whether the observed patterns, imagined with the "blurry" techniques of optical recording or 2-DG, are in fact real.[†] This question will be returned to briefly below, following a further discussion of the nature of the "vortex" pattern.

One of the key predictions of this paper is that the center of the vortex region, where all orientations are represented, should have an amplitude which approaches zero, because in the vicinity of the vortex center, where all orientations must be (smoothly) represented, the amplitude of the field must go to zero.

This interpretation has two aspects: if the vortex structure is due to a "true" neuronal correlation function, then the central region of near-zero amplitude will likely be "filled in" by some other neural system, rather than merely "waste" the cortical patch. We propose that the "puff–extra puff" architecture of V1 is the architectural correlate of this observation. And, to the extent that a particular vortex pattern is artifactual, i.e., caused by the intrinsic smoothing of some measurement technique, then an observation of small amplitude at the center of the vortex would be expected in any case.

In previous work (Schwartz and Rojer, 1992), we have shown that our simulations are very similar in appearance to experimental data from cat area 18 (Bonhoeffer and Grinvald, 1991), in which "pinwheel" structure has been described in cat visual cortex. These "pinwheels" are essentially identical, in both their local and global structure, to the vortex patterns shown in Fig. 8. In earlier work, we made a similar point with respect to macaque orientation column patterns, and bandpass-filtered orientation maps (Rojer and Schwartz, 1990a), where we concluded that orientation column pattern, at least as presented in the optical dye recordings of Blasdel (1992), is best modeled by an isotropic bandpass filter applied to random orientation patterns.

The ease with which hypercolumn structures may be constructed suggests that the brain need not specify "developmental rules" to create hypercolumns: it only needs to provide a local correlation function for initially random orientation-tuned neurons. Any developmental mechanism (or theoretical model) which provides a smoothing of initially random neuronal orientation preference is sufficient to produce vortexlike "hypercolumns."

In summary, both the ocular dominance and orientation column patterns appear to be consistent with the types of patterns which are created by spatial

---

*The question of the "resolution" of optical dye has been little studied. For example, it is not stated how the "resolution" is measured, nor even what the term "resolution" is taken to mean in Frostig *et al.* (1990).

[†]Verification of "vortex" and "pinwheel" structure in cortical hypercolumns has never been published. Hubel and Weisel (1974) only reported local correlation of orientation. Recent optical dye experiments (Bonhoeffer and Grinvald, 1991; Blasdel, 1992) mention "verification" of the vortex patterns observed by single-unit recording, but have not published any data demonstrating this verification. This is a nontrivial point, given the difficulty in the earlier and extensive experiments of Hubel and Wiesel to definitively establish the details of cortical hypercolumn pattern via single-unit recording.

filtering of noise patterns. Thus, the parameters of the filter serve to characterize the cortical patterns. This reduces a description of the cortical columnar patterns to the specification of a small set of parameters which fix the frequency, bandwidth, and orientation of the generating filters. The implication of this observation is that the dramatic patterns associated with orientation and ocular dominance are simply the results of local correlation of the neuronal features: the correlation function or, equivalently, filter function of the cortex is thus the primary quantity which must be measured, and the details of this correlation function are apparently sufficient to characterize the observed patterns of cortical columns.

3. Cortical Topography: One- and Two-Dimensional Models

In the previous section, columnar structure was addressed, independent of topographic structure. In the present section, a review of the current understanding of topographic structure of V1, independent of the details of columnar structure, is provided.

3.1. One-Dimensional Models of Cortical Magnification Factor

Virtually all psychophysical approaches to cortical magnification factor have been based on the use of a one-dimensional map function, of the form

$$M_{\text{Cortex}}(\theta) = \frac{k}{\theta + a} \tag{1}$$

where $M_{\text{Cortex}}(\theta)$ represents cortical magnification factor, i.e., the differential change in cortical position with respect to retinal eccentricity θ; k and a are two constants which specify the fit.

It is important to realize that magnification factor is essentially a "derivative." It represents the ratio of small changes in cortical location to corresponding small changes in retinal location. The question naturally arises: what is it the derivative of? We should first note that since there are two possible directions at each location in the cortex, and two independent directions in the retina, it is not immediately clear how to specify the ratio of cortical difference to retinal difference as a scalar function. This represents a mathematical difficulty which has largely been ignored in physiological and psychophysical studies which involve the retinotopic map. The simple use of a scalar magnification factor, which is nearly universal in this context, can only hold under limiting circumstances, which amount to an assumption of local isotropy. Since a map which is locally isotropic is also, by definition, a conformal map, there are very significant consequences to the usual assumption of scalar magnification factor. On the other hand, the isotropic nature of the map is an experimental question, which is not automatically true! In order to clarify this issue, we will embark on a brief but necessary digression concerning the nature of the derivative of a map from a planar region to a planar region, following an earlier exposition of this subject (Schwartz, 1984), in this context.

3.2. Two-Dimensional Models of the Cortical Map Function

In Schwartz (1976, 1977a, 1980) it was shown that if one represents the cortical map function in complex variables as

$$w = k \cdot \log(z + a) \tag{2}$$

then the magnitude of the derivative of this map function is

$$|w'| = \left| \frac{k}{z + a} \right| \tag{3}$$

Figure 9 shows the geometry of this mapping and Fig. 10 shows approximations using this function to V1 topography in several different species.

A function of a complex variable is by definition analytic, or conformal, which is equivalent to the assumption of local isotropy. This is the only condition under which the usual scalar magnification function is sufficient to characterize the cortical map.

The constants k and a may be determined by plotting inverse magnification factor against eccentricity, which is linear: intercept = a/k and slope = $1/k$. In this two-parameter fit, only the constant a is of qualitative significance to the "shape," or the "foveal" extent of the map. The constant k merely provides a normalization. However, if one is interested in the numerical value of magnification factor in units such as mm/deg, then both parameters are significant.

The physical interpretation of the two constants may be understood as follows. The constant k is essentially a normalization factor, which for most purposes is uninteresting. The constant a determines the extent of the linear

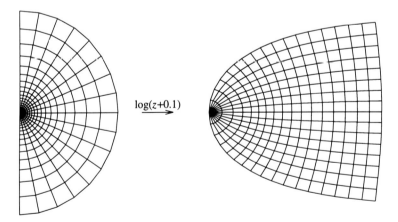

$\log(z+0.1)$

Figure 9. The conformal mapping of the unit half disk by $log\,(z + a)$, shown here with $a = 0.1$. The vertical meridian in the domain (half-circle as a model for the retina) is mapped to the curved boundary at left in the range (model of the cortex). In the central, or foveal, region, the map is asymptotically linear, turning to a logarithmic geometry in the region $> a$. From Rojer and Schwartz (1990b).

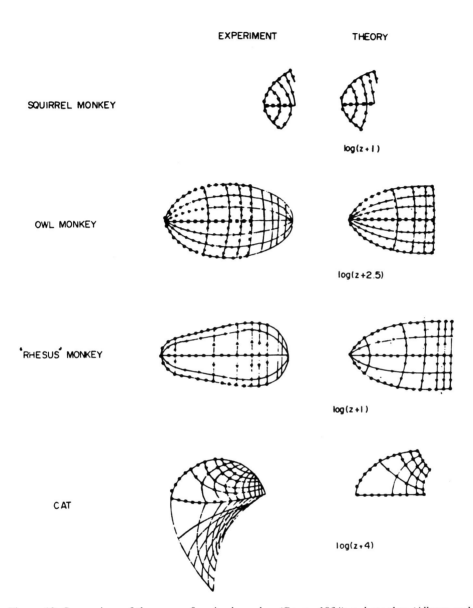

Figure 10. Comparison of the maps of squirrel monkey (Cowey, 1964), owl monkey (Allman and Kaas, 1971), rhesus monkey (Daniel and Whitteridge, 1961), and cat (Tusa *et al.*, 1978) with complex logarithmic approximations. The original experimental mappings are shown in the column labeled "experiment," and the complex logarithmic mappings in the column labeled "theory," where the constant *a* has been chosen to give the best "visual" fit to the experimental data. From Schwartz (1980).

regime of the cortical map, which we suggest should provide the definition of the term *fovea*. Consider the asymptotic forms of the map function $\log(z + a)$:

$$\log(z + a) \simeq \log(a) + \frac{z}{a}; \qquad z < a \tag{4}$$

$$\log(z + a) \simeq \log(z); \qquad z > a \tag{5}$$

For $z > a$ the map is essentially logarithmic, while for $z < a$ the map is essentially linear. Current estimates for the magnitude of a are in the range of 0.3 to 0.8°, which correspond roughly to the usual definition of the "fovea." It has been proposed (Schwartz, 1980) that this change in the behavior of the cortical representation, from linear to logarithmic, be used as a quantitative estimate of the size of the "fovea." In any case, the constant a is best understood in terms of the location of change in the geometric behavior of the cortical map.

In the psychophysical literature, another interpretation is in current use for these parameters. Cortical magnification factor is represented as (Wilson *et al.*, 1990)

$$M_{\text{Cortex}}(\theta) = \frac{M_{\text{Cortex}}(0)}{1 + E/E_2} \tag{6}$$

In this parameterization, E_2 is the eccentricity at which magnification has fallen by a factor of 2. Both descriptions are equivalent, since they are both linear functions.

Thus,

$$a = E_2 \tag{7}$$

$$k = M_{\text{Cortex}}(0) \cdot E_2 \tag{8}$$

Here, the $[k, a]$ parameter set will be used. It explicitly makes the connection to cortical map geometry via the interpretation of a. The "halving" of cortical magnification factor in the interpretation of the $[M(0), E_2]$ parameter set suggests no physiological interpretation. However, both descriptions are equivalent, since they model inverse cortical magnification factor as a linear, scalar function.

3.2.1. Geometry of Complex Log Function

Now, if one asserts that a scalar function is sufficient to model cortical magnification factor, then one has, implicitly, assumed* that the two-dimensional structure of the cortical map is given by a function of the form

$$w = k \cdot \log(z + a) \tag{9}$$

---

*If magnification is a scalar, then it is independent of direction, hence is isotropic, hence represents a conformal map. The unique conformal map is specified by finding the unique conformal map whose derivative is the presumed scalar magnification function. In the case that this function is linear, then a complex log map is the obvious solution.

whose derivative (in absolute magnitude) is

$$|w'| = \left| \frac{k}{z + a} \right| \qquad (10)$$

In fact, functions of the complex log type do provide a fairly good fit to all existing data. For example, as shown in Schwartz (1980), the data of both Talbot and Marshall (1941) and Daniel and Whitteridge (1961) are approximately fit to map function of the form $\log(z + a)$, with the constant $a \simeq 1°$, as shown in Fig. 10. Dow *et al.* (1981, 1985) found that a function of this form provided a good fit to their data, using a constant $a \simeq 0.3°$. The data of Tootel *et al.* (1982) are fit, to "good approximation," by the same complex log function with $a \simeq 0.3°$ (Schwartz, 1985b). van Essen *et al.*, (1984) have reported an inverse linear magnification function, with constant $a \simeq 0.8°$, and have stated that the complex log model is a good approximation, although there may be localized departures from isotropic mapping near the representation of the lower vertical meridian.

It appears that there is a rough consensus that primate visual cortex is, to a first approximation, a complex logarithmic structure characterized by a "foveal" constant $a \simeq 0.3$ to $a \simeq 0.9°$ (Wilson *et al.*, 1990).

The complex log mapping may be regarded as a warping of a conventional scene. This is illustrated in Fig. 11.

3.2.2. Two-Parameter Mapping

One problem with the simple complex log fit is that it begins to fail around 15–20°, where the actual shape of the cortex requires that the boundaries of the map begin to close in to make a closed, vaguely elliptical surface (see Fig. 18 for the shape of the central 10° of cortex, and Figs. 14 and 17 for the shape of the entire cortex). Also, cortical magnification is somewhat sublinear beyond 20–40° of visual field. Both of these problems can be simply addressed, however, by the use of a second logarithmic function, which requires one additional parameter, as suggested in Schwartz (1983):

$$w = k \cdot \frac{\log(z + a)}{\log(z + b)} \qquad a \simeq 0.3°, \, b \simeq 50° \qquad (11)$$

Thus, simple two- and three-parameter fits provide a good approximation to the topographic map structure in primate V1. However, these fits, as well as the numerical conformal fit that is to be presented shortly, are based on the existence of "isotropy": the local cortical magnification must be a scalar, i.e., independent of local direction.

In order to fully understand the nature of "cortical magnification," we need to take a brief diversion into calculus. It is necessary to take a close look at the nature of the "derivative" of a cortical map function (Schwartz, 1983). Although both obscure (and elementary in a mathematical sense), it is impossible to understand the details of topographic mapping without grasping the following argument, which has been largely ignored in both the psychophysical and physiological literatures.

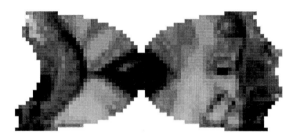

Figure 11. Texture mapping of digitized image with logarithmic map. From Rojer and Schwartz (1990b).

3.2.3. Regular Mappings of 2-D Surfaces into 2-D Surfaces

For a regular map $F: \mathbb{R}^2 \to \mathbb{R}^2$; $F: (x,y) \mapsto (f(x,y),g(x,y))$, the Jacobian is:

$$J = \begin{bmatrix} \partial f/\partial x & \partial f/\partial y \\ \partial g/\partial x & \partial g/\partial y \end{bmatrix} = \begin{bmatrix} f_x & f_y \\ g_x & g_y \end{bmatrix}$$

As mentioned in earlier work (Schwartz, 1984), it is instructive to rewrite the Jacobian in terms of its symmetric and antisymmetric parts:

$$2J = S + A = (J + J^T) + (J - J^T) = \begin{bmatrix} 2f_x & f_y + g_x \\ f_y + g_x & 2g_y \end{bmatrix} + \begin{bmatrix} 0 & f_y - g_x \\ g_x - f_y & 0 \end{bmatrix}$$

It is further useful to rewrite the symmetric part as the sum of a diagonal and a traceless matrix:

$$S = \begin{bmatrix} 2f_x & f_y + g_x \\ f_y + g_x & 2g_y \end{bmatrix} = D + T = \begin{bmatrix} \text{tr} & 0 \\ 0 & \text{tr} \end{bmatrix} + \begin{bmatrix} 2f_x - \text{tr} & f_y + g_x \\ f_y + y_x & 2g_y - \text{tr} \end{bmatrix}$$

where tr represents the trace of the Jacobian, i.e., $\text{tr} = f_x + g_y$.

The traceless component of the Jacobian can be diagonalized by an orthogonal transformation, R, since it is symmetric and positive definite, which we write explicitly:

$$RTR^{-1} = \begin{bmatrix} \lambda & 0 \\ 0 & -\lambda \end{bmatrix}$$

The differential structure of a regular mapping has been dissected into the sum of three geometric invariant components:

- A rotation (the antisymmetric part A)
- A dilation (the diagonal part D) and
- A shear (the traceless part T)

The shear component is a compression along one axis and an expansion of equal magnitude along another axis, of magnitude equal to λ. The direction of the principal axes are given by the rotation matrix R. In the case that the mapping has no shear, then T (traceless, symmetric part of the Jacobian) is zero and the differential structure of the mapping is characterized by only two numbers: the rotation and the dilation. The dilation is the "magnification factor," and this is the single number which usually emerges from physiological or anatomical experimental work.

In the literature, the apparent compression of the cortical magnification factor (CMF) perpendicular to the axes of ocular dominance columns is sometimes expressed as a ratio of gain along the two axes. In the above analysis, if we consider only the symmetric portion

$$RSR^{-1} = RDR^{-1} + RTR^{-1} = D + RTR^{-1} = \begin{bmatrix} \mathrm{tr} + \lambda & 0 \\ 0 & \mathrm{tr} - \lambda \end{bmatrix}$$

and examine the ratio of the two nonzero terms, we obtain the "compression ratio" of the mapping.

Having provided the necessary analysis, we conclude by pointing out that the characterization of a neural map by microelectrode measurements of "magnification," i.e., by comparison of small changes in one neural layer compared with small changes in another, requires four such measurements at each point, in order to retrieve the Jacobian matrix. This has never been reported in any physiological experiment, and would be technically difficult to accomplish. However, obtaining only a single magnification measurement at each point, as in conventional magnification experiments, is only sufficient if the mapping is conformal, i.e., if the "shear" component is zero. In the absence of a direct measurement of the Jacobian of the cortical map, our strategy has been to use numerical and analytic conformal maps to model the cortex. If these work, then the shear component is negligible, and we don't, in fact, need to perform the very difficult experiment that would be required to provide the Jacobian. This strategy, which seems to be effective for primate V1, will now be reviewed.

3.2.4. Numerical Conformal Mapping

We have performed (Rosenbluth *et al.*, 1992) a series of 2-DG studies of primate retinotopic mapping, using computer reconstruction and computer brain flattening, together with numerical conformal mapping, to provide an isotropic map fit to primate topography in V1. Before presenting some preliminary results of this work, we provide some background material on the properties of isotropic map functions and conformal mapping.

3.3. Application of Conformal Mapping to Cortical Topography

Conformal mappings may be defined in a number of equivalent ways, which emphasize different aspects of their geometric or analytic properties (Ahlfors, 1966):

- Complex analytic functions $f(z) = (u(x,y), v(x,y))$, for

$$\frac{df}{dz} \neq 0$$

represent conformal mappings.
- The real and imaginary parts of the map function satisfy the Cauchy–Riemann equations:

$$\frac{\partial u}{\partial x} = \frac{\partial v}{\partial y}$$

and

$$\frac{\partial v}{\partial x} = -\frac{\partial u}{\partial y}$$

- A conformal mapping is locally isotropic. This means that an infinitesimal area element is magnified equally in all directions. In the language of the preceding section, conformal mappings have no shear component.
- Infinitesimal angles are preserved by conformal mapping.
- The real and imaginary parts of the map function are harmonic conjugate functions, i.e., they satisfy the Laplace equation,

$$\nabla^2 u(x,y) = \nabla^2 v(x,y) = 0$$

and intersect orthogonally.

This property provides important practical application to areas of potential theory (electrostatics, fluid mechanics, etc.) where the Laplace equation occurs.

The Riemann mapping theorem [see Frederick and Schwartz (1990) for discussion in this context and Ahlfors (1966) for general discussion] guarantees the existence and uniqueness of conformal mappings between regions. It is of fundamental importance, since it states that given any two planar brain regions (e.g., retina and cortex), there is a unique conformal mapping between them which is specified by the "shape" of the regions, a single point correspondence and angle specifying the relative orientation of the domain and range regions.

Riemann Theorem. Given a region, there exists a conformal mapping of this region onto the unit disk (Ahlfors, 1966). The mapping is made unique by fixing the mapping of a single point in the region onto the center of the unit disk, and fixing the orientation of the unit disk.

The use of this theorem in the present context consists in mapping two regions into the unit disk, which then implicitly provides the desired map function.

The Riemann theorem is perhaps surprising, in that it does not seem intuitively possible that all of the details of a two-dimensional map could be specified by a single point and orientation. However, this may be understood better by considering the one-dimensional analogy. If one specifies the boundary points of a curve (i.e., two points), and states that the curve must satisfy the equation

$$\frac{d^2y}{dx^2} = 0$$

the curve must be a straight line (i.e., zero curvature), and we then know all of the internal points only from the boundary conditions, which are simply the endpoints of the line. In two dimensions, the Laplace equation condition that is one of the possible defining characteristics of a conformal mapping, as summarized above, is a direct generalization of the specification of a straight line in one dimension. Conformal mappings thus are the generalizations to mappings of $R^2 \to R^2$ of the straight line as a mapping of $R^1 \to R^1$! This is the reason that they are so widely used as a modeling tool in engineering, physics, and mathematics. Conformal mappings are the simplest possible regular mapping in two dimensions.

3.3.1. Computer Simulation via Texture Mapping

In the case of visual cortex, there is considerable experimental evidence that the mapping of the retina to the surface of primary visual cortex is approximately isotropic. Thus, in order to model the representation of a visual image on the surface of the cortex, we need to construct a conformal approximation to this map, and to perform a conformal texture map of given visual field images. In order to illustrate this process, we show a numerical flattening that we have performed recently of the surface of primary visual cortex of the monkey.

In this work, we were able to identify a single point (the representation of the blind spot, or optic disk) in the eye, and an orientation (the orientation of the horizontal meridian). These observations, together with the flattened representation (and its boundary), were sufficient to generate the cortical map function (Weinshall and Schwartz, 1987).

The agreement between this method of determining the cortical map, and direct microelectrode measurements of the cortical map function, is excellent. Figure 12 shows a natural scene, mapped via this conformal approximation.

3.3.2. Numerical Conformal Mapping of 2-DG Computer-Flattened Cortex

During the long history of attempts at measurement and modeling of the primate V1 topographic map, there has never been, to my knowledge, an attempt to assess the "error."

In fact, there has been a very wide range in the reported values of the constant a in the linear magnification fit: Dow *et al.* (1981) reported the smallest value

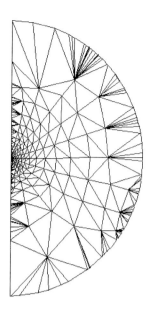

Figure 12. Texture mapping of high-resolution scene representing nearly full range of human (3 min of arc maximum resolution to 100° of field). At left is shown the "wire frame" which triangulated the retinal and cortical regions. A numerical conformal mapping estimated from an ocular dominance column reconstruction, as described in the text, was used to map the nodes of the "wire frame" model, and then the interiors of each triangular patch were texture mapped from retina to cortex, using a local bilinear approximation, resulting in the mapping shown at right. On the left page is a lab scene that was digitized to an effective resolution of 16,000 × 16,000. On the right is the cortical model of this data, which extended over more than 100° of visual field, and which resolved

(about 0.3°), while Rovamo and Virsu (1979) and Tootell *et al.* (1988b) reported values in the range of 1.5°–2.0°, and Hubel and Freeman (1977) reported 4°! Is this range of variance due to normal differences between animals, or to error in experimental procedure? And, more to the point, how is it possible to compare different measures, such as retinal ganglion cell density, visual acuity, or stereo fusion area, with cortical magnification factor, when one or both of the derived curves has no error analysis supplied?

In order to provide a high-precision estimate of the two-dimensional cortical map function, we set out several years ago to provide the following data and computational procedures.

3.3.3. Computer Flattening

In order to mathematically model the cortex, it is convenient to work in a planar model. Numerical methods for conformal mapping have only been published for planar domains. Differential geometry is best avoided in numerical problems, and the cortex is relatively flat. [We have measured the mean and Gaussian curvature of primate V1 and find that curvature of cortex does not preclude an accurate flattening (Schwartz and Merker, 1985, 1986).] We have found that it is possible to flatten cortex with a mean local error of roughly 5% (Schwartz *et al.*, 1987, 1988, 1989; Wolfson and Schwartz, 1989) and have ap-

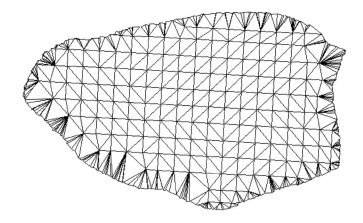

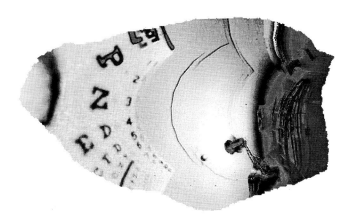

Figure 12. (*Continued*) about 3 arc-min in the center. The retinal view (left) contained a Snellen Chart, which is not visible in this low-resolution reconstruction, but which was resolvable in the digital image. The Snellen Chart image in the cortex, seen on the right, occupies nearly one-third of the entire cortical surface. This simulation is unique, in that it represents a simultaneous wide-angle and high-resolution-image model of the cortex. This necessitated the use of an extremely high-resolution input frame (effectively 16,000 × 16,000). From Schwartz *et al.* (1988).

plied this method to the reconstruction of primate ocular dominance columns and to primate topography in V1. Figure 13 shows the wire-frame model of a flattened cortex, reconstructed from serial sections. This experiment was from a macaque which had one eye enucleated, and whose brain was subsequently stained for CO, revealing the structure of the ocular dominance column pattern, which was computer mapped from digitized images of the serial sections into the flattened cortical representation (Schwartz *et al.*, 1988). Note the representation of the blind spot in Fig. 14. The blind spot (optic disk) in macaques is located at about 17° of eccentricity on the horizontal meridian, as shown in Fig. 14. We will return to the significance of this particular type of data shortly.

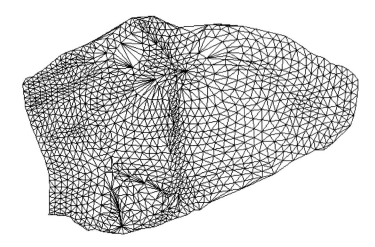

Figure 13. Wire frame showing flattened polyhedral model of cortex, digitized from serial sections, and numerically flattened with the algorithm of Wolfson and Schwartz (1989) and Schwartz *et al.* (1989). From Wolfson and Schwartz (1989).

3.3.4. Numerical Map Fitting

As outlined above, we have developed several fitting methods using complex logarithmic functions, and also a more general numerical conformal mapping method. We have also outlined the conditions under which local shear in the map function might be modeled, although we have not performed any numerical analysis of sheared mappings to date. Using the numerical conformal mapping procedure outlined above, and described in detail in Frederick and Schwartz (1990), we have modeled the details of a 2-DG experiment which will now be described.

3.3.5. 2-DG Measurements of V1 Topography

2-DG is a metabolic marker. When radioactively labeled, it can be injected, and then, following some paradigm for stimulation, the brain can be imaged to detect the relative pulse density of neurons [that is, under the common assumption that 2-DG density is proportional to metabolic activity which in turn is presumed proportional to neuronal pulse density (!)], following the stimulation.

2-DG has been used in humans, using PETT scanning to reconstruct the pattern of firing (Schwartz, 1981; Schwartz *et al.*, 1983). The stimulus pattern in this case was a logarithmically structured stimulus, similar to that shown in Fig. 15.

A similar use of this technique was provided by Tootell *et al.* (1982), who used monkeys and physical brain flattening (i.e., pressing the opercular part of monkey visual cortex between glass slides). In this, and subsequent accounts of this research, these workers modeled their data with a simple, one-dimensional magnification curve. Initially they claimed contradictions with the complex logarithmic model and their data (Tootell *et al.*, 1982). However, a simple complex log fit to their data was subsequently published (Schwartz, 1985b), followed by agreement that the complex log provided a "good" fit to their data (Tootell *et al.*, 1985). This data, and a complex log model superimposed over it, is shown in Fig. 16. This log map had an *a* of 0.3°, in agreement with the data of Dow *et al.* It appears to provide a good fit to the data of Tootel *et al.*, with the exception of a region near the intersection of the inferior occipital sulcus and the calcarine cortex, where the 2-DG images seem to be significantly sheared. This coincides with a region in which van Essen *et al.* (1984) reported observing anisotropies (shear) in V1 topography, and also coincides with some of our recent 2-DG computer-flattened models, as shown in Fig. 17.

We are currently in the final stages of analyzing a series of 2-DG experiments, using computer flattening and conformal mapping. The result from one experiment is shown in Fig. 17.

In this experiment, monkeys were shown stimuli like that of Fig. 15, subsequent to paralysis, eye position plotting, and injection with 2-DG labeled with [14]C. Following about 40 min of stimulation, the monkeys were sacrificed, their brains removed, and one hemisphere was physically flattened while the other hemisphere was blocked for subsequent coronal section on a cryostat. The autoradiography of the physically flattened hemisphere is shown in Fig. 18. The computer-flattened, reconstructed other hemisphere is shown in Fig. 17. Both of these figures correspond to the stimulus shown in Fig. 15.

Using a single intersection of the ring and ray pattern as one of the constraining data needed to apply the Riemann mapping theorem to construct a conformal model of this data, as outlined above, and in Frederick and Schwartz (1990), the conformal map of the stimulus of Fig. 15 is shown in Fig. 19. The superposition of this conformal model and the computer-reconstructed autoradiographic data is shown in Fig. 20. It is clear that the agreement between the conformal model and the data is excellent. However, there appears to be evi-

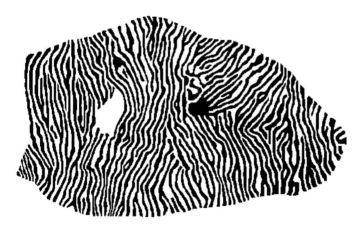

Figure 14. Using the flattened cortical model of the type shown in Fig. 13, the density of cytochrome oxidase was texture mapped from the original serial sections (obtained from a one-eye enucleated monkey). This shows the ocular dominance column pattern. Note the representation of the optic disk, which is the dark region along the horizontal meridian. This data was hand-traced after computer plotting to provide a "black–white" representation.

Figure 15. An example of the 2-DG visual stimulus used in the 2-DG experiment discussed in this chapter.

dence of some shear in the lower corner of the map, which is at the intersection of the inferior occipital sulcus and the lip of the calcarine sulcus. This is the same phenomenon seen in the data of Tootel *et al.* shown above (Fig. 16), and has also been described by van Essen *et al.* (1984). It is remarkable, however, that this feature, which appears to be common in macaques, is quite localized to the lip of the lower calcarine sulcus: we call this feature the calcarine anomaly: it represents a localized departure from the otherwise excellent fit of an isotropic (conformal) map to V1 topography.

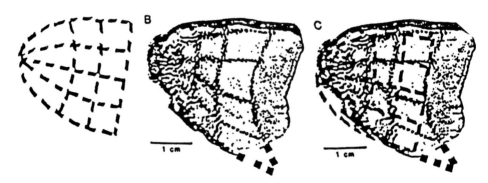

Figure 16. The 2-DG data of Tootell *et al.* (1982), overlaid with a complex log mapping of constant *a* = 0.3. From Schwartz (1985b).

In this study, we performed a regression, in order to find the best-fit conformal map to our data. In the regression, we allowed both eye position and cyclotorsion to be the free variables, and the error function for the fit was the summed square difference of data landmarks (e.g., intersections of rings and rays) with corresponding conformal map landmarks. We used a nonparametric regression procedure [Nelder simplex method (Press *et al.*, 1988)], and were able to determine the best fit to the eye position and cyclotorsion of the paralyzed monkeys from this procedure. It appears that this procedure has provided the excellent fit shown in Fig. 20. There is a significant error associated with plotting the retinal landmarks of paralyzed monkeys. van Essen *et al.* (1984) estimated this error to be about 0.5° typically. In the data shown in Fig. 17, we found a 1° error, and we also obtained an estimate of 11° for cyclotorsion from the regression. This is in good agreement with the cyclotorsion of paralyzed monkeys, and with our measured value for cyclotorsion in this particular case. We therefore feel that the regression procedure that we used has allowed us to null out the significant error due to imprecision of retinal plotting in these procedures.

In the near future, we plan to publish analyses for several monkeys, and to provide an estimate of "error" for the whole procedure. At the present time, we have hand-sketched "error bars" in Fig. 20, which represent the distance between the computer-predicted feature, and the actual measured feature in the map (features are corners of "boxes," intersections of rings and rays). It is evident that the average error is considerably less than 1 mm, over perhaps 10–15 mm of cortex. Since the error of brain flattening is roughly 5%, we feel that this error is about as small as possible to achieve under these conditions. In summary, the average error in this fit is considerably less than 10%, although

Figure 17. Computer-flattened macaque V1, left hemisphere of monkey DG-14, texture mapped with 2-DG pixel data following stimulation of monkey with the Fig. 15 stimulus. The image of the boxes, rings, and rays of the stimulus is visible in the data.

there is an error of 4–5 mm (perhaps 30%) localized in the lower left corner of the operculum in Fig. 20, in the region of the calcarine anomaly. Put another way, if we have a technique which reliably indicates the visual field coordinates of a single point in the cortex, then (if we have access to the anatomical "shape" of the cortical boundary), we can reproduce the entire cortical map to within an accuracy of roughly 1 mm. This somewhat counterintuitive result illustrates the power of these techniques.

What is the cause of the calcarine anomaly? We have not found any correlation to excessive Gaussian curvature of cortex in that region, although the intersection of the lip of the calcarine and the inferior occipital cortex is one of the most severely folded parts of the macaque visual cortex. Perhaps a shear may be

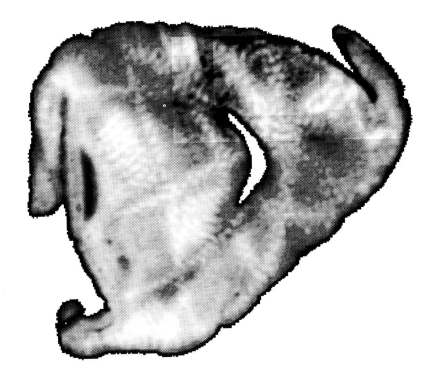

Figure 18. Same as above, but the operculum of V1, rather than the whole extent of V-1, is shown. This is the right hemisphere of monkey DG-14, whose left hemisphere is shown in computer-flattened format in Fig. 17. This was prepared by physical flattening, rather than computer flattening. Note the eye-offset that is apparent in this figure: there is a central point with six rays emanating from it that is apparently about 1° away from the nominal location of the fovea, which is at the intersection of the lunate and inferior occipital sulci (the "pointy" end of the operculum). This eye-offset error is in the expected range, due to error in plotting the center of the fovea. Note the large "cortical" effect of this relatively small retinal error. We have regressed out this eye position error, and the regression reports an error which is about 1°. Independently, the regression reported a cyclotorsion of the eye of about 11°, which is consistent with the cyclotorsion measured during the experiment. Thus, we believe that we have been able, by studying left and right hemispheres from the same monkey, to "null" out the significant error introduced in the map estimate by errors in retinal landmark plotting.

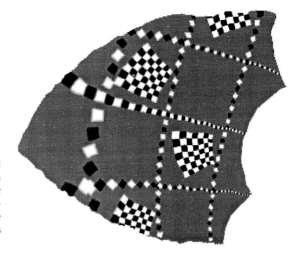

Figure 19. A numerical conformal mapping, prepared by choosing a single point correspondence in the data of Fig. 17 to construct a numerical conformal mapping. Only the opercular extent of the mapping is shown.

introduced in the developmental events leading to the creation of the sulcal pattern of the cortex. Perhaps an idea suggested by Tootell *et al.* (1982), that in this region there is a less than 2:1 compression of the ocular dominance column pattern, is the cause. It is difficult to say, although the location of this anomaly in the region of major sulcus folding, and its topographic location in the vicinity of about 8° on the upper vertical meridian, make it unlikely that this anomaly is driven by any functional considerations.

In summary, we offer this preliminary data as an estimate of the reliability

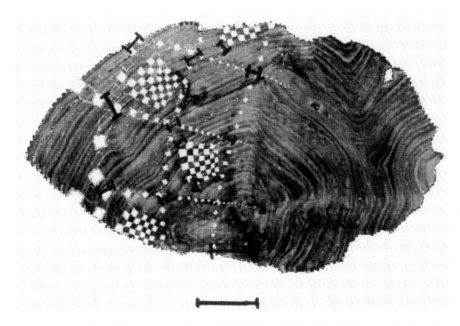

Figure 20. An overlay of the computer-flattened 2-DG data of Fig. 17 and the numerical conformal mapping of Fig. 19. "Error" bars have been sketched in to show, in a preliminary fashion (this data is not yet fully analyzed), an estimate of the reliability of the map function. Bar = 5 mm.

of conformal approximation of cortical topography in V1. We believe that this is the first time that a regression or error analysis has been performed on the two-dimensional structure of V1 topography.

3.3.6. Application to Reconstruction of Human V1 Topography

One major application of the use of numerical conformal mapping, which has been validated by the monkey data shown in Fig. 20, is that it provides an entirely new method for estimating topographic map structure in primates, as pointed out in Weinshall and Schwartz (1987). In order to reconstruct a conformal approximation to a topographic mapping, it is only necessary to have access to the following data:

1. A single point correspondence
2. The relative orientation of the two domains
3. A flattened representation of the boundary of the cortex

This follows directly from the Riemann mapping theorem, and is shown by construction in Fig. 20, since the computer model in that figure was constructed from a single point (ring and ray intersection), and the flattened cortical representation. Note that the alternative to this procedure is to perform an exhaustive (physically and figuratively!) point-by-point physiological mapping of differential magnification factor measurements. Difficult and errorprone as this is in monkey research, it is literally impossible (for legal and ethical reasons) in human research. However, anatomical data on the V1 boundary are readily available in both monkey and human (magnetic resonance images are now capable of noninvasively providing submillimeter resolution of tissue density, which is sufficient to model the boundaries of human cortex, and to flatten human cortex using our flattening software, to excellent precision).

Now, in the human, physiological mapping methods have been applied in only a few, very limited experiments. PETT scanning, even with current resolutions in the range of 5–7 mm, is much too coarse to provide a reliable estimate of the human map. Human ocular dominance columns from postmortem, one-eyed patients have been reconstructed (Horton and Hedley-White, 1984) using the same CO stain used in the monkey experiment summarized in Fig. 14.

In order to demonstrate the feasibility of this method, we have used the columnar pattern, as shown in Fig. 14, to measure the position of the optic disk, which is clearly evident in this figure. This provides us with a landmark at 17° on the horizontal meridian. We then constructed the unique conformal mapping that corresponds to this constraining data point. The result is shown in Fig. 21.

By numerical differentiation, we have found that the magnification function obtained from this map is essentially indistinguishable from microelectrode estimates of V1 cortical magnification (Weinshall and Schwartz, 1987). Therefore, this approach provides a novel route to estimating cortical magnification factor, and one which is applicable to humans. Thus, by either reconstructing the ocular dominance column in postmortem one-eyed humans, or, perhaps, by using echo-planar MRI techniques to find a single visual landmark, followed by computer flattening of the MRI-measured sections of human cortex, we would be able to reconstruct a highly reliable estimate of human V1 topography. Moreover, were this done in an MRI paradigm, we would have the added benefit of a

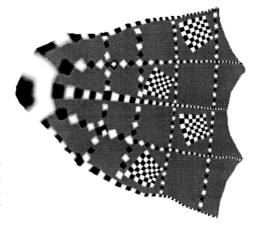

Figure 21. A numerical conformal map, made from a computer-flattened ocular dominance column preparation (Fig. 14), using the known location of the optic disk to constrain the complete conformal map.

living subject, thus allowing psychophysical study of the same individual whose anatomy has been characterized. An experiment of this kind is currently being planned. The basic calibration of the method, together with the necessary algorithms, has been provided in the context of the monkey work outlined above. If successful, it may be possible to provide the first reliable estimate of human V1 topography. At the present time, we see no other viable route to these extremely important data.

3.3.7. Psychophysical–Anatomical Parallels

A major theme of the literature on human retinocortical topography has always been the possible existence of scaling laws between psychophysical measurements and retinal and cortical anatomical measurements. Daniel and Whitteridge (1961) were likely the first to make the point that human minimum angle of resolution has a similar functional form to inverse cortical magnification. In later work, Schwartz (1977b) suggested that retinal ganglion cell density and cortical topography might be related by a particularly simple developmental rule: the retina (actually LGN) fiber input spreads across the cortex in the manner of an isotropic fluid: fixing a single pint (and orientation) would be sufficient to provide a unique map.

During the 1980s there were many papers published which sought to link acuity, stereo acuity, vernier acuity, motion thresholds, and retinal receptor and retinal ganglion cell densities to cortical magnification factor. There are two major problems with this literature, however.

- There has usually been no error analysis performed on any of the data. This is particularly problematic given the wide spread in the measurement of all the quantities involved. For example, different workers have found values of the constant a, which characterizes the cortical map, to be between 0.3 and 4°. This is a variation of more than 1000%. And none of these experiments provides an estimate of error.
- Human cortical magnification has never been reliably measured. Existing estimates of it tend to be in the range of $a = 1.7$ to $a = 6°$, but these estimates are based on extremely poor-quality data, and are many hun-

dreds of percent higher than that found in monkey. The only reliable monkey measurements have been made with exhaustive microelectrode plotting and with 2-DG analysis of serial sections. These techniques are not possible in human subjects since careful microelectrode exploration is both illegal and unethical. Therefore, human magnification factor is currently known by guesswork, not by measurement. This fact greatly compromises the comparisons of human psychophysics to a largely unknown human magnification factor, or of monkey magnification factor to largely unknown monkey psychophysical data!

A recent review of this subject (Wilson *et al.*, 1990) attempts to rationalize this situation by suggesting that there are really two *a* parameters, a retinal one in the range of 1.5 to 4.0° and a cortical one in the range of 0.3 to 0.8°. In this chapter, a summary of psychophysical measurements, and their relative *a* values (called E_2 herein) are shown (in slightly modified form) in Table I.

The attempt to cluster these data is important, but it seems to be the case that the measurements in this field are spread over an extremely large range. For example, Wilson *et al.* (1990) cite the data of Tootell *et al.* (1982) as providing support for a value of 0.8 for *a*. But recently, it appears that the same data have been presented with an analysis of 1.5 for *a* by the same research group (Tootell *et al.*, 1988b), which would move this measurement from one group of Wilson *et al.* to the other. And, recently, Wassle *et al.* (1989) have provided evidence that the classical fits of retinal ganglion cell density have been contaminated by a background of displaced amacrine cells that have tilted the linear regression curves toward values of *a* that are considerably too high. The new reanalyzed values of retinal ganglion cell density of Wassle *et al.* are in agreement with the

Table I. Values of *a* from Psychophysical and Anatomical Sources[a]

$a \in [1.5^o, 4^o]$	Species	References
Retinal cones	Human	Osterberg (1935)
Pβ ganglion cell density	Macaque	Perry and Cowey (1985), Schein and de Monasterio (1987)
Ganglion cell density	Macaque	Wassle *et al.* (1989)
Inverse cortical magnification	Macaque	Tootell *et al.* (1988b)
Cortical receptive field size	Macaque	Dow *et al.* (1981)
Minimum angle of resolution	Human	Wertheim (1894), Weymouth (1958), Westheimer (1979), Levi and Klein (1982), Coletta and Williams (1987)
Contrast sensitivity	Human	Rovamo *et al.* (1978), Koenderink *et al.* (1978), Watson (1987)
Slow velocity discrimination	Human	McKee and Nakayama (1984), Wright (1987)
$a \in [0.3^o, 0.9^o]$		
Inverse cortical magnification	Macaque	Dow *et al.* (1981), Tootell *et al.* (1982), van Essen *et al.* (1984)
Optimal 2-dot vernier acuity	Human	Westheimer (1982)
Abutting vernier acuity	Human	Bourdon (1902), Levi *et al.* (1985)
Vernier crowding zone	Human	Levi and Klein (1982)
Optimal 3-dot bisection	Human	Yap *et al.* (1987), Klein and Levi (1987)
Stereoacuity	Human	Fendick and Westheimer (1983)

[a]After Wilson *et al.* (1990).

recent a of 1.5° from Tootell *et al.* (1988b), although evidently not in agreement with the value of a of 0.8° from the same investigators using the same data set (Tootell *et al.*, 1982), and is in significant disagreement with the data of Dow *et al.* (1981, 1985).

There is clearly a problem here, whose source is easy to see: virtually all data in the visual system, from retina, through LGN, to V1, V2, and on to MT and V4, scale in a very roughly linear fashion, as do most psychophysical data. Given the wide range of reported measurements, it is not hard to make arguments that sets of phenomena do, or do not match, or even cluster into several groups. It appears that this situation will remain somewhat confused until such time as investigators begin to publish error bars, or confidence limits on their data, so that a comparison of two curves can be made statistically. In the near future, we expect to publish such a result, based on the analysis of data like those shown in Fig. 20. Also, the progress made by the Wassle group in obtaining more reliable retinal data is an important step toward clarifying the retinal component of this problem.

4. Cortical Protomaps: Joint Columnar and Topographic Structure in V1

The entire preceding discussion has been predicated on the notion that visual cortex represents a "regular" mapping of the visual field, i.e., a well-behaved, continuous map whose Jacobian is nonzero and finite. In fact, the situation is considerably more complex than this. In the input layers of V1 (layer IV), there are two complete maps, one for each eye, interlaced in the form of thin strips, or columns. We have shown a computer-flattened texture mapping of this ocular dominance column pattern from our lab, in Fig. 14.

In addition, there are at least two more independent columnar systems in V1: one for orientation, and the so-called "cytochrome oxidase" puffs, which tend to occupy the singular regions where all orientations in cortex come together.

Recently, Landau and Schwartz (1992) have developed computational methods to represent the cortex, including arbitrary columnar substructure. This algorithm will be reviewed here, along with some computer simulations of recent binocular 2-DG data, which are well modeled by this approach.

4.1. Protocolumns and Protomaps

The details of the protocolumn algorithm are not elementary, and depend on the manipulation of several data structures from computational geometry. But the general idea of it is simple: we imagine that the submodalities of the cortex (e.g., left eye input, or orientation input at some specific angle) constitute a complete map of retinal space. We call these maps "protomaps," i.e., they are the "regular" map before it has been "cut" into columns, which are then "squeezed" and packed together to make up the real, observable cortex. For ocularity, there are two protomaps: a left and right eye protomap. For orientation, there are as many protomaps as distinct orientation values which are to be modeled. Thus, if we discretize orientation in terms of angular bins of 15° each, we would need to construct 12 protomaps for the orientation system.

The advantage of this conceptualization is that the protomaps are "regular" maps which can be described by conventional continuum approaches, as we have outlined above.

To proceed with the algorithm, we then define "protocolumns." The proto-columns are constructed from the observed individual columns in the brain, by a warping transformation which expands the observed columns until they smoothly fill the cortex. This is sometimes called a "grow" operation in computer graphics; we have used a Voronoi diagram method to actually compute the protocolumn pattern. Thus, the protocolumn pattern represents a form of "proximity" diagram. For a given "real" column, its unique protocolumn is the locus of pints nearest to that real column. Thus, for a typical single ocular dominance column in cortex, its protocolumn is a somewhat "fatter" version of itself, whose detailed shape is a function of that column and its near neighbors (see Fig. 22). This construction is designed so that the complete set of proto-columns perfectly tessellates the cortex. We then are able to image-warp from a protocolumn to its actual column, and thereby construct a two-dimensional sim-ulation of the topography and columnar structure of V1.

One consequence of this construction is that for "zebra-skin"-like patterns, such as the V1 ocular dominance column system, there is a roughly 2:1 compres-sion between a column and its protocolumn. This agrees with an early estimate of the experimental "shear" within a cortical ocular dominance column (LeVay *et al.*, 1975), but is in significant disagreement with the recent estimate of Tootell *et al.*, (1988b) that there is a 1:1 compression within a column, resulting in a 2:1 compression in the overall cortical map.

In Fig. 22 (top), we show an example of the computation of a single proto-column from its observed ocular dominance column, and at the bottom of the fig-ure show the entire protomap of the left eye. In Fig. 23, we show a simulation of the protocolumn algorithm and compare it with a recent binocular 2-DG experi-ment of Tootell *et al.* (1988a). The agreement is good, but it should be stressed that few quantitative data are available at the present time to constrain this fit.

Tootell *et al.* (1988a), in fact, favor a model in which the protomap, to use our term, is actually isotropic, and the "real" cortex is 100% anisotropic, and the local shear within an ocular dominance column is therefore something like 1:1. We favor the opposite model, in which the protomap is isotropic, the "real cortex" is isotropic, and there is a shear of 2:1 within an individual ocular dominance column.

One reason that we favor this type of model is that it is very difficult to believe that there could be as good a fit as we have shown in Fig. 20 of an isotropic model to a highly anisotropic cortex. Tootel *et al.* in fact have not provided any two-dimensional modeling to support, or to clarify, their conjecture.

And, to avoid confusing the reader at this point with notions of columnar shear, we point out that virtually all methods of measuring cortical magnification factor operate on a scale which is larger than the size of individual cortical columns, so that the small-scale columnar structure is essentially averaged out. However, if there was anything much different from a 2:1 shear within individu-al columns, this effect would summate coherently over many columns* across the

---

*Ocular dominance columns appear to be coherent, in the sense of running roughly parallel, over regions of the cortex which vary between several millimeters to perhaps 1 cm.

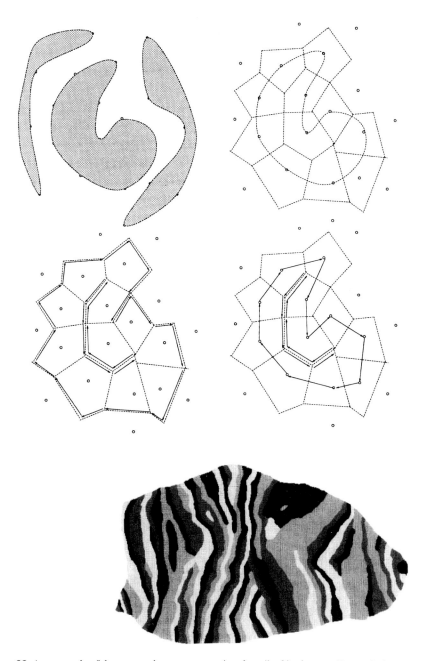

Figure 22. An example of the protocolumn construction described in the text. On top is shown some of the steps of the algorithm which finds Voronoi polygons from the experimental column data. On the bottom left is the thresholded image of the ocular dominance column, and on the right is the corresponding protocolumn pattern. Each protocolumn is shaded with a different shade of gray, in order to help visualize them. Note that the protocolumns surround their corresponding columns, are about twice the thickness, and form a smooth, jigsaw-like tessellation of the cortex, which property follows from the definition of the Voronoi polygon construction used to create this protomap. From Landau and Schwartz (1992).

Figure 23. Using the protocolumn algorithm and numerical conformal mapping, a binocular 2-DG experiment (Tootell *et al.*, 1988b) is simulated, and compared with the data. On the left is a computer simulation, using the protocolumn algorithm to model the columnar structure, and a conformal mapping to model the topographic structure, of a line that was imaged on the left and right retinas of a monkey, whose cortex was then physically flattened, and autoradiographically "developed." Experimental data from Tootell *et al.* (1988a). There is a close resemblance between the protocolumn simulation on the left, and the experimental data on the right. Note the "rippling" of the cortical image, due to the ocular dominance column pattern. This is apparently because the images of the line are more than a single column pair apart on the cortex, i.e., the cortical binocular disparity was greater than the size of a single hypercolumn. From Landau and Schwartz (1992).

cortex in such a way as to have a large macroscopic effect. We see no evidence of this, with the exception of the calcarine anomaly, and therefore assert that the most likely state of the local cortical shear is roughly 2:1 in layer IV, with the direction parallel to the column boundaries having a magnification factor that is twice that in the direction perpendicular to the local column boundary. This shear is then apparently "relaxed" by binocular mixing in the surrounding laminae, leading to the observation of an isotropic, or conformal cortex, as observed by a number of recent experiments (Dow *et al.*, 1981, 1985; van Essen *et al.*, 1984; Daniel and Whitteridge, 1961; Rosenbluth *et al.*, 1992), but not by Tootell *et al.* (1988b). To conclude this section, we present in Fig. 24 a computer simulation, using a numerical conformal mapping and the protocolumn algorithm of Landau and Schwartz (1992) to simulate a binocular spatial mapping, at the level of layer IVC of V1, using a natural scene which has been digitized to span the range from 0 to 100°, with a maximum resolution which corresponds to about 3 min of arc.

5. Applications: Machine Vision and Computational Anatomy

In the introduction of this chapter, it was stated that the interplay of computational methods and neuroscience was bidirectional. Up to this point, the use of techniques which have been developed in computer science and mathematics has been applied to the description of patterns in the nervous system. Now, the opposite direction of application will be illustrated. The space-variant or foveating architecture of the primate visual system has recently begun to be applied to the construction of high-performance machine vision systems. This area of research will now be briefly reviewed.

5.1. Vertical–Horizontal Pyramids

Variable resolution in spatial vision has been one of the principal themes in vision research for more than 30 years. Early in this era, this field bifurcated between "spatial frequency" enthusiasts and "space domain" partisans. There

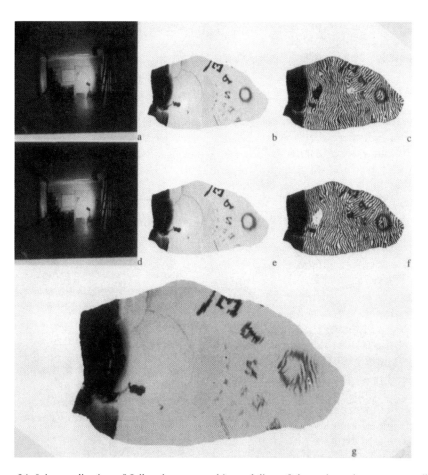

Figure 24. Joint application of full-scale topographic modeling of the retinostriate system, as illustrated in Fig. 12, with the protocolumn algorithm. This illustrates the concept of "polymap," as described in the text. At the top left (a, d) is shown a pair of (low-resolution) 100° scenes of a hallway. Panels b and e show the corresponding left and right eye topographic mappings. A disparity of 5 arc min has been introduced (simple lateral shift of one image, relative to the other, by 5 arc min). Panels c and f show the left and right eye protomaps, with the image data from the center mapped into the ocular dominance column pattern. In panel g, the left and right eye images are "or"ed together to form a single image, which represents a simulation of a binocular stimulus, on the surface of the cortex, when a small binocular disparity is present. Note that a 5 min of arc disparity is within the range of Panum's area, and is perceptually significant. The effect of the disparity on the cortical image is to produce a "ripple" whose magnitude is proportional to disparity, and whose "carrier" frequency is determined by the ocular dominance column spacing (see the "O" of the Snellen Chart). Farther into the periphery, the small disparity begins to have a negligible effect on the binocular "cortical image."

was a time when the Fourier transform was reified in vision, and it should be pointed out that a global convolution of this form is, in a certain sense, the antithesis of a spatial map. There was also much contention between physiologists who preferred a spatial frequency interpretation to the space-domain receptive field interpretation originated by Hubel and Wiesel. This contention reached its peak, if not its reductio ad absurdum, in a paper (DeValois and DeValois, 1988) which showed that cortical simple cells were linear in their response to gratings, but that sinusoidal gratings were "better" stimuli, since a sinuosoidal grating of a given amplitude had a larger cortical response than a slit of the same amplitude. It was soon pointed out that the normalization of a square slit and a sinusoid would be expected, on "energy" grounds, to differ by a factor of $\sqrt{2}$!, just as electrical waveforms which are sinusoidal have $\sqrt{2}$ more energy than square waves of the same amplitude. The choice of normalization is thus arbitrary. This showed how the visual system, which was asserted to be linear, could also show "a better response to" sinusoidal waveforms compared with square (i.e., edge) waveforms!

More recently, this contention has faded away. There are relatively few who still argue that a Fourier transform is somehow performed in the brain. And, the waning of this form of argument makes an interesting historical counterpoint to the area of cortical topography.

Thus, it has been pointed out (Schwartz, 1985a) that the variation of receptive field size across the surface of the cortex might be considered a form of "horizontal" multiple resolution representation, while the variation in receptive field size at a single location in the visual field might be considered a form of "vertical" multiple resolution. If so, then the two major features of spatial vision could be viewed from a unified perspective.

Donald Kelly has extensively explored this idea (Kelly, 1990) and has studied multiple resolution representation using Gabor functions, together with complex log mapping to represent the spatial structure of the visual field. This work represents the only computational attempt to date to unify the spatial frequency description of cortical receptive fields with the space-variant facts of cortical topography.

The theme of "vertical–horizontal" space-variance in vision provides a major thread in the context of machine vision. In fact, it can be argued that the only efficient approaches to real-time practical machine vision have been and, more to the point, will be based on space-variant architectures.

For example, Burt (Burt and Adelson, 1981; Anderson *et al.*, 1985) and his collaborators have provided "pyramid" algorithms for machine vision for more than a decade. By and large, multiscale, or pyramid approaches to vision appear to be the only practical means of realizing real-time performance. This is because most early vision processing can be performed on a coarse-to-fine grid, greatly decreasing the processing required, so that refinement at high acuity only needs to process a relatively small area of the sensor.

In the early work of Burt, the operative metaphor was that of "spatial frequency." A full image sensor was subsampled to provide perhaps four levels of a pyramid, with increasingly larger "receptive fields." Burt showed that this results in a total amount of pixels that was (Σ 1 + $^1/_4$ + $^1/_{16}$ + $\cdots$) = $^4/_3$ larger than the original single resolution frame. Therefore, a multiresolution pyramid "costs" little more than a conventional image, but provides a potentially large speedup on processing.

However, the original image itself is of order N^2, where N is a large number in the range of 256 to 1000 pixels. $^{4}/_{3} \cdot N \cdot N$ is a large number if $N \cdot N$ is a large number! This is the principal bottleneck in machine vision. Thirty frames per second of 0.256 million pixels is more than 6,000,000 pixels/sec, and each pixel must be processed in real time with many hundreds of machine instructions to complete a machine vision task.

More recently, Burt (1988a,b) and others have begun to use a so-called "truncated" pyramid, in which the higher-resolution levels of the pyramid are computed on smaller regions of the sensor. The highest-resolution level of this truncated pyramid thus exists on the smallest segment of the sensor area. In other words, this is a discrete approximation to a foveal architecture!

Yeshurun and Schwartz (1989) used the term "space-variant vision" to refer to machine vision performed with a sensor whose resolution varies smoothly across its surface, like that of the human visual system. The use of such sensors is rapidly becoming an important factor in machine vision: Burt's truncated pyramid is one example, but more recently many other applications of space-variant vision have begun to emerge. In the next section, we will review some work of our laboratory in this area, because it illustrates the potential that space-variant architectures have in machine vision, and also because it may shed important insight into the nature of the constraints and architectural principles that are operative in human vision.

5.2. Cortical Architecture and Machine Vision

The nature of computational function in visual cortex is, at present, almost completely unknown. Despite an impressive amount of knowledge at all levels of the nervous system, from the level of ion channels and membrane biophysics, neuron trigger features, columnar and topographic architectures, and multiple areas of visual function throughout extrastriate cortex, there is no confident answer, at the present time, to the question:

What are the computational functions of the visual cortex?

However, we can speculate on several aspects of visual computation which are supported by the particular form of space-variant mapping which occurs in the cortex. We will review several of these proposals, which have been advanced over the years, but will begin with one, which is certain beyond any reasonable doubt, which follows directly from the analysis presented in this chapter, and which has had practical consequences to the recent design of machine vision systems.

5.2.1. Ten Thousand Pounds of Brain

The human visual system is able to cover a wide visual field, and achieve high maximum resolution, without the need for an unreasonably large number of spatial channels. The foveating, or space-variant construction of V1 provides a dramatic form of data compression. Just how dramatic this compression is can be seen from the following simple estimate. Suppose we wish to "cover" a solid angle which is comparable to human vision (let us say about 100 × 100°), with a maximum resolution of 1 min of arc. If we were to attempt this with conventional video sensor technology, we would require our sensor to have 6000 × 2 × 6000 × 2 pixels (the factor of 2 × 2 is for sampling). In Fig. 12, we have simulated a 16,000 × 16,000 pixel scene with the map function estimate from monkey V1, which is represented by a simulated cortical representation of 16,000 pixels in total. In other words, we have achieved, in Fig. 12, a compression of about 16,000:1, by the use of a space-variant sensing strategy similar to that used in primate vision (Schwartz *et al.*, 1988).

In a more careful analysis of this compression factor, Rojer and Schwartz (1990b) have defined a measure of sensor quality, which was termed F/R quality, defined as the ratio of sensor field of view to maximum resolution, as outlined above. This is a measure of the spatial dynamic range of a sensor. Estimating a primate visual field of 140° (vertical) and 200° (horizontal), the number of "pixels" in a complex log sensor such as the human retina is estimated to be about 150,000. This number is consistent with the number of fibers in the optic tract (about 1,000,000), since we have not accounted for color, on–off and off–on pathways, noncortical afferents in the optic tract, and redundancy of sampling. We believe that a count of about 10^5 "pixels," or "sampling units" or "spatial degrees of freedom" is consistent both with cortical topography and with the number of fibers in the optic tract.

Nakayama (1990) has also provided an estimate of the number of "pixels" required to encode contrast. He obtained an estimate of 25,000 "pixels," somewhat lower than ours, but he did not provide any details of his calculation.

The number of pixels in a conventional, space-invariant sensor (e.g., a TV sensor) of the same F/R ratio is 600,000,000.* These estimates for the space-variant and space-invariant pixel burden of vision sensors suggest that compression ratios of between 3500:1 and 10,000:1 are achieved.

Since the primate cortex is roughly 50% (exclusively) visual, and the human brain weighs about 3 lb, we estimate that our brains would weigh many thousands of pounds if we were to maintain the same spatial dynamic range, but used a space-invariant, or nonfoveal architecture.

Since wide-angle vision with high acuity would appear to be of great selective advantage, and since a brain which weighs 5000–30,000 lb is not, it appears that we have identified at least one indisputable functional correlate of visual cortex spatial architecture.

5.2.2. Space-Variant Active Vision

As we began to outline above, the use of space-variant architectures in machine vision has come to be an area of increasing importance. Starting with the pyramid algorithms of Burt, in the early 1980s, there has been an increasing realization in the machine vision community that multiresolution sensing may be critical to achieve high-performance, real-time machine vision.

---

*Shostak (1992) has reported the following "pixel" estimates for the full visual field, using a space-invariant (nonfoveal) architecture:

- Solid angle of human vision: 15,000°² [180° (horizontal) × 135° (vertical)]
- Maximum resolution: 0.5 min of arc
- Sampling factor: 2
- Space-invariant sensor size: 36,000 × 28,000 = 1,000,000,000 pixels

Shostak's estimate is larger than ours because he used an assumed 0.5 min of arc, rather than a 1 min of arc maximum resolution.

During the past several years, with support from DARPA's Artificial Neural Network Technology program, we have built a machine vision system which utilizes the complex log geometry as its sensing strategy. The system that we have constructed has established extremely high performance on certain measures, which we will now review.

5.2.3. Cortex-I

We have constructed a miniature space-variant active vision system, using a complex logarithmic sensing strategy (Bederson *et al.*, 1992b; Wallace *et al.*, 1993), which we call CORTEX-I. The benchmark application for this system was to acquire moving targets (automobiles), track them with the camera, and to use pattern recognition techniques to read the license plates of the cars as they drove past the camera system. In order to solve this problem, a series of hardware and algorithmic problems were solved. At the hardware level, a novel actuator design was produced and implemented, called the spherical pointing motor. This design produced a two-degree-of-freedom camera pointing device which was fast, very compact, and extremely inexpensive to produce. Additional hardware innovations were the production of subminiature camera and lens systems, and custom VLSI sensor and DSP-based image processing hardware. At the algorithmic level, it was necessary to develop attentional algorithms that were capable of locating an object of interest (e.g., a license plate) in a complex scene, containing moving objects, in real time, to track the object of interest, and to perform image processing and pattern recognition on the tracked object. This work is fully described in a recent series of papers (Bederson *et al.*, 1992a,b; Wallace *et al.*, 1993).

The license plate reading benchmark was achieved (Bederson *et al.*, 1992b) with a hardware system that occupied less than 0.5 cubic feet, weighed less than 10 lb, and cost roughly $2000 in parts to build, inclusive of video camera, lenses, motors, and computer system. A picture of CORTEX-I is shown in Fig. 25. A picture of the custom "eye" spherical motor and camera system is shown in Fig. 26, and a sample of the license plate reading task is shown in Fig. 27.

The most notable aspect of this system was its ability to perform a difficult machine vision task, in real time, with the support of only 12 MIPS of processing power. This system is roughly 10–100 times smaller, cheaper, and computationally less expensive (in MIPS) than other contemporary machine vision systems. The reason for this economy can be traced directly to the use of a space-variant sensing strategy. Our system processed only 1400 pixels per frame, instead of the usual 64,000 to 256,000 pixels common in machine vision. In effect, we exploited the same type of leverage, outlined above for the human visual system (although not quite the same magnitude). And, the scaling down of our system's cost and size must be understood to apply not only to the sensor, but also to the memory, the CPU power, and to almost all other aspects of the system.*

We are currently building a second-generation system, which will provide 200 MIPS or more, and which will be mounted on a miniature robot vehicle. We expect the performance of this system to affirm the practical significance of space-variant image architectures.

---

*Like the human eye, the log map sensor does not require high-quality optics off-axis, as conventional cameras do. This made possible very small and light lenses, which in turn allowed actuators to be very small and light. There are a number of synergistic benefits which followed from the complex log sensor geometry.

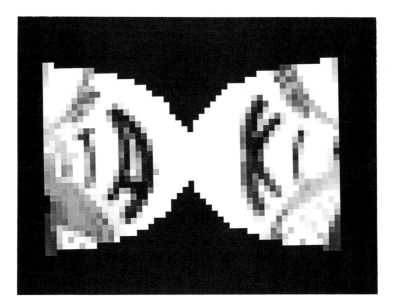

Figure 25. The computer engine and camera of the machine vision system CORTEX-I. On the top of the rack is mounted a miniature active camera system, shown in more detail in Fig. 26 The rack shown contains three DSPs, one microprocessor, and several boards of custom electronics, including motor control, communications, and camera control. At the time of this writing, this system is the smallest functional active vision system that has yet been constructed, by at least an order of magnitude. On the bottom is shown the output of this system, a license plate image. The system was successfully demonstrated to acquire moving targets (vehicles), to locate the license plates on the moving vehicles using a model-based "attentional algorithm," and to perform optical character recognition of the license plates, in real time while tracking the moving target.

Figure 26. A "spherical pointing motor" built to control the active vision system camera of CORTEX-I. The camera is shown in the center, and is about 5 × 10 *mm.* The whole system is about 1.5 inches on a side.

Figure 27. A frame from the final pattern recognition stage of CORTEX-I. This frame shows the OCR algorithm classifying the characters on a stationary license plate.

5.2.4. Commodity Robotics

Machine vision and robotics have been disappointing in the marketplace. This is not because they have failed to perform. It is because the relatively modest performance that contemporary machine vision provides is not cost-effective. Almost all vision and robotic tasks are more cheaply performed, at the present, by human workers. We have recently argued that the availability of extremely low-cost robotic and machine vision hardware, in the consumer price range, and with the functionality provided by the prototype CORTEX-I, will have a major impact on industrial and military applications, which we have likened to the impact of the low-cost PC on computation in general. We have called this (so far hypothetical) phenomenon the "PC metaphor" for machine vision, or the coming of "commodity robotics." For a number of years, we have predicted that high-performance vision robots will ultimately use space-variant, active vision architectures: the factors of hundreds or thousands in data reduction that this provides are simply too great to imagine any competition from a conventional space-invariant system. Whether this comes to pass remains to be seen. However, judging by the rapidly increasing interest in machine vision groups to build active vision systems in general, and space-variant or foveating systems in particular (e.g., Sandini and Dario, 1989; Sandini *et al.*, 1989; van der Spiegel *et al.*, 1989; Weiman, 1988b, 1990; Baloch *et al.*, 1991), we feel confident

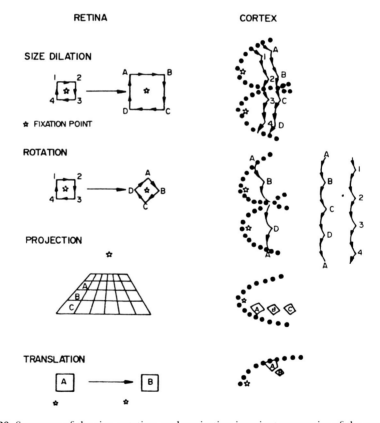

Figure 28. Summary of the size, rotation, and projection invariant properties of the complex logarithm. From Schwartz, (1980).

that this trend will be realized within the next decade. And if does, at least one area of biological vision will have made an indisputable contribution to high-performance machine vision.

5.2.5. Special Geometric Properties of the Complex Logarithm

One area of great activity has been the application of the special geometric properties of the complex log function to visual computation. These properties were known prior to the realization that primate cortex possessed complex log geometry, but were linked to the Fourier transform, via the "Mellin transform" (Brousil and Smith, 1967; Casasent and Psaltis, 1976). In the space domain, it was pointed out by Schwartz (1976, 1977c) and by Weiman and Chaikin (1979) that the complex log map possessed some intriguing size, rotation, and projection invariant properties. These are summarized in pictorial form in Fig. 28.

In a sense, the favorable size, rotation, and projection invariant properties of the complex log map are bought at the expense of the severe breaking of translation symmetry that is implied by a space-variant mapping. At the present time, it is not known whether the symmetry properties of the complex log are significant. For one thing, they follow only a complex log approximation to cortical topography that is valid outside of the foveal region (i.e., for visual angles greater than a, where our best guess for a is about 0.5°). Most likely, the geometric properties of the complex log mapping will find application in the simplification of optical flow, as sketched out in Fig. 28 (Schwartz, 1980). These aspects of the complex log have been most intensively investigated by Chaikin and Weiman (1980), Weiman (1988a), Sandini *et al.* (1989), and Jain (1986; Jain *et al.*, 1987).

6. Conclusion

In this chapter, a review has been presented of attempts to model, and to measure, the topographic and columnar structure of visual cortex. Since columnar structure appears to occur jointly with topographic structure in cortical sensory systems, we have briefly reviewed constructions, such as the protomap, protocolumn, and cortical polymap (Landau and Schwartz, 1992) which have been introduced recently to make algorithmic sense of structures, such as V1, which are both topographic and multimodal in a columnar fashion. Having introduced these ideas, we have a framework in which to understand visual cortex in terms of the simpler mathematical idea of "regular map," i.e., a map with a well-behaved Jacobian, and we have reviewed the history, the mathematics (both analytic and numerical), and the current experimental status of modeling such neural maps. Preliminary data have been presented which suggest that the topographic map of visual cortex is in fact quite well approximated by a conformal map whose structure is very close to the simple analytic model of the complex logarithm.

In order to fully present this idea, a fairly large amount of mathematical detail has been presented. The justification for doing this has been the primacy of this form of structure for understanding vision, and also the fact that the original papers presenting this work are scattered across several different literatures (computer graphics, physiology, neural modeling) in such a way that makes it very difficult for the interested reader to obtain a full understanding of these methods. The present chapter represents an attempt to summarize all of this material in one place, and the result is, clearly, quite a lot to expect of the reader. However, should one be interested in understanding the nature of spatial vision, and spatial patterns of functional architecture in visual cortex, it would appear that coming to terms with the material presented here is, in the long run, an essential step in understanding the basic nature of spatial representation in primary visual cortex.

ACKNOWLEDGMENT. This manuscript was prepared under support from NIMH No. 45969, which is gratefully acknowledged, as is earlier support from the System Development Foundation, and from the Air Force Office of Scientific Research, Life Sciences Directorate.

7. References

Ahlfors, L., 1966, *Complex Analysis*, McGraw–Hill, New York.

Allman, J. M., and Kaas, J. H., 1971, Representation of the visual field in striate and adjoining cortex of the owl monkey (*Aotus trivirgatus*), *Brain Res.* **35**:89–106.

Anderson, C. H., Burt, P. J., and van der Wal, G. S., 1985, Change detection and tracking using pyramid transform techniques, *Proc. SPIE Conf. Intelligent Robots and Computer Vision*, pp. 72–78.

Baloch, A. A., 1991, Visual learning: Adaptive expectations, and behavioural conditioning of the mobile robot mavin, *Neural Networks* **4**:271–302.

Bederson, B., Wallace, R., and Schwartz, E., 1992a, Two miniature pan-tilt devices, *IEEE International Conference on Robotics and Automation*, in press.

Bederson, B., Wallace, R. S., and Schwartz, E. L., 1992b, A miniaturized active vision system, in: *11th IAPR International Conference on Pattern Recognition*, Specialty Conference on Pattern Recognition Hardware Architecture, The Hague.

Blasdel, G. G., 1992, Differential imaging of ocular dominance and orientation selectivity in monkey striate cortex, *J. Neurosci.* **12**:3115–3138.

Blasdel, G., and Salama, G., 1986, Voltage sensitive dyes reveal a modular organisation in monkey striate cortex, *Nature* **321**:579–585.

Bonhoeffer, T., and Grinvald, A., 1991, Iso-orientation domains in cat visual cortex are arranged in pinwheel-like patterns, *Nature* **353**:429–431.

Bourdon, B., 1902, *La Perception Visuelle de l'Espace*, Scheicher, Paris.

Brousil, J. K., and Smith, D. R., 1967, A threshold-logic network for shape invariance, *IEEE Trans. Comput.* **16**:818–828.

Burt, P. J., 1988a, Algorithms and architectures for smart sensing, *Proc. DARPA Image Understanding Workshop*, pp. 139–153.

Burt, P. J., 1988b, Smart sensing within a pyramid machine, *IEEE Proc.* **76**(8):1006–1015.

Burt, P., and Adelson, T., 1981, A Laplacian pyramid for data compression, *IEEE Trans. Commun.* **8**:1230–1245.

Casasent, D., and Psaltis, D., 1976, Position, rotation and scale-invariant optical correlation, *Appl. Opt.* **15**:1793–1799.

Chaikin, G. M., and Weiman, F. R., 1980, Conformal computational geometry for machine vision, in: *Proceedings 5th International Conference on Pattern Recognition*.

Chinn, W. G., and Steenrod, N. E., 1966, *A First Course Topology*, Mathematical Association of America, Washington, D.C.

Coletta, N., and Williams, D. R., 1987, Psychophysical estimate of extrafoveal cone spacing, *J. Opt. Soc. Am. A* **4**:1503–1513.

Cowey, A., 1964, Projection of the retina on to striate and prestriate cortex in the squirrel monkey, *Saimiri sciureus*, *J. Neurophysiol.* **27**:366–293.

Daniel, M., and Whitteridge, D., 1961, The representation of the visual field on the cerebral cortex in monkeys, *J. Physiol. (London)* **159**:203–221.

DeValois, R. L., and DeValois, K. K., 1988, *Spatial Vision,* Oxford University Press, London.

Dow, B. M., Snyder, A. Z., Vautin, R. G., and Bauer, R., 1981, Magnification factor and receptive field size in foveal striate cortex of monkey, *Exp. Brain Res.* **44:**213–228.

Dow, B., Vautin, R. G., and Bauer, R., 1985, The mapping of visual space onto foveal striate cortex in the macaque monkey, *J. Neurosci.* **5:**890–902.

Fendick, M., and Westheimer, G., 1983, Effects of practice and the separation of test targets on foveal and peripheral stereoacuity, *Vision Res.* **23:**145–150.

Fox, P. T., Miezin, F. M., Allman, J. M., Essen, D. C. V., and Raichle, M. E., 1987, Retinotopic organization of human visual cortex mapped with positron emission tomography, *J. Neurosci.* **7:**913–922.

Frederick, C., and Schwartz, E. L., 1990, Conformal image warping, *IEEE Comput. Graphics Appl.* **March:**54–61.

Frostig, R. D., Lieke, E. E., Ts'o, D. Y., and Grinvald, A., 1990, Cortical functional architecture and local coupling between neuronal activity and the microcirculation revealed by in vivo high-resolution optical imaging of intrinsic signals, *Proc. Natl. Acad. Sci. USA* **87:**6082–6086.

Horton, J. C., and Hedley-White, E. T., 1984, Cytochrome oxidase studies of human visual cortex, *Philos. Trans. R. Soc. London Ser. B* **304:**255–272.

Hubel, D. H., and Freeman, D. C., 1977, Projection into the visual field of ocular-dominance columns in macaque monkey, *Brain Res.* **122:**336–343.

Hubel, D. H., and Livingstone, M. S., 1985, Complex-unoriented cells in a subregion of primate area 18, *Nature* **315:**325–327.

Hubel, D. H., and Wiesel, T. N., 1974, Sequence regularity and geometry of orientation columns in the monkey striate cortex, *J. Comp. Neurol.* **158:**267–293.

Jain, R., 1986, Complex log mapping and the focus of expansion, in: *ACM Siggraph/Sigart Interdisciplinary Workshop,* pp. 137–142. Elsevier/North-Holland, Amsterdam.

Jain, R., Bartlett, S. L., and O'Brien, N., 1987, Motion stereo using ego-motion complex logarithmic mapping, *PAMI* **3:**356–369.

Kelly, D. H., 1990, Retinocortical processing of spatial patterns, *SPIE Transactions on Human Vision and Electronic Imaging,* 1249.

Klein, S. A., and Levi, D. M., 1987, Position sense of the peripheral retina, *J. Opt. Soc. Am. A* **4:**1543–1553.

Koenderink, J., van de Grind, A. W., and Bouman, M. A., 1978, Perimetry of contrast detection thresholds and moving sine wave patterns, *J. Opt. Soc. Am.* **68:**845–865.

Landau, P., and Schwartz, E. L., 1992, Computer simulation of cortical polymaps: A proto-column algorithm, *Neural Networks* **5:**187–206.

LeVay, S., Hubel, D. H., and Wiesel, T. N., 1975, The pattern of ocular dominance columns in macaque visual cortex revealed by a reduced silver stain, *J. Comp. Neurol.* **159:**559–576.

Levi, D., Klein, S. A., and Aitsebaomo, A., 1985, Vernier acuity, crowding and cortical magnification, *Vision Res.* **25:**963–977.

Levi, D. M., and Klein, S. A., 1982, Vernier acuity, crowding and amblyopia, *Invest. Ophthalmol. Visual Sci.* **23:**398–407.

McKee, S., and Nakayama, K., 1984, The detection of motion in the peripheral visual field, *Vision Res.* **24:**25–32.

Miller, K. D., Keller, J. B., and Stryker, M. P., 1989, Ocular dominance column development: Analysis and simulation, *Science* **245:**605.

Nakayama, K., 1990, The iconic bottleneck and the tenuous link between early visual processing and perception, in: *Vision: Coding and Efficiency* (C. Blakemore, ed.), Cambridge University Press, London, pp. 411–422.

Osterberg, G., 1935, Topography of the layer of rods and cones in the human retina, *Acta Ophthalmol.* Supp. **6:**1–103.

Perry, V. H., and Cowey, A., 1985, The ganglion cell and cone distributions in the monkeys retina: Implications for central magnification factor, *Vision Res.* **25:**1795–1810.

Press, W. H., Flannery, B. P., Teukolsky, S. A., and Vetterling, W. T., 1988, *Numerical Recipes in C: The Art of Scientific Computing,* Cambridge University Press, London.

Rojer, A., and Schwartz, E. L., 1990a, Cat and monkey cortical columnar patterns modeled by bandpass-filtered 2d white noise, *Biol. Cybern.* **62:**381–391.

Rojer, A. S., and Schwartz, E. L., 1990b, Design considerations for a space-variant visual sensor with complex-logarithmic geometry, *10th International Conference on Pattern Recognition,* Volume 2, pp. 278–285.

Rosenbluth, D., Munsiff, A., Albright, T., and Schwartz, E., 1992, Computer reconstruction from 2dg serial sections of the topographic map of macaque visual cortex, *Soc. Neurosci. Abstr.* **18:**742.

Rovamo, J., and Virsu, V., 1979, An estimation and application of human cortical magnification factor, *Exp. Brain Res.* **37:**495–510.

Rovamo, J., Virsu, V., Laurinen, P., and Hyvarinen, L., 1978, Cortical magnification factor predicts the photopic contrast sensitivity of peripheral vision, *Nature* **271:**54–56.

Sandini, G., and Dario, P., 1989, Active vision based on space-variant sensing, *Int. Symp. Robotics Res.*

Sandini, G., Bosero, F., Bottino, F., and Ceccherini, A., 1989, The use of an anthropomorphic visual sensor for motion estimation and object tracking, *Proc. OSA Topical Meeting on Image Understanding and Machine Vision.*

Schein, S. J., and de Monasterio, F. M., 1987, Mapping of retinal and geniculate neurons onto striate cortex of macaque, *J. Neurosci.* **7:**996–1009.

Schwartz, E. L., 1976, Analytic structure of the retinotopic mapping and relevance to perception, *6th Annual Meeting of the Society for Neuroscience Abstracts* **6:**1636.

Schwartz, E. L., 1977a, Afferent geometry in the primate visual cortex and the generation of neuronal trigger features, *Biol. Cybernetics* **28:**1–24.

Schwartz, E. L., 1977b, The development of specific visual projections in the monkey and the goldfish: Outline of a geometric theory of receptotopic structure, *J. Theor. Biol.* **69:**655–685.

Schwartz, E. L., 1977c, Spatial mapping in primate sensory projection: Analytic structure and relevance to perception, *Biol. Cybern.* **25:**181–194.

Schwartz, E. L., 1980, Computational anatomy and functional architecture of striate cortex: A spatial mapping approach to perceptual coding, *Vision Res.* **20:**645–669.

Schwartz, E. L., 1981, Positron emission tomography studies of human visual cortex, *Soc. Neurosci. Abstr.* **7:**367.

Schwartz, E. L., 1983, Cortical anatomy and size invariance, *Vision Res.* **18:**24–58.

Schwartz, E. L., 1984, Anatomical and physiological correlates of human visual perception, *IEEE Trans. Syst. Man Cybern.* **14:**257–271.

Schwartz, E. L., 1985a, Image processing simulations of the functional architecture of primate striate cortex, *Invest. Ophthalmol. Visual Res. (Suppl.)* **26(3):**1–64.

Schwartz, E. L., 1985b, On the mathematical structure of the retinotopic mapping of primate striate cortex, *Science* **227:**1066.

Schwartz, E. L. (ed.), 1990, *Computational Neuroscience,* MIT Press, Cambridge, Mass.

Schwartz, E. L., and Merker, B., 1985, Flattening cortex: An optimal computer algorithm and comparisons with physical flattening of the opercular surface of striate cortex, *Soc. Neurosci. Abstr.* 15.

Schwartz, E. L., and Merker, B., 1986, Computer-aided neuroanatomy: Differential geometry of cortical surfaces and an optimal flattening algorithm, *IEEE Comput. Graphics Appl.* **6(2):**36–44 (March).

Schwartz, E. L., Christman, D. R., and Wolf, A. P., 1983, Human primary visual cortex topography imaged via positron-emission tomography, *Brain Res.* **104:**104–112.

Schwartz, E. L., and Rojer, A. S., 1992, A computational study of cortical hypercolumns and the topology of random orientation maps, *Soc. Neurosci. Abstr.* **18:**742.

Schwartz, E. L., Shaw, A., and Weinshall, D., 1987, Flattening visual cortex at image resolution: Quantitative computer reconstruction of the macaque ocular dominance column pattern, *Neurosci. Abstr.* p. 1293.

Schwartz, E. L., Merker, B., Wolfson, E., and Shaw, A., 1988, Computational neuroscience: Applications of computer graphics and image processing to two and three dimensional modeling of the functional architecture of visual cortex, *IEEE Comput. Graphics Appl.* **8(4):**13–28 (July).

Schwartz, E. L., Shaw, A., and Wolfson, E., 1989, A numerical solution to the generalized mapmaker's problem, *IEEE Trans. Pattern Anal. Mach. Intell.* **11:**1005–1008.

Shostak, S., 1992, The ultimate motion imaging system: What- and when? *Advanced Imaging* pp. 39–41.

Swindale, N. V., 1980, A model for the formation of ocular dominance column stripes, *Proc. R. Soc. London Ser. B* **208:**243–264.

Swindale, N. V., 1982, A model for the formation of orientation columns, *Proc. R. Soc. London Ser. B* **215:**211–230.

Talbot, S. A., and Marshall, W. H., 1941, Physiological studies on neural mechanisms of visual localization and discrimination, *Am. J. Ophthalmol.* **24:**1255–1263.

Tootell, R. B., Silverman, M., Switkes, E., and DeValois, R., 1982, Deoxyglucose analysis of retinotopic organization in primate striate cortex, *Science* **218**:902–904.

Tootell, R. B., Silverman, M. S., Switkes, E., and DeValois, R., 1985, Deoxyglucose retinotopic
mapping and the complex log model in striate cortex, *Science* **227**:1066.

Tootell, R. B. H., Hamilton, S. L., Silverman, M. S., and Switkes, E., 1988a, Functional anatomy of
macaque striate cortex. I. Ocular dominance, binocular interactions, and baseline conditions, *J.
Neurosci.* **8**:1531–1568.

Tootell, R. B. H., Switkes, E., Silverman, M. S., and Hamilton, S. L., 1988b, Functional anatomy of
macaque striate cortex. II. Retinotopic organization, *J. Neurosci.* **8**:1569.

Tusa, R. J., Palmer, L. A., and Rosenquist, A. C., 1978, The retinotopic organization of area 17
(striate cortex) in the cat, *J. Comp. Neurol.* **177**:213–236.

van der Spiegel, J., Kreider, F., Claeys, C., Debusschere, I., Sandini, G., Dario, P., Fantini, F., Belluti,
P., and Soncini, G., 1989, A foveated retina-like sensor using ccd technology, in: *Analog VLSI
Implementations of Neural Networks* (C. Mead and M. Ismail, eds.), Kluwer, Boston.

van Essen, D. C., Newsome, W. T., and Maunsell, J. H. R., 1984, The visual representation in striate
cortex of the macaque monkey: Asymmetries, anisotropies, and individual variability, *Vision Res.*
24:429–448.

Wallace, R., Ong, P.-W., Bederson, B., and Schwartz, E., 1993, Space variant image processing, *Int. J.
Mach. Vision,* in press.

Wassle, H., Grunert, U., Rohrenbeck, J., and Boycott, B. B., 1989, Cortical magnification factor and
the ganglion cell density of the primate retina, *Nature* **341**:643–646.

Watson, A. B., 1987, Estimation of local spatial scale, *J. Opt. Soc. Am. A* **4**:1579–1582.

Weiman, C. F. R., 1988a, 3-d sensing with polar exponential sensor arrays, *SPIE Conf. Digital and
Optical Shape Representation and Pattern Recognition.*

Weiman, C. F. R., 1988b, Exponential sensor array geometry and simulation, *SPIE Conf. Digital and
Optical Shape Representation and Pattern Recognition.*

Weiman, C. F. R., 1990, Video compression via log polar mapping, *SPIE Symposium on OE/Aerospace
Sensing,* pp. 1–12.

Weiman, C. F., and Chaikin, G., 1979, Logarithmic spiral grids for image-processing and display,
Computer Graphics and Image Processing **11**:197–226.

Weinshall, D., and Schwartz, E. L., 1987, A new method for measuring the visuotopic map function
of striate cortex: Validation with macaque data and possible extension to measurement of the
human map, *Soc. Neurosci. Abstr.,* p. 1291.

Wertheim, T., 1894, Uber die indirekte sehscharfe, *Z. Psychol. Physiol. Sinnesorg.* **7**:172–189.

Westheimer, G., 1979, The spatial sense of the eye, *Invest. Ophthalmol. Visual Sci.* **18**:893–912.

Westheimer, G., 1982, The spatial grain of the perifoveal visual field, *Vision Res.* **22**:157–162.

Weymouth, F. W., 1958, Visual sensory units and the minimal angle of resolution, *Am. J. Ophthalmol.*
46:102–113.

Wilson, H., Levi, D., Maffei, L., Rovamo, J., and DeValois, R., 1990, The perception of form in:
Visual Perception: The Neurophysiological Foundations, Academic Press, New York.

Winfree, A. T., 1987, *When Time Breaks Down,* Princeton University Press, Princeton, N.J.

Wolfson, E., and Schwartz, E. L., 1989, Computing minimal distances on arbitrary polyhedral surfaces, *IEEE Trans. Pattern Anal. Mach. Intell.* **11**:1001–1005.

Wright, M. J., 1987, Spatiotemporal properties of grating motion detection in the center and the
periphery of the visual field, *J. Opt. Soc. Am. A* **4**:1627–1633.

Yap, Y. L., Levi, D. M., and Klein, S. A., 1987, Peripheral hyperacuity: Three dot bisection scales to a
single factor from 0 to 10 degrees, *J. Opt. Soc. Am. A* **4**:1557–1561.

Yeshurun, Y., and Schwartz, E. L., 1989, Shape description with a space-variant sensor: Algorithms
for scan-path, fusion and convergence over multiple scans, *IEEE Trans. Pattern Anal. Mach. Intell.*
11:1217–1222.

10

Motion Processing in Monkey Striate Cortex

GUY A. ORBAN

1. Introduction

Motion processing, i.e., the processing of retinal image movement, is of great importance for primates (for review, see, e.g., Nakayama, 1985). In fact, motion processing could be considered fundamental to vision since retinal images are always moving as a result of micro eye movements, essential for visual perception. However, retinal image motion, whether generated by micro or macro eye movements, including pursuit and saccades, contains no information about the outside world. This is not the case for retinal image motion generated by the subject's own movements. The spatiotemporal changes in the retinal light distribution induced by relative movement between the observer and the environment, generated either by object motion or by self motion, are referred to as optic flow. Optic flow is a rich source of information about the outside world. It provides information about the 3-D trajectory of moving objects of the moving subject as well as about the 3-D structure of the environment. Furthermore, motion is a clear signal for image segmentation and perceptual grouping. In addition to its many perceptual uses, retinal motion also contributes to the control of eye movements, saccades as well as pursuit and optokinetic nystagmus. The term *motion processing* generally refers to the analysis of retinal image motion inasmuch as this leads to control of eye position and to extraction of information about the outside world.

GUY A. ORBAN • Laboratorium voor Neuro- en Psychofysiologie, K.U. Leuven, Medical School, B-3000 Leuven, Belgium.
Cerebral Cortex, Volume 10, edited by Alan Peters and Kathleen S. Rockland. Plenum Press, New York, 1994.

An early step in motion processing is to build a topographic representation of local velocity, i.e., the computation of the direction and speed of retinal image motion in every position of the visual field. By this I mean that a group of cortical neurons tuned for direction or speed represent the direction and speed of motion occurring at a retinal locus by virtue of their distributed activity. It seems that striate cortex (V1) contributes substantially to this process. A number of extrastriate areas involved in motion processing, such as MT (Rodman *et al.*, 1989; Girard *et al.*, 1992) and V3 (Girard *et al.*, 1991), receive their major input from V1, even though they also receive some direct input from the retina bypassing V1. Striate cortex (area 17) is the largest of all cortical visual areas (Van Essen *et al.*, 1992) and contains the most detailed representation of the visual field. Thus, the representation of local velocities in V1, while suffering many shortcomings (see below), has at least the enormous advantage of having a very fine spatial grain.

The contribution of V1 to lower-order motion processing, i.e., to building a local velocity representation, can be investigated with simple stimuli such as bars, gratings, or random dot fields. These are the stimuli which have been generally used in the study of area V1, and it is fair to say that most of these studies were restricted to the representation of local directions, neglecting more or less the representation of local speed. While it is tempting to conclude on the basis of these studies that V1 contributes only to low-level motion processing, it may be premature to do so, since the choice of stimuli may have biased the results. This bias is gradually being corrected since more and more studies have started to investigate the possible contribution of V1 to higher motion processing, using more complex stimuli and stimulus paradigms. Therefore, this chapter will be divided into three parts: (1) lower-order motion processing: direction selectivity, (2), lower-order motion processing: speed sensitivity, and (3) higher-order motion processing. Earlier qualitative studies revealed a number of important aspects of motion processing in V1 such as the occurrence of direction-selective cells (Hubel and Wiesel, 1968), the relatively low incidence of direction-selective cells, their laminar pattern (Dow, 1974), and the clear preference of V1 cells for slow speeds (Dow, 1974). The present review is based on more recent, quantitative studies, most of which were done in the Old World macaque monkey.

2. Lower-Order Motion Processing: Direction Selectivity

This is by far the most extensively studied aspect of motion processing in V1. Most of these studies have used elongated stimuli, either light or dark bars and edges introduced by Hubel and Wiesel (1959) or gratings introduced by Campbell *et al.* (1968). Only recently have random dot patterns been used.

2.1. Direction Selectivity Tested with Elongated Stimuli

The first quantitative study (Schiller *et al.*, 1976a) used moving edges, but most subsequent studies used either moving slits, i.e., light bars (Albright, 1984; Mikami *et al.*, 1986; Orban *et al.*, 1986), or drifting gratings (De Valois *et al.*,

1982; Foster *et al.*, 1985; Hawken *et al.*, 1988). All of these studies were done in the anesthetized, paralyzed preparation, except for the Mikami *et al.* (1986) and Poggio *et al.* (1977) studies and some of the experiments in the Schiller *et al.* (1976a) study. As shown by this latter study, there seems to be little difference between the two preparations with respect to direction selectivity.

2.1.1. Direction Selectivity and Orientation Tuning

Testing a neuron with an elongated stimulus, a line or a grating, moving in different directions (Fig. 1) entails changes not only in direction but also in orientation. Interpretation of the results is difficult but can be addressed in several ways. Orientation selectivity can be investigated in isolation using stationary stimuli. These, contrary to a claim often made, are as effective in driving V1

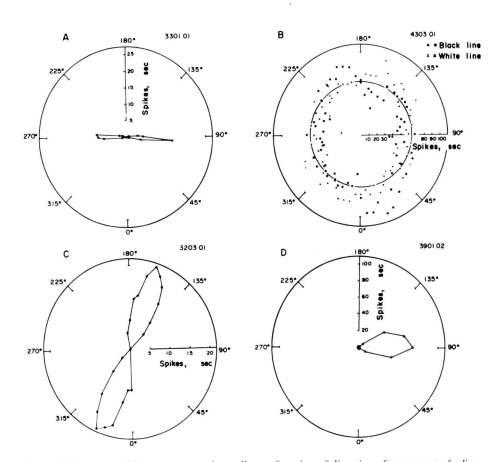

Figure 1. Responses of four macaque striate cells as a function of direction of movement of a line. The angle on the polar plot represents the direction while the radial dimension represents the response in spikes per second. The direction of the line or the grating is always at right angles to its orientation. Angles from 0 to 180° correspond to lines moving from right to left and/or from down to up; angles from 180 to 360° denote the same orientations but with movements from left to right and/or up to down. The first three cells (A, B, C) are non-direction-selective and vary in degree of orientation tuning from very narrow (A) through average (C) to no tuning (B). The cell in D was both orientation- and direction-selective. Cells in A and B had simple RFs, those in C and D complex RFs. From De Valois *et al.* (1982) with permission.

cells as moving stimuli (Albright, 1984). Direction-selective mechanisms can be studied in isolation with stimuli lacking overt orientation such as moving spots or random dot fields. Two arguments are commonly put forward in support of the view that in V1, but not necessarily in other areas (see Albright, 1984), orientation selectivity contributes to directional tuning measured using elongated stimuli. First, in V1 the optima obtained with moving slits and stationary slits are generally similar, although the width of tuning is not exactly the same (Albright, 1984). Second, the direction tuning obtained with moving spots, which have no orientation as such (Fig. 2), is wider than that obtained in the same cells with moving slits, which possess an intrinsic orientation (Schiller *et al.*, 1976b).

In V1 cells, directional tuning for elongated stimuli results from the operation of both orientation- and direction selective mechanisms (Fig. 1). The directional tuning benefits from the contribution of orientation-selective mechanisms in terms of increased narrowness of tuning. There is, however, a manner in which orientation and direction selectivity may be separated, even when using elongated stimuli. A given orientation corresponds to only a single axis of motion, usually taken orthogonal to the orientation, but along that axis it corresponds to two directions. Hence, orientation selectivity cannot contribute to the difference in response to opposite directions. Indeed, directional differences remain the same whether moving bars or spots are used (Schiller *et al.*, 1976b). Therefore, any difference in response to opposite directions, measured on the optimal axis (Fig. 3), may be taken as a specific measure of the direction-selective mechanisms operating in a V1 cell tested with elongated stimuli. This difference in response is referred to as direction selectivity (or directional preference). This is generally contrasted with directional tuning, which is the tuning as a function of direction, a parameter spanning a range of 360°. In conclusion, elongated stimuli drive both direction and orientation selectivity mechanisms. And thus when using such stimuli, only direction selectivity should be used as a measure of the contribution of V1 cells to the representation of local directions.

2.1.2. Incidence of Direction-Selective Cells

Direction selectivity is not an all-or-nothing property. All studies except perhaps that of De Valois *et al.* (1982) agree that the distribution of direction selectivity among V1 cells is unimodal. Two measures of direction selectivity have been used: either the simple ratio of net responses in opposite directions, termed here the direction ratio (DR) (Schiller *et al.*, 1976a; De Valois *et al.*, 1982; Foster *et al.*, 1985; Hawken *et al.*, 1988), or the direction index (DI). The DR, depending on whether or not the preferred direction is in the denominator, can be either a fraction, as in the Schiller *et al.* (1976a) study, or an integer (De Valois *et al.*, 1982; Foster *et al.*, 1985; Hawken *et al.*, 1988). In the first case, strong direction selectivity corresponds to near-zero values, in the second case to large values.

The direction index was introduced by Baker *et al.* (1981) and Orban *et al.* (1981b). It was defined as $DI = 1 - R_{np}/R_p$ (Baker *et al.*, 1981) or as $DI = 100 (1 - R_{np}/R_p)$ (Orban *et al.*, 1981b), where R_p and R_{np} are net responses in the preferred and nonpreferred direction, respectively. The response is either the average firing rate evaluated over a relatively long interval (Albright, 1984) or the maximum firing rate within a relatively narrow bin (Orban *et al.*, 1986). Since small responses can produce spuriously large DIs, it is advisable to require responses to exceed some criterion, such as the significance level (Orban *et al.*, 1981b; Mikami *et al.*, 1986). The DI can range from 0 to 1, or from 0 to 100, depending on the exact definition. When the responses are measured in average firing rate over a relatively large interval (Albright, 1984), and motion in the nonpreferred direction inhibited spontaneous firing, the DI can exceed 1 or 100. A DI of zero indicates no direction selectivity and a large DI, close to 1 or 100 depending on the definition, indicates strong direction selectivity. The DI is more stable than the DR when responses in the nonpreferred direction become

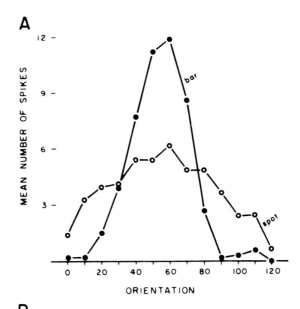

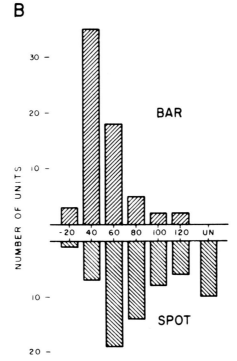

Figure 2. (A) Comparison of selectivity for orientation, and direction which covaries, tested with a moving bar and a moving spot. (B) Distribution of selectivity for orientation (or direction) with bars and spots. The cell population for each distribution is the same. From Schiller *et al.* (1976b) with permission.

very small and allows one to take into account inhibition. These seem to be strong arguments favoring the use of the DI to measure direction selectivity.

Irrespective of the exact criterion used to define direction selectivity, there seems to be a general agreement across the studies that about one-quarter to one-third of the V1 cells are direction selective. Using a DR of 2/1 as the criterion for direction selectivity, and gratings as stimuli, De Valois *et al.* (1982) reported a proportion of 29% direction-selective V1 cells. With the same technique, Foster *et al.* (1985) reported a proportion of 20%. In a third study with a comparable definition and stimulus, Hawken *et al.* (1988) reported a value of 28%. In an earlier study, the only one with unanesthetized monkeys, Poggio *et al.* (1977) had reported 26% of the cells responding to gratings to be direction-selective but they did not explicitly mention their criterion for direction selectivity.

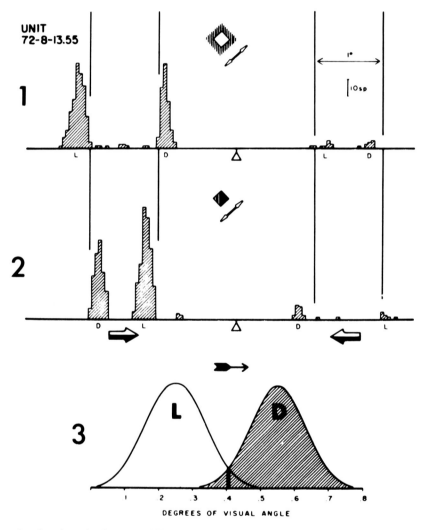

Figure 3. Direction-selective (or unidirectional) S2-type cell with separate subfields for light- and dark-edge. (1) Response to a 1° light square moving to and fro on the optimal axis. (2) Response to a 1° dark square. (3) Schematic drawing of the receptive field showing size, separation, and directionality of the two subfields. From Schiller *et al.* (1976a) with permission.

Still using a DR of 2/1 as criterion, but with edges as stimuli, Schiller *et al.* (1976a) obtained a much larger fraction (48%) of direction-selective cells. Albright (1984) reported a proportion of 31% direction-selective cells employing light bars as stimuli and using as criterion a DI of 66, corresponding to a 3/1 ratio. Orban *et al.* (1986) observed 27% direction-selective cells employing the same stimuli and the same DI of 66, but averaged over different speeds (see below). Again with a DI of 66 but now with random dot fields as stimuli, Snowden *et al.* (1992) reported 32% of the cells to be direction-selective (see below). While these studies reflect a rather nice agreement among studies in different laboratories, they also suggest that different types of stimuli might yield slightly different proportions of direction-selective cells. It seems that edges yield the largest proportion, and gratings the smallest.

2.1.3. Receptive Field Organization and Direction Selectivity

In all four studies (Schiller *et al.*, 1976a; Poggio *et al.*, 1977; Orban *et al.*, 1986; Hawken *et al.*, 1988) which have addressed the question, there is a small tendency for simple cells or simplelike cells to be more direction-selective than complex or complexlike cells. This result was obtained regardless of the criteria used to distinguish among receptive field organizations. Schiller *et al.* (1976a) and Orban *et al.* (1986) used the separation of discharge peaks for moving edges or moving bars of opposite polarity. The two other studies relied on the modulation of the response to drifting gratings as the criterion. Neither study demonstrated a large effect but the difference between simple and complex cells seems genuine because of the consistency of the findings across studies.

2.1.4. Laminar Dependence of Direction Selectivity

The first study to report that direction selectivity showed a laminar dependence was that of Dow (1974) who reported that many layer 4B neurons were direction-selective. Livingstone and Hubel (1984) also reported a high incidence of direction-selective cells in layers 4B and 6.

The first quantitative study, that of Poggio *et al.* (1977), reports that direction selectivity was more prevalent in layers 4, 5, and 6 than in layers 2 and 3. This was confirmed by two subsequent studies. Orban *et al.* (1986), using light bars, reported a higher incidence of direction-selective cells in layers 4B and 6, both of which project to MT, than in other layers. Hawken *et al.* (1988), using gratings, reported a high incidence of direction-selective cells in layers 6, 4Cα, 4A, and 4B in order of decreasing incidence (Fig. 4). They noted that layer 4A is very thin and that the allocation of cells to 4A and 4B is not always straightforward. Taken together, these results suggest that there is a concentration of direction-selective cells in the two horizontal bands of striate cortex which receive magnocellular geniculate input and project to MT: laminae 4B and 4Cα in the middle of the cortical depth, and lamina 6 lying at the bottom. It is worth noting that in these two bands reside smooth dendritic neurons whose axons spread horizontally over long distances (Lund, 1988). These cells resemble the "basket" cells of cat visual cortex and may play a role in the generation of direction selectivity.

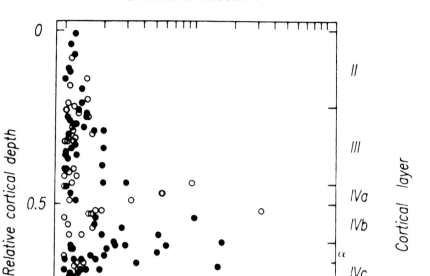

Figure 4. Direction selectivity (or directional preference) of 147 macaque striate cells is plotted as a function of laminar position. Direction selectivity is measured by the direction ratio (DR) calculated as the ratio of peak response to gratings drifting in the preferred direction to the response in the nonpreferred direction. Only cells in layer 6 and middle layer 4 show pronounced direction selectivity. (○) 54 complex cells plus 3 nonlinear nonoriented cells; (●) 81 simple cells plus 9 linear nonoriented cells. From Hawken *et al.* (1988) with permission.

2.1.5. Eccentricity Dependence of Direction Selectivity

Relatively little has been published about the eccentricity dependence of direction selectivity. Most studies, particularly quantitative studies, have been performed in the parts of V1 representing foveal or parafoveal vision.

Orban *et al.* (1986) compared direction selectivity in two samples. In one sample, the central sample, neurons had receptive fields within 2° from the fixation point. In the peripheral sample, neurons had receptive fields between 15 and 25° from the fixation point. The proportion of direction-selective cells dropped from 27% in the central sample to 5% in the peripheral one. On the other hand, Hawken *et al.* (1988) found the proportion of direction selective cells to be very similar in their foveal sample (eccentricity < 1.5°) and perifoveal sample (eccentricity 1.5–3.5°). Clearly, more work is required to understand the changes in direction selectivity with eccentricity.

2.1.6. Speed Dependence of Direction-Selective Cells

421

MOTION
PROCESSING
IN MONKEY
STRIATE CORTEX

It is well established that in the cat, the direction selectivity of area 17 and 18 neurons is speed dependent (Movshon, 1975; Orban *et al.*, 1981b; Duysens *et al.*, 1987; Orban, 1991). In many cortical neurons, direction selectivity decreases at lower speed. This is usually interpreted as a decrease in the sequential effects between neighboring inputs resulting from increased temporal separation between these inputs. A motion sequence contains both sequential or second-order effects, resulting from stimulation of neighboring retinal loci in sequence, and local or first-order effects, resulting from stimulation of each single locus on its own (Orban, 1986). The cortical neurons continue to respond at low speed since the local factors still produce responses. These responses will lack direction selectivity because of the failure of the sequential effects. At fast speeds, first-order factors fail because of the short duration of stimulation. Sequential effects also fail since the maximum spatial separation which allows interactions will be exceeded. Thus, direction selectivity and response both vanish together. A similar, clear-cut speed dependence of direction selectivity was noted by Orban *et al.* (1986) in V1 of the monkey. As shown in the examples taken from this study (Fig. 5), the increase in direction selectivity with increasing speed can be related either to decreased response in the nonpreferred direction or to an increase in response in the preferred direction.

On average, V1 cells lost their direction selectivity below 0.2°/sec in the central sample and below 0.7°/sec in the peripheral sample (Fig. 5). Mikami *et al.* (1986) have shown that for V1 cells the maximum Δt (temporal summation) between inputs measured with stroboscopic stimuli was 114 msec and did not depend on eccentricity. This suggests that central V1 cells receive input from neighboring points as close as 1.4 min arc in the central sample and 4.8 min arc in the peripheral sample.

Given the speed dependence of direction selectivity, it is difficult to characterize the direction selectivity by means of the DI at a single speed. Therefore, Orban *et al.* (1986) have advocated the use of the mean direction index (MDI; Orban *et al.*, 1981b) which is a weighted average of DI at different velocities, the strength of responses at each speed being the weighting factors. This index captures the behavior of the neuron over the whole range of speeds over which it is responsive.

2.1.7. Contrast Dependence of Direction Selectivity

The contrast dependence of direction selectivity is a way to test how invariant the direction-selective property is at the level of V1. It is also of considerable interest with respect to the mechanisms producing direction selectivity. One possible mechanism in the case of a light bar is the synergy between leaving an OFF area and entering an ON area (Albus, 1980). The synergy in the case of the light bar arises from the longer latency of responses to a light bar leaving an OFF area than responses entering an ON area (Orban *et al.*, 1985). In the opposite direction, this synergy would be absent and the response would be weaker. If direction selectivity for a light bar was based on such a mechanism, it should not be present for a dark bar.

Schiller *et al.* (1976a) investigated the contrast dependence of direction selectivity for edges and concluded that there was a significant interaction be-

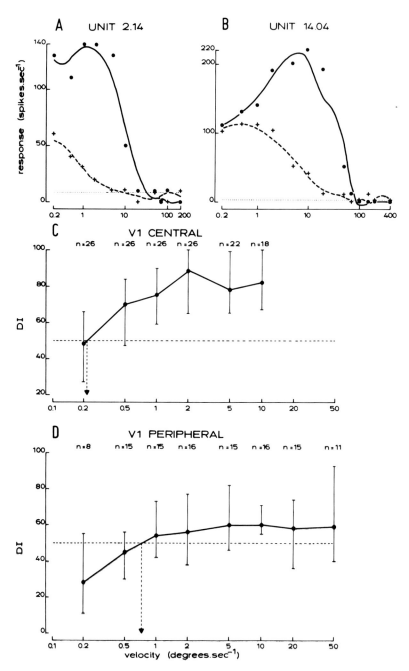

Figure 5. Loss of direction selectivity at low speeds. (A, B) Speed response curves (light bars) in preferred (dots, solid lines) and nonpreferred (crosses, dashed lines) direction of a cell of the central V1 sample (A) and peripheral V1 sample (B). (C, D) The DI of cells with MDI > 40 plotted as a function of stimulus speed (velocity) for the central V1 sample (C) and peripheral V1 sample (D). The median (dots) and first and third quartiles are indicated. Arrowheads indicate the speed at which on average V1 cells lose direction selectivity (DI = 50). Number of cells responsive at each speed are indicated above the graphs. From Orban *et al.* (1986) with permission.

tween these two properties in about half of the simple cells but not in complex cells.

Using both light and dark bars as stimuli, we have made similar observations. For a number of cells, direction selectivity depends on contrast polarity but for about an equal number this is not the case. As shown in Fig. 6, many V1 cells are non-direction-selective for either polarity, but 7 cells (N = 44) have an MDI exceeding 50 for both polarities and 6 cells have an MDI exceeding 50 for one polarity but not for the other. This is very different from area MT (Fig. 6) where most cells are direction-selective for both polarities (Lagae *et al.*, 1993).

Figures 7 and 8 show the speed–response curves for opposite directions of motion for light and dark bars in two V1 cells. For one neuron (Fig. 7), an end-stopped complex cell, the direction selectivity is invariant for contrast polarity. The other neuron (Fig. 8), an end-stopped simple cell, is direction-selective for the dark bar but not for the light bar. Notice that for this cell the above-mentioned synergy between leaving an OFF area and entering an ON area could explain the small difference in response to opposite directions for the light bar, but not the marked selectivity for the dark bar.

Given the contrast independence of MT direction selectivity, we have proposed the MDI averaged for light and dark bars, as a composite number charac-

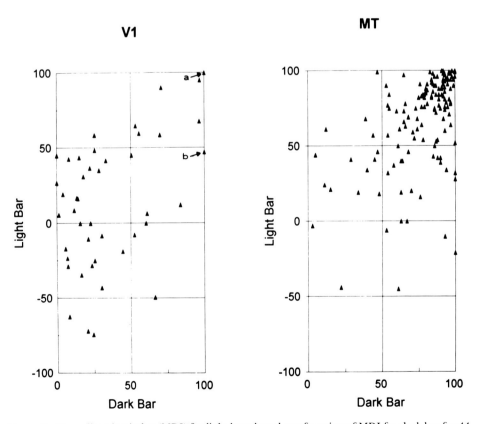

Figure 6. Mean direction index (MDI) for light bar plotted as a function of MDI for dark bar for 44 striate (V1) cells and 147 MT cells. "a" and "b" indicate the data points of the cells shown in Figs. 7 and 8, respectively. The MT cells are from Lagae *et al.* (1993). The striate cells are from Gulyás, Lagae, Raiguel, and Orban (unpublished).

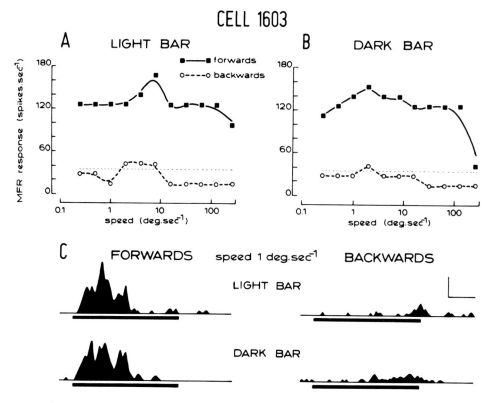

Figure 7. Striate neuron (HC cell of layers 2–3, RF eccentricity 5.7°) with direction selectivity invariant for contrast polarity. (A, B) Speed–response curves in preferred (forwards) and non-preferred (backwards) directions for (A) light bar and (B) dark bar. (C) Peristimulus time histograms for light and dark bars moving at 1°/sec. In A and B, the response is maximum firing rate measured in 8-msec bins; horizontal lines indicate significance level. In C, horizontal bars below histograms indicate motion duration; calibration bars indicate 50 spikes/sec and 500 msec. Gulyás, Lagae, Raiguel, and Orban (unpublished).

terizing direction selectivity for bars (Lagae *et al.*, 1993). The distribution of this average measure of direction selectivity is very different in V1 and MT (Fig. 9).

2.1.8. Properties of Direction-Selective Cells

Given the association of motion perception with the magnocellular stream (e.g., Maunsell and Newsome, 1987), Hawken *et al.* (1988) have tested whether direction-selective cells share some of the magnocellular properties. These authors observed that direction-selective cells are more contrast sensitive and tuned to lower spatial frequencies than nonselective cells. This was in line with the earlier observation of Foster *et al.* (1985) who also reported that the direction-selective V1 cells are tuned to lower spatial frequencies than non-direction-selective cells.

2.1.9. Mechanisms Producing Direction Selectivity

Schiller *et al.* (1976a) have argued convincingly that mechanisms producing direction selectivity differ from those producing orientation selectivity. The

CELL 1618

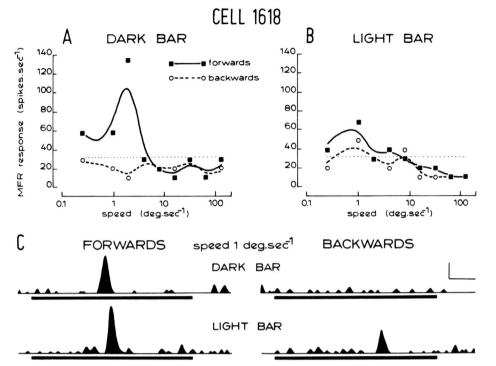

Figure 8. Striate neuron (layer 2–3 HS cell, RF eccentricity 5.8°) which is direction-selective for the dark bar but not the light bar. (A, B) Speed–response curves in preferred (forwards) and non-preferred (backwards) directions for (A) dark bar and (B) light bar. (C) Peristimulus time histograms representing responses to light and dark bars moving at 1°/sec. Same conventions as in Fig. 7, except calibration bars indicate 25 spikes/sec and 500 msec. Gulyás, Lagae, Raiguel, and Orban (unpublished).

nature of these mechanisms has, however, received relatively little attention in the monkey. Given the fact that most direction-selective V1 cells are also orientation-selective (Schiller *et al.*, 1976a; De Valois *et al.*, 1982), the mechanisms that have been proposed for the cat visual cortex are relevant here. Contrary to speed sensitivity, which depends on both first-order or local factors and

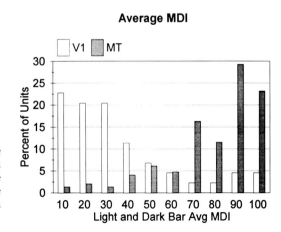

Figure 9. Distribution of the average MDI (averaged for light and dark bar) in 44 V1 and 147 MT cells. The MT cells are from Lagae *et al.* (1993). The V1 cells are from Gulyás, Lagae, Raiguel, and Orban (unpublished).

sequential factors, direction selectivity can depend only on sequential factors (Orban, 1986). These sequential factors can be either linear or nonlinear, i.e., resulting from interactions between neighboring inputs. An example of a linear mechanism is the superposition or synergy between leaving an area of one polarity and entering one of opposite polarity discussed earlier. Another example, which is probably more important, is the orientation of the RF in the space-time domain (Reid *et al.*, 1987). This refers to a systematic increase in latency, within an ON or OFF subfield, with changes in position in one direction. The relative contribution of linear and nonlinear sequential factors to direction selectivity is still undetermined. A complicating factor, as argued by Hamilton *et al.* (1989), is that some of the nonlinear interactions measured between neighboring points could reflect the distortion introduced into the output of the linear mechanism by the threshold nonlinearity of the spike generation.

The speed dependence of direction selectivity can be explained by the two types of sequential factors. In the linear case, the speed dependence reflects the degree of correspondence between stimulus speed and the speed specified by the RF orientation in space-time. In the nonlinear case, the speed dependence arises from the coincidence of the speed with the optimal spatial and temporal separation of the interacting inputs.

The importance of second-order interactions was demonstrated by the study of Mikami *et al.* (1986) who determined the maximum spatial and temporal separation between inputs allowing direction selectivity with stroboscopically illuminated moving light bars. They observed that in V1 the maximum temporal separation (Δt) was independent of eccentricity and averaged 114 msec. The maximum spatial separation (Δx) increased with RF eccentricity from 0.1° at 2° eccentricity to 1° at 25° eccentricity.

2.2. Directional Tuning Investigated with Random Dot Patterns

Only one recent study (Snowden *et al.*, 1992) has been published in which random dot stimuli were used in V1 of the alert macaque. We (Marcar *et al.*, 1992) are using similar stimuli in the anesthetized, paralyzed preparation. At this stage it seems worthwhile to draw attention to the need to describe this type of stimulus carefully. A random dot pattern as used by Snowden *et al.* (1992) should be distinguished from a random textured stimulus as introduced by Hammond and MacKay (1975) and Orban *et al.* (1975). In the latter case each pixel is randomly set to black or white and the ratio of black to white is exactly 50/50. In the former case, discrete elements, white (dark) dots, or points, are randomly positioned over a background of the opposite polarity. In this case the white/black ratio generally differs from 50/50, and the contrast polarity of the background dominates. This sort of stimulus has been used by Newsome and Paré (1988) since the signal-to-noise ratio can be easily manipulated by having only a fraction of dots moving coherently. For these random patterns, the pixel size or the dot size are important stimulus specifications, as these variables determine the power spectrum of the stimuli, which should be specified wherever possible. We have recently shown that MT cells respond well to a random dot stimulus with 6 minarc dot size (Marcar *et al.*, 1991). It is the same random dot stimulus we have begun to use in V1 (Marcar *et al.*, 1992).

As mentioned above, it is important to distinguish directional tuning from direction selectivity, which refers just to a difference in response to opposite directions on the optimal axis. Unfortunately, Snowden *et al.* (1992) report only the direction selectivity of V1 cells, and not their directional tuning when tested with random patterns. Their results (Fig. 10) confirm that V1 cells are indeed much less direction-selective than MT cells (see also Albright, 1984). They also measured the variability of responses, which determines neuronal discrimination capacity as effectively as does the degree of selectivity or response strength (Vogels and Orban, 1990), and they plotted the variance as a function of mean response at optimal direction. As was the case for responses at optimum orientation (Vogels *et al.*, 1989), the log variance increased linearly with log mean response, with an intercept close to zero and a slope of one.

Thus, there is little information on the directional tuning of V1 cells measured with random stimuli, We have begun to explore this issue, and two examples are shown in Fig. 11. Figure 12 shows examples of three MT cells for comparison. Although the V1 sample is still small, two observations emerge. First, V1 cells show little selectivity and the tunings are often erratic with multiple peaks (Fig. 11). Second, the directional tuning in V1 seems to depend much more on speed than that in area MT. Indeed, the direction and speed of motion seem to be separable in area MT (see Orban, 1992; Rodman and Albright, 1987). This, together with the V1–MT difference in contrast dependence, suggests that the purpose of the processing taking place in area MT is not so much to increase the degree of selectivity for speed or direction as to separate these two variables and make them less dependent on other stimulus attributes such as contrast polarity (Lagae *et al.*, 1993) or form-cue (Albright, 1992; Olavarria *et al.*, 1992).

The degree of neuronal selectivity is usually captured by the bandwidth of the tuning measured, e.g., as the width at half height (Henry *et al.*, 1974). Alternatively, a Gaussian can be fitted to the tuning curve, in which case the selectivity is given by the standard deviation (SD) of that distribution (Snowden *et al.*, 1992). These measures are unsatisfactory for several reasons. First, they fit only the response in the preferred direction and one then needs two numbers with which to specify the directional tuning: the DI and the bandwidth or SD. Second, these numbers are difficult to compute for irregular tunings. Third, no bandwidth can be calculated for cells with shallow tuning in which the response never decreases below 50% of the maximum. Therefore, we suggest that in order to measure the degree of selectivity in the directional tuning one should use the selectivity index derived from circular statistics: a single number characterizes both the width of tuning and the direction selectivity. The SI is given by the formula

$$\text{SI} = \frac{\sqrt{[\Sigma_x R_x \sin(x)]^2 + [\Sigma_x R_x \cos(x)]^2}}{\displaystyle\sum_{x-1}^{n} R_x}$$

and ranges from 0 (no selectivity) to 1 (response to only one direction). The formula shows that the SI is calculated on the actual data points and not on a curve fitted to those points, which is an additional advantage. The selectivity index first depends on direction selectivity: to exceed a value of 0.3–0.5, the response has to be reasonably direction-selective (i.e., more response for the optimal direction than the opposite); further increase in SI requires the response also to be narrowly tuned around the optimum direction (see Fig. 12). On

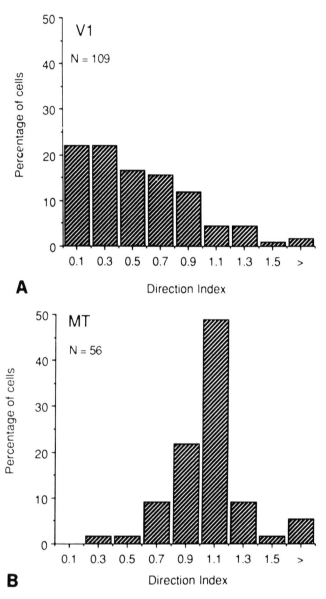

Figure 10. Distribution of direction index (DI) for cells in area V1 and MT. From Snowden *et al.* (1992) with permission.

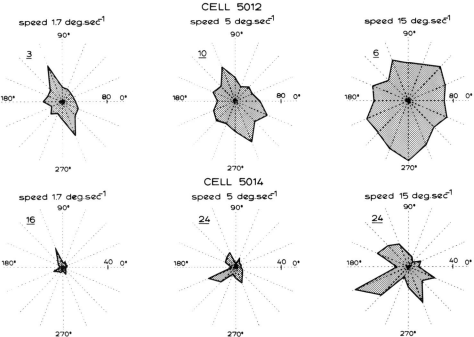

Figure 11. Direction tuning of two striate cells obtained with a random textured stimulus moving at three different speeds. The number underlined on the left of each plot indicates the selectivity index for direction (multiplied by 100). Notice the change in tuning with speed. From Marcar, Xiao, Raiguel, and Orban (unpublished).

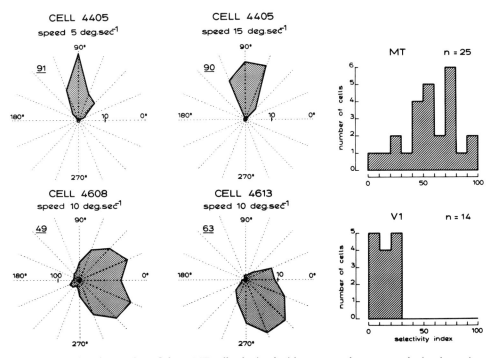

Figure 12. Direction tuning of three MT cells obtained with same random textured stimulus as in Fig. 11. Cell 4405 was tested at two speeds. Underscored numbers indicate the selectivity index (multiplied by 100). On the right are shown distributions of selectivity index for direction in 25 MT and 14 V1 cells. From Marcar, Xiao, Raiguel and Orban (unpublished).

the right-hand side of Fig. 12 we have plotted this selectivity index for 14 V1 cells and 25 MT cells tested with the same random dot stimulus and same speed (5°/sec). As predicted from previous studies, the direction SI is much larger in area MT than in V1.

3. Lower-Order Processing: Speed Sensitivity

Relatively little work has been done on speed selectivity of V1 neurons. Except for an unpublished study by Connolly and Van Essen mentioned in Van Essen (1985), there is as far as I am aware only one published study (Orban *et al.*, 1986) based on a collaboration between Leuven and Lyon. In Leuven, we have gone on to collect some data on contrast dependence of speed sensitivity which I will briefly mention here. Again a point of nomenclature must be made. In earlier studies (Movshon, 1975; Orban *et al.*, 1981a), the term *velocity* was used for changes in speed along a straight line. It is preferable to use speed for this scalar parameter and to keep velocity for the vector quantity which has a direction and a speed. In this way, speed and direction are the elementary parameters specifying local translation.

3.1. Motion Processing versus Processing of Images from Stationary Scenes

As mentioned earlier, it is important to realize, when using a paralyzed preparation, that not all neurons responsive to motion on the retina are involved in analysis of motion signals arising from object or subject motion. Indeed, during fixation there are micro eye movements by which the images of stationary objects are shifted over the retina at relatively low speeds (Skavenski *et al.*, 1975). Thus, cells responsive to slow motion in the paralyzed preparation will be activated by stationary stimuli in the alert preparation. In fact, many V1 cells, especially those in the part devoted to foveal vision, are responsive to very slow movements as noted before by Dow (1974). They correspond to what we have labeled velocity low pass (VLP) cells (Orban *et al.*, 1981a), defined as having a large response (exceeding 50% of maximum) at slow speeds and an upper cut-off speed (i.e., the speed beyond the optimal at which the response is decreased to 50% of maximum) below 20°/sec (Fig. 13). More than two-thirds of the foveal V1 cells belong to this group (Orban *et al.*, 1986).

The finding that there is a group of VLP neurons in the part of the visual area devoted to central vision is a very general one, since it has been observed in three areas of cat visual cortex: areas 17, 18, and 19 (Orban *et al.*, 1981a; Duysens *et al.*, 1982) and in three monkey visual cortical areas: V1, V2 (Orban *et al.*, 1986), and MT (Lagae *et al.*, 1993). However, the largest proportions of VLP cells occurred in area 17 of the cat and in V1 and V2 of the monkey (Fig. 13). The responses of these VLP V1 cells start to decline at about 3°/sec, and Orban *et al.* (1986) have suggested that these cells, in particular the non-direction-selective ones, underlie the acuity functions in vision. This would explain why in humans, acuity is equally good for slowly moving targets and for stationary ones: because the same velocity low pass cells are active in the two conditions. It would also

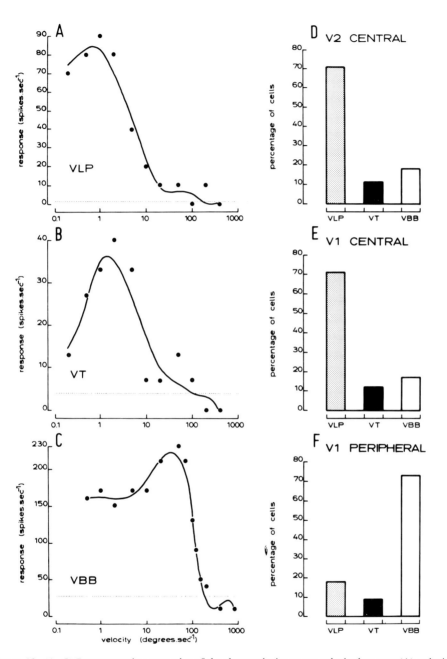

Figure 13. (A–C) Representative examples of the three velocity types: velocity low pass (A), velocity tuned (B), and velocity broadband (C); (D–F) proportion of the three velocity types in the three experimental samples: V2 subserving central vision (0–2° eccentricity) (D), V1 with same eccentricity (E), and V1 subserving peripheral vision (15–25°) (F). From Orban *et al.* (1986) with permission.

account for the fact that acuity decreases for speeds over $3°/\text{sec}$: because the population of VLP cells begins to be less responsive.

3.2. Measurement of Retinal Speed

It is quite clear that VLP cells are in fact not narrowly tuned, since their firing is generally constant between 0.2 and $2°/\text{sec}$. This is also the case for another group of cells which have been labeled velocity broadband (VBB) cells (Orban *et al.*, 1981a). These cells respond well over a wide range (100-fold or so) of speeds (Fig. 13; see also Fig. 7). There is, however, in the monkey a third group of cells, labeled velocity tuned (VT) cells, which are narrowly tuned for speed (tuning widths below 50). These cells are relatively uncommon (10%) in V1 (Orban *et al.*, 1986) but occur frequently in MT (Lagae *et al.*, 1993). I have suggested (Orban, 1985) that these cells underlie speed discriminations. This is supported by the fact that lesions of MT severely disrupt this ability in the monkey (Vandenbussche *et al.*, 1991).

3.3. Factors Influencing Speed Sensitivity

There are marked dependences of speed sensitivity on laminar and eccentricity location as reported by Orban *et al.* (1986). In foveal V1, the two layers, 4B and 6, projecting to MT contain far fewer (44%) VLP cells than other layers (84%). Thus, the speed characteristics of V1 cells in the layers projecting to MT are much more similar to those of MT cells than the characteristics of the overall V1 population. In addition, there is a profound influence of position of the RF in the visual field on the speed sensitivity of V1 neurons. The upper cut-off speed increases markedly with eccentricity so that cells with more peripheral RFs respond to faster speeds (Orban *et al.*, 1986). This is also a general finding for all five cortical visual areas—areas 17, 18, 19 of the cat and V1 and MT of the monkey—which have been studied so far. This represents a fundamental adaptation of the visual cortex to a behavioral pattern in which the animal frequently moves through the environment and thereby generates an optic flow pattern on its retina: this flow contains increasingly faster speeds toward the edge of the visual field.

Recently we have started to look at the contrast dependence of speed sensitivity in V1 neurons, and to compare it with that of MT cells. Although the V1 sample is still small, it is clear that the speed characteristics depend on contrast polarity much more in V1 than in MT (Fig. 14). In V1 cells there is reasonably good correlation between response strength for light and dark bars. The correlation between optimal or upper cut-off speeds for light and dark bars is far less in V1 than in area MT. This suggests that, as for direction selectivity, the processing in area MT does not so much increase the selectivity for speed but that it decreases the dependency of this selectivity on other attributes.

3.4. Mechanisms Underlying Speed Sensitivity

Hardly any research has addressed the mechanisms underlying speed sensitivity. Contrary to direction selectivity, both first-order (local stimulation) and

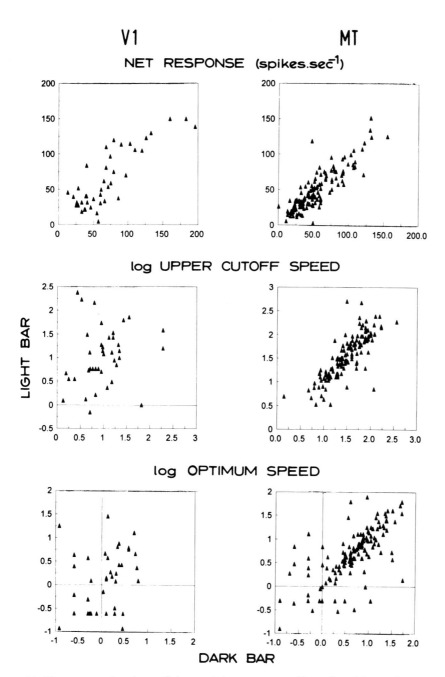

Figure 14. Net response (maximum firing rate), log upper cut-off speed, and log optimum speed compared for light and dark bars in 44 V1 cells and 147 MT cells. Correlation coefficients were 0.84 and 0.91 for responses in area V1 and MT, respectively, 0.2 and 0.81 for log upper cut-off speeds in area V1 and MT; and 0.33 and 0.69 for log optimum speeds in area V1 and MT. The MT cells are from Lagae *et al.* (1993); the V1 cells from Gulyás, Lagae, Raiguel, and Orban (unpublished).

second-order (sequence of stimulations) effects could contribute to speed sensitivity. The latter ones can be either linear or nonlinear. There is at least some indication that indeed first-order characteristics do contribute to speed sensitivity. In particular, it has been shown in the cat (Duysens *et al.*, 1985) that the speed dependence of VLP cells reflects a dependence on duration of illumination. A similar dependence on duration, but shifted to short durations, has been noted in monkey V1 (Orban, Bullier, and Kennedy, unpublished).

4. Higher-Order Motion Processing

Relatively few studies have been devoted to this question. Since most of these were comparative studies involving other areas, generally MT, the V1 samples were relatively small. These studies are, however, important for determining whether the specialization of V1 for lower-order processing is real or simply induced by the bias in the choice of stimuli.

4.1. Retinal Motion versus Stimulus Motion

As mentioned earlier, retinal motion does not necessarily correspond to object or even stimulus motion. This issue has been addressed experimentally by Galletti *et al.* (1984), who compared responses of V1 cells to a bar moving while the monkey fixated a stationary target and to the same bar moved over the RF by a pursuit eye movement in the opposite direction. These authors observed that in 90% of the V1 cells, responses were indistinguishable in the two cases. Ten percent of the V1 cells responded markedly less during pursuit than to stimulus movement. The role of these cells is not completely clear since motion perception arises both with the eyes being stationary (afferent motion perception) and during pursuit (efferent motion perception). It is noteworthy that in the two examples illustrated, direction-selective cells were tested with slow tracking eye movements (1°/sec). Perhaps these cells distinguish between object motion and retinal image motion induced by slow movements during fixation. The nature of the signal producing the difference in response is still unclear. The signal seems not to arise from an antagonistic surround (see below), since the background used in the experiments was an unstructured background unlikely to drive such a surround.

4.2. The Aperture Problem

Movshon *et al.* (1985) noted that a one-dimensional pattern such as a grating or a line is ambiguous with respect to its direction of motion: when viewed through a circular aperture, only motion orthogonal to the orientation can be perceived. This perceptual effect is referred to as the aperture problem. These authors went on to reason that V1 cells, because of the orientation selectivity of their receptive field, should suffer from a similar limitation: they should respond only to patterns moving in a direction orthogonal to their direction. Indeed, authors such as Schiller had reported that the preferred direction of a

V1 cell was orthogonal to its preferred orientation (Schiller *et al.*, 1976a). In order to test this idea, Movshon *et al.* (1985) compared the directional tuning for gratings and for plaids made by the superposition of two gratings at right angles (or at 135°). They reasoned that a cell which suffered from the aperture problem would be "component direction-selective" and yield a bimodal directional tuning for plaids. On the contrary, a cell capable of resolving the aperture problem would be "pattern direction-selective" and give a directional tuning with a single maximum for plaids just as it did for gratings.

These authors devised a quantitative method based on a comparison of partial correlation coefficients with the predictions for component and pattern selectivity, in order to distinguish between the two types of cells. This procedure left a substantial number of cells unclassified. The main observation, however, was that a quarter of the 108 MT cells tested fell into the "pattern direction-selective" category, while none of the 69 cells recorded in cat area 17 or monkey V1 were so categorized. Unfortunately, the number of V1 cells actually recorded was not specified. Nor was it specified what proportion of end-stopped cells were included in the V1 sample. Some caution is required in interpreting these results in that end-stopped cells can be direction-selective (Fig. 7) and might suffer less from the aperture problem than end-free cells. In a subsequent study, reported only in abstract form (Movshon and Newsome, 1984), these authors went on to show that the V1 neurons which actually projected to area MT, as shown by antidromic activation, were also "component direction-selective," just as were the other V1 cells. For obvious technical reasons, the numbers of cells tested in this second study were small.

4.3. Bar Texture Interactions and Antagonistic Surrounds

Antagonistic surrounds can be demonstrated in two ways. One way is to test a neuron with random dot or random textured stimuli of increasing diameters: beyond a certain diameter the response will decrease, indicating the presence of an antagonistic surround. These area summation curves are the 2-D extension of length response curves used to demonstrate end-stopping (Kato *et al.*, 1978). The alternative is to study the influence of a moving random background stimulus on a bar moving in the classical RF. These two techniques have been used in owl monkey MT (Allman *et al.*, 1985) and macaque MT (Tanaka *et al.*, 1986; Lagae *et al.*, 1989, 1990). For V1 neurons, which often respond only weakly to random stimuli, the second technique is the method of choice, and has been applied by Allman *et al.* (1990) in the owl monkey V1, and by our group in macaque V1 (Gulyás *et al.*, 1987; Orban *et al.*, 1989).

There seems to be a genuine species difference between the owl monkey and the macaque, as many V1 cells in the owl monkey have a surround while few do in the macaque. Allman *et al.* (1990) have reported that in the owl monkey, 4/5 of 21 cells recorded from in V1 were influenced by the textured background motion. On the contrary, such an interaction was absent in the majority of the 47 V1 cells recorded in the macaque (Orban *et al.*, 1989). Interestingly, two of the 6 macaque V1 cells which were influenced by the textured background motion reversed their preferred direction with the direction of the background motion (Fig. 15). We have labeled such cells antiphase cells (Orban *et al.*, 1987). Such cells have also been recorded in the monkey superior colliculus (Bender and David-

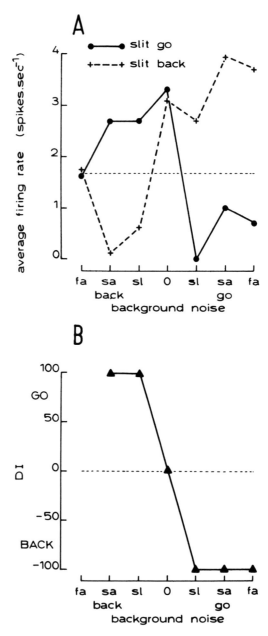

Figure 15. (A) Response of a striate neuron (S cell layer 2, RF eccentricity 5.4°) to a bar moving in forwards (go) and backwards (back) direction plotted as a function of the motion of a textured background: the background moved either forwards (go) or backwards (back) or was stationary (0), its speed was either faster (fa), the same (sa), or slower (sl) than the bar. Horizontal dashed line: significance level. (B) DI derived from data in A plotted as a function of background motion: this plot illustrates nicely the reversal in preferred direction with reversal in direction of background motion. From Gulyás, Lagae, Raiguel, and Orban (unpublished).

son, 1986) and we have also seen them in monkey V2 and V4t (Orban *et al.,* 1989) but not in area MT (Lagae *et al.,* 1989). Similar cells were observed by Allman *et al.* (1990) in V1 of the owl monkey: 7 of 21 cells behaved in this manner. Another 10 cells in owl monkey V1 were suppressed equally well by all directions of background motion.

The species difference between owl monkeys and macaques might be related to the nocturnal/diurnal life-style of the two species and the weakness of color vision in the owl monkeys. It has been suggested that in the cat, an animal with weak color vision (Loop *et al.,* 1987), horizontal intracortical connections (Gilbert and Wiesel, 1983) are implicated in the creation of these motion surrounds (Orban *et al.,* 1987). Given the importance of color vision in the macaque, comparable horizontal connections in V1 might be used to create color surrounds such as those of double opponent cells (Livingstone and Hubel, 1984) and modified type II cells (Ts'o and Gilbert, 1988) rather than motion surrounds. This anatomy would fit with the laminar distribution of direction selectivity in macaque striate cortex: direction selectivity is almost absent in laminae 3 and 5 in which horizontal connections are abundant. Horizontal connections also occur in layers 4B (Rockland and Lund, 1983) and 6 (Lund, 1973) and these might underlie some of the motion surrounds. Since the owl monkey has little color vision (Jacobs and Deegan, 1992), the horizontal connections in area 17 of this species, just as in the cat, might instead be used to generate motion surrounds.

4.4. Responses of V1 to Transparent Stimuli

One recent study by Snowden *et al.* (1991) has investigated the responses of V1 cells to transparent stimuli. These authors found that V1 cell responses to a set of random dots were very little affected by the addition of a second set of dots moving in the opposite direction. In contrast, MT cells were much more strongly influenced. This might be related to the difference in direction selectivity between V1 and MT (see above), since the inhibitory influences necessary to create the direction selectivity will be triggered by the additional set of dots.

4.5. Orientation Selectivity for Kinetic Boundaries

Neurons at later stages in the ventral pathway (V4 and IT) are selective for orientation of kinetic boundaries (Logothetis and Charles, 1990; Vogels *et al.,* 1992). On the other hand, MT cells are not selective for the orientation of kinetic boundaries (Marcar *et al.,* 1991), although lesions of MT seem to impair the perception of shapes defined by kinetic boundaries (Marcar and Cowey, 1992). In an effort to find the earliest cortical level at which this selectivity arises, we are presently investigating areas V1 and V2 (Marcar *et al.,* 1992). Up to now we have tested 14 V1 neurons and have observed no cells selective for the orientation of kinetic boundaries. In this case, as in many of the studies of higher-order motion processing, further work is required to elucidate the specific role of V1.

5. Conclusion

Motion processing is a relatively little-studied aspect of visual processing in V1. Striate cortex contains an extremely detailed map of the visual field. Only a fraction of the neurons in this map have the capacity to represent the two main parameters of translation: direction and speed. These direction- and speed-selective neurons occur mainly in layers projecting to area MT, an area largely devoted to motion processing. The motion representation in V1, although very detailed, is not invariant for other parameters, such as contrast polarity, and represents only the velocity component orthogonal to an elongated contour. Many of these shortcomings in the V1 representation are corrected at the next level, i.e., in area MT. Areas MT and V1 are active almost simultaneously (Raiguel *et al.*, 1989) and I propose that these areas act in concert to achieve both the precision and generality required of motion processing. At least in the macaque, there has been little evidence for any processing beyond lower-order processing primarily directed at building a local representation of speed and direction.

6. References

Albright, T. D., 1984, Direction and orientation selectivity of neurons in visual area MT of the macaque, *J. Neurophysiol.* **52:**1106–1130.

Albright, T. D., 1992, Form-cue invariant motion processing in primate visual cortex, *Science* **255:**1141–1143.

Albus, K., 1980, The detection of movement direction and effects of contrast reversal in the cat's striate cortex, *Vision Res.* **20:**289–293.

Allman, J., Miezin, F., and McGuinness, E., 1985, Direction- and velocity-specific responses from beyond the classical receptive field in the middle temporal visual area (MT), *Perception* **14:**105–126.

Allman, J., Miezin, F., and McGuinness, E., 1990, Effects of background motion on the responses of neurons in the first and second cortical visual areas, in: *Signal and Sense: Local and Global Order in Perceptual Maps* (G. M. Edelman, W. E. Gall, and W. M. Cowan, eds.), Wiley–Liss, New York, pp. 131–141.

Baker, J., Petersen, S., Newsome, W., and Allman, J., 1981, Visual response properties of neurons in four extrastriate visual areas of the owl monkey (Aotus trivirgatus): A quantitative comparison of medial, dorsomedial, dorsolateral and middle temporal areas, *J. Neurophysiol.* **45:**397–416.

Bender, D. B., and Davidson, R. M., 1986, Global visual processing in the monkey superior colliculus, *Brain Res.* **381:**372–375.

Campbell, F. W., Cleland, B. G., Cooper, G. F., and Enroth-Cugell, C., 1968, The angular selectivity of visual cortical cells to moving gratings, *J. Physiol. (London)* **198:**237–250.

De Valois, R. L., Yund, E. W., and Hepler, N., 1982, The orientation and direction selectivity of cells in macaque visual cortex, *Vision Res.* **22:**531–544.

Dow, B. M., 1974, Functional classes of cells and their laminar distribution in monkey visual cortex, *J. Neurophysiol.* **37:**927–946.

Duysens, J., Orban, G. A., van der Glas, H. W., and de Zegher, F. E., 1982, Functional properties of area 19 as compared to area 17 of the cat, *Brain Res.* **231:**279–291.

Duysens, J., Orban, G. A., Cremieux, J., and Maes, H., 1985, Velocity selectivity in the cat visual system. III. Contribution of temporal factors, *J. Neurophysiol.* **54:**1068–1083

Duysens, J., Maes, H., and Orban, G. A., 1987, The velocity dependence of direction selectivity of visual cortical neurones in the cat, *J. Physiol. (London)* **387:**95–113.

Foster, K. H., Gaska, J. P., Nagler, M., and Pollen, D. A., 1985, Spatial and temporal frequency selectivity of neurones in visual cortical areas V1 and V2 of the macaque monkey, *J. Physiol. (London)* **365:**331–363.

Galletti, C., Squatrito, S., Battaglini, P. P., and Maioli, M. G., 1984, 'Real-motion' cells in the primary visual cortex of macaque monkeys, *Brain Res.* **301:**95–110.

Gilbert, C. D., and Wiesel, T. N., 1983, Clustered intrinsic connections in cat visual cortex, *J. Neurosci.* **3:**1116–1133.

Girard, P., Salin, P. A., and Bullier, J., 1991, Visual activity in areas V3a and V3 during reversible inactivation of area V1 in the macaque monkey, *J. Neurophysiol.* **66:**1493–1503.

Girard, P., Salin, P. A., and Bullier, J., 1992, Response selectivity of neurons in area-MT of the macaque monkey during reversible inactivation of area-V1, *J. Neurophysiol.* **67:**1437–1446.

Gulyás, B., Orban, G. A., and Spileers, W., 1987, A moving noise background modulates responses of striate neurones to moving bars in the cat but not in the monkey, *J. Physiol. (London)* **390:**28P.

Hamilton, D. B., Albrecht, D. G., and Geisler, W. S., 1989, Visual cortical receptive fields in monkey and cat: Spatial and temporal phase transfer function, *Vision Res.* **29:**1285–1308.

Hammond, P., and MacKay, D. M., 1975, Differential responses of cat visual cortical cells to textured stimuli, *Exp. Brain Res.* **22:**427–430.

Hawken, M. J., Parker, A. J., and Lund, J. S., 1988, Laminar organization and contrast sensitivity of direction-selective cells in the striate cortex of the Old World monkey, *J. Neurosci.* **8:**3541–3548.

Henry, G. H., Bishop, P. O., and Dreher, B., 1974, Orientation, axis and direction as stimulus parameters for striate cells, *Vision Res.* **14:**767–777.

Hubel, D. H., and Wiesel, T. N., 1959, Receptive fields of single neurones in the cat's striate cortex, *J. Physiol. (London)* **148:**574–591.

Hubel, D. H., and Wiesel, T. N., 1968, Receptive fields and functional architecture of monkey striate cortex, *J. Physiol. (London)* **195:**215–243.

Jacobs, G. H., and Deegan, J. F., 1992, Cone photopigments in nocturnal and diurnal procyonids, *J. Comp. Physiol.* **171:**351–358.

Kato, H., Bishop, P. O., and Orban, G. A., 1978, Hypercomplex and the simple/complex cell classification in cat striate cortex, *J. Neurophysiol.* **41:**1071–1095.

Lagae, L., Gulyás, B., Raiguel, S., and Orban, G. A., 1989, Laminar analysis of motion information processing in macaque V5, *Brain Res.* **496:**361–367.

Lagae, L., Raiguel, S., Xiao, D., and Orban, G. A., 1990, Surround properties of MT neurons show laminar organization, *Soc. Neurosci. Abstr.* **16:**6.

Lagae, L., Raiguel, S., and Orban, G. A., 1993, Speed and direction selectivity of macaque middle temporal (MT) neurons, *J. Neurophysiol.* **69:**19–39.

Livingstone, M. S., and Hubel, D. H., 1984, Anatomy and physiology of a color system in the primate visual cortex, *J. Neurosci.* **4:**309–356.

Logothetis, N. K., and Charles, E. R., 1990, V4 responses to gratings defined by random dot motion, *Invest. Ophthalmol. Visual Sci.* **31**(4):90.

Loop, M. S., Millican, C. L., and Thomas, S. R., 1987, Photopic spectral sensitivity of the cat, *J. Physiol. (London)* **382:**537–553.

Lund, J. S., 1973, Organization of neurons in the visual cortex, area 17, of the monkey (Macaca mulatta), *J. Comp. Neurol.* **147:**455–496.

Lund, J. S., 1988, Anatomical organization of macaque monkey striate visual cortex, *Annu. Rev. Neurosci.* **11:**253–288.

Marcar, V. L., and Cowey, A., 1992, The effect of removing superior temporal cortical motion areas in the macaque monkey: II) Motion discrimination using random dot displays, *Eur. J. Neurosci.* **4:**1228–1238.

Marcar, V. L., Raiguel, S. E., Xiao, D., Maes, H., and Orban, G. A., 1991, Do cells in area MT code the orientation of a kinetic boundary? *Soc. Neurosci. Abstr.* **17:**525.

Marcar, V. L., Raiguel, S. E., Xiao, D., Maes, H., and Orban, G. A., 1992, Do cells in area V2 respond to the orientation of kinetic boundaries? *Soc. Neurosci. Abstr.* **18:**1275.

Maunsell, J. H. R., and Newsome, W. T., 1987, Visual processing in monkey extrastriate cortex, *Annu. Rev. Neurosci.* **10:**363–401.

Mikami, A., Newsome, W. T., and Wurtz, R. H., 1986, Motion selectivity in macaque visual cortex. II. Spatiotemporal range of directional interactions in MT and V1, *J. Neurophysiol.* **55:**1328–1339.

Movshon, J. A., 1975, The velocity tuning of single units in cat striate cortex, *J. Physiol. (London)* **249:**445–468.

Movshon, J. A., and Newsome, W. T., 1984, Functional characteristics of striate cortical neurons projecting to MT in the macaque, *Soc. Neurosci. Abstr.* **10:**933.

Movshon, J. A., Adelson, E. H., Gizzi, M. S., and Newsome, W. T., 1985, The analysis of moving visual patterns, in: *Pattern Recognition Mechanisms* (C. Chagas, R. Gattass, and C. Gross, eds.), Pontifical Academy of Sciences, Vatican City, pp. 117–151.

Nakayama, K., 1985, Biological image motion processing: A review, *Vision Res.* **25**:625–660.

Newsome, W. T., and Paré, E. B., 1988, A selective impairment of motion perception following lesions of the middle temporal visual area (MT), *J. Neurosci.* **8**:2201–2211.

Olavarria, J. F., DeYoe, E. A., Knierim, J. J., Fox, J. M., and Van Essen, D. C., 1992, Neural responses to visual texture patterns in middle temporal area of the macaque monkey, *J. Neurophysiol.* **68**:164–181.

Orban, G. A., 1985, Velocity tuned cortical cells and human velocity discrimination, in: *Brain Mechanisms and Spatial Vision* (D. J. Ingle, M. Jeannerod, and D. N. Lee, eds.), Nijhoff, The Hague, pp. 371–388.

Orban, G. A., 1986, Processing of images in the geniculocortical pathway, in: *Visual Neuroscience* (J. D. Pettigrew, K. J. Sanderson, and W. R. Levick, eds.), Cambridge University Press, London, pp. 121–141.

Orban, G. A., 1991, Quantitative electrophysiology of visual cortical neurones, in: *Vision and Visual Dysfunction,* Volume 4 (J. Cronly-Dillon, gen. ed., and A. G. Leventhal, ed.), Macmillan & Co., London, pp. 173–222.

Orban, G. A., 1992, The analysis of motion signals and the nature of processing in the primate visual system, in: *Artificial and Biological Vision Systems* (G. A. Orban and H. H. Nagel, eds.), Springer-Verlag, Berlin, pp. 24–56.

Orban, G. A., Callens, M., and Colle, J., 1975, Unit responses to moving stimuli in area 18 of the cat, *Brain Res.* **90**:205–219.

Orban, G. A., Kennedy, H., and Maes, H., 1981a, Response to movement of neurons in areas 17 and 18 of the cat: Velocity sensitivity, *J. Neurophysiol.* **45**:1043–1058.

Orban, G. A., Kennedy, H., and Maes, H., 1981b, Response to movement of neurons in areas 17 and 18 of the cat: Direction selectivity, *J. Neurophysiol.* **45**:1059–1073.

Orban, G. A., Hoffmann, K. -P., and Duysens, J., 1985, Velocity selectivity in the cat visual system. I. Responses of LGN cells to moving bar stimuli: A comparison with cortical areas 17 and 18, *J. Neurophysiol.* **54**:1026–1049.

Orban, G. A., Kennedy, H., and Bullier, J., 1986, Velocity sensitivity and direction selectivity of neurons in areas V1 and V2 of the monkey: Influence of eccentricity, *J. Neurophysiol.* **56**:462–480.

Orban, G. A., Gulyás, B., and Vogels, R., 1987, Influence of a moving textured background on direction selectivity of cat striate neurons, *J. Neurophysiol.* **57**:1792–1812.

Orban, G. A., Lagae, L., Raiguel, S., Gulyás, B., and Maes, H., 1989, Analysis of complex motion signals in the brain of cats and monkey, in: *Models of Brain Function* (R. M. J. Cotterill, ed.), Cambridge University Press, London, pp. 151–165.

Poggio, G. F., Doty, R. W., Jr., and Talbot, W. H., 1977, Foveal striate cortex of behaving monkey: Single-neuron responses to square-wave gratings during fixation of gaze, *J. Neurophysiol.* **40**:1369–1391.

Raiguel, S. E., Lagae, L., and Orban, G. A., 1989, Response latencies of visual cells in macaque areas V1, V2 and V5, *Brain Res.* **493**:155–159.

Reid, R. C., Soodak, R. E., and Shapley, R. M., 1987, Linear mechanisms of directional selectivity in simple cells of cat striate cortex, *Proc. Natl. Acad. Sci. USA* **84**:8740–8744.

Rockland, K. S., and Lund, J. S., 1983, Intrinsic laminar lattice connections in primate visual cortex. *J. Comp. Neurol.* **216**:303–318.

Rodman, H. R., and Albright, T. D., 1987, Coding of visual stimulus velocity in area MT of the macaque, *Vision Res.* **27**:2035–2048.

Rodman, H. R., Gross, C. G., and Albright, T. D., 1989, Afferent basis of visual response properties in area MT of the macaque. I. Effects of striate cortex removal, *J. Neurosci.* **9**:2033–2050.

Schiller, P. H., Finlay, B. L., and Volman, S. F., 1976a, Quantitative studies of single-cell properties in monkey striate cortex. I. Spatiotemporal organization of receptive fields, *J. Neurophysiol.* **39**:1288–1319.

Schiller, P. H., Finlay, B. L., and Volman, S. F., 1976b, Quantitative studies of single-cell properties in monkey striate cortex. II. Orientation specificity and ocular dominance, *J. Neurophysiol.* **39**:1320–1333.

Skavenski, A. A., Robinson, D. A., Steinman, R. M., and Timberlake, G. T., 1975, Miniature eye movements of fixation in rhesus monkey, *Vision Res.* **15**:1269–1273.

Snowden, R. J., Treue, S., Erickson, R. G., and Andersen, R. A., 1991, The response of area MT and V1 neurons to transparent motion, *J. Neurosci.* **11**:2768–2785.

Snowden, R. J., Treue, S., and Andersen, R. A., 1992, The response of neurons in areas V1 and MT of the alert rhesus monkey to moving random dot patterns, *Exp. Brain Res.* **88**:389–400.

Tanaka, K., Hikosaka, H., Saito, H., Yukie, Y., Fukada, Y., and Iwai, E., 1986, Analysis of local and wide-field movements in the superior temporal visual areas of the macaque monkey, *J. Neurosci.* **6**:134–144.

Ts'o, D. Y., and Gilbert, C. D., 1988, The organization of chromatic and spatial interactions in the primate striate cortex, *J. Neurosci.* **8**:1712–1727.

Vandenbussche, E., Saunders, R. C., and Orban, G. A., 1991, Lesions of MT impair speed discrimination performance in the Japanese monkeys (Macaca fuscata), *Soc. Neurosci. Abstr.* **17**:8.

Van Essen, D. C., 1985, Functional organization of primate visual cortex, in: *Cerebral Cortex* (A. A. Peters and E. G. Jones, eds.), Plenum Press, New York, pp. 259–329.

Van Essen, D. C., Anderson, C. H., and Felleman, D. J., 1992, Information processing in the primate visual system: An integrated systems perspective, *Science* **255**:419–423.

Vogels, R., and Orban, G. A., 1990, How well do response changes of striate neurons signal differences in orientation: A study in the discriminating monkey, *J. Neurosci.* **10**:3543–3558.

Vogels, R., Spileers, W., and Orban, G. A., 1989, The response variability of striate cortical neurons in the behaving monkey, *Exp. Brain Res.* **77**:432–436.

Vogels, R., Sáry, G., and Orban, G. A., 1992, Responses of inferotemporal units to luminance, kinetic and texture boundaries, *Invest. Ophthalmol. Visual Sci.* **33**(4):1131.

Temporal Codes for Colors, Patterns, and Memories

11

JOHN W. McCLURKIN,
JENNIFER A. ZARBOCK,
and LANCE M. OPTICAN

1. Introduction

We can see and understand complicated visual images without conscious effort. Other physical abilities, such as balancing, walking, and talking, are also effortless. One major difference between sensory and motor activity is that we can decompose our movements into a series of very small motions. This conscious decomposition allows our introspection to help us understand how we move. Unfortunately, we cannot decompose our visual perceptions into serial elements of seeing. Thus, introspection cannot help us understand how we see. The present work attempts to help us understand how the brain sees by decomposing vision into elements below our conscious perception: the activity of individual neurons.

In the first part of this chapter we will discuss how neurons in an early part of the visual system, the striate cortex or V1, encode information about the color and pattern of a stimulus. However, perception depends not only on the visual scene, but also on the goals and expectations of the observer. Thus, to fully understand vision, it is necessary to understand both the effects of stimuli and behavior on visual neurons. In the second part of this chapter we will consider how the act of discriminating among stimuli on the basis of color or pattern affects this encoding process.

JOHN W. McCLURKIN, JENNIFER A. ZARBOCK, and LANCE M. OPTICAN • Laboratory of Sensorimotor Research, National Eye Institute, National Institutes of Health, Bethesda, Maryland 20892

Cerebral Cortex, Volume 10, edited by Alan Peters and Kathleen S. Rockland. Plenum Press, New York, 1994.

2. Neuronal Encoding of Visual Information

Classical neurophysiology characterizes visual neurons by the area of the brain in which they are found, the size of their receptive field, and their selectivity for different stimulus features. However, it is well known that individual neurons are influenced by many different stimulus features (such as color and pattern), and even by the behavioral task the animal is performing. It is usually thought that these many different influences end up confounded within the activity of individual neurons. It would then be necessary to monitor the activity of a large population of neurons to sort out all of the different influences. We contend that this approach to characterizing neuronal function does not elucidate the neuronal mechanisms underlying visual perception.

The problem faced by the brain is not to sort out confounded features. Instead, because neurons in V1 have small receptive fields, the brain must ultimately join together local features into a global percept. This function would best be served if the individual neurons in V1 encoded unambiguous information about all of the visual features within their receptive fields in some way that would make them easily available in later visual areas. We suggest in this chapter that the temporal modulation of a neuron's activity may be used to encode unambiguous information about multiple stimulus features. This multiplex code hypothesis states that codes for different features (e.g., color, pattern, texture) coexist within the activity of a single neuron without confusion. Temporal correlation and filtering provide simple ways for neurons in later areas to decode or process these temporal messages about local visual features.

2.1. Mean Activity Confounds Features

The current theory of cortical visual processing states that neurons in the primary visual cortex are used to encode information about only the pattern of a stimulus or about only the color of a stimulus (Livingstone and Hubel, 1984). This theory is based in part on electrophysiological studies which have shown that neurons sensitive to color tend to be broadly tuned for orientation and spatial frequency. Conversely, those neurons which exhibit narrow orientation and/or spatial frequency tuning tend to have little color sensitivity (Dow and Gouras, 1973; Lennie *et al.*, 1990; Livingstone and Hubel, 1984). Further support for this theory was provided by studies using metabolic markers, which showed that achromatic stimuli produced different foci of labeling within area V1 than did chromatic stimuli (Tootell *et al.*, 1988a–c). Finally, neurons with similar color sensitivities and similar orientation preferences are grouped together (Michael, 1981; Ts'o and Gilbert, 1988). This theory is also consistent with the idea that neurons can encode unambiguous information about only a single stimulus parameter or a single combination of stimulus parameters (Barlow, 1972, 1985).

One limitation common to both the electrophysiological studies and the metabolic marker studies is that both consider only the strength of the neuronal responses. Electrophysiological studies usually measure only the number of spikes elicited by the stimuli, and the uptake of metabolic markers depends solely on mean activity level. If several different stimulus features, such as color and pattern, each modulate neuronal activity, then the mean level of activity they elicit must confound those features. In fact, neurons in V1 encode stimulus-related information in the distribution as well as in the number of spikes in their responses (Cattaneo et al., 1981; Eckhorn and Pöpel, 1975; Richmond and Optican, 1990). Further, we have shown that V1 neurons are capable of encoding unambiguous information about the pattern, the luminance, and the duration of a stimulus (Gawne et al., 1991b). We sought to extend this finding by determining whether the functional division of color and pattern processing defined on the basis of spike count would persist when the distribution of spikes was considered as well.

2.2. Responses to Colored Walsh Patterns

The usual method used to study the color and pattern sensitivity of visual neurons is to find some optimal pattern and then to present that pattern in a number of colors; or to find some optimal color and then to present a number of patterns in that color (Dow and Gouras, 1973; Lennie et al., 1990). (Classically, an optimal stimulus is simply the one that evokes the largest number of spikes in the response.) This technique can only provide a complete representation of neuronal function if the effects of color and pattern on the neurons are linearly separable. We chose not to assume linear separability of pattern and color, because our previous work showed that pattern and luminance were not linearly separable in complex cells recorded in primary visual cortex (Gawne et al., 1991b). Therefore, we used a factorial design to construct a set of 36 stimuli by combining six different patterns, based on Walsh functions,* with six different colors. The six colors were matched for apparent brightness (35.35 cd/m²). The black pixels of the patterns had a brightness of 0.087cd/m², and the background luminance of the screen was 18.00 cd/m².

The responses of one neuron (a complex cell) to these 36 stimuli are shown in Fig. 1 as continuous functions of neuronal activity. The curves in Fig. 1 give the probability that the neuron will generate a spike at that point in time after the stimulus onset (the spike density function) (Richmond et al., 1990). It is clear that the number of spikes in this neuron's response (proportional to the area under the curves) depended on the stimulus presented. However, it is equally clear from Fig. 1 that the distribution of spikes in the responses also depended on the stimulus presented.

To determine the relative importance of the number versus the distribution of spikes in the response, we computed the amount of information transmitted by the neuron using Shannon's information theory (Optican et al., 1991). Since the intrinsic code for transmitting visual information is not known, we computed information based on two codes. The first code was a univariate code based on

---

*Walsh functions are a set of orthogonal functions, varying in sequency, the number of times the function changes sign, that only take the values of -1 or 1 (Ahmed and Rao, 1975). The Walsh functions we used ranged in sequency from 0 to 7. Eight one-dimensional spatial patterns can be created from this set of Walsh functions by assigning black to -1 and a color to 1. Sixty-four two-dimensional patterns can be created by multiplying, pixel by pixel, a set of vertical and a set of horizontal one-dimensional patterns. The six patterns we used in these experiments were drawn from this set of 64 possible patterns. Because each Walsh function is orthogonal to all others, stimuli based on Walsh functions can be considered letters in an alphabet for patterns (Richmond et al., 1987).

the number of spikes, and the other was a multivariate code based on the number and temporal distribution of spikes in the response. We constructed the spike count code by counting the number of spikes in the 128-msec portion of the response defined by the thick, horizontal bar under the spike density functions. We constructed the temporal code from the first four components of the Karhunen–Loève (K-L) transform of the spike density function (Richmond and Optican, 1987). The K-L transform is similar to the Fourier transform in that both represent responses as the sum of a series of basis waveforms. The two transforms differ in that the basis waveforms of the K-L transform are the principal components of the data set rather than sines and cosines. We chose the K-L over the Fourier transform because the coefficients of the principal component waveforms are uncorrelated, and thus allowed us to determine if information carried by the spike distribution was independent of that carried by the spike count. The first component of this temporal code reflected the number of

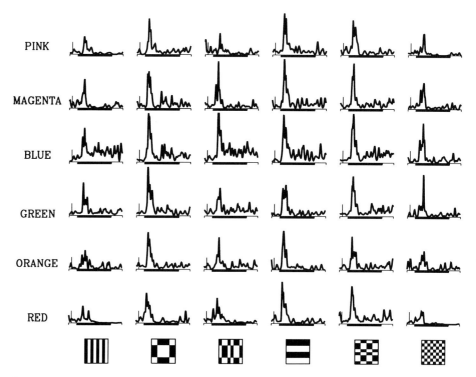

Figure 1. Responses of a neuron recorded in monkey primary visual cortex to the 36 stimuli used in these experiments. The pattern of each stimulus is indicated by the icon below each column of responses. The color of the white pixels is indicated by the label to the left of each row of responses. The intensity of each color was 35 cd/m², the intensity of the black pixels was 0.087 cd/m², and the background intensity was 18 cd/m². The responses are represented as spike density functions (a continuous estimate of the probability of spike occurrence throughout the response). The thin, vertical bars to the left of each spike density function indicate the time of stimulus onset; this bar's height corresponds to a probability of 0.1. The thick, horizontal bars indicate the 128-msec portion of the response used in the analyses. Our analysis began 20 msec after stimulus onset to allow for the neuron's response latency. The monkey's reaction time was approximately 200 msec after stimulus onset. Our analysis included only 128 msec of the neuronal activity to avoid any effects of eye movement. Note that the activity of this neuron was modulated in a complex way by both the color and pattern of the stimulus.

spikes in the response, and the higher components reflected aspects of the distribution of spikes in the response that were independent of the spike count.

For the neuron shown in Fig. 1, the code based on the spike count carried 0.310 ± 0.035 SE bit of information about the 36 stimuli. However, the temporal code carried 0.880 ± 0.088 SE bit of information. For all neurons in our sample, the spike count code carried an average of 0.250 ± 0.050 SE bit of information about the 36 stimuli in our set. In comparison, the temporal code carried an average of 0.980 ± 0.101 SE bit of information. Thus, the fluctuations in the waveforms of the spike density functions shown in Fig. 1 are not random, but contain stimulus-related information, indeed, more information than the spike count alone. Further, the information carried by the temporal distribution of spikes is independent of that carried by the spike count.

Previous studies have suggested that color and pattern have independent effects on V1 neuronal responses (Lennie *et al.*, 1990; Michael, 1978a,b). However, an examination of Fig. 1 shows that this neuron's responses to color depend on pattern and vice versa. For example, the differences among the six red Walsh patterns (bottom row) are larger than the differences among the six blue Walsh patterns. Further, the responses to patterns one and six (leftmost and rightmost columns) vary considerably with color, but the responses to patterns two and four do not vary much with color. Therefore, to obtain a measure of a neuron's ability to encode information about color independently of pattern, we pooled the responses to each color across all six patterns. Similarly, to obtain a measure of a neuron's ability to encode information about pattern independently of color, we pooled the responses to each pattern across all six colors. The pooled responses to the six colors and the six patterns for one neuron are presented as rasters and spike density functions in Fig. 2. It is clear from the rasters and spike density functions in Fig. 2 that the number of spikes in these pooled responses vary with color but not pattern. This figure also reveals differences in the temporal distribution of spikes among the responses to pattern.

We again used Shannon's information theory to quantify the degree to which single neurons could transmit information about both the color and the pattern of a stimulus, and how that information was distributed in the response. The temporal codes based on the K-L transform described above allowed us to determine that the information carried by the distribution of spikes was independent of that carried by the number of spikes in the response, but did not allow us to trace the development of this information over time in the response. Therefore, we constructed time-domain temporal codes by dividing each response into sixty-four 2-msec epochs, activity in each of which we considered as one letter in a temporal code (Chee-Orts and Optican, 1993). As with the temporal codes based on the K-L transform, the time-domain temporal codes reflected both the number and distribution of spikes in the response. For the neuron shown in Fig. 2, the spike count code carried more information about color than about pattern (Fig. 3, Spikes). In contrast, the 64-component temporal code carried nearly equal amounts of information about the color and the pattern of the stimulus (Fig. 3, Waveform).

To quantify the phenomenon illustrated in Figs. 2 and 3, that neurons appear to encode information about only one stimulus parameter when a spike count code is used but about both parameters when a temporal code is used, we computed a ratio of the relative amounts of color and pattern information a neuron transmitted by dividing the smaller value by the larger for both the spike count and 64-component temporal codes. This resulted in a series of ratios between zero and one. Smaller ratios indicate the tendency to encode only one stimulus parameter, whereas larger ratios indicate the tendency to encode both parameters. The average of the ratios for the spike count code was 0.33 ± 0.048 SE and the average of the ratios for the temporal code was 0.85 ± 0.025 SE. Thus, when only the number of spikes is considered, neurons in area V1 appear to encode information about pattern only or color only, but when the number and temporal distribution of spikes are considered, these neurons appear to encode information about both pattern and color. This result calls into question the idea that color and pattern information are separated into different streams in V1.

The neurons in our sample all encoded information about both the color and pattern of a stimulus in the temporal distribution of spikes, but information

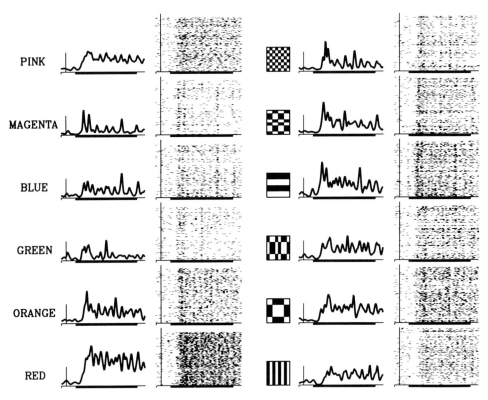

Figure 2. Pooled responses to pattern and color. On the left are the responses of a neuron to each color (indicated by the label) averaged across all six patterns. On the right are the responses of the same neuron to each pattern (indicated by the icon) averaged across all six colors. Each response is represented both as a spike density function and as a raster diagram (each dot represents a spike and each line of spikes represents the response to one trial). As with the spike density functions, the thin vertical bar indicates the time of stimulus onset and the thick horizontal bar indicates the portion of the response used in our analyses. Examination of the rasters indicates that there is clearly a greater difference in magnitude among the averaged responses to color than among the averaged responses to pattern. However, examination of the spike density functions reveals differences in spike distribution among the responses to the different patterns as well as among the responses to the different colors. The rasters of the averaged responses show that these differences in spike distribution among the different levels of each parameter are consistent across trials and across the levels of the other parameter.

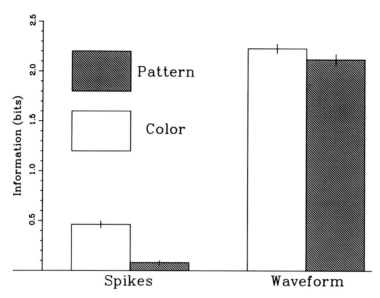

Figure 3. Color and pattern information. This graph shows the amount of color and pattern information transmitted by the neuron shown in Fig. 2, assuming a spike count code (Spikes) and a 64-component temporal code (Waveform). The spike count code carried 5.7 times as much information about color as about pattern (0.463 ± 0.04 SE bit versus 0.081 ± 0.02 SE bit), making it appear that this neuron is only encoding color information. However, the temporal code carried nearly equal amounts of information about color and pattern (2.227 ± 0.05 SE bits versus 2.118 ± 0.6 SE bits). In general, temporal codes carried much more information than did strength codes. Furthermore, the amounts of information about color and pattern were more nearly equal in the temporal codes than in the strength codes.

about these parameters did not develop at the same rate. To analyze the dynamics of color and pattern encoding, we constructed a series of time-domain temporal codes consisting of increasing numbers of 2-msec epochs of the response. The first code was based on the first epoch alone, the second code was based on the first two epochs considered jointly, etc. The last code was based on all 64 epochs considered jointly. These 64 information measures were averaged across all of the neurons in our sample and plotted as a function of time (Fig. 4). It is apparent that the information about pattern develops somewhat more rapidly than does the information about color early in the response, though the total information about each parameter is nearly the same.

2.3. Structure of the Color and Pattern Codes

The information measurements demonstrate that temporal modulation of the neuronal responses carries unambiguous information about both color and pattern, but they give no indication of how this information is encoded. Each neuronal response might be an instance of a unique, inseparable code for the color/pattern combination of the associated stimulus. Alternately, each neuronal response might be the result of multiplexing together a code for color with a code for pattern. Multiplexed codes would have the advantage of allowing complex stimuli to be encoded with far fewer code words. For example, our 36

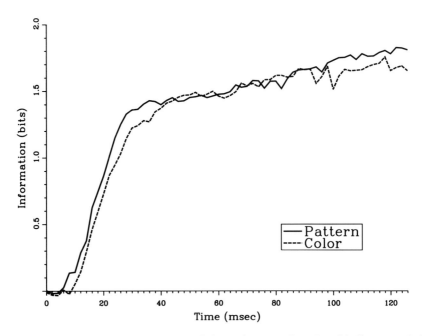

Figure 4. Development of color and pattern information over time. For this figure, we built 64 temporal codes by considering increasing numbers of consecutive 2-msec epochs in the response of the neuron. Thus, the first point represents the amount of information in the first 2 msec of the response, the information at 20 msec represents that carried by a temporal code composed of the first ten 2-msec epochs considered jointly, etc. Pattern information first appears 6 msec after the beginning of the response, grows rapidly for the next 24 msec, and then grows more slowly through the rest of the response. Color information first appears 10 msec after the beginning of the response, grows more slowly than pattern information for the next 20 msec, and only equals pattern information after 40 msec. Thereafter, information about both pattern and color grow at nearly the same rate for the duration of the response.

stimuli could be represented by a separable code with just 12 elements (six for color and six for pattern), but would require 36 inseparable codes. A separable code might also require only small populations of neurons to decode complex stimuli, rather than the large populations that would be required if inseparable codes were used.

To test the multiplex code hypothesis, we formulated the following model for the responses of V1 neurons:

$$R_{ij}(k) = C_i(k) \times P_j(k) + \bar{R}(k) \qquad (1)$$

where k represents the 64 time samples in the temporal response, $R_{ij}(k)$ is the waveform of the response to the stimulus with the ith color and jth pattern, $C_i(k)$ is the waveform representing the code for the ith color, $P_k(k)$ is the waveform representing the code for the jth pattern, and $\bar{R}(k)$ is the average of all the response waveforms elicited by all the stimuli. Thus, $C_i(k)$ and $P_j(k)$ represent that part of the response which is unique to the color or pattern of the stimulus. This model is illustrated in Fig. 5.

We fit our model to the data using nonlinear regression implemented by a three-layer neural network (Zarbock, 1992). This procedure was similar to the one we used for decomposing neuronal responses of neurons in inferior temporal cortex into pattern and memory codes (Eskandar *et al.*, 1992a). The network (Fig. 6) had two channels, one for stimulus color and one for stimulus pattern, which were combined in a single output layer. The first layer of each channel had six nodes, one for each of the six colors or six patterns. The hidden layer of each channel had 64 nodes, one for each point in the temporal codes for color or pattern [$C_i(k)$ or $P_j(k)$]. Within one channel, each of the six input nodes was connected by adjustable weights to each of the 64 hidden nodes. This allowed each color or each pattern to affect all 64 time samples of the response, thereby forming a temporal code for color or pattern. The two channels were combined into a single output layer, consisting of 64 nodes, by multiplying each point in the color channel's hidden layer with the corresponding point in the pattern channel's hidden layer. There were no adjustable weights between the two hidden layers and the output layer. Each of the 64 nodes of the output layer had an adjustable bias which converged to the average response of the neuron, $\bar{R}(k)$.

If the multiplex code hypothesis is correct, this model should be able to fit the individual responses very well, because the model has 12 degrees of freedom to match those of the response data. However, if the inseparable code hypothesis is correct, the data would have 36 degrees of freedom, and the model [Eq. (1)] could represent at most one-third of the responses, or just some average of the

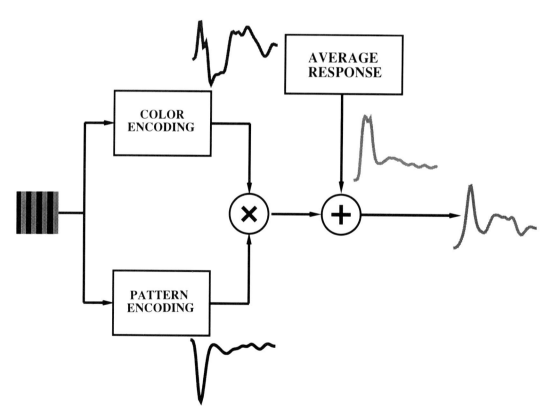

Figure 5. Multiplex temporal encoding hypothesis. According to this hypothesis, separate codes for the color and the pattern of a stimulus are generated by a neuron. These codes are multiplied together and then added to the average response of the neuron to arrive at the response to a particular stimulus.

responses. Thus, a pattern of uniformly high correlations between a neuron's responses and the model's fit to them would support the multiplex code hypothesis. On the other hand, a bimodal distribution of high and low correlations would support the inseparable code hypothesis. If the multiplex code hypothesis is correct, then our network should also be able to predict the responses to other stimuli made up of combinations of the same colors and patterns. However, if the inseparable code hypothesis is correct, then the network should not be able to predict the responses to novel stimuli. To test our hypothesis, we trained the network using 30 of the 36 stimuli (Fig. 7). The correlations between the network's output and the neuronal responses on which it was trained were uniformly high (Fig. 8). In addition, the correlations between the network's output and those responses not used in training were also high (Fig. 9). This ability both to fit all the neuronal responses and to predict responses to novel feature combinations provides support for the multiplex code hypothesis.

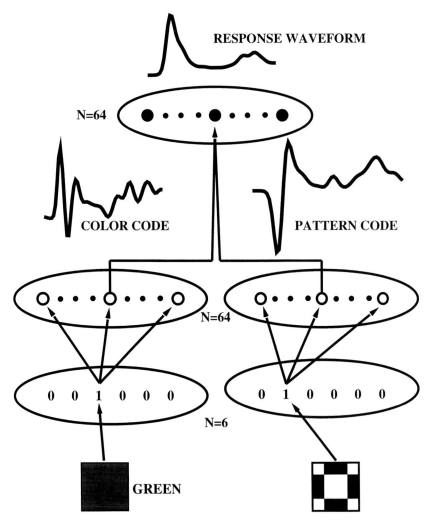

Figure 6. Neural net for nonlinear regression. We used a modified neural network to extract the color and pattern waveforms required by our model as described in Eq. (1). The network had two channels, one for the stimulus's color and one for its pattern. See text for explanation of the network.

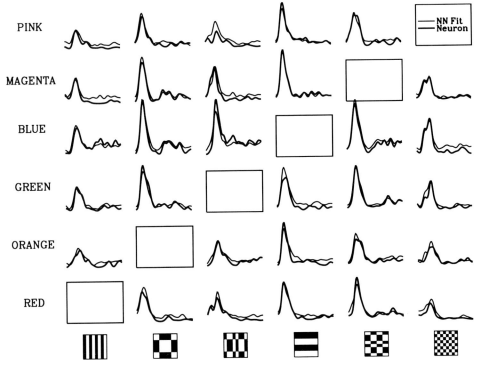

Figure 7. Model fits to the neuronal data for pattern and color of the target. The thick lines represent the responses of the neuron that were used to train the network, and the thin lines represent the responses of the network after training. The icons indicate the pattern, and the labels indicate the color of the stimulus that elicited each response. The empty boxes indicate combinations that were not used in training the network. This figure shows that, after training, the network was able to fit the neuronal responses quite well.

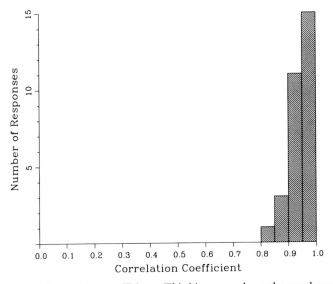

Figure 8. Frequency of correlation coefficients. This histogram shows the number of responses in each range of values for the data shown in Fig. 7. The correlations between the model's output and the neuron's responses to the 30 stimuli that were used to train the network are very high and are unimodally distributed. This supports our hypothesis that the neuronal responses can be decomposed into a small set of temporal codes for color and pattern.

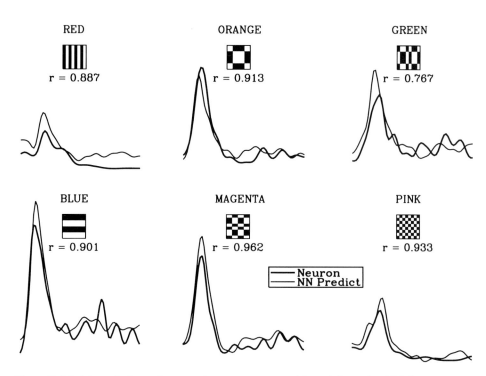

Figure 9. Model predictions of neuronal responses not used in training. The thick lines represent the responses of the neuron that were not used to train the network, and the thin lines represent the model's predictions of those responses. The label and icon above each set of functions indicate the stimulus that elicited the neuronal response. The correlation coefficients between the network's prediction and the neuron's response are also given above each pair of curves. All of these coefficients are statistically significant beyond the 0.001 level. Thus, the model embodied in Eq. (1) is a valid representation of the way that V1 neurons encode color and pattern.

The success of the model represented by Eq. (1) suggests that neurons encode information about stimuli by multiplexing together a small number of basic waveforms that represent the various stimulus parameters. To determine what these color and pattern code waveforms look like, we presented each of the six colors and patterns to the network after it had been trained and examined the pattern of activation across the 64 nodes of the hidden layers of each channel. An example of the color and pattern code waveforms for one complex cell is presented in Fig. 10. Note that many of these temporal codes for the colors or patterns are clearly differentiable on the basis of their waveform. The extent to which these waveforms can be differentiated from each other is what makes them good temporal codes for pattern or color.

3. Neuronal Encoding of Remembered Information

Previous studies that have examined the effect of behavior on visual processing have reported that, whereas neurons in cortical area V1 exhibit task-related effects, neurons in V1 do not (Moran and Desimone, 1985; Wurtz, 1969). However, these studies measured only the number of spikes in the neuronal responses. Because the distribution of spikes is clearly an important aspect of the response, we wanted to determine if V1 neurons could be shown to exhibit task-

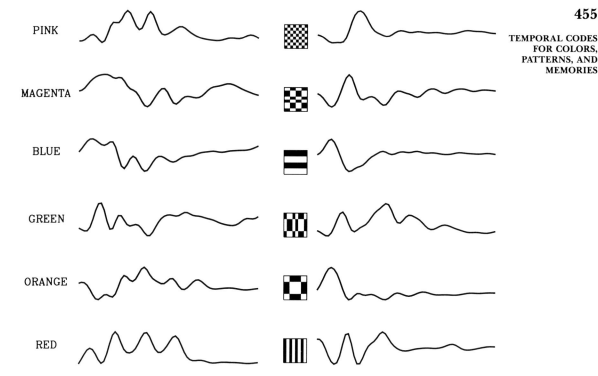

Figure 10. Temporal codes for color and pattern. The waveforms on the left are estimates of neuronal codes for the six colors (indicated by the labels), and the waveforms on the right are estimates of the neuronal codes for the six patterns (indicated by the icons) used in these experiments. These waveforms were obtained by examining the pattern of activation in the hidden layers of the color and pattern channels after the network had been trained.

related effects when the distribution of spikes was taken into consideration. We chose to examine the effects of a discrimination task on visual processing, because discriminating among a set of objects is one of the most important visual tasks in which an animal engages.

For the discrimination task used in these experiments, the monkey was given a foveal cue after he had fixated a small spot, and was then required to make a saccade to one of three peripheral targets while we recorded the activity of single V1 neurons. One of the targets was centered on the neuron's receptive field, and the other two were at equidistant points on a circle surrounding the fixation point (Fig. 11). The cues were colored squares or black-and-white Walsh patterns, and the targets were colored Walsh patterns. When color was the cue, the monkey's task was to make a saccade to the target that had the same color as the cue (Fig. 11, left). When pattern was the cue, the monkey's task was to make a saccade to the target that had the same pattern as the cue (Fig. 11, right). The location of the correct target was randomized so that the monkey could not use spatial cues to perform the task.

3.1. Responses to Remembered Colors and Patterns

To examine the effects of the cue on target responses, the responses of one neuron to the six stimulus colors (pooled across pattern) preceded by each of the six cue colors that were recorded while the monkey performed the discrimination task are shown in Fig. 12. The responses in each column were evoked by target stimuli having the color indicated below, and the responses in each row were preceded by a cue having the color indicated at the left. If this neuron had been insensitive to the cue, then all of the responses in each column would have been identical. However, there are clear differences among the responses in each column. To quantify the effect of the cue, we pooled the responses within each row, formed 64-component temporal and spike count codes as above, and calculated how much cue-related information each neuron transmitted using these two codes. The neuron shown in Fig. 12 transmitted no information about either the color (0.006 ± 0.02 SE bit) or the pattern (-0.003 ± 0.016 SE bit, not shown) of the cue in the spike count code, but this neuron did transmit information about both the color (1.65 ± 0.038 SE bits) and the pattern (1.74 ± 0.042 SE bits, not shown) of the cue in the temporal code.

Note that the cue did not fall within the receptive field of the neuron, and the neuron did not respond to the onset or offset of the cue. Nonetheless, the cue affected the responses of the neuron to the targets. What possible mechanisms could mediate a cue/target interaction? Long-range lateral connections have been demonstrated in the striate cortex. Therefore, in a discrimination task, any task-related effects could be the result of passive spread of visual–visual interactions within V1. To test the ability of this mechanism to explain our results, we conducted the discrimination task using two different cue–target sequences. For eight neurons, the cue was presented for 300 msec, then the

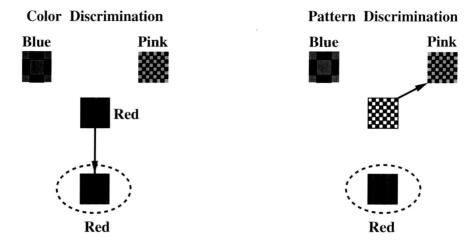

Figure 11. The discrimination tasks. The left side of the figure represents the response that the monkey should make when discriminating on the basis of color, and the right side represents the response that the monkey should make when discriminating on the basis of pattern. The dashed circle indicates the location of the neuron's receptive field. The arrow indicates the direction of a "correct" saccade. In both discrimination tasks, the monkey was first required to fixate a 0.2° black square (not shown). This square was then replaced by the central cue. The cue did not fall on the classical receptive field. In simultaneous discrimination tasks, the three peripheral stimuli were presented 300 msec after the cue turned on. The cue and stimuli then remained on until the monkey made a saccade. In sequential discrimination tasks, the cue was shown for 300 msec and then turned off. The peripheral stimuli were turned on 67 msec after the cue went off, and remained on until the monkey made a saccade.

targets were also turned on. In this case, the cue and the targets were on the screen together, so the task was a simultaneous matching of a current stimulus and a current cue. For 11 neurons, the cue was presented for 300 msec, but was turned off 67 msec before the targets were presented. In this case, the task was the sequential matching of the current stimulus with a previous cue. If task-related effects are caused by visual–visual interactions, then these effects should be weaker, and/or earlier in the sequential task than in the simultaneous task.

Across the eight neurons recorded in the simultaneous discrimination task, there was essentially no information about either the color (0.008 ± 0.005 SE bit) or the pattern of the cue (0.006 ± 0.003 SE bit) in the spike count code. This finding is consistent with previous studies which reported no task-related effects in V1 when only the number of spikes was considered. In contrast, there was cue-related information in the temporal code. The eight neurons recorded in the simultaneous discrimination task transmitted an average of 1.97 ± 0.34 SE bits of information about the color of the cue and 1.41 ± 0.18 SE bits of information about the pattern of the cue in the temporal code. The 11 neurons recorded in the sequential discrimination task also did not transmit any cue-related information in the spike count code (color: 0.004 ± 0.002 SE bit; pattern: 0.003 ± 0.004 SE bit), but did transmit cue-related information in the temporal code

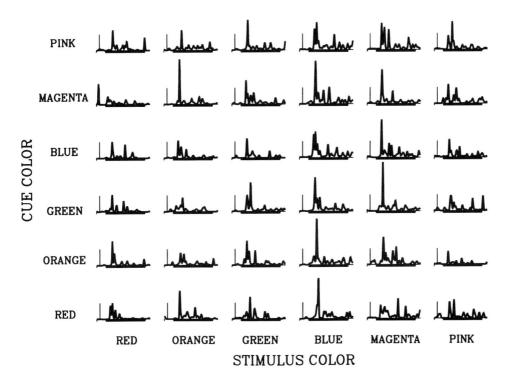

Figure 12. Responses to stimulus color as a function of cue color. These spike density functions represent the responses of a neuron to stimulus color (indicated by the labels below each column) averaged across stimulus pattern as a function of the color of the cue which preceded the stimulus (indicated by the labels to the left of each column). If the cue had had no effect on this neuron, the six responses in each column would have been the same. However, there are obvious differences in the response waveforms within a column, indicating that the foveal cue affected the response to the peripheral stimulus.

(color: 1.23 ± 0.17 SE bits; pattern: 1.36 ± 0.12 SE bits). The neurons in the simultaneous task did transmit significantly more information about the color of the cue than did the neurons in the sequential task ($t = 2.09$, $df = 17$, $p = 0.026$), but there was no difference in the amount of information transmitted about the pattern of the cue ($t = 0.73$, $df = 17$, $p = 0.239$). Thus, some of the information about the color of the cue was probably carried by lateral projections within V1. However, the large amount of information about the color of the cue transmitted by the neurons in the sequential task, and the similarity in the amounts of information transmitted about the pattern of the cue by the neurons in the simultaneous and sequential tasks argues that the effects of the cue on the responses of these neurons cannot be explained entirely by the lateral spread hypothesis.

A second method of determining whether the cue-related effects on the temporal code represent a passive spread of information laterally in the cortex, rather than the requirements of the task, is to examine where in the response the cue-related effects occur. In the simultaneous task, the cue was turned on 300 msec before the stimulus was turned on, but in the sequential task, the cue was turned on 367 msec before the stimulus to allow for the cue–stimulus gap. If the cue-related effects represent a passive, lateral spread of information, we would expect to see cue-related information earlier in the sequential task than in the simultaneous task. If, on the other hand, the cue-related effects are distributed in a similar fashion through the responses of both groups of neurons, we could conclude that these effects reflect the requirements of making the discrimination. To measure the time course of the cue-related effects, we measured the amount of cue-related information present in successive 2-msec epochs of the response. The results of this analysis are shown in Fig. 13 for a neuron recorded in the simultaneous discrimination task, and for one recorded in the sequential discrimination task. For both pattern and color cues, the cue-related information is distributed in a very similar fashion in the responses of both neurons. This suggests that the cue-related effects on the responses of these neurons are not the result of a passive spread of information via the lateral projection within V1, but rather represent a memory trace synchronized to the onset of the target stimulus. This memory trace may be provided through feedback connections from higher cortical areas.

3.2. Structure of the Memory Trace Codes

In the first part of this chapter, we presented evidence that supports the hypothesis that neurons represent compound stimuli by multiplexing together temporal codes for color and pattern. Now we have shown that the response to the target stimulus also contains information about the color or pattern of a preceding cue. If color and pattern are represented in the brain as temporal patterns of neuronal activity, then perhaps there is also a temporal code for the cue information. Indeed, recent studies in inferior temporal cortex have found temporal codes representing the pattern of a remembered stimulus and the pattern of the current stimulus multiplexed together (Eskandar *et al.*, 1992a,b). Thus, we hypothesize that temporal codes for the color and pattern of the cue are multiplexed with those for the target stimuli.

To test this hypothesis, we formulated the following models of the cue effects on the responses of V1 neurons:

$$RP_{ih}(k) = CP_i(k) \times Cq_h(k) + \bar{R}(k) \tag{2}$$

and

$$R^c_{jg}(k) = P^c_j(k) \times Pq_g(k) + \bar{R}(k) \tag{3}$$

where k represents the 64 points in the temporal code, $CP_i(k)$ are the temporal codes for color (averaged over pattern) and pattern (averaged over color), and $\bar{R}(k)$ is the average response waveform [see Eq. (1)]. $RP_{ih}(k)$ represents the response to the ith target color, averaged over all target patterns, when the hth color was the cue, and thus $Cq_h(k)$ represents the code for the hth cue color. $R^c_{jg}(k)$ represents the response to the jth target pattern, averaged over all target colors, when the gth pattern was the cue, and thus $Pq_g(k)$ represents the code for the gth cue pattern. Because each of the six stimulus patterns, independent of color, could be preceded by any of the six cue patterns, there were a total of 36

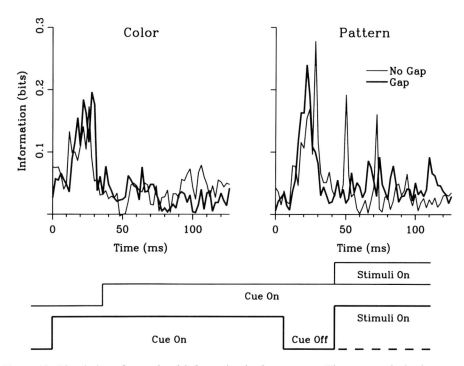

Figure 13. The timing of cue-related information in the response. The two graphs in the upper panel show the amount of color cue-related (left) and pattern cue-related (right) information at each 2-msec interval in the neuronal response. The thin lines plot the cue-related information transmitted by a neuron in the simultaneous discrimination task, and the thick lines plot the cue-related information transmitted by a different neuron in the sequential discrimination task. The traces at the bottom describe the temporal sequence of events in these two tasks. The cue-related information is distributed in the same way in the responses of the two neurons even though the cue itself was turned on 67 msec earlier in the sequential discrimination task. This suggests that cue-related information is maintained by some type of memory function and delivered when needed.

stimulus-pattern, cue-pattern combinations. Similarly, there were a total of 36 stimulus-color, cue-color combinations.

We fit these models to the data using the same nonlinear regression technique we used to determine the codes for stimulus color and pattern (Fig. 6). In this case, however, one channel of the network represented either the color or pattern of the stimulus and the other represented the color or pattern of the cue. If the variations among the responses to stimulus color as a function of cue color represent random fluctuations, then this regression technique might be able to fit a set of neuronal responses but would not be able to predict novel combinations of cue color and stimulus color. Therefore, to test our model for cue pattern, we trained the network using 30 of the $Rp_{ih}(k)$ responses, and then asked the network to predict the remaining six $Rp_{ih}(k)$. Similarly, to test our model for cue color, we trained the network using 30 of the $Rc_{jg}(k)$ responses, and then asked the network to predict the remaining six $Rc_{jg}(k)$. The ability of the network to fit the 30 neuronal responses to stimulus color by cue color is shown in Fig. 14. The correlations between the neuronal responses and the network outputs were uniformly high across these 30 responses (Fig. 15). The network was also able to predict the responses to novel cue-color combinations that had not been used in

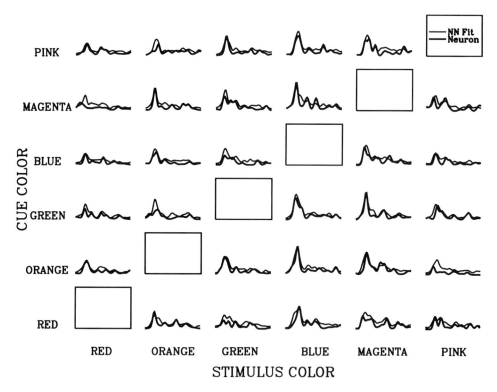

Figure 14. Model fits to the neuronal data for cue and target colors. The thick lines represent the responses of the neuron that were used to train the network, and the thin lines represent the responses of the model after training. The labels under each column of responses indicate the color of the stimulus, and the labels to the left of each row of responses indicate the color of the cue. The boxes indicate combinations that were not used in training the network. This figure shows that, after training, the network was able to reproduce both the magnitude and the temporal waveform of the neuronal responses.

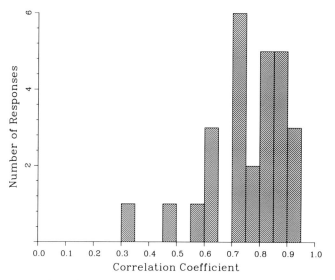

Figure 15. Frequency of correlation coefficients. This histogram shows the number of responses in each range of coefficient values for the data shown in Fig. 14. The correlations between the model's output and the neuron's responses to the 30 colors that were used to train the network are high and are unimodally distributed. This supports our hypothesis that the neuronal responses can be decomposed into a small set of temporal codes for cue color and target color.

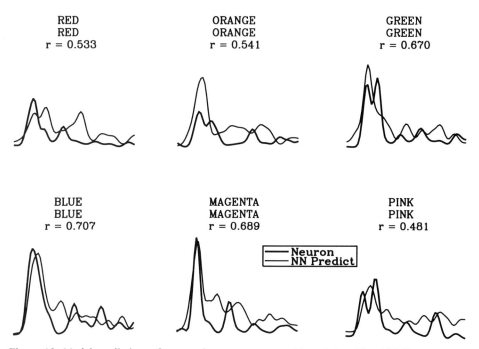

Figure 16. Model predictions of neuronal responses not used in training. The thick lines represent the responses of the neuron that were not used to train the network, and the thin lines represent the network's predictions of those responses after training. The labels above each set of functions indicate the color of the cue and the target that elicited the neuronal response. The correlation coefficients between the network's prediction and the neuron's response are also given above each pair of plots. All of these coefficients are statistically significant beyond the 0.001 level. Thus, the model embodied in Eq. (3) is a valid representation of the effect of a cue on the processing of color by V1 neurons.

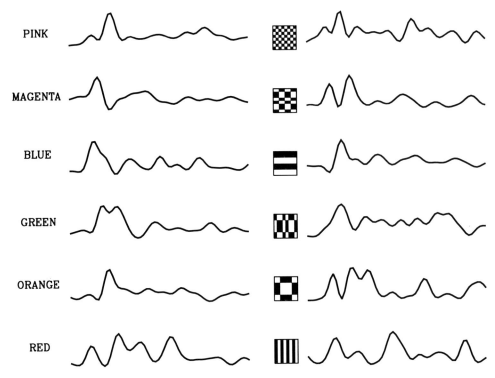

Figure 17. Codes for cue color and cue pattern. The waveforms on the left are our estimates of the temporal codes for the six cue colors (indicated by the labels), and the waveforms on the right are our estimates of the neuronal codes for the six cue patterns (indicated by the icons) used in these experiments. These waveforms were obtained by examining the pattern of activation in the hidden layers of the color and pattern channels after the network had been trained.

training (Fig. 16). This suggests that the cue-related response variations reflect information about the cue, and are not just random fluctuations. Examples of the code waveforms for the color of the cue and the pattern of the cue for one complex cell are presented in Fig. 17. Many of the codes representing the color and pattern of the cue stimuli are differentiable on the basis of their waveforms, qualifying them to act as temporal codes for cue color or pattern.

4. Functional Roles for Temporal Codes

Temporally encoded information within the visual system has now been seen in every area from the retina to the inferior temporal cortex (McClurkin *et al.*, 1991). The question remains as to whether the visual system makes any use of this ubiquitous code. It would be very difficult to answer this question with current neurophysiological techniques. Nonetheless, there are several possible ways in which temporal codes would prove advantageous.

4.1. Efficient Encoding and Decoding

Many experiments have now shown that neurons in primary visual cortex convey information about the parameters of a visual stimulus in the temporal modulation of their responses. The work reported here shows that this information may be carried by temporal codes which are separable, and thus can be built up from simple codes for independent stimulus features. Presumably, information about pattern, luminance, color, texture, disparity, etc., could all be multiplexed together in the responses of individual neurons. Thus, only relatively small numbers of neurons would be required to encode and decode visual information. This suggests that when considering the role of different areas of visual cortex, attention must be focused on some function other than simply unconfounding visual information. The neurons in primary visual cortex, for example, are not simply forming a "neuronal image" of the visual world. Instead, they can be thought of as sending messages about the visual scene, rich with information about local details.

4.2. Role of Memory Traces

The discovery in V1 neurons of a memory trace, specific enough to encode a cue's color and pattern, suggests a new role for neurons in primary visual cortex. Either V1 neurons can store detailed visual memories for a very short time, or remembered information flows back into V1 from extrastriate areas. Since the information in the memory trace does not appear in the response until at least 35 to 40 msec after stimulus onset, it seems unlikely that the memory trace arises locally in V1. Perhaps it comes from an area much farther away, such as inferior temporal cortex or the limbic system. A direct anatomical link from inferior temporal cortex to area V1 has been demonstrated that could serve as the path to carry this signal (Douglas and Rockland, 1992).

4.2.1. Optimum Channel Utilization

This raises the important question: what function would V1 be performing that requires a memory trace? Since neurons have a limited information capacity, the neuronal representation of all the visual information in a scene must be dispersed across many neurons. Yet, one can suppose that the processing required by a discrimination task would be more efficient if the information needed to make the judgment were represented as compactly as possible. Perhaps V1 neurons adapt dynamically to the requirements of a specific visual task, so that the maximum amount of useful information is available to later areas in a concentrated form. Feedback from these later areas would then show up as cue-related information, and might be used by neurons in V1 to partition their available channel capacity in the most efficient way.

4.2.2. Spatial Localization

An alternative hypothesis deals with the issue of localizing the correct target. This is an essential visual function, since the monkey must respond by making a saccade to the target with the feature that matches that of the cue. The information about where this target resides is certainly available in V1, since it is retinotopically mapped. Suppose the cue was the color red. If the temporal code for the color red were sent to the entire visual cortex, like a "broadcast" message, it might function to enhance the activity of those neurons which also contained codes for the color red. Thus, a local map in V1 would be built up that represents the distribution of the cue feature in space.

4.2.3. Memory/Feature Linking

The presence of temporal codes for current and remembered visual information in V1 is also consistent with a recent hypothesis about the role of memory in visual perception. This highly speculative hypothesis states that the visual system is organized as an associative memory for small image fragments (called icons), coupled with a multiresolution pyramid for visual feature extraction (Nakayama, 1990). A schematic rendering of this idea is shown in Fig. 18. In this hypothesis, a visual object is represented by the aggregation of many small icons in the associative memory, corresponding to attentional fixations at different levels (and hence, different scales) within the feature pyramid. Visual perception relies on the interaction of the associative memory's icons with the visual feature pyramid through a narrow-bandwidth channel. This interaction builds up a chain of related icons that forms the visual perception.

This model can be adapted to our results by specifying that their mechanisms involve temporal modulation. The information carried by one neuron with its multiplexed temporal codes, or a small group of neurons carrying related codes, might define the icon patch in Nakayama's hypothesis. Interaction between the associative memory and the feature pyramid would be achieved by broadcasting temporal messages from the associative memory and correlating them with the temporal messages within the feature pyramid. The channel between the memory and the pyramid would have a limited bandwidth because you cannot broadcast more than one message at a time without making the correlations impossible to interpret.

For example, suppose the task is to match the red cue with the color of a target stimulus. The associative memory would broadcast the temporal code for "red" back to the feature pyramid. Temporal correlation (essentially a multiplication and low-pass filtering operation) would then be used to find the neurons whose activity also contained the code for red. Thus, the need to connect the associative memory with the feature pyramid to build up a complete percept of a visual object would explain why the cue information was present, even in V1 neurons. The cue information would be interrogating the feature information in the pyramid to determine those icons that had to be linked to form a percept of the object.

5. The Multiplex Code Hypothesis

Previously, Optican and Richmond (1987) proposed a multiplex filter hypothesis that stated that responses of individual neurons could be represented as multiple spatial–temporal channels, each one contributing to the overall response of the neuron. Although models based on this hypothesis have been successful at predicting responses of neurons in the lateral geniculate nucleus (Gawne *et al.*, 1991a) and primary visual cortex (Richmond *et al.*, 1989), the hypothesis does not explain how neurons within a large population cooperate to encode visual information.

A new hypothesis can now be formed that extends the multiplex filter hypothesis by explaining how individual neurons must cooperate within a group to represent visual information. We propose the *multiplex code hypothesis:* that

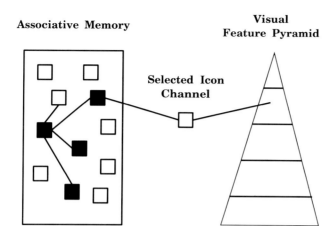

Figure 18. Schematic representation of an associative memory/visual pyramid for perception (after Nakayama, 1990). Squares represent icons, small visual fragments that form the basis of an associative memory. The visual scene is represented at multiple scales within the feature pyramid. A narrow-bandwidth channel connects the memory with the pyramid. A percept is built up by linking many icons together (dark squares). The links are based on information from the visual pyramid.

individual neurons encode information about multiple stimulus features by multiplexing together invariant temporal components to form a separable code. Under the multiplex code hypothesis, many neurons would necessarily share common temporal components. Some support for this prediction was found among the neurons in this study. Furthermore, the new hypothesis suggests that individual neurons within a population cooperate by increasing the bandwidth

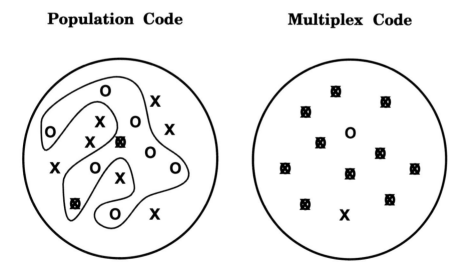

Figure 19. Representations of visual images by population and multiplex codes. In the standard population code model (based on response strength), neurons responding to one object (X) would form a different set from neurons responding to another object (O). Differences between these sets would be confounded by neurons responding to both objects (⊗). In the multiplex code model, neurons in the same set could respond to different objects with different temporal codes (⊗) without confounding the messages.

available for sending information. Thus, if neurons are lost because of injury or disease, the performance of the system would degrade gracefully by lowering the available bandwidth. This loss of bandwidth would cause a degradation of overall performance, without creating a strict dichotomy between spared and lost abilities.

Finally, the multiplex code hypothesis implies a revised role for neuronal populations in encoding and decoding visual information. Under the classic hypothesis that individual neurons within a population encode unique combinations of stimulus features, different stimuli would lead to activity in different subpopulations of neurons (Fig. 19 left). Decoding these messages would then require some other part of the brain to "observe" the two-dimensional distribution of activity within the population. Under the multiplex code hypothesis, the same neurons within a population would encode many different combinations of stimulus features with a separable code. Thus, different stimuli would lead to different activities within the same subpopulation of neurons (Fig. 19 right). Decoding these messages would then only require some other part of the brain to separate the signals coming from the individual neurons into their component elements.

6. Conclusion

This chapter has reviewed some of the advantages of using temporal modulation to encode messages about stimulus features. The difficulty is in showing that the brain actually makes use of such codes. In the future, we will address this problem by looking at the feature codes used by neurons in many visual areas, and by neurons from different animals. Similar temporal codes for a given pattern or color across different visual areas or across different animals would be highly suggestive that such codes were meaningful to the brain. Commonality of these codes would suggest that they may be fundamental elements of visual perception, and thus could act as a foundation for understanding how we see.

7. References

Ahmed, N., and Rao, K. R., 1975, *Orthogonal Transforms for Digital Signal Processing*, Springer-Verlag, Berlin.

Barlow, H. B., 1972, Single units and sensation: A neuron doctrine for perceptual psychology, *Perception* **1**:371–394.

Barlow, H. B., 1985, The twelfth Bartlett memorial lecture: The role of single neurons in the psychology of perception, *Q. J. Exp. Psychol.* **37A**:121–145.

Cattaneo, A., Maffei, L., and Morrone, C., 1981, Patterns in the discharge of simple and complex visual cortical cells, *Proc. R. Soc. London B Ser.* **212**:279–297.

Chee-Orts, M. -N., and Optican, L. M., 1993, Cluster method for analysis of transmitted information in multivariate neuronal data, *Biol. Cybern.* **69**:29–35.

Douglas, K. L., and Rockland, K. S., 1992, Extensive visual feedback connections from ventral inferotemporal cortex, *Soc. Neurosci. Abstr.* **18**:390.

Dow, B. M., and Gouras, P., 1973, Color and spatial specificity of single units in the rhesus monkey foveal striate cortex, *J. Neurophysiol.* **36**:79–100.

Eckhorn, R., and Pöpel, B., 1975, Rigorous and extended application of information theory to the afferent visual system of the cat. II. Experimental results, *Kybernetik* **17**:7–17.

Eskandar, E. N., Optican, L. M., and Richmond, B. J., 1992a, Role of inferior temporal neurons in visual memory. II. Multiplying temporal waveforms related to vision and memory, *J. Neurophysiol.* **68**:1296–1306.

Eskandar, E. N., Richmond, B. J., and Optican, L. M., 1992b, Role of inferior temporal neurons in visual memory. I. Temporal encoding of information about visual images, recalled images, and behavioral context, *J. Neurophysiol.* **68:**1277–1295.

Gawne, T. J., McClurkin, J. W., Richmond, B. J., and Optican, L. M., 1991a, Lateral geniculate neurons in behaving primates: III. Response predictions of a channel model with multiple spatial-to-temporal filters, *J. Neurophysiol.* **66:**809–823.

Gawne, T. J., Richmond, B. J., and Optican, L. M., 1991b, Interactive effects among several stimulus parameters on the responses of striate cortical complex cells, *J. Neurophysiol.* **66:**379–389.

Lennie, P., Krauskopf, J., and Sclar, G., 1990, Chromatic mechanisms in striate cortex of macaque, *J. Neurosci.* **10:**649–669.

Livingstone, M. S., and Hubel, D. H., 1984, Anatomy and physiology of a color system in the primate visual cortex, *j. Neurosci.* **4:**309–356.

McClurkin, J. W., Optican, L. M., Richmond, B. J., and Gawne, T. J., 1991, Concurrent processing and complexity of temporally encoded neuronal messages in visual perception, *Science* **253:**675–677.

Michael, C. R., 1978a, Color vision mechanisms in monkey striate cortex: Simple cells with dual opponent-color receptive fields, *J. Neurophysiol.* **41:**1233–1249.

Michael, C. R., 1978b, Color-sensitive complex cells in monkey striate cortex, *J. Neurophysiol.* **41:**1250–1266.

Michael, C. R., 1981, Columnar organization of color cells in monkey's striate cortex, *J. Neurosci.* **46:**587–604.

Moran, J., and Desimone, R., 1985, Selective attention gates visual processing in extrastriate cortex, *Science* **229:**782–784.

Nakayama, K., 1990, The iconic bottleneck and the tenuous link between early visual processing and perception, in: *Vision: Coding and Efficiency* (C. Blakemore, ed.), Cambridge University Press, London, pp. 411–422.

Optican, L. M., and Richmond, B. J., 1987, Temporal encoding of two-dimensional patterns by single units in primate inferior temporal cortex. III. Information theoretic analysis, *J. Neurophysiol.* **57:**162–178.

Optican, L. M., Gawne, T. J., Richmond, B. J., and Joseph, P. J., 1991, Unbiased measures of transmitted information and channel capacity from multivariate neuronal data, *Biol. Cybern.* **65:**305–310.

Richmond, B. J., and Optican, L. M., 1987, Temporal encoding of two-dimensional patterns by single units in primate inferior temporal cortex: II. Quantification of response waveform, *J. Neurophysiol.* **57:**147–161.

Richmond, B. J., and Optican, L. M., 1990, Temporal encoding of two-dimensional patterns by single units in primate primary visual cortex. II. Information transmission, *J. Neurophysiol.* **64:**370–380.

Richmond, B. J., Optican, L. M., Podell, M., and Spitzer, H., 1987, Temporal encoding of two-dimensional patterns by single units in primate inferior temporal cortex: I. Response characteristics, *J. Neurophysiol.* **57:**132–146.

Richmond, B. J., Optican, L. M., and Gawne, T. J., 1989, Neurons use multiple messages encoded in temporally modulated spike trains to represent pictures, in: *Seeing Contour and Colour* (J. J. Kulikowski and C. M. Dickinson, eds.), Pergamon Press, Elmsford, N.Y., pp. 701–710.

Richmond, B. J., Optican, L. M., and Spitzer, H., 1990, Temporal encoding of two-dimensional patterns by single units in primate primary visual cortex. I. Stimulus–response relations, *J. Neurophysiol.* **64:**351–369.

Tootell, R. B. H., Hamilton, S. L., and Switkes, E. 1988a, Functional anatomy of macaque striate cortex. IV. Contrast and magno-parvo streams, *J. Neurosci.* **8:**1594–1609.

Tootell, R. B. H., Silverman, M. S., Hamilton, S. L., De Valois, R. L., and Switkes, E., 1988b, Functional anatomy of macaque striate cortex. III. Color, *J. Neurosci.* **8:**1569–1593.

Tootell, R. B. H., Silverman, M. S., Hamilton, S. L., Switkes, E., and De Valois, R. L., 1988c, Functional anatomy of macaque striate cortex. V. Spatial frequency, *J. Neurosci.* **8:**1610–1624.

Ts'o, D. Y., and Gilbert, C. D., 1988, The organization of chromatic and spatial interactions in the primate striate cortex, *J. Neurosci.* **8:**1712–1727.

Wurtz, R. H., 1969, Response of striate cortex neurons during rapid eye movements in the monkey, *J. Neurophysiol.* **32:**975–986.

Zarbock, J. A., 1992, Temporally encoded messages about visual features are invariant across primate cortices V1, V2, V3, V4, M.Sc. thesis, Johns Hopkins University, Baltimore.

12

The Human Primary Visual Cortex

RODRIGO O. KULJIS

1. Introduction

The human primary visual (striate) cortex was one of the first specialized regions identified in the cerebral cortex (Gennari, 1782). Because of its unique anatomical organization and the striking clinical deficits that result from lesions to it (Rizzo, this volume), this region has received nearly continuous attention for over two centuries, making it the best understood of all cortical areas in the human brain.

   In recent years, many exciting findings have been made that are beginning to transform the way we think about and study the human visual cortex. For example, functional mapping with positron emission tomography (PET) is being used with increasing frequency not only to map out the exact boundaries of the visual areas, but also to study the functional role of the striate and parastriate cortices (e.g., Fox *et al.*, 1986, 1987; Kushner *et al.*, 1988; Mora *et al.*, 1989; Zeki *et al.*, 1991; Woods *et al.*, 1991). Other, still evolving strategies include quantitative imaging of cerebral blood flow using magnetic resonance imaging (MRI) techniques (Belliveau *et al.*, 1991), and nuclear magnetic resonance spectroscopy. The latter method can be used to study changes in the concentration of various types of molecules *in vivo* during physiologic activity, so that processes such as stimulus-induced glycolysis in the visual cortex can be monitored (Prichard *et al.*, 1991). Although these techniques are still in their infancy, they offer consider-

RODRIGO O. KULJIS • Department of Neurology, The University of Iowa, Iowa City, Iowa 52242-1053.

Cerebral Cortex, Volume 10, edited by Alan Peters and Kathleen S. Rockland. Plenum Press, New York, 1994.

able promise to increase both the breadth and the speed of our understanding of normal and disordered function in the visual cortex.

These potentially revolutionary neuroimaging methods require integration and comparison with what has been learned from more established methodologies. This is especially important to validate and to determine the accuracy of the new methods, and to carefully assess the findings that result from their application. For example, it will be necessary to establish precise correlations between the increasingly more detailed MRI images and those familiar to us from light microscopy. Such an effort is already under way, with modifications of conventional MRI parameters that offer the promise of resolving gross cytoarchitectonic landmarks *in vivo* (Damasio *et al.*, 1991), including the stria of Gennari in the visual cortex (Clark *et al.*, 1992).

In addition to the new radiological approaches to the study of the visual cortex, there has been a resurgence of interest in the study of this region exploiting novel postmortem tract-tracing methods based on carbocyanines (e.g., Burkhalter and Bernardo, 1989; Burkhalter, 1991). These methods are limited to short-range pathways, but are especially helpful to study the human brain, in which axonal transport tract-tracing methods cannot be employed. Coupled with renewed interest in the use of silver impregnation of degenerating pathways (e.g., Clarke and Miklossy, 1990), these methods will be of considerable help in the continuing effort to analyze the connectivity of the visual cortex in humans. Furthermore, there is an increasing interest in understanding the mode of involvement of the striate cortex in patients with Alzheimer's disease (e.g., Beach *et al.*, 1989; Braak *et al.*, 1989; Bell and Ball, 1990; Kuljis and Van Hoesen, 1991; Kuljis, 1992a). The latter effort has capitalized on the relatively well-known histological organization of the striate cortex, permitting insights into the pathophysiology of the illness which cannot be obtained by examining other less well understood cortices.

This chapter reviews some salient aspects of the organization, development, and histopathology of the human striate cortex. The text is focused in the primary visual cortex, without attempting to make significant inroads into parastriate visual cortices, since these are less well characterized in the human brain. Furthermore, because most of the present volume is dedicated to the nonhuman primate visual cortex, I have limited discussions of animal models and refer instead to pertinent chapters in this volume dealing primarily with nonhuman primates.

2. Structure

The primary visual cortex is the most clearly demarcated area of the cerebral cortex. In sections, its boundaries can be ascertained even with the naked eye because of a unique anatomical feature, the stria, line, band, or stripe of Gennari. The stria was named after its discoverer, the Italian anatomist Francesco Gennari (1782), who reported seeing this feature for the first time on 2 February 1772. It is a white line situated about midway through the thickness of the cortical gray matter. The stria extends throughout the primary visual cortex and ends abruptly at the boundary with surrounding cortical areas (Fig. 1D). Because of this unique stria, the primary visual cortex is frequently referred

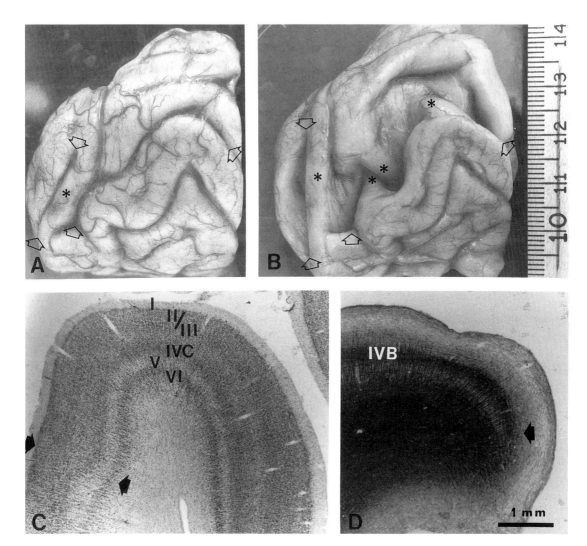

Figure 1. (A, B) Photomacrographs of the medial aspect of the left occipital lobe from a 72-year-old man. (A) Undisturbed specimen, showing the curved pattern of the calcarine fissure (ends marked with hollow arrows), which is interrupted by a cuneolingual gyrus (asterisk). (B) After opening part of the calcarine fissure, its complex pattern of folding can be appreciated. This contains about two-thirds of the primary visual cortex. Three additional small cuneolingual gyri can be appreciated within the calcarine fissure (asterisks), in addition to the more posterior cuneolingual gyrus that was visible in the interhemispheric fissure prior to dissection (asterisk in A), (C, D) Photomicrographs of cresyl violet and myelin (Gallyas) stains, respectively, cut perpendicularly to the lip of the calcarine fissure. They illustrate low-power views of the striking laminar organization of the striate cortex, and especially its distinctive layer IV, which can be easily seen to end abruptly at the boundary with the prestriate cortex (area 18 of Brodmann, 1903), indicated by the solid arrows. Roman numerals designate cortical layers described in detail in the text.

to as the "striate" cortex. The stripe was discovered independently by Félix Vicq d'Azyr (1786; Glickstein and Rizzolatti, 1984; Glickstein, 1988), and is therefore often named after Gennari and/or Vicq d'Azyr, although the latter is now archaic and therefore infrequently used. The stria corresponds to the external band of Baillarger (1840), and to layer IVB (Fig. 1C,D) of modern nomenclature (H. Braak, 1976; E. Braak, 1982), described in detail below (see also Peters, this volume).

2.1. Macroscopic Organization

The human primary visual cortex is situated in the medial aspect and the pole of the occipital lobe, in the banks and lips of one of the primary furrows of the cerebral cortex: the "calcarine" fissure. The fissure was given this name because of its proximity with the *calcar avis* ("the bird's spur"), an elevation in the medial wall of the posterior horn of the lateral ventricle. The fissure was described and named by T. H. Huxley (1861) and was also called the "striate fossa" or "pit" by G. Elliot Smith (cited in Polyak, 1958). The calcarine fissure is a long furrow extending on the medial surface of the occipital lobe more or less horizontally. It is classically described as originating from the isthmus of the fornicate gyrus below the splenium of the corpus callosum and extending posteriorly to the occipital pole (Fig. 1A,B). The anterior portion of the fissure is rather shallow, while the posterior stretch is usually deeper and contains a substantial amount of cortex.

The configuration of the calcarine fissure is variable, both among individuals and between the left and right cerebral hemispheres in most persons (Smith, 1904, 1907a,b). This individual variability is important, since it presents difficulties in the precise identification of this region. These difficulties have both diagnostic and therapeutic implications, which are discussed in the next subsection. In general, the fissure is roughly horizontal or gently curved. Its height within the medial aspect of the hemisphere is variable, such that the size of the cuneus may vary by a factor of two or three among individuals (Stensaas *et al.*, 1974). Usually, the calcarine fissure meets the parieto-occipital fissure at about the junction of its anterior two-thirds, before reaching the splenium of the corpus callosum. The posterior end is often forked near the occipital pole, but this feature can be highly variable (Fig. 1A,B). Two lips are distinguishable: the upper cuneal or cuneofornicate (dorsal labium) and the lower or lingual (ventral labium). The lips are frequently separated from the cuneus and the lingual gyrus by secondary sulci that are more or less parallel to the calcarine fissure, and are referred to as the paracalcarine sulci (Fig. 1B shows such a sulcus in the lingual gyrus). In some cases, the calcarine fissure may be interrupted by one or more cuneolingual gyri that span between the homonymous gyri and are visible on the surface of the interhemispheric fissure (asterisk in Fig. 1A). More frequently, one or more cuneolingual gyri are present in the depth of the calcarine fissure, but they become visible only after the fissure is opened (Fig. 1B).

The calcarine artery, the most distal branch of the posterior cerebral artery, is usually situated in the depth of the calcarine fissure. This artery often gives a dorsal branch to the parieto-occipital sulcus, and another branch to the collateral sulcus which spreads over the ventral aspect of the occipital lobe. Like the cortical territory it supplies, the configuration of the calcarine artery is highly variable among individuals. It bears so little relation to the configuration of the striate cortex that angiographic studies have proven unreliable in predicting its exact location (Smith and Richardson, 1966). The calcarine artery supplies at least most of the primary visual cortex in average individuals (Polyak, 1958), although exhaustive angio- and cytoarchitectonic comparative studies appear not to have been performed on this issue. It is possible that given the highly variable anatomy of the occipital pole, the calcarine artery, and the pattern of folding of the calcarine cortex, distal branches of the middle cerebral artery may feed a somewhat laterally displaced margin of the visual cortex (Polyak, 1958; Smith and Richardson, 1966; Hoyt and Margolis, 1970; Hoyt and Newton, 1970). Individual anatomical variability may be adduced, for example, to explain some cases of sparing of the foveal/macular region of the visual field after massive stroke in the posterior cerebral artery territory (Polyak, 1958; Miller, 1982). It should be emphasized, however, that preservation of the vascular supply to the macular representation in the striate cortex is but one of several mechanisms invoked to explain the clinical observation of macular visual field sparing after occipital lesions. Other proposed mechanisms for this phenomenon include defective or compensatory eccentric fixation, which, because of technical artifacts during visual field perimetry, may give the false impression of macular sparing (Verhoeff, 1943; see Rizzo, this volume).

2.2. Topographic Variability in the Localization of the Striate Cortex

Several authors have painstakingly demonstrated that there is a highly variable topographic relationship between the calcarine fissure and other gross anatomical landmarks on the surface of the occipital pole (Bolton, 1900; Brodmann, 1918; Filimonoff, 1932; Brindley, 1972; Stensaas et al., 1974). This variability occurs in several aspects: (1) the total area of this region can vary from 1284 to 3702 mm² among individuals; (2) the amount of striate cortex exposed in the surface of the interhemispheric fissure can vary from 359 to 1308 mm², and (3) the pattern of folding of the striate cortex in the hemispheres of a single individual is also highly variable and asymmetric (Stensaas et al., 1974). These factors, along with differences in methodology, explain in part the considerable range of variation in the total surface of striate cortex reported among various studies. For example, Putnam (1926) reported an average area of 20.7 cm² of striate cortex per hemisphere, whereas Brodmann provided a figure of 34.5 cm². The "correct" figure is probably closer to that reported by Putnam (1926) and Stensaas et al. (1974), namely, about 21.34 cm². Collectively, these investigators have studied the largest number of hemispheres and sampled them at the closest intervals. In general, it can be said that most of the striate cortex is buried in the calcarine fissure and its branches: 67% according to the study carried out by Stensaas et al. (1974). Therefore, only about one-third of the striate cortex is exposed on the surface of the hemisphere.

This variability in the topography of the striate cortex has several important implications: (1) gross landmarks on the surface of the occipital lobe and angiographic studies cannot be used to locate the striate cortex precisely; (2) topical resections, stimulation, and any other kind of surgical interventions that require accurate localization of the striate cortex are considerably hampered by the inability to localize it precisely in vivo; (3) since topographic variability is so amply documented in the distribution of the primary visual cortex, it is to be expected that most other areas of the cortex are similarly variable. This poses severe limitations to the accurate localization of cortical areas and their functions in humans. This is especially true whenever such localization relies principally on gross anatomical landmarks and lacks cytoarchitectonic confirmation. Fortunately, recent modifications in the protocols for magnetic resonance imaging are beginning to resolve regional differences in cortical architecture (Damasio *et al.*, 1991). Such methods have been applied to the striate cortex, where the stria of Gennari can be resolved and visualized (Clark *et al.*, 1992), and thus they may eventually provide a satisfactory *in vivo* alternative to classical cytoarchitectonic parcellations using postmortem material.

The above considerations also have therapeutic implications. For example, the topographic variability in the location of the striate cortex is conspicuous among the impediments to develop a useful prosthetic device to restore visual function in the blind. Such devices rely on the ability to produce visual sensations similar to phosphenes by electrical stimulation of the surface of the striate cortex (Brindley and Lewin, 1968; Button and Putnam, 1962; Dobelle *et al.*, 1974). One strategy calls for the insertion of an array of electrodes apposed to the surface of the cortex that can be driven by a computer using data gathered by a video camera. This may be feasible because discernible spatiotemporal stimulation patterns in the array of electrodes have already helped to provide useful "visual" cues to a blind person (Dobelle *et al.*, 1974; Girvin, 1988). Two of the significant problems affecting the potential for success of this method are that: (1) only the portion of striate cortex exposed on the interhemispheric fissure (i.e., about one-third of its surface) can be stimulated, and (2) there is no reliable method to precisely localize the borders of the striate cortex *in vivo*, so that it is not possible to ensure that the prosthesis is applied precisely to the primary visual cortex.

2.3. Cellular Anatomy

The human striate cortex exhibits a distinct pattern of lamination when examined with cell and fiber stains, as well as stains for lipofuchsin deposits (Fig. 2). Several different schemes of lamination have been proposed, some of which are summarized in Fig. 3. In general, these schemes tend to follow the notion of a six-layered cerebral isocortex first described by Berlin (1858), which was elaborated upon by Brodmann (1903, 1906, 1909, 1918). Within this six-layered prototype, the main layers are designated by Roman numerals following the recommendations of Vogt and Vogt (1919), and two or more sublayers are often recognized, most commonly within layers III, IV, and sometimes in layer V. The most notable alternate scheme is that of Ramón y Cajal (1890, 1899, 1900, 1909–1911), which was followed closely by his pupil Lorente de Nó (1938). The scheme proposed by Ramón y Cajal uses Arabic numerals to designate nine different layers, without distinguishing sublayers. Despite this rather unusual nomenclature, Ramón y Cajal's divisions have approximate equivalents in most of the schemes proposed by other authors (see Fig. 3). Therefore, his scheme does not differ fundamentally except in that it does not endorse the concept of a six-layered cortex. At present, most authors describing the striate cortex tend to follow Lund's nomenclature (1973; see also Lund and Boothe, 1975), which was

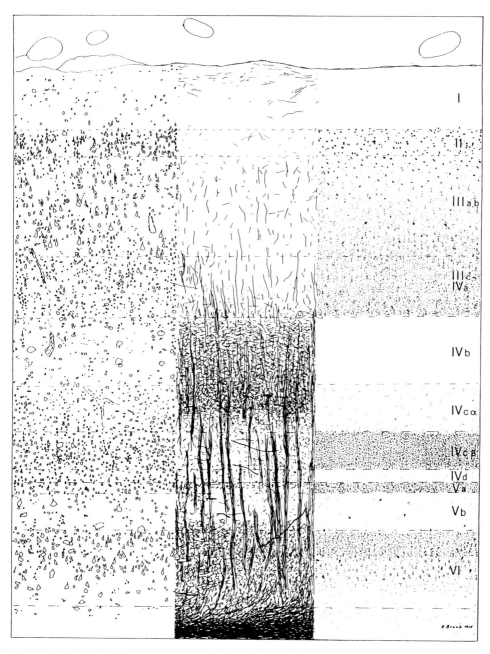

Figure 2. Drawing of the arrangement of cell bodies (left column), myelinated fibers (middle column), and lipofuchsin deposits (right column) in the human striate cortex, as seen in transverse sections perpendicular to the pial surface. Cortical layers are designated per Braak's nomenclature in the right column. Reproduced from Braak (1976) with permission. Courtesy of Professor Heiko Braak.

developed for the macaque monkey (see Peters and Lund *et al.*, both in this volume). In the present account, I will rely principally on H. Braak's (1976) nomenclature, which was formulated specifically for the human striate cortex, following Lund's scheme closely while acknowledging some minor species differences.

The pattern of lamination in the human striate cortex is the result of different packing densities of neurons, astrocytes, microglial cells, and blood vessels throughout its depth. The principal determinant elements of lamination are the neuronal somata (Fig. 4), with other types of elements playing a minor role (E. Braak, 1982). As shown in Figs. 2 and 4, layer I is a molecular layer that contains very few neurons, and has a distinct boundary with the more cellular layer II. Most of the cells in layer I are fibrillary astrocytes, along with some oligodendrocytes and microglial cells. Layer I neurons have ovoid perikarya, from which few processes emerge, and, in general, have not been extensively characterized. These neurons can be arbitrarily subdivided into large, so-called Cajal–Retzius neurons (Ramón y Cajal, 1909–11), and neurons with smaller perikarya that are probably local circuit neurons. Little is known about the latter. Cajal–Retzius neurons have been observed in fetal material, although many investigators are unable to find them in adult humans (Ramón y Cajal, 1929), raising the possibility that they die postnatally, or that they are transformed into other cell types. However, at least some Cajal–Retzius cells may survive into adulthood (Marín-Padilla and Marín-Padilla, 1982).

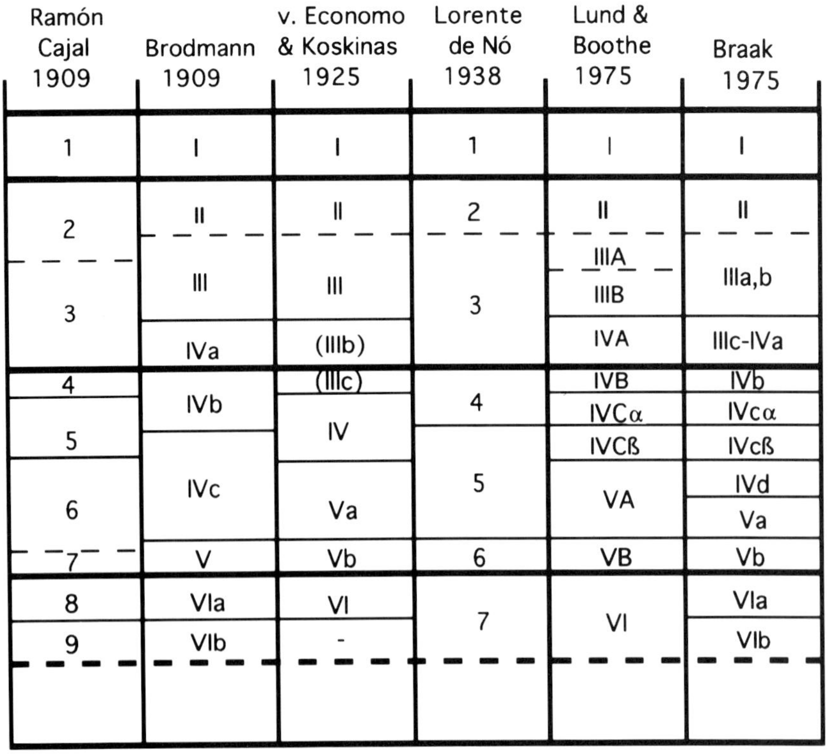

Ramón Cajal 1909	Brodmann 1909	v. Economo & Koskinas 1925	Lorente de Nó 1938	Lund & Boothe 1975	Braak 1975
1	I	I	1	I	I
2	II	II	2	II	II
3	III	III	3	IIIA / IIIB	IIIa,b
	IVa	(IIIb)		IVA	IIIc-IVa
4	IVb	(IIIc)	4	IVB	IVb
5		IV		IVCα	IVcα
				IVCß	IVcß
6	IVc	Va	5	VA	IVd / Va
7	V	Vb	6	VB	Vb
8	VIa	VI	7	VI	VIa
9	VIb	-			VIb

Figure 3. Sketch of the lamination pattern of the striate cortex according to the nomenclature of several of the authors discussed in the text. Modified from Braak (1976).

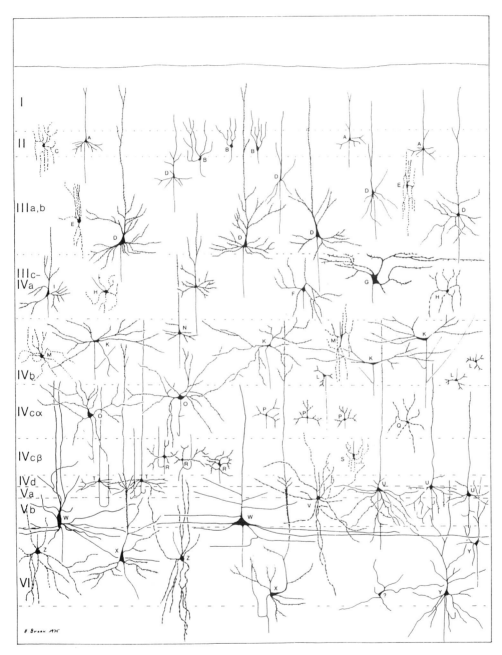

Figure 4. Drawing of the different types of neurons that populate the striate cortex. Reproduced from Braak (1976), with permission. Courtesy of Professor Heiko Braak.

Layers II–VI contain both pyramidal and nonpyramidal neurons (Fig. 4). Nonpyramidal neurons in the form of spiny stellate cells predominate in layers IV and—to a lesser extent—VIB, while pyramidal neurons predominate in layers II/III and V. There is considerable variability in perikaryal size in both populations of spiny neurons. The largest spiny stellate neurons are about 30 μm in diameter, while the diameter of pyramidal neurons varies from about 8 μm (the smallest pyramidal cells of layer II) to 50 μm (largest Meynert–Cajal neurons in layers IVb and V).

Layer II contains small pyramidal neurons with perikaryal diameters of about 8–11 μm, and short apical dendrites that extend into layer I (Fig. 4). They have short apical dendrites that receive numerous small boutons with spherical vesicles that make predominantly asymmetrical synapses. Contrary to previous descriptions (e.g., Polyak, 1958), Braak (1976) has found that layer II contains no spiny stellate cells. The boundary with the subjacent layer III is poorly marked or imperceptible, justifying the common practice of designating these layers together as "layers II/III." Layer III contains predominantly medium-sized and small pyramidal neurons. The largest of these tend to have apical dendrites that branch and end in a tuft near the layer I/II border. Their descending axons branch profusely in layer V before entering the white matter (Braak, 1976). In monkeys, these cells have been shown to project to the prestriate cortex (area 18 of Brodmann), with which they participate in reciprocal corticocortical connections (Felleman and Van Essen, 1991). Differences in cell packing density, lipofuchsin deposits, and the distribution of histochemical markers reveal two to three sublaminae within layer III (Figs. 2 and 3).

Layer IV can also be subdivided into several sublayers based on the packing density of cells and the distribution of histochemical markers, such as the mitochondrial enzyme cytochrome oxidase (see below and Wong-Riley, this volume). Traditional descriptions of this layer maintain that it is devoid of pyramidal neurons, and largely made up of spiny stellate neurons. Golgi studies indicate that small neurons indistinguishable from "classical" pyramidal neurons are situated in this layer (e.g., "I" and "N" in Fig. 4; H. Braak, 1976; E. Braak, 1982). These pyramidal neurons are, however, present in small numbers in layer IV, predominantly in the interface with layer III, and virtually absent from layers IVB and IVC. In addition to the so-called "classical" pyramidal and nonpyramidal neurons, H. Braak and E. Braak have felt that a third class of "polygonal" neurons is also situated in layer IV. These cells do not conform to strict morphological criteria for pyramidal or stellate neurons. Similar cells have been identified in animals, and have been variably designated as "spiny stellate cells" (Garey, 1971; LeVay, 1973; Lund, 1973) or "stellate cells" (Ramón y Cajal, 1909–11; Valverde, 1971) and "star pyramids" (Lorente de Nó, 1938), among other terms. It should be emphasized that these various types of neurons are not necessarily equivalent, and that the terms that describe them are not freely interchangeable. As a result, considerable controversy often arises about the names that should be given to neurons. The polygonal neurons are probably modified pyramidal neurons, since they share a number of morphological features with them, including: (1) dendritic spines, (2) a gradual transition of cytoplasmic features from the soma to the dendrites, and (3) distal dendritic bifurcations. In addition, their axons arise from the basal aspect of the perikaryon, and descend into the white matter, which is also atypical for "classical" spiny stellate neurons.

Layers IIIc–IVb contain the largest polygonal neurons, which correspond to the Meynert–Cajal solitary neurons (Meynert, 1872; Ramón y Cajal, 1899), and an array of smaller polygonal neurons. The former are known also as outer Meynert cells, to distinguish them from Meynert cells in layers V and VI (Clark, 1942; Chan-Palay *et al.*, 1974; Peters, this volume). Layer IVCα contains a mixture of small and medium-sized polygonal neurons, while layer IVCβ has predominantly small polygonal, spiny neurons.

Layer V has some small and medium-sized polygonal neurons, although small typical pyramidal cells predominate along the interface of layers IVd/Va of Braak (1976). Layer VB has larger pyramidal cells, including the large Meynert–Cajal cells. In general, pyramidal neurons have ascending dendrites. Those with the largest perikarya tend to have the longest ascending dendrites. For example, the tiniest pyramidal cells in layer VA often have delicate, unbranched apical dendrites ending in layer IV, while the dendrites of the Meynert–Cajal neurons reach the interface between layers I and II before bifurcating. In monkeys, some of the cells of layer V have been shown to project to the superior colliculus and pulvinar (Lund *et al.*, 1975).

Layer VI, also known as the multiform layer, contains several varieties of cells. These include: (1) small pyramidal neurons with apical dendrites that end in layer V; (2) medium-sized pyramidal neurons with apical dendrites that end in layer IVB without ramification; (3) small and medium-sized triangular neurons with asymmetrically disposed dendritic branches that are covered by dendritic spines; and (4) neurons with atypical features, such as two or more apical dendrites, and major dendrites that originate laterally. In monkeys, cells in this layer have been shown to project to the lateral geniculate nucleus (Lund *et al.*, 1975).

Some recent work has focused on the nonpyramidal neurons of the striate cortex that contain unique markers, such as the calcium-binding protein parvalbumin (Blümcke *et al.*, 1990) and the neuroactive peptide known as neuropeptide Y (Berman and Fredrickson, 1992), which can be detected immunocytochemically (see Jones *et al.*, this volume). Provided there is suitably short postmortem autolysis, preparations can be obtained that approach the quality of perfusion-fixed nonhuman primate material. This has permitted cross-species comparisons, and helped characterize these neurons in some detail. For example, parvalbumin-containing neurons belong to several morphologically distinct classes of nonpyramidal neurons that cocontain GABA; they are situated in layers II–VI, and appear to be densest in supragranular layers (Blümcke *et al.*, 1990). Neuropeptide Y-containing neurons are less numerous than parvalbumin-positive cells, appear to belong to a less wide variety of nonpyramidal neuron types, and are present predominantly in the white matter immediately subjacent to the cortical gray (Berman and Fredrickson, 1992). We expect to see more studies of these neurons in the near future. This will help reduce a considerable gap in our knowledge of nonpyramidal neurons compared with pyramidal neurons in the human striate cortex, especially in layers other than layer IV.

2.4. Organization of Intrinsic Connections

The precise pattern of inter- and intralaminar intrinsic connectivity of the human striate cortex remains a mystery, since virtually all that is known about its connections relies on fortuitous lesions examined with degeneration methods. The latter methods are best suited to determine gross features of long-range projections, but unfortunately they are not ideal for analyzing intrinsic connections. Obviously, the method of choice for detailed connectivity (i.e., application of tracers that depend on axonal transport *in vivo*) cannot be employed in humans because of technical and ethical limitations, but this situation has improved with the advent of postmortem pathway tracers, especially the carbocyanines. Burkhalter and Bernardo (1989) have used this method to study the organization of corticocortical connections both within the striate cortex, and

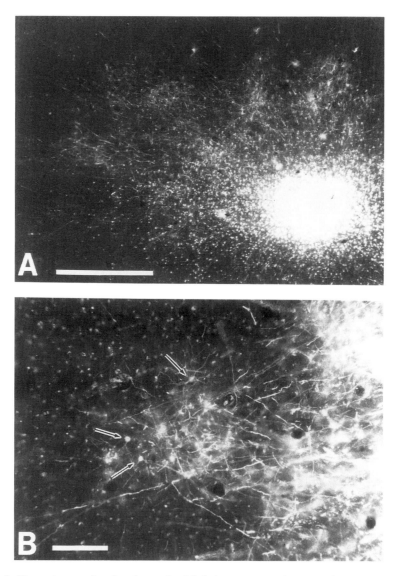

Figure 5. Photomicrographs of carbocyanine-labeled postmortem human striate cortex. (A) Tangential section through layer III showing four clustered projections within ~2-mm radius from the dye application site (lower right corner of the panel). (B) Higher-power view of one of these clusters. Bars: A = 1 mm, B = 0.2 mm. Reproduced from Burkhalter and Bernardo (1989) with permission. Courtesy of Dr. Andreas Burkhalter.

between the striate and immediately adjacent parastriate cortex (area 18 of Brodmann). The method still has limitations, among them the long intervals (weeks) that must elapse between tracer application and its useful diffusion, and the short range of connections that can be traced (up to 6 mm). Nevertheless, useful information has been gained, and the results have been similar to those obtained with axonal transport methods in nonhuman primates (see Rockland and Lund *et al.*, this volume).

In human striate cortex, there is a distinctive pattern of interlaminar connectivity, such that carbocyanine applications restricted to supragranular layers result in bands of labeling immediately underneath the injection site in layers IV and V. There are also bundles of descending axons that stream into the white matter, and probably participate in corticocortical connections with area 18. Tracer applications centered midway through the depth of the cortex result in more horizontally dispersed labeling of layer IVB, and in a very narrow columnlike zone of labeling in the supragranular layers. Patterns of connectivity in the tangential domain (i.e., parallel to the pial surface) appear to be also very selective. For example, tangentially cut sections systematically reveal patches, or clusters, of presumed terminals at discrete intervals near the tracer application site (Fig. 5). Such clusters are very similar to the "periodic intrinsic connections" in the striate and parastriate cortices of other mammals, including nonhuman primates (Rockland and Lund, 1982, 1983; Rockland, 1985). Such periodic intrinsic connections are considered to be an important feature of the connectivity of the striate cortex, and they may be involved in a variety of integrative processes carried over large portions of the visual field (Gilbert *et al.*, 1990).

Burkhalter and Bernardo (1989) examined their carbocyanine material together with sections counterlabeled for the histochemical localization of the mitochondrial enzyme cytochrome oxidase (CO). This enzyme is a convenient marker of functionally and anatomically distinct compartments within individual hypercolumns, or postulated functionally distinct "units" described in the striate cortex of macaque monkeys (for reviews, see Livingstone and Hubel, 1984; Kuljis and Rakic, 1989, 1990; and Wong-Riley, this volume). Carbocyanine applications centered in CO-rich patches revealed connections preferentially with nearby CO puffs. The projections tended to avoid the intervening CO-poor interpuffs. In contrast, applications centered in interpuffs revealed preferentially connections with nearby interpuffs, and they tended to avoid the puffs.

2.5. Patterns of Cytochrome Oxidase Enzymatic Activity and Evidence for Modular Organization

Histochemical labeling for the mitochondrial enzyme CO has proven a useful method for labeling regions of the human striate and parastriate cortices. These areas exhibit differential levels of high oxidative metabolic activity (Horton and Hedley-Whyte, 1984; Horton *et al.*, 1990; see also Allman and Zucker, 1990, and Wong-Riley, this volume). In humans, a high level of CO activity is present predominantly in the neuropil of layers II–IV and in large pyramidal perikarya. Labeling is particularly intense in layer IVC, where it is continuous along the tangential plane, and exhibits a texture felt to closely reflect the pattern of geniculocortical terminals preterminals. Labeling in layers II–IVB is less intense and is discontinuous in the tangential plane (Horton and Hedley-Whyte,

1984). In these layers, CO activity is more or less regularly distributed in zones of the neuropil variably referred to as "patches," "puffs," "dots," "spots," and "blobs." The puffs are roughly oval in shape when viewed in sections cut tangentially to the pial surface. They are most distinct in layer III and tend to fade on each site toward layer IVB, and toward layer II. Puffs measure about 400 by 250 μm in diameter in tissue sections cut parallel to the pial surface. The density of puffs is roughly one per 0.6–0.8 mm² of cortex, and they are organized in rows spaced about 1 mm apart that intersect the border between striate and parastriate cortex at a right angle. Other oxidative enzymes and acetylcholinesterase exhibit similar labeling patterns.

Because of its resemblance to a nearly identical pattern of CO labeling in the macaque monkey striate cortex (Horton, 1984; Kuljis and Rakic, 1989; Wong-Riley, this volume), histochemical labeling for this enzyme provides a cogent basis for comparison between the striate cortices in humans and monkeys. For example, oxidative enzyme methods reveal no trace of a honeycomblike band between layers III and IVB in humans, and this is consistent with the apparent lack of layer IVA in Nissl stains of human tissue.

Another apparent interspecies difference is in the labeling intensity and laminar location of the puffs: human puffs are paler than those in macaques and are located deeper in the cortex. Consequently, human puffs are very hard to recognize in layer II, except in unusually good transverse sections, and they extend into layer IVB. Macaque puffs, by contrast, are more obvious in layer II, and do not extend into layer IVB.

A third difference pertains to the size and spacing of the puffs, which are about twice as large (450–250 versus 240–125 μm) and exhibit about twice the interrow distance (1 mm versus 450 μm) in humans compared with macaque monkeys. These histochemically revealed compartments have been felt to provide a convenient marker for functionally distinct cell assemblies, or units known as hypercolumns in macaque striate cortex (Horton, 1984; Livingstone and Hubel, 1984; Kuljis and Rakic, 1989, 1990). If they reveal similar functionally distinct compartments in humans, it would seem that hypercolumns are twice as large in humans compared with macaques. This obviously raises important questions about their phylogenetic and functional implications, which have not been resolved (Florence and Kaas, 1992).

In accord with the notion that CO reveals functionally meaningful compartments in human cortex is the fact that enzymatic activity decreases in response to monocular deprivation and retinal or ocular ablation. Monocular enucleation and large retinal lesions result in a pattern of regular, alternating dark and light bands of CO labeling in layer IVC (Fig. 6A), very similar to those observed in monkeys after comparable procedures (Horton, 1984; Horton and Hedley-Whyte, 1984; Horton et al., 1990), except that they are twice as wide. The eye dominance columns demonstrated by the CO method also resemble those observed using a modified Glees stain (Hitchcock and Hickey, 1980). The relative simplicity of the enzymatic method permits mapping the columns over a large portion of the striate cortex (Horton et al., 1990; Fig. 6B). These columns form a mosaic of irregular parallel stripes 500–1000 μm wide, oriented at right angles to the boundary of the striate cortex. They are wider near this border, where the central portion of the retina is represented (see below), and the columns served by the ipsilateral eye become progressively fragmented toward the peripheral representation. Finally, the columns disappear in the region representing the

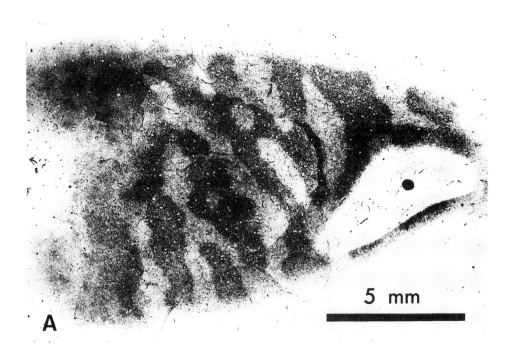

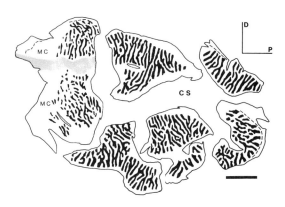

Figure 6. Pattern of cytochrome oxidase activity in layer IVC of patients who have undergone unilateral eye removal. (A) Photomontage of semitangential sections from a small patch of calcarine cortex. These demonstrate the distribution of the enzyme in stripes, also known as eye dominance columns, with frequent anastomoses and bifurcations. (B) Drawing of the distribution of enzymatic activity in six tissue blocks from the same patient, covering nearly the entire extent of the striate cortex. MC, monocular crescent; CS, calcarine sulcus; D, dorsal, P, posterior. Bar = 10 mm. Reproduced from Horton and Hedley-Whyte (1984) and Horton *et al.* (1990), respectively, with permission. Courtesy of Dr. Jonathan Horton.

monocular crescent (Fig. 6B). As might be expected, this arrangement of eye dominance columns is similar to that in the macaque monkey (LeVay *et al.*, 1975, 1980), except for the differences in size mentioned previously.

3. Development

The ontogenesis of the human striate cortex provides a cogent model for understanding the developement of less well understood cortices. It is also of interest because of its direct relationship to the emergence of visual capabilities in infants. The relative wealth of information about the striate cortex makes it a nearly ideal focal point for developmental studies aimed at elucidating the mechanisms that control the development of the cortex in general. The following paragraphs review some of the salient aspects on this topic, including observations on patients with congenital anophthalmia and an experimental model of this condition in monkeys.

3.1. Normal Development

It is not known precisely when the development of the striate cortex begins, since the unique cytoarchitectonic features that distinguish it among all other cortical areas do not appear until late in intrauterine life (Vignal, 1888). Therefore, until specific markers of phenotypes unique to this area become available, the beginnings of the striate cortex can be traced, along with the rest of the cortical mantle, to about day 41 of gestation. According to Marín-Padilla (1983), fibers from a primitive internal capsule reach the cerebral vesicle at this stage, heralding the development of the cerebral cortex. Others would set the birth of the cortex at about day 54 of gestation, when the cortical plate appears (His, 1904; Bartelmez and Dekaban, 1962; reviewed in Kuljis, 1992b). In any event, there is practically nothing to distinguish the future striate cortex from any other region on the surface of the telencephalic vesicle at such early stages of development, so the earliest tentative identification of the prospective striate cortex has traditionally relied on the ability to recognize the incipient calcarine fissure.

The earliest reliable sign of the development of the visual cortex has been traced to about week 8 gestation (term = 40 weeks), when a barely discernible, shallow sulcus appears in the posterior aspect of the telencephalic vesicle, identifying the future location of the calcarine fissure (Zheng *et al.*, 1989; Masood *et al.*, 1990). At this early stage, the five major layers of the telencephalic vesicle (from the pial surface to the ventricular surface: marginal zone, cortical plate, intermediate zone, subventricular zone, and ventricular zone) are distinguishable in the *anlage* of the striate cortex. It is at this stage that the germinal epithelium of the ventricular zone exhibits its highest mitotic rate. About week 8–9 of gestation, however, the mitotic rate in the subventricular zone is very low, and it is not until week 13–15 that significant levels of mitotic activity become evident in this region. This implies that many of the cells that will occupy deeper positions within the cortical plate are generated in the ventricular zone, while those destined to occupy more superficial layers of the cortical plate are generated in both the ventricular and subventricular zones. This has important implications in terms of the origin of individual cortical laminae, and perhaps also in terms of their selective patterns of connectivity in the adult. Autoradiographic studies in every mammalian species examined to date indicate that neurons situated in deeper layers are born early in gestation, while neurons destined to more superficial layers are generated progressively later, according to the so-called inside-out gradient of histogenesis (Angevine and Sidman, 1961; Berry and Rogers, 1965; Hicks and D'Amato, 1968; Shimada and Langman, 1970; Fernández and Bravo, 1974; Rakic, 1974). It has been postulated that a combination of innate cues and epigenetic factors related to the time of birth of cortical neurons may contribute to specify the phenotypic differentiation of the neurons that make up the cerebral cortex (Rakic, 1988). This may include the differential connectivity among neurons situated in various cortical layers (Angevine and Sidman, 1961), although synaptic contacts form well after the migration of neurons to the cortical plate has taken place (Rakic *et al.*, 1986).

As a result of the process of neuronal migration and subsequent differentiation outlined above, the organization of the striate cortex undergoes progressive elaboration. Using traditional cytoarchitectonic methods, His (1904) and Filimonoff (1929) believed that they were able to recognize the striate cortex unequivocally by $3\frac{1}{2}$ months of gestation, the earliest stage that they examined. However, the full array of layers found in adults becomes detectable only by month 6 of gestation according to Brodmann (1906).

Many complex processes of differentiation continue well after the histogenesis of the striate cortex. They extend into the postnatal period, and perhaps even into the second decade of life (Huttenlocher and de Courten, 1987). Neuronal density in the striate cortex is over 10×10^6 neurons/mm³ at 21 weeks of gestation (term = 40 weeks), but at birth it decreases steeply to about 9×10^4 neurons/mm³. The decline continues until about 4 months of age, when it drops to about 4×10^4 neurons/mm³ and eventually becomes stable at a mean density of 3.5×10^4 neurons/mm³ in young adults. Thereafter, there do not seem to be any appreciable changes with age (Leuba and Garey, 1987, 1989). These figures, when corrected for tissue shrinking during processing and for the increase in the volume of the striate cortex that takes place up to postnatal month 8 imply about 31% overall cell loss in this region (Klekamp *et al.*, 1991) during development. Such cell loss is similar to the loss of 16% described in the macaque monkey (O'Kusky and Colonnier, 1982), 12% in the hamster (Finlay and Slattery, 1983), and 31% in the mouse (Heumann *et al.*, 1978). Neuronal loss in the striate cortex is presumably related to remodeling phenomena that occur in the visual system, such as the competitive elimination of overproduced retinal ganglion cell axons demonstrated in both monkeys (Rakic and Riley, 1983) and humans (Provis *et al.*, 1985), and associated phenomena that help refine the organization of the visual system under the influence of postnatal epigenetic cues.

An additional important aspect of the development of the striate cortex pertains to synapse formation and elimination. Huttenlocher and de Courten (1987) have found that synaptogenesis in the striate cortex of human infants is most rapid between 2 and 4 months of age, with a subsequent loss of about 40% of synapses between 8 months and 11 years of age. Thereafter, synapse numbers become stable, at least into the sixth decade. Similar observations have been made in the macaque monkey by Rakic *et al.* (1986), indicating that the phenomenon is generally valid for primates at least, and that synapse loss probably has a role in the refinement of cortical organization and function. For example, it has been proposed that the sudden increase in visual capacities in human infants of 4–5 months of age is related to the peak of synaptic density that occurs at about the same time throughout the cerebral cortex. The subsequent reduction in synapse number presumably reflects refinements in striate cortex circuitry that mediate subsequent improvements in visual function (Huttenlocher and de Courten, 1987).

Much remains to be investigated about postnatal developmental phenomena in the striate cortex to attain a comprehensive understanding of the relationship between structural refinement and function. For example, Burkhalter (1991) has shown a temporal difference in the establishment of extrinsic connections between areas, as opposed to intrinsic connections within the striate cortex. Using carbocyanines for postmortem pathway tracing, he has found that fibers interconnecting the striate cortex with the prestriate cortex are present at birth near the boundary between these regions. In contrast, fibers interconnecting different regions within the striate cortex are absent at birth, and do not appear until 7 months of age. This is an important finding, indicating that anatomically and functionally distinct circuits in the striate cortex develop at different stages in ontogenesis. This may help to probe the mechanisms that mediate the development of visual capabilities in humans. For example, Bronson (1974) proposed that early vision in humans depends primarily on subcortical circuits. However, Braddick and Atkinson (1988) made behavioral observations they believe indicate cortical involvement in early visual behavior in human infants. Burkhalter's findings add support to the proposal of Braddick and Atkinson, since the sequential development of different types of connections in the striate cortex may underlie some aspects of the maturation of visually dependent behavior.

3.2. Development of the Striate Cortex in Patients with Congenital Anophthalmia

The complexity in the organization of the striate cortex poses significant questions on the role of genetic versus epigenetic cues in its specification. Not surprisingly, this specific aspect of the nature versus nurture problem has been difficult to approach experimentally, yet recent observations have begun to shed some light on the subject. This subsection is concerned with the role of retinal cues in the development of the striate cortex, a subject that can be approached by analyzing patients that are born without eyes, but who nevertheless do have a striate cortex. Important additional information relevant to this topic has been obtained recently in monkeys subjected to eye ablation prenatally, which address questions that could not be answered in anophthalmic mice (reviewed in Kuljis, 1991). Together, these paradigms provide a powerful approach to understand some of the mechanisms that specify the development of the striate cortex.

3.2.1. General Findings in Congenitally Anophthalmic Humans

Perhaps one of the most interesting aspects of the development of the striate cortex pertains to the role of retinal cues in the specification of its laminar and modular architecture. One of the cleanest ways to address this experimentally would be to completely prevent the development of the eyes. This is obviously a difficult experimental paradigm, although there are several animal models that approach this ideal (reviewed in Kuljis, 1991). In humans, there is a particularly instructive experiment of nature that addresses this question: true congenital anophthalmia (Fig. 7). This is a rare condition in which no vestige of eye tissue can be demonstrated in the orbits, even after extensive histological examination of serial sections of the orbital content (Bolton, 1900; Cosmetattos, 1931; Recordon and Griffiths, 1938; Duckworth and Cooper, 1964; Haberland and Perou,

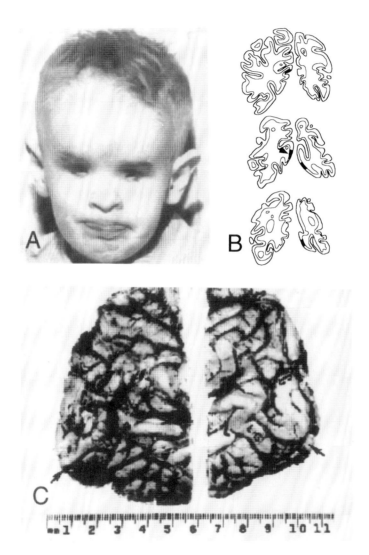

Figure 7. Characteristic features of patients with true congenital anophthalmia. (A) Appearance of the patient, with sunken orbital contents in which no microscopic remnants of eye tissue were found. (B) Sketch of three coronal sections from the occipital lobe, where a cytoarchitectonically distinct striate cortex was present (solid black lines on the right), although reduced in total surface area. A "transitional cortex between striate and parastriate areas" was also found, illustrated by the gray hatching on the left (arrowhead). (C) View of the medial aspect of the occipital lobes, which display an anomalous pattern of gyration that includes a short calcarine fissure. Modified from Haberland and Perou (1968) with permission.

1968; Brunquell *et al.*, 1984). Perhaps surprisingly, these patients have a histologically well-developed striate cortex, which can reportedly be distinguished unequivocally from adjacent cortical regions on cytoarchitectural grounds (e.g., Bolton, 1900; Haberland and Perou, 1968; Brunquell *et al.*, 1984). The striate cortex in human anophthalmia, however, may not be completely normal. Although it appears to be of normal thickness and retains most of the layers that characterize its normal counterpart, there is some dispute, in particular about layer IVB. Some authors have reported that the *stria* of Gennari is absent (Duckworth and Cooper, 1964; Haberland and Perou, 1968). Others maintain that the stria is well developed (Cosmetattos, 1931; Recordon and Griffiths, 1938; Brunquell *et al.*, 1984). Bolton (1900) also reported that this layer is present, but found that it is reduced to about half the normal thickness. Reports are variable as to other minor changes, suggesting that there may be subtle modifications in cell numbers and/or density, as well as in the demarcation of individual laminae. However, most of the remaining cytoarchitectural features of the striate cortex are not substantially affected. There is more agreement that the surface area of the striate cortex is nearly always reduced (Fig. 7B), a feature perhaps consistent with the nearly unanimous finding that the calcarine fissure is short. This is associated with changes in the gyri of the occipital lobe (Fig. 7C).

A reduction in the surface area of the striate cortex has also been documented in a primate model of anophthalmia developed by Rakic (1988) (see also Dehay *et al.*, 1989). This model also generated the intriguing finding that a novel cytoarchitectonic area, termed "area X," is interposed between the striate and prestriate cortices, and that patches of area X are found as "islands" surrounded by cytoarchitectonically typical striate cortex (Rakic *et al.*, 1991). Area X-like regions have been so far identified in only one human case, examined by Haberland and Perou (1968; Fig. 7B). Their report indicates the presence of a "transitional cortex" situated between areas 17 and 18 of Brodmann. It is not known if area X-like patches occur within the striate cortex in anophthalmic humans. This is the result principally of the lack of awareness of the phenomenon, and of the inability to examine the problem adequately in routinely prepared autopsy material. Because of these difficulties, such islands could have been present in Haberland and Perou's case, and in patients studied by other investigators, but may have gone undetected. Thus, the findings in anophthalmic monkeys not only confirm but also extend the observations in humans, emphasizing the need for more extensive cytoarchitectonic analysis in humans, and for additional cross-species comparisons in future clinical cases. Rakic (1988) has correlated the reduced area of the striate cortex with the reduced size of the lateral geniculate nucleus, which also fails to develop its characteristic lamination. He concluded that the area subtended by the striate cortex depends strongly on the number of axons supplied to it by the lateral geniculate nucleus.

It should be emphasized that the autopsy findings in most of the anophthalmic humans reviewed here were collected from patients without evidence of even vestigial remnants of eyes in their orbits. This is important, since microphthalmia is a more frequent condition than true anophthalmia. In the former condition, eyes do develop that could provide cues, which in turn may instruct the development of the striate cortex. Nevertheless, these eyes may be so small that microscopic examination of the orbit is necessary to demonstrate them (Pritkin, 1980). Since the development of the striate cortex in the patient reports summarized above presumably occurred without any cues from the retina throughout gestation, it is reasonable to assume that such cues are not essential for the specification of the main cytoarchitectonic features of the striate cortex. Retinal information may nevertheless be important, to refine the architecture of the visual cortex in postnatal life. For example, lateral (or horizontal, i.e., parallel to the pial surface) intrinsic connections in the striate cortex develop much later (i.e., postnatally) than vertical intrinsic connections, which are present before birth (Burkhalter, 1991). Presumably, such horizontal connections depend more heavily on visual experience and may not be implemented in the absence of retinal input. This is consistent with the reports that the stria of Gennari is underdeveloped or absent in some anophthalmic patients. Such a possibility is testable in the nonhuman primate model of anophthalmia, and perhaps also in future cases of human anophthalmia.

3.2.2. What Do Models of Anophthalmia Teach Us about the Mechanisms That Specify Modular Architecture in the Cerebral Cortex?

The fact that the striate cortex can develop in the complete absence of retinal cues has been confirmed not only in human anophthalmia, but also in similar rodent and primate models of this condition (Kuljis and Rakic, 1990; reviewed in Kuljis, 1991). A particularly interesting observation was made in the primate model of anophthalmia: hypercolumns are present and appear to have normal dimensions, as revealed by the presence of CO puffs. As in controls (Kuljis and Rakic, 1989), there is a strong tendency for neuropeptide Y (NPY)-containing neurons to be situated preferentially outside the puffs (Kuljis and Rakic, 1990). Taken together, both features indicate that the striate cortex can develop not only normal lamination, but also a modular compartmentalization, in the absence of cues from photoreceptors. It is not known if this is the case in humans, since the puffs have been sought in only one of the few published cases (Brunquell *et al.*, 1984). Unfortunately, the puffs could not be found in this case, and it is unclear whether this reflects an actual absence or a technical artifact. It is hard to decide between these two possibilities, since: (1) details on the length of postmortem autolysis and fixation parameters are not available; (2) the patient died at 27 years of age, after suffering from retardation of psychomotor development, a seizure disorder, hypothyroidism, and hypoadrenalism, all of which could have had adverse effects on cortical organization, in addition to the anophthalmia; (3) brain weight was low (1050 g), the frontal lobes were small, and there was a right parietal protuberance, all features suggestive of cerebral maldevelopment beyond the confines of the visual cortices, that could have had additional adverse effects on the striate cortex; and (4) it is not known whether the puffs present in the striate cortex of congenitally anophthalmic monkeys can be maintained for decades after birth. For these reasons, it is still likely that humans with pure congenital anophthalmia will eventually be found to have CO puffs, as do monkeys subjected to prenatal retinal ablation (Kuljis and Rakic, 1990).

The presence of hypercolumns in anophthalmic primates would imply that substantial aspects of cortical organization can be specified in the absence of peripheral receptor cues. Such a hypothesis is tantamount to proposing that intrinsic cortical, or thalamocortical programs are capable of instructing the development of the cortex, which does not necessitate peripheral instructive or modulatory cues (Rakic, 1988; Kuljis and Rakic, 1990). This hypothesis appears diametrically opposed to that formulated from experiments in the rodent somatosensory system, where ablation of sinus hair follicles, or their nerves, immediately after birth, results in developmental failure of their corresponding modular representation (the so-called "barrels") in the contralateral somatosensory cortex (reviewed in Kuljis, 1991). This elegant line of research has been traditionally interpreted as indicating that peripheral receptor cues (from vibrissal sinus hair follicles) have an essential instructive role. In this interpretation, receptor cues "imprint" a pattern of modular organization in the somatosensory cortex, such that there is a one-to-one correspondence between the number and relative topographic location of vibrissae and their representation in the cortex. Thus, the two experimental models (primate versus murine) appear to support opposite conclusions. The findings in the rodent somatosensory system would be most consistent with a tabula rasa model of cortical specification, in which peripheral cues imprint a pattern of modular organization onto the cortex. In contrast, models of true congenital anophthalmia indicate that the visual cortex can be specified in the absence of cues from photoreceptors, if not from the retina as a whole (see Kuljis and Rakic, 1990, for a discussion of this topic), and that the specification of this region may be instructed by a protomap contained within the brain itself (Rakic, 1988).

Several factors may account for what superficially seems like a discrepancy between the models of specification in the rodent somatosensory versus the primate visual cortices. Careful experimental testing of these factors will probably contribute to address the validity of the tabula rasa versus protomap hypotheses of cortical specification in the near future. In the first place, the comparison between the two cortices is not straightforward, since their modular organizations are not readily comparable. For example, the most easily recognized markers of modules are in different layers: barrels are situated in layer IV in rodents, while puffs are situated mainly in layers II/III in monkeys. Furthermore, barrels are markers of spatially discontinuous, entirely self-contained receptors in individual sinus hair follicles, while puffs are distributed along the representation of a continuous sheet of photoreceptors. Differences such as these raise the possibility that the mechanisms that specify functionally and anatomically different cortices may be different. This needs to be rigorously tested.

An additional complication is whether the developmental events that lead to the failure of barrel development (after *postnatal* receptor ablation) are comparable to those when the retina is surgically ablated *in utero* (Kuljis and Rakic, 1990), or fails to develop completely, as in humans with true anophthalmia. The experimental paradigms in rodents rely on postnatal surgery that occurs well after many circuits have developed in the somatosensory system. Thus, the barrels may fail to form because the ablation has a profoundly destructive effect on a sensory system that is already substantially "wired up." The striate cortex in anophthalmic humans, by contrast, develops without ever receiving retinal cues, as is the case for monkeys subjected to retinal ablation *in utero*. In the primate experiments, photoreceptor cues are never made available to the brain because these receptors were removed before they established synaptic contacts. Furthermore, the retinal ablations are made before the development of the superficial layers of the cortex, where puffs are situated (Kuljis and Rakic, 1990).

3.2.3. Is There a Functional Role for the Striate Cortex in Congenitally Anophthalmic Patients?

What functional role, if any, may the striate cortex of anophthalmic humans be playing? It is known that the visual cortex of sighted individuals has a high background level of oxidative metabolic activity (Fig. 8A), and a characteristic dominant alpha rhythm (i.e., 8–12 Hz), which is detectable electroencephalographically when the eyes are closed. Eye closure reduces the metabolic activity in this region, although the average metabolic rate and blood flow remain high compared with most of the cortex (Fig. 8B). Blindness acquired in adulthood results in a loss of the alpha rhythm, and in a clear-cut reduction in metabolic

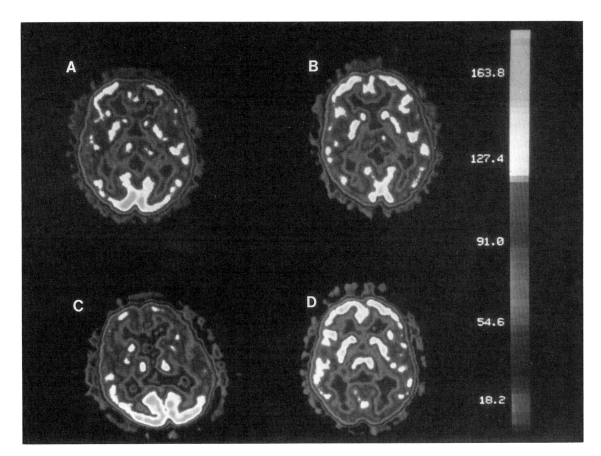

Figure 8. Positron emission tomographic scans of a normal subject with the eyes opened (A) and closed (B), from a patient with early onset blindness (C), and from a patient who became blind in adulthood (D). The scans represent horizontal brain slices that include the visual cortex, and reveal a high level of glucose utilization in the occipital lobe (lower portion of the scan in all panels) in the normal, sighted subject (A). Closing the eyes results in a decrease of glucose utilization in the occipital lobe (B). Interestingly, congenitally blind patients, as well as patients who lose vision early in postnatal life, exhibit levels of glucose utilization at least comparable to those of normal subjects with their eyes open (C). By contrast, patients who lost vision as adults exhibit low levels of glucose utilization in the visual cortex (D), which are clearly below those of controls with their eyes closed (B). The scale to the right represents percents of mean glucose utilization. Modified from Veraart *et al.* (1990) and reproduced with permission. Courtesy of Drs. Claude Veraart and André Goffinet.

activity in the visual cortex, below the levels observed in normals with their eyes closed (Wanet-Defalque *et al.*, 1988; Veraart *et al.*, 1990; Fig. 8D). Surprisingly, patients who become blind early in infancy or prenatally exhibit high levels of metabolic activity in the presumed region of the visual cortex (Wanet-Defalque *et al.*, 1988; Veraart *et al.*, 1990; Fig. 8C). This would be consistent with our observation of a normal topographic pattern of CO activity in the striate cortex of macaque monkeys subjected to prenatal retinal ablation (Kuljis and Rakic, 1990), and with Allman and Zucker's (1990) proposal that "the surviving population of neurons in the lateral geniculate nucleus in monkeys with their eyes removed well before birth might be sufficient to provide a tonic drive to the blobs [a.k.a. "puffs"] and thus engage their metabolic machinery." Congenitally blind, as well as early blind patients display no alpha rhythm, but they do have ongoing slow negative cortical potentials similar to those in sighted controls (Noebels *et al.*, 1978). The above observations imply that the striate cortex of congenitally blind subjects is far from inactive, despite the lack of retinal input. It is not known, however, if the ongoing spontaneous activity in the striate cortex of congenitally blind patients has any beneficial, adaptive role, since it may simply reflect residual, functionally irrelevant activity.

Similar issues arise in the case of congenitally deaf individuals. In such patients, it is somewhat clearer that the auditory cortex may still have a functionally relevant role. That is, congenitally deaf patients exhibit a spread of visually evoked electrocortical activity into their presumed "auditory" cortices, suggesting that these cortices may be used for visual information processing (Neville, 1990). This intriguing possibility may underlie the folk notion, which has some experimental support (Ryugo *et al.*, 1975), that spared sensory modalities are enhanced in deaf or blind individuals. In a strictly speculative vein, the cross-modality utilization of residual visual cortex in congenitally blind individuals may be susceptible to exploitation in helping them use sensory substitution strategies (Veraart and Wanet, 1986; Bachy-Rita, 1972).

Finally, because of the relatively intact architecture and modular organization of the visual cortex in congenitally blind patients, it may be theoretically possible to provide them with vision by means of prosthetic devices capable of directly stimulating their striate cortex (Brindley, 1972; Button and Putnam, 1962; Dobelle *et al.*, 1974) and thus engaging their residual visual structures. This hypothetical prospect will require, among other advances, a more comprehensive experimental understanding of the factors that control the development, specification, and maintenance of visual structures in both animals and humans.

4. Involvement of the Striate Cortex in Alzheimer's Disease

Alzheimer's disease is a progressive dementia of unknown cause that afflicts over four million persons in the United States. Neuropathologically, the disease is characterized by the presence of large numbers of senile plaques (with variable amounts of dystrophic neurites and amyloid protein deposits), diffuse interstitial amyloid deposits without dystrophic neurites (also known as "diffuse plaques"), neurofibrillary tangles within neuronal perikarya, diffusely scattered dystrophic neurites (also known as "neuropil threads"), and cell and synapse loss (Khachaturian, 1985; Mirra *et al.*, 1991).

It was first thought that the visual symptoms in patients with this condition did not correlate with anatomical changes in the visual system. This seeming paradox can be traced to early descriptions of the topographic distribution of the macro- and microscopic changes that emphasized the relative paucity of cortical atrophy, as well as microscopic stigmata, in the striate cortex (Blessed *et al.*, 1968; Brun and Gustafson, 1976; Brun and Englund, 1981). This provided the basis for what is still referred to as the "sparing" or "relative sparing" of this area in Alzheimer's disease (Mirra *et al.*, 1991). As a result, visual dysfunction in Alzheimer's patients has often been explained mainly on the basis of the relatively high density of lesions (predominantly neurofibrillary tangles) in association cortices, including the visual association cortices. However, many recent reports indicate that the notion about sparing of the striate cortex in Alzheimer's disease may not be entirely correct. One of the issues that remains to be addressed is the observation that both the total volume and neuronal size in the striate cortex do not change appreciably with age, at least up to the seventh decade (Haug, 1984). This is intriguing, considering that a small but macroscopically noticeable degree of sulcal widening and gyral narrowing is found in virtually all normal elderly individuals. Such changes are commonly referred to as age-associated "brain atrophy," and incorrectly thought of as more or less uniformly distributed throughout all regions of the cerebral cortex. Given the relative resistance of the striate cortex to age-associated brain shrinkage, it is possible that pathologic processes that result in cortical atrophy in the elderly are less apparent in this region. Since this feature is shared with the motor cortex, and since both cortices display less relative shrinkage than other cortical areas in normal elderly individuals and in patients with Alzheimer's disease, the presumed resiliency of these regions to the effects of Alzheimer's disease may be more apparent than real.

It has been recently recognized that retinal and optic nerve involvement may play a role in the visual defects observed in patients with Alzheimer's disease. These include anomalies of color vision, spatial contrast sensitivity, and susceptibility to visual masks (Hinton *et al.*, 1986; Blanks *et al.*, 1989; Katz and Rimmer, 1989). Electrophysiological assessment, using tools such as electroretinography and cortical visual evoked potentials, has supported the reports of retinal involvement, while also indicating that there is primary, as well as association visual cortical dysfunction (Katz *et al.*, 1989; Cronin-Golomb *et al.*, 1991). For example, pattern-evoked electroretinograms are often abnormal in patients with Alzheimer's disease, while flash-evoked electroretinograms are less affected (Katz *et al.*, 1989). In contrast, cortical potentials evoked by flashes of light are more commonly affected than potentials elicited by shifting patterns. There is also a host of psychophysical abnormalities that include impaired contrast sensitivity (predominantly at low spatial frequencies), tritanomalous defects in color discrimination and deficits in stereoacuity which, in the absence of ocular abnormalities on clinical examination, suggest a primary cortical basis for these manifestations (Cronin-Golomb *et al.*, 1991). Obviously, there needs to be an awareness of the probable involvement of the striate cortex in the visual defects associated with Alzheimer's disease.

Several independent groups of investigators have reassessed the precise mode of involvement of the striate cortex in Alzheimer's disease, and there is considerable agreement in their observations. For example, immunolabeling for the abnormal cytoskeletal protein A68 and Gallyas's silver impregnation method for Alzheimer's cytoskeletal lesions reveal a stereotyped pattern of involvement that consists of lesions concentrated in layers I–III and V, and which essentially spares layers IV and VI (Fig. 9; Beach *et al.*, 1989; Braak *et al.*, 1989; De Carlos *et al.*, 1990; Kuljis and Van Hoesen, 1991). This bilayered distribution of lesions turns out to be a general feature of the entire cerebral cortex, that may reflect a common mode of involvement related to the basic uniformity in structure of the cortex (Kuljis and Van Hoesen, 1991, and unpublished results). This basic bilaminar pattern is only quantitatively modified among regions, with increasing numbers and density of lesions in the areas traditionally described as severely affected. This view modifies classical notions about the selective regional distribution of lesions in Alzheimer's disease, in that it suggests a gradient of pathology in which the striate cortex is a distinct participant. A study of the distribution of astrocytic gliosis, revealed by immunolabeling of glial fibrillary acidic protein, also shows a distinct laminar pattern. Lesions are concentrated in superficial and deep layers, and tend to avoid layer IV. Despite some problems with the identification of individual layers, the pattern of involvement of nonneuronal cells appears similar to that revealed by methods selective for neuronal cytoskeletal lesions (Beach and McGeer, 1988).

Interestingly, and in agreement with previous observations, the recent studies indicate that the lesions in the striate cortex in Alzheimer's disease consist mainly of senile plaques, neuropil threads, and neuropil threadlike labeling with antibodies to A68 (this antibody labels cells and processes that do not have cytoskeletal abnormalities detectable by other methods, in addition to some cytoskeletal lesions; Fig. 9C). However, neurofibrillary tangles are virtually absent from the striate cortex, although large numbers of tangles in association cortices are necessary to fulfill the current diagnostic criteria for Alzheimer's disease (Khachaturian, 1985; Mirra *et al.*, 1991). This is an important difference between primary and association cortices, since, until recently, much emphasis was placed on the *density of tangles*, and not of senile plaques, in assessing the severity of Alzheimer's disease (Blessed *et al.*, 1968; Tomlinson *et al.*, 1970). However, Arnold *et al.* (1991) have shown a consistent tendency for a reciprocal relation between tangles and plaques. In the association cortices of patients with Alzheimer's disease, neurofibrillary tangles are present in high densities, but the density of senile plaques is low. Their primary cortices, by contrast, have low densities of tangles, in agreement with classical teaching, but quite high densities of senile plaques. The latter observation had apparently escaped the attention of earlier investigators.

The preceding findings raise a new perspective about the involvement of the striate cortex in Alzheimer's disease. This area of the cortex can no longer be conceived of as "spared," since it exhibits a density of senile plaques that is among the highest of all cortical areas (Arnold *et al.*, 1991). This observation, coupled with the striking laminar distribution of the cytoskeletal lesions described above, and the recent finding of a spatially discontinuous, periodic distribution of plaques in the tangential domain (Tikoo and Kuljis, 1991, and in preparation), call for a renewed effort in reassessing the precise mode of involvement of the striate cortex in Alzheimer's disease. Such an undertaking may help to understand the pathophysiology of cortical dysfunction in this condition. For example, while the precise impact (or lack thereof) of neurofibrillary tangles and

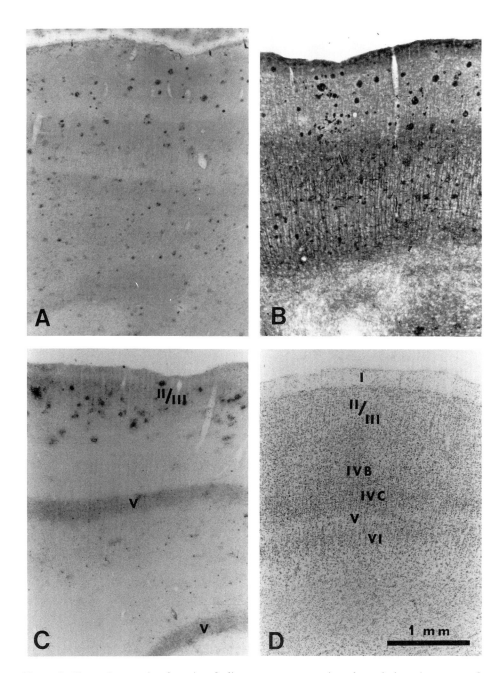

Figure 9. Photomicrographs of a series of adjacent transverse sections through the striate cortex of a 79-year-old woman with Alzheimer's disease, labeled, impregnated, or stained according to the modified Bielschowsky method (A), the method of Campbell *et al.* (1987) (B), A68-like immunoreactivity (C), and cresyl violet (D, for cytoarchitectonic orientation). Note the large numbers of senile plaques (A) and diffuse amyloid plaques (B) present in this region, as well as the characteristic bilaminar pattern of immunoreactivity to A68 (C). These features are described in detail in the text. Combinations of Roman numerals and letters designate cortical layers in panels C and D, according to the conventions illustrated in Figs. 1–4.

senile plaques on cortical function is often hotly debated, a solution to the problem has remained elusive to experimental assessment. This is why the analysis of the striate cortex may contribute to the elucidation of some aspects of the pathophysiology of Alzheimer's disease, since the lesions that this region contains are predominantly senile plaques, and they are distributed in a highly stereotypic, laminar fashion. As discussed in the preceding sections, the laminar connectivity of the striate cortex is highly selective and known in great detail in macaques, which, incidentally, only develop plaques and not tangles as they age. This knowledge, extrapolated cautiously to humans, may help to determine whether there is selective vulnerability among the parallel pathways (and, by extension, the functions) served by the striate cortex. Selective vulnerability is plausible since the visual deficits in Alzheimer's disease may depend, at least in part, on the sites of termination or origin of pathways in which the plaques are situated. Recent studies of this kind have tentatively identified corticocortical and corticocollicular pathways as potential candidates for involvement (Kuljis, 1992a). For example, the presence of a band of dystrophic neurites and A68-containing processes in layer V (Fig. 9) in patients with Alzheimer's disease suggests that the outputs from neurons situated in this layer, to the pulvinar and to the superior colliculus, may be affected. This hypothesis is consistent with defects in extraocular movements (colliculus) and visual attention (pulvinar), which have been well documented in patients with Alzheimer's disease (Katz and Rimmer, 1989; Cronin-Golomb *et al.,* 1991; Kuljis, 1993). Obviously, the presence of newly recognized lesions in some of the target structures themselves (e.g., Kuljis, 1993) are likely contributors to the defects that may result from lesions in the striate cortex, and an effort needs to be made to sort out their role in the visual symptoms in Alzheimer's disease. Should this connectivity-biased approach to the pathophysiology of Alzheimer's disease be successful, it may represent a virtually unique opportunity to understand the precise impact of senile plaques in cortical function. This approach, augmented by psychophysical assessment, may be applicable to several other diseases and conditions affecting this region, since there is a large body of detailed knowledge about the organization and function of the striate cortex.

5. Conclusions

The human primary visual (striate) cortex has enjoyed the attention of investigators for over 200 years. As a result, it is perhaps the best understood of all areas in the human cerebral cortex. Recent improvements in neuroimaging methods, postmortem pathway tracing techniques, psychophysical methods, and new developmental paradigms promise ongoing progress in elucidating the substrates of visual capabilities in humans. Together with data gathered from the study of other sensory systems, we may soon see the emergence of clinical applications of this knowledge to alleviate conditions such as congenital and acquired blindness. This may take the form of an implantable prosthesis or extracranial stimulation device, coupled with high-resolution video cameras connected to a portable computer. Such a device may be capable of providing spatiotemporally discernible stimuli to the striate cortex and allow a blind patient to navigate safely in space.

Another important development is the application of our understanding of the anatomical and physiological organization of the visual cortex to analyze its precise mode of involvement in specific pathological conditions. Perhaps the prototype of this approach is the analysis of the substrate of visual impairment in Alzheimer's disease, which will hopefully serve as a model for approaches to understand several other conditions.

ACKNOWLEDGEMENTS. I would like to thank Drs. Heiko Braak, Andreas Burkhalter, Jonathan Horton, André Goffinet, and Claude Veraart for granting permission to reproduce illustrations from some of their work. I am also grateful to Drs. K. S. Rockland and A. Peters for their editorial guidance, to Drs. M. Rizzo and M. Wall for helpful criticisms on the manuscript, and to Mrs. Shirley Knapp for assistance with histological and photographic processing. Material from my laboratory was prepared with support from the American Federation for Aging Research, Grant RR 05372 from the Biomedical Research Support Branch, Division of Research Facilities and Resources, National Institutes of Health, and PHS-NIH grant NS29856.

6. References

Allman, J., and Zucker, S., 1990, Cytochrome oxidase and functional coding in primate striate cortex: A hypothesis, *Cold Spring Harbor Symp. Quant. Biol.* **55**:979–982.

Angevine, J. B., and Sidman, R. L., 1961, Autoradographic study of cell migration during histogenesis of cerebral cortex in the mouse, *Nature* **192**:766–768.

Arnold, S. E., Hyman, B. T., Flory, J., Damasio, A. R., and Van Hoesen, G. W., 1991, The topographical and neuroanatomical distribution of neurofibrillary tangles and neuritic plaques in the cerebral cortex of patients with Alzheimer's disease, *Cere. Cortex* **1**:103–116.

Bach-y-Rita, P., 1972, *Brain Mechanisms in Sensory Substitution,* Academic Press, New York.

Baillarger, J. G. F., 1840, Recherches sur la structure de la couche corticale des circonvolutions du cerveau, *Mem. Acad. Med.* **8**:149–183.

Bartelmez, G. W., and Dekaban, A. S., 1962, The early development of the human brain, *Contrib. Embryol.* (Publ. 253) **37**:13–32.

Beach, T. G., and McGeer, E. G., 1988, Lamina-specific arrangement of astrocytic gliosis and senile plaques in Alzheimer's disease visual cortex, *Brain Res.* **463**:357–361.

Beach, T. G., Walker, R., and McGeer, E. G., 1989, Lamina-selective A68 immunoreactivity in primary visual cortex of Alzheimer's disease patients, *Brain Res.* **501**:171–174.

Bell, M. A., and Ball, M. J., 1990, Neuritic plaques and vessels of visual cortex in aging and Alzheimer's dementia, *Neurobiol. Aging* **11**:359–370.

Belliveau, J. W., Kennedy, D. N., McKinstry, R. C., Buchbinder, B. R., Weiskoff, R. M., Cohen, M. S., Vevea, J. M., Brady, T. J., and Rosen, B. R., 1991, Functional mapping of the human visual cortex by magnetic resonance imaging, *Science* **254**:716–719.

Berlin, R., 1858, Beitrag zur Structurlehre der Broßhirnwindungen, Inauguraldissertation, Junge, Erlangen.

Berman, N. E. J., and Fredrickson, E., 1992, Morphology and laminar distribution of neuropeptide Y immunoreactive neurons in the human striate cortex, *Synapse* **11**:20–27.

Berry, M., and Rogers, A. W., 1965, The migration of neuroblasts in the developing cerebral cortex, *J. Anat.* **99**:691–709.

Blanks, J. C., Hinton, D. R., Sadun, A. A., and Miller, C. A., 1989, Retinal ganglion cell degeneration in Alzheimer's disease, *Brain Res.* **501**:364–372.

Blessed, G., Tomlinson, B. E., and Roth, M., 1968, The association between quantitative measures of dementia and of senile change in the grey matter of elderly subjects, *Br. J. Psychiatry* **114**:797–811.

Blümcke, I., Hof, P. R., Morrison, J. H., and Celio, M. R., 1990, Distribution of parvalbumin immunoreactivity in the visual cortex of Old World monkeys and humans, *J. Comp. Neurol.* **301**:417–432.

Bolton, J. S., 1900, The exact histological localisation of the visual area of the human cerebral cortex, *Philos. Trans. R. Soc. London Ser. B* **193**:165–222.

Braak, E., 1982, *On the Structure of the Human Striate Area*, Springer-Verlag, Berlin.

Braak, H., 1976, On the striate area of the human isocortex. A Golgi and pigmentarchitectonic study, *J. Comp. Neurol.* **166**:341–364.

Braak, H., Braak, E., and Kallus, P., 1989, Alzheimer's disease: Areal and laminar pathology in the occipital isocortex, *Acta Neuropathol.* **77**:494–506.

Braddick, O., and Atkinson, J., 1988, Sensory selectivity, attentional control, and cross-channel integration in early visual development, *Minn. Symp. Child Psychol.* **20**:105–143.

Brindley, G. S., 1972, The variability of the human striate cortex, *J. Physiol. (London)* **225**:1–3P.

Brindley, G. S., and Lewin, W. S., 1968, The sensations produced by electrical stimulation of the visual cortex, *J. Physiol. (London)* **196**:479–493.

Brodmann, K., 1903, Beiträge zur histologischen lokalisation der Großhirnrinde. Der Calcarina-typus, *J. Psychol. Neurol.* **2**:133–159.

Brodmann, K., 1906, Beiträge zur Histologischen Lokalisation der Großhirnrinde. V. Über den allgemeinen Bauplan des Cortex palii bei den Mammaliern und zwei homologe Rindenfelder im besonderen. Zugleich ein Betrag zur Furchenlehre, *J. Psychol. Neurol.* **6**:275–400.

Brodmann, K., 1909, Vergleichende Lokalisationslehre der Großhirnrinde, Barth, Leipzig.

Brodmann, K., 1918, Individuelle variationen der sehsphare und ihre bedeutung fur die klinik der hinterauptschusse, *Allg. Z. Psychiatr.* (Berlin) **74**:564–568.

Bronson, G., 1974, The postnatal growth of visual capacity, *Child Dev.* **45**:873–890.

Brun, A., and Englund, E., 1981, Regional pattern of degeneration in Alzheimer's disease: Neuronal loss and histopathological grading, *Histopathology* **5**:549–564.

Brun, A., and Gustafson, L., 1976, Distribution of cerebral degeneration in Alzheimer's disease, *Arch. Psychiatr. Nervenkr.* **223**:15–33.

Brunquell, P. J., Papale, J. H., Horton, J. C., Williams, R. S., Zgrabik, M. J., Albert, D. M., and Hedley-Whyte, E. T., 1984, Sex-linked hereditary bilateral anophthalmos. Pathologic and radiologic correlation, *Arch. Ophthalmol.* **102**:108–113.

Burkhalter, A. 1991, Developmental status of intrinsic connections in visual cortex of newborn humans, in: *The Changing Visual System* (P. Bagnoli and W. Hodos, eds.), Plenum Press, New York, pp. 247–254.

Burkhalter, A., and Bernardo, K. L., 1989, Organization of corticocortical connections in human visual cortex, *Proc. Natl. Acad. Sci. USA* **86**:1071–1075.

Button, J., and Putnam, T., 1962, Visual responses to cortical stimulation in the blind, *J. Iowa Med. Soc.* **52**:17–21.

Campbell, S. K., Switzer, R. C., and Martin, T. L., 1987, Alzheimer's plaques and tangles: A controlled enhanced silver staining method, *Soc. Neurosci. Abstr.* **13**:678.

Chan-Palay, V., Palay, S. L., and Billings-Gagliardi, S. M., 1974, Meynert cells in the primate visual cortex, *J. Neurocytol.* **3**:631–658.

Chugani, H. T., Phelps, M. E., and Mazziotta, J. C., 1987, Positron emission tomography study of human brain functional development, *Ann. Neurol.* **22**:487–497.

Clark, V. P., Courchesne, E., and Grafe, M., 1992, *In vivo* myeloarchitectonic analysis of human striate cortex using magnetic resonance imaging, *Cereb. Cortex* **2**:417–424.

Clark, W. E. L. G., 1942, The cells of Meynert in the visual cortex of the monkey, *J. Anat.* **76**:369–376.

Clarke, S., and Miklossy, J., 1990, Occipital cortex in man: Organization of callosal connections, related myelo- and cytoarchitecture, and putative boundaries of functional visual areas, *J. Comp. Neurol.* **298**:188–214.

Cosmetattos, G. F., 1931, De la structure du centre visuel cérébral chez les anophthalmes congenitaux, *Arch. Ophthalmol.* **48**:282–289.

Cronin-Golomb, A., Corkin, S., Rizzo, J. F., Cohen, J., Growdon, J. H., and Banks, K. S., 1991, Visual dysfunction in Alzheimer's disease: Relation to normal aging, *Ann. Neurol.* **29**:41–52.

Damasio, H., Kuljis, R. O., Yuh, W., Van Hoesen, G. W., and Ehrhardt, J., 1991, Magnetic resonance imaging of human intracortical structure *in vivo*, *Cereb. Cortex* **1**:374–379.

De Carlos, J. A., López-Mascaraque, L., and Valverde, F., 1990, Morphological characterization of Alz-50 immunoreactive cells in the developing neocortex of kittens, in: *The Neocortex: Oxtogeny and Phylogeny* (B. L. Finlay, G. Innocenti, and H. Scheich, eds.), Plenum Press, New York, pp. 193–197.

Dehay, C., Horsburgh, G., Berland, M., Killackey, H., and Kennedy, H., 1989, Maturation and connectivity of the visual cortex in monkey is altered by prenatal removal of retinal input, *Nature* **337**:265–267.

Dobelle, W. H., Mladejovsky, M. G., and Girvin, J. P., 1974, Artificial vision for the blind. Electrical stimulation of visual cortex offers hope for a functional prosthesis, *Science* **183**:440–444.

Duckworth, T., and Cooper, E. R. A., 1964, A study of anophthalmia in an adult, *Acta Anat.* **63**:509–522.

Economo, C. F. von, and Koskinas, G. N., 1925, Die Cytoarchitektonik der Hirnirinde des erwachsenen Menschen, Springer-Verlag, Berlin.

Felleman, D. J., and Van Essen, D. C., 1991, Distributed hierarchical processing in the primate cerebral cortex, *Cere. Cortex* **1**:1–47.

Fernández, V., and Bravo, H., 1974, Autoradiographic study of development of the cerebral cortex in the rabbit, *Brain Behav. Evol.* **9**:317–332.

Filimonoff, I. N., 1929, Zur embryonalen und postembryonalen Enwicklung der Großhirnirinde des Menschen, *J. Psychol. Neurol.* **39**:323–389.

Filimonoff, I. N., 1932, Uber die variabilitat der grosshirnrindenstruktur. Regio occiptalis beim erwachsenen menschen, *J. Psychol. Neurol.* **44**:1–96.

Finlay, B. L., and Slattery, M., 1983, Local differences in the amount of early cell death in neocortex predict local specialization, *Science* **219**:1349–1351.

Fishbein, D. S., Chrousos, G. A., Di Chiro, G., Wayner, R. E., Patronas, N. J., and Larson, S. M., 1987, Glucose utilization of visual cortex following extra-occipital interruptions of the visual pathways by tumor. A positron emission tomographic study, *J. Clin. Neuro-ophthalmol.* **7**:63–68.

Florence, S. L., and Kaas, J. H., 1992, Ocular dominance columns in area 17 of Old World macaque and talapoin monkeys: Complete reconstruction and quantitative analyses, *Visual Neurosci.* **8**:449–462.

Fox, P. T., Mintun, M. A., Raichle, M. E., Miezin, F. M., Allman, J. M., and Van Essen, D. C., 1986, Mapping human visual cortex with positron emission tomography, *Nature* **323**:806–809.

Fox, P. T., Miezin, F. M., Allman, J. M., Van Essen, D. C., and Raichle, M. E., 1987, Retinotopic organization of human visual cortex mapped with positron-emission tomography, *J. Neurosci.* **7**:913–922.

Garey, L. J., 1971, A light and electron microscopic study of the visual cortex of the cat and monkey, *Proc. R. Soc. London Ser. B* **179**:21–40.

Gennari, F., 1782, *De Peculiari Structura Cerebri Nonnulisque Ejus Morbus,* Ex Regio, Parmae.

Gilbert, C. D., Hirsch, J. A., and Wiesel, T. N., 1990, Lateral interactions in visual cortex, *Cold Spring Harbor Symp. Quant. Biol.* **55**:663–677.

Girvin, J. P., 1988, Current status of artificial vision by electrocortical stimulation, *Can. J. Neurol. Sci.* **15**:58–62.

Glickstein, M., 1988, The discovery of the visual cortex, *Sci. Am* **259**:118–127.

Glickstein, M., and Rizzolatti, G., 1984, Francesco Gennari and the structure of the cerebral cortex, *Trends Neurosci.* **7**:464–467.

Glickstein, M., and Whitteridge, D., 1987, Tatsuji Inouye and the mapping of the visual fields in the human visual cortex, *Trends Neurosci.* **10**:350–353.

Haberland, C., and Perou, M., 1968, Primary bilateral anophthalmia, *J. Neuropathol. Exp. Neurol.* **28**:337–351.

Haug, H., 1984, Macroscopic and microscopic morphometry of the human brain and cortex. A survey in the light of new results, in: *Brain Pathology* (G. Pilleri and F. Tagliavini, eds.), **1**:123–149.

Heumann, D., Leuba, G., and Rabinowicz, T., 1978, Postnatal development of the mouse cerebral neocortex. IV. Evolution of the total cortical volume, of the population of neurons and glial cells, *J. Hirnforsch.* **19**:385–393.

Hicks, S. P., and D'Amato, C. J., 1968, Cell migrations to the isocortex of the rat, *Anat. Rec.* **160**:619–634.

Hinton, D. R., Sadun, A. A., Blanks, J. C., and Miller, C., 1986, Optic nerve degeneration in Alzheimer's disease, *N. Engl. J. Med.* **315**:485–487.

His, W., 1904, *Die Entwickelung des Menschlichen Gehirns wärend der ersten Monate,* Hirzel, Leipzig, pp. 176–180.

Hitchcock, P. F., and Hickey, T. L., 1980, Ocular dominance columns: Evidence for their presence in humans, *Brain Res.* **182**:176–179.

Hockfield, S., Tootell, R. B., and Zaremba, S., 1990, Molecular differences among neurons reveal an organization of human visual cortex, *Proc. Natl. Acad. Sci. USA* **87**:3027–3031.

Holmes, G., and Lister, W. T., 1916, Disturbances of vision from cerebral lesions with special reference to the cortical representation of the macula, *Brain* **39**:34–73.

Horton, J. C., 1984, Cytochrome oxidase patches: A new cytoarchitectonic feature of monkey visual cortex, *Philos. Trans. R. Soc. London Ser. B* **304**:199–253.

Horton, J. C., and Hedley-Whyte, E. T., 1984, Mapping of cytochrome oxidase patches and ocular dominance columns in human visual cortex, *Philos. Trans. R. Soc. London Ser. B* **304**:255–272.

Horton, J. C., Dagi, L. R., McCrane, E. P., and de Monasterio, F. M., 1990, Arrangement of ocular dominance columns in human visual cortex, *Arch. Ophthalmol.* **108**:1025–1031.

Hoyt, W. F., and Margolis, W. T., 1970, Arterial supply of the striate cortex: Angiographic changes with occlusion of the posterior cerebral artery, *Excerpta Med. Int. Congr.* 1323–1332.

Hoyt, W. F., and Newton, T. M., 1970, Angiographic changes with occlusion of arteries that supply the visual cortex, *N.Z. Med. J.* **72**:310–317.

Huttenlocher, P. R., and de Courten, C., 1987, The development of synapses in striate cortex of man, *Hum. Neurobiol.* **6**:1–9.

Huxley, T. H., 1861, On the brain of *Ateles paniscus, Proc. Zool. Soc. London* p. 247.

Katz, B., and Rimmer, S., 1989, Ophthalmologic manifestations of Alzheimer's disease, *Surv. Ophthalmol.* **34**:31–43.

Katz, B., Rimmer, S., Iragui, V., and Katzman, R., 1989, Abnormal pattern electroretinogram in Alzheimer's disease: Evidence for retinal ganglion cell degeneration? *Ann. Neurol.* **26**:221–225.

Khachaturian, Z. S., 1985, Diagnosis of Alzheimer's disease, *Arch. Neurol.* **42**:1097–1105.

Klekamp, J., Riedel, A., Harper, C., and Kretschmann, H. J., 1991, Quantitative changes during the postnatal maturation of the human visual cortex, *J. Neurol. Sci.* **103**:136–143.

Kuljis, R. O., 1991, Development of the visual cortex deprived prenatally of retinal cues, in: *The Changing Visual System* (P. Bagnoli and W. Hodos, eds.), Plenum Press, New York, pp. 255–267.

Kuljis, R. O., 1992a, Differential distribution of Alzheimer's disease lesions in the striate cortex and visual nuclei, *Neurology* **42**(Suppl.):444 (1007P).

Kuljis, R. O., 1992b, Development of the human brain: The emergence of the neural substrate for pain and conscious experience, in: *The Beginnings of Individual Human Life: Medical, Ethical and Legal Issues* (F. Beller, R. F. Weir, and H. M. Sass, eds.), Kluwer, in press.

Kuljis, R. O., 1993, Lesions in the pulvinar of patients with Alzheimer's disease, in: *Alzheimer's Disease and Related Disorders. Advances in the Biosciences,* Vol. 87 (M. Nicolini, P. F. Zatta, and B. Corain, eds.). Pergamon, London, pp. 191–192.

Kuljis, R. O., and Rakic, P., 1989, Neuropeptide Y-containing neurons are situated outside cytochrome oxidase puffs in macaque visual cortex, *Visual Neusosci.* **2**:57–62.

Kuljis, R. O., and Rakic, P., 1990, Hypercolumns in primate visual cortex can develop in the absence of cues from photoreceptors, *Proc. Natl. Acad. Sci. USA* **87**:5303–5306.

Kuljis, R. O., and Van Hoesen, G. W., 1991, Pancortical, regionally distinct bilaminar distribution of A68 immunoreactivity in Alzheimer's disease, *Neurology* **41**(Suppl.):377 (909S).

Kushner, M. J., Rosenquist, A., Alavi, A., Rosen, M., Dann, R., Fazekas, F., Bosley, T., Greenberg, J., and Reivich, M., 1988, Cerebral metabolism and patterned visual stimulation: A positron emission tomographic study of the human visual cortex, *Neurology* **38**:89–95.

Leuba, G., and Garey, L. J., 1987, Evolution of neuronal numerical density in the developing and aging human visual cortex, *Hum. Neurobiol.* **6**:11–18.

Leuba, G., and Garey, L. J., 1989, Comparison of neuronal and glial numerical density in primary and secondary visual cortex in man, *Exp. Brain Res.* **77**:31–38.

LeVay, S., 1973, Synaptic patterns in the visual cortex of the cat and monkey. Electron microscopy of Golgi preparations, *J. Comp. Neurol.* **150**:53–86.

LeVay, S., Hubel, D. H., and Wiesel, T. N., 1975, The pattern of ocular dominance columns in macaque visual cortex revealed by a silver stain, *J. Comp. Neurol.* **159**:559–576.

LeVay, S., Wiesel, T. N., and Hubel, D. H., 1980, The development of ocular dominance columns in normal and visually deprived monkeys, *J. Comp. Neurol.* **191**:1–52.

Livingstone, M. S., and Hubel, D. H., 1984, Specificity of intrinsic connections in primate primary visual cortex, *J. Neurosci.* **4**:2830–2835.

Lorente de Nó, R., 1938, The cerebral cortex: Architecture, intracortical connections and motor projections, in: *Physiology of the Nervous System* (J. F. Fulton, ed.), Oxford University Press, London, pp. 291–321.

Lund, J. S., 1973, Organization of neurons in the visual cortex, area 17, of the monkey (*Macaca mulatta*), *J. Comp. Neurol.* **147**:455–496.

Lund, J. S., and Boothe, R. G., 1975, Interlaminar connections and pyramidal organization in the visual cortex, area 17, of the macaque monkey, *J. Comp. Neurol.* **159**:305–334.

Lund, J. S., Lund, R. D., Hendrickson, A. H., Bunt, A. H., and Fuchs, A. F., 1975, The origin of efferent pathways from the primary visual cortex, area 17, of the macaque monkey as shown by retrograde transport of horseradish peroxidase, *J. Comp. Neurol.* **164**:287–304.

McIntosh, H., and Parkinson, D., 1990, GAP-43 in adult visual cortex, *Brain Res.* **518**:324–328.

Marg, E., Adams, J. F., and Rutkin, B., 1968, Receptive fields of cells in the human visual cortex, *Experientia* **24**:348–350.

Marín-Padilla, M., 1983, Structural organization of the human cerebral cortex prior to the appearance of the cortical plate, *Anat. Embryol.* **168**:21–40.

Marín-Padilla, M., 1987, The chandelier cell of the human visual cortex: A Golgi study, *J. Comp. Neurol.* **256**:61–70.

Marín-Padilla, M., and Marín-Padilla, M. T., 1982, Origin, prenatal development and structural organization of layer I of the human cerebral (motor) cortex, *Anat. Embryol.* **164**:161–206.

Masood, F., Wadhwa, S., and Bijlani, V., 1990, Early development of visual cortex in human fetuses, *Arch. Ital. Anat. Embryol.* **95**:1–10.

Meynert, T., 1867, Der Bau der Großhirnrinde und seine örtlichen Verschiedenheiten, nebst einem pathologisch-anatomischen Corollarium, *V. Jahresschr, Psychiatr.* **1**:77–93.

Meynert, T., 1868, Der Bau der Großhirnrinde und seineörtlichen Verschiedenheiten, nebst einem pathologisch-anatomischen Corallarium, *V. Jahresschr. Psychiatr.* **2**:88–113.

Meynert, T., 1872, in *Sticker's Handbuch D. Gewebelehre*, Volume II (quoted in Chan-Palay *et al.*, 1974).

Miller, N. R., 1982, *Walsh and Hoyt's Clinical Neuro-Opthalmology*, 4th ed., Williams & Wilkins, Baltimore, pp. 146–147.

Mirra, S. S., Heyman, A., McKeel, D., Sumi, S. M., Crain, B. J., Brownlee, L. M., Vogel, F. S., Hughes, J. P., van Belle, G., and Berg, L., 1991, The Consortium to Establish a Registry for Alzheimer's Disease (CERAD). Part II, Standardization of the neuropathologic assessment of Alzheimer's disease, *Neurology* **41**:479–486.

Mora, B. N., Carman, G. J., and Allman, J. M., 1989, *In vivo* functional localization of the human visual cortex using positron emission tomography and magnetic resonance imaging, *Trends Neurosci.* **12**:282–284.

Neville, H., 1990, Intermodal competition and compensation in development. Evidence from studies of the visual system in congenitally deaf adults, *Acad. Sci. Ann. N.Y.* **208**:71–91.

Noebels, J. L., Roth, W. T., and Kopell, B. S., 1978, Cortical slow potentials and the occipital EEG in congenital blindness, *J. Neurol. Sci.* **37**:51–58.

O'Kusky, J., and Colonnier, M., 1982, Postnatal changes in number of neurons and synapses in visual cortex (area 17) of macaque monkey: A stereological analysis in normal and monocularly deprived animals, *J. Comp. Neurol.* **210**:291–306.

Phillipson, O. T., Kilpatric, I. C., and Jones, M. W., 1987, Dopaminergic innervation of the primary visual cortex in the rat, and some correlations with human visual cortex, *Brain Res. Bull.* **18**:621–633.

Polyak, S. L., 1958, *The Vertebrate Visual System*, The University of Chicago Press, Chicago.

Prichard, J., Rothman, D., Novotny, E., Petroff, O., Kuwabara, T., Avison, M., Howseman, A., Hanstock, C., and Shulman, R., 1991, Lactate rise detected by ^{1}H NMR in human visual cortex during physiologic stimulation, *Proc. Natl. Acad. Sci. USA* **88**:5829–5831.

Pritkin, R. I., 1980, The rarity of true congenital bilateral anophthalmos, *Meta. Pediatr. Ophthalmol.* **4**:165–167.

Provis, J. M., van Driel, D., Billson, F. A., and Russell, P., 1985, Human fetal optic nerve: Overproduction and elimination of retinal axons during development, *J. Comp. Neurol.* **238**:92–100.

Putnam, T. J., 1926, Studies of the central visual connections: General relationship between external geniculate body, optic radiation and visual cortex in man, *Arch. Neurol. Psychiatry* (Chicago) **16**:566–596.

Rakic, P., 1974, Neurons in rhesus monkey visual cortex: Systematic relation between time of origin and eventual disposition, *Science* **183**:425–427.

Rakic, P., 1988, Specification of cerebral cortical areas, *Science* **241**:170–176.

Rakic, P., and Riley, K. P., 1983, Overproduction and elimination of retinal axons in the fetal rhesus monkey, *Science* **219**:1441–1444.

Rakic, P., Bourgeois, J. -P., Eckenhoff, M. F., Zecevic, N., and Goldman-Rakic, P. S., 1986, Concurrent overproduction of synapses in diverse regions of primate cerebral cortex, *Science* **232**:232–235.

Rakic, P., Suñer, I., and Williams, R. W., 1991, A novel cytoarchitectonic area induced experimentally within the primate visual cortex, *Proc. Natl. Acad. Sci. USA* **88**:2083–2087.

Ramón y Cajal, S., 1890, Textura de las circunvoluciones cerebrales de los mamíferos inferiores. Nota preventiva, Gac. Méd. Catalana, Dec. 15.

Ramón y Cajal, S., 1899, Estudios sobre la corteza cerebral humana. Corteza visual, *Rev. Trim. Microgr.* **4**:1–63.

Ramón y Cajal, S., 1900, *Studien über die Hirnrinde des Menschen. 1. Die Sehrinde,* Barth, Leipzig.

Ramón y Cajal, S., 1909–1911, Histologie du Système Nerveux de L'Homme et des Vertébrés. Maloine, Paris. (Reprinted 1952–1955 by the Consejo Superior de Investigaciones Científicas, Madrid.)

Ramón y Cajal, S., 1929, *Studies on Vertebrate Neurogenesis* (L. Guth, trans.), Thomas, Springfield, Ill.

Recordon, E., and Griffiths, G. M., 1938, A case of primary bilateral anophthalmia (clinical and histological report), *Br. J. Ophthalmol.* **22**:353–360.

Rockland, K. S., 1985, A reticular pattern of intrinsic connections in primate area V2 (area 18), *J. Comp. Neurol.* **235**:457–478.

Rockland, K. S., and Lund, J. S., 1982, Widespread periodic intrinsic connections in the tree shrew visual cortex (area 17), *Science* **215**:1532–1534.

Rockland, K. S., and Lund, J. S., 1983, Intrinsic laminar lattice connections in primate visual cortex, *J. Comp. Neurol.* **216**:303–318.

Ryugo, D. K., Ryugo, R., Globus, A., and Killackey, H. P., 1975, Increased spine density in auditory cortex following visual or somatic deafferentation, *Brain Res.* **90**:143–146.

Sauer, B., Kammradt, G., Krathhausen, I., Kretschmann, H. J., Lange, H. W., and Wingert, W., 1983, Qualitative and quantitative development of the visual cortex in man, *J. Comp. Neurol.* **214**:441–450.

Shimada, M., and Langman, J., 1970, Cell proliferation, migration and differentiation in the cerebral cortex of the golden hamster, *J. Comp. Neurol.* **139**:227–244.

Smith, C. G., and Richardson, W. F. G., 1966, The course and distribution of the arteries supplying the visual (striate) cortex, *Am. J. Ophthalmol.* **612**:1391–1396.

Smith, G. E., 1904, The morphology of the occipital region of the cerebral hemisphere in man and the apes, *Anat. Anz.* **24**:436–451.

Smith, G. E., 1907a, New studies on the folding of the visual cortex and the significance of the occipital sulci in the human brain, *J. Anat.* **41**:198–207.

Smith, G. E., 1907b, A new topographical survey of the human cerebral cortex, being an account of the distribution of the anatomically distinct cortical areas and their relationship to the cerebral sulci, *J. Anat. Physiol. London* **41**:237–254.

Stensaas, S. S., Eddington, D. K., and Dobelle, W. H., 1974, The topography and variability of the primary visual cortex in man, *J. Neurosurg.* **40**:747–755.

Teuber, H. L., Battersby, W. S., and Bender, M. B., 1960, *Visual Defects after Penetrating Missile Wounds of the Brain,* Harvard University Press, Cambridge, Mass.

Tikoo, R. K., and Kuljis, R. O., 1991, Selective distribution of Alzheimer's disease lesions in the visual cortex, *Soc. Neurosci. Abstr.* **17**:351 (147.1).

Tomlinson, B. E., Blessed, G., and Roth, M., 1970, Observations on the brains of demented old people, *J. Neurol. Sci.* **11**:205–242.

Valverde, F., 1971, Short axon neuronal subsystems in the visual cortex of the monkey, *Int. J. Neurosci.* **1**:181–197.

Veraart, C., and Wanet, M. -C., 1986, Sensory substitution of vision by audition, in: *Electronic Spatial Sensing for the Blind* (D. H. Warren and E. R. Strelow, eds.), Nijhoff, The Hague, pp. 217–238.

Veraart, C., De Volder, A. G., Wanet-Defalque, M. C., Bol, A., Michel, C., and Goffinet, A. M., 1990, Glucose utilization in human visual cortex is abnormally elevated in blindness of early onset but decreased in blindness of late onset, *Brain Res.* **510**:115–121.

Verhoeff, F. H., 1943, A new answer to the question of macular sparing, *Arch. Ophthalmol.* **30**:421–425.

Vicq d'Azyr, F., 1786, *Traité d'Anatomie et de Physiologie Avec des Planches Coloriées. Représentant au Naturel les Divers Organes de l'Homme et des Animaux,* Didot l'Aîné, Paris.

Vignal, W., 1888, Recherches sur le développement des éléments des couches corticales du cerveau et du cervelet chez l'homme et les mammifères, *Arch. Physiol. Norm. Pathol. (Paris) Sec. IV* **2**:228–254.

Vogt, O., and Vogt, C., 1919, Allgemeinere Ergebnisse unserer Hirnforschung, *J. Psychol. Neurol.* **25:**279–462.

von Monakow, C., 1889, Experimentalle und pathologisch-anatomische untersuchungen uber die optischen zentren und bahnen, *Arch Psychiatr. Nervenkr.* **20:**714–787.

Wanet-Defalque, M. C., Veraart, C., De Volder, A., Metz, R., Michel, C., Dooms, G., and Goffinet, A., 1988, High metabolic activity in the visual cortex of early blind human subjects, *Brain Res.* **446:**369–373.

Woods, S. W., Hegeman, I. M., Zubal, I. G., Krystal, J. H., Koster, K., Smith, E. O., Heninger, G. R., and Hoffer, P. B., 1991, Visual stimulation increases technetium-99m-HMPAO distribution in human visual cortex, *J. Nucl. Med.* **32:**210–215.

Zeki, S., Watson, J. D. G., Leuck, C. J., Friston, K. J., Kennard, C., and Frackowiack, R. S. J., 1991, A direct demonstration of functional specialization in human visual cortex, *J. Neurosci.* **11:**641–649.

Zheng, D. R., Guan, Y. L., Luo, Z. B., and Yew, D. T., 1989, Scanning electron microscopy of the development of layer I of the human visual cortex, *Dev. Neurosci.* **11:**1–10.

Zilles, K., Werners, R., Büsching, U., and Schleicher, A., 1986, Ontogenesis of the laminar structure in areas 17 and 18 of the human visual cortex, *Anat. Embryol.* **174:**339–353.

The Role of Striate Cortex
Evidence from Human Lesion Studies

13 (large numeral, chapter number)

MATTHEW RIZZO

1. Introduction

Much of what is known about the role of human striate cortex in vision comes from the study of dysfunction in patients with specific lesions of visual pathways, from the retina to the occipital lobe and the adjoining temporal and parietal regions. That evidence depends on neuro-ophthalmological, neuropsychological, and psychophysical techniques. The neuroanatomy is provided, in vivo, by modern neuroimaging techniques such as magnetic resonance imaging (MRI) (Damasio and Damasio, 1989; Damasio and Frank, 1992), and (less often) at autopsy. Positron emission tomography (PET) studies offer another window on regional localization and visual function. Comparative anatomical studies on the functional organization of the visual system, particularly in the monkey, also provide insights into the corresponding organization in the human. This chapter examines these converging lines of evidence in the context of what they tell us about the role of human striate cortex [which is also referred to as the primary visual cortex, the calcarine cortex, area 17 of Brodmann (1909), and more recently as area V1] (see Fig. 1). The emphasis in this chapter is on human brain lesion studies. In the introductory section, we start with a definition of human visual areas and deficits, and follow with comments on the human brain lesion method applied to vision, on human–monkey homologies, and on subcortical inputs to V1.

MATTHEW RIZZO • Division of Behavioral Neurology and Cognitive Neuroscience, Department of Neurology, The University of Iowa College of Medicine, Iowa City, Iowa 52242.

Cerebral Cortex, Volume 10, edited by Alan Peters and Kathleen S. Rockland. Plenum Press, New York, 1994.

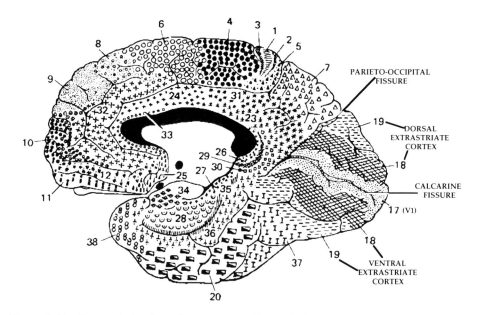

Figure 1. Brodmann depicted the human cytoarchitectonic fields on the mesial surface of the human right hemisphere. The striate or calcarine cortex, also known as area 17 or V1, is depicted as symmetrically distributed about the calcarine fissure. The peripheral fields are represented anteriorly, in the depths of the calcarine fissure. The cortical representation of the retinal fovea is located posteriorly toward the occipital pole. The exact position of V1 varies among individuals and between hemispheres, and sometimes extends slightly beyond the occipital pole onto the lateral surface of the hemisphere, not shown here. The parieto-occipital fissure is a reliable anterior border for dorsal area 17. An average of 92% of calcarine cortex lies posterior to the intersection point formed by the parieto-occipital and calcarine fissure, and the amount of intracalcarine area 17 averages 62% of total striate cortex (Rademacher *et al.,* 1992). V1 is concentrically surrounded in its ventral and dorsal aspects by extrastriate regions, areas 18 and 19, which should correspond to the monkey's V2/V3. V2/V3 are mostly mesial, but they do extend onto the lateral surface of the hemisphere. Ventromesial areas 18 and 19 may also correspond, in part, to a human homologue of the monkey's area V4 complex (see Fig. 6). Preliminary evidence suggests that a human homologue of the simian MT complex is present on the lateral surface of the brain, which is not shown here. The occipital cortex is primarily supplied by branches of the posterior cerebral artery, which arises from the vertebrobasilar vascular system in most individuals. Occlusion of these branches results in structural brain lesions in V1, V2/V3, and V4 causing visual deficits. Lesions in putative human MT complex, however, may depend more on occlusion of posterior temporal and posterior parietal branches of the middle cerebral artery.

1.1. Definition of Human Visual Areas and Deficits

A focal lesion of the human visual cortex results in a corresponding visual field defect: an area of visual loss surrounded by preserved vision known as a scotoma (*skotos,* Greek for "darkness"). Scotomata are operationally defined by the inability to report the presence of targets of specific size and luminance in various portions of the visual field. The questions that concern us are: what do patients "see," or at least, how well do they perform, with such islands of defective vision? Also, how can we link that psychometric performance with models of the underlying neurophysiological process that relate to V1? [See Brindley (1970) and Teller (1984) on "linking propositions".]

Patients with acquired lesions of V1, or of the white matter connections to and from V1, may report the experience of "darkness," "blackness," or a "hole," in their vision. [See Kolb (1990) for a self-report on the experience of his own occipital stroke; the issue of residual vision in a V1 scotoma is addressed in Section 5.] These defects are located in the visual fields opposite to the side of the lesion and are homonymous, that is, they occupy the same hemifield in each eye, an arrangement which follows from the reversal of real-world images by the lens, and from the decussation of the nasal fibers of the optic nerve in the optic chiasm. They are also congruent, meaning the defects in the two eyes are near identical when they are superimposed. The specific size, shape, and location of a V1 scotoma within the visual fields depend on the locus of the lesion in visual cortex. Inouye (1909) and Holmes and Lister (1916) were aware of this special geometry, and used it to construct a retinotopic map of V1 (see Section 3).

Damage to the macular representation in V1 is especially troublesome since it may interfere with ocular fixation, visual scanning, and the ability to process visual details. Visual field defects at or near the fixation point are called, respectively, "central" or "paracentral" scotomata. A visual field defect that is restricted to the upper or lower quadrant of a hemifield is known as a quadrantanopia. A lesion of V1 below the calcarine fissure results in an upper quadrantanopia while a V1 lesion above the calcarine fissure causes a lower quadrantanopia. A visual field defect that occupies both the upper and lower visual portions of the same hemifield of both eyes is known as a hemianopia (or hemianopsia, or homonymous hemianopia).

A homonymous hemianopia may include the entire foveal representation on one side, a condition known as foveal splitting; alternatively, the few degrees of vision near the fixation point may be preserved, so that there is foveal sparing (anatomical explanations are explored in Section 4). A homonymous hemianopia may spare the peripheralmost sector of the temporal visual fields of the eye contralateral to an occipital lobe lesion. This spared peripheral sector of vision is referred to as the "monocular temporal crescent" and it depends on unpaired peripheral fibers of the nasal retina and their representation in contralateral V1 (see Section 3).

Homonymous hemianopias may be bilateral, leading to a severe loss of peripheral vision; if there is foveal sparing, a "keyhole" of vision around the fixation point remains. Complete cortical blindness is rare, and it is generally associated with considerable damage that extends well beyond V1. Affected patients generally appear confused and are difficult to test.

Human V1 visual field defects can be compared and contrasted with those caused by extrastriate (prestriate) lesions. Classically, human extrastriate cortex has been divided into two large fields, areas 18 and 19 of Brodmann. The human extrastriate areas contain additional functional representations equivalent to the simian areas V2, V3, V4, and MT (V5) (see chapter by Van Essen in Volume 3 and chapter by Casagrande and Kaas in this volume). One specific scheme for homologies was proposed by Clark and Miklossy (1990). Visual field defects caused by lesions in V2 and V3, however, may be difficult to distinguish from those caused by lesions of V1 or of the optic radiations (see Section 7). In contrast, lesions in the ventral visual association cortex, or related white matter connections, in the fusiform and lingual gyri are associated with comparatively distinctive deficits which accord with damage or undercutting of a human homologue of the monkey's area V4. The resultant deficit is quite different from a V1 deficit, because the loss of vision is more selective. It may include a relative defect of color processing, so-called central achromatopsia, and pattern processing (see Section 8). Lesions in the dorsolateral visual association cortex that should include a human homologue of the monkey's area MT complex may cause defective processing of visual motion.

While the focus of this chapter is on human lesion data emphasizing localized damage, we note that PET studies (see Section (9) point to a highly interactive, synthetic organization in which multiple regions are activated by a single task or stimulus. This underscores the idea that during normal vision the striate and prestriate regions interact strongly with each other, as well as with other cortical and subcortical structures. The visually relevant network includes, cortically, frontal (area 8 and parts of 6) and parietal (areas 5, 7, and 39) eye and hand fields, and the visual association cortex in the temporal lobe (e.g., area 37); and subcortically, the pulvinar, caudate, superior colliculus, dorsolateral pontine nucleus, and vermis of the cerebellum. Consequently, a lesion of visual cortex is liable to cause dysfunction ("diaschesis") in several connected areas in the same hemisphere, and across the callosum. A lesion is unlikely to produce isolated effects in any single hypothetical cortical module, such as a color or motion module.

1.2. Comment on the Human Brain Lesion Method Applied to Vision

Human lesion studies provide many of the data from which critical inferences on the role of human V1 derive. Therefore, it is important to understand the perspective of these studies, the nature of the information they offer, and the behavioral factors that affect their interpretation.

Humans with occipital lobe lesions can make sophisticated visual judgments on a wide variety of psychophysical procedures. Unlike monkeys in experimental situations, they are not limited to lever pressing or eye movement responses on rudimentary perceptual tasks as the main expression of their visual abilities. They can generally report verbally on what they do or do not see. Thus, their behavior may be compared with their descriptions of actual visual experience.

Verbal descriptions of dysfunction along the psychological dimensions of color, shape, depth, or motion can provide insights on the visual defect in a patient, which in turn helps to motivate further studies. An example is the pursuit of a color processing defect in patients with extrastriate lesions who complain that their world is now seen only in gray. Patients' descriptions may provide the dominant clues for diagnosing certain visuoperceptive defects, the existence of which might not otherwise be detected. Such defects include metamorphopsia, described by some patients as a carnival-mirror-like effect; monocular polyopia, characterized by multiple ghostly images of a single object which are not caused by distortions of the ocular media or surface; micropsia or macropsia, objects looking smaller or larger than they should, but without any retinal damage; palinopsia, the persistence of visual afterimages; and hallucinations (Walsh and Hoyt, 1969). These human phenomena have physiological underpinnings, such as stroke, migraine, epilepsy, neurodegenerative disease, or drug effects at specific receptor sites. They probably reflect specific patterns of activation or damage within visual cortex including V1 (Anderson and Rizzo, 1992).

Not all patients are able to provide lucid descriptions of their visual experience. Some have neural damage in language-related structures and cannot express their idea in verbal terms. In color anomia, for example, reliance on verbal responses may lead to the erroneous conclusion that a patient has abnormal color vision, when in fact colors are simply misnamed. Another problem is that some individuals may not be aware of their perceptual defect and consequently do not report it. This falls in the category of anosognosia (a neg. + Greek nosos disease + gnosis knowledge). "Anton's syndrome" (1899), the denial of blindness following extensive damage in the visual cortex, is an example of anosognosia related to vision. So is the denial of a V1 scotoma in a patient who also has left spatial hemineglect caused by a lesion of the right lateral parietal lobe.

On the other hand, patients with cerebral lesions may in fact retain perception, but deny it. They are not malingerers, but are simply unaware of their residual capacities. Knowledge without awareness (Tranel and Damasio, 1985) can be dramatically demonstrated, for example, in prosopagnosia (Greek prosopon face + agnosia not knowing). Patients no longer consciously recognize previously familiar faces, or are unable to learn new faces following bilateral lesions in the inferior visual association cortex (Damasio, 1985a,b). Nevertheless, they may still discriminate between familiar and unfamiliar faces on forced choice tasks, and discrimination is also indicated by galvanic skin response (Tranel and Damasio, 1985), or eye movement criteria (Rizzo *et al.*, 1987). In a similar vein (but at a "lower" level), subjects with area V1 lesions have been reported to perform reasonably well on simple forced choice detection tasks, and on localization tasks in which the accuracy of finger pointing or eye movements are taken as the index of perception for targets presented in their scotoma. However, these patients have no conscious experience of the items they localize or detect. This phenomenon has been described under the oxymoron "blindsight" (Weiskrantz, 1987, 1990; Cowey and Stoerig, 1991; see Section 5).

Finally, we note that lesions of the human visual cortex occur without a guiding plan, at least not one that is easily discerned, and without respect for human convenience, either the patients' or researchers'. The strategy of Damasio and colleagues at Iowa has been to establish a standing registry of brain-damaged subjects (Palca, 1990). From this registry, which now includes a base of 2000 individuals, including over 200 with lesions of visual cortex, it is possible to select cooperative individuals with lesions in an anatomical territory of interest so that hypothesis-driven experiments can be conducted. Cerebrovascular lesions, especially in the distribution of the posterior cerebral artery (Marinkovic *et al.*, 1987; Pessin *et al.*, 1987), are the most common etiology for visual cortical lesions. Other underlying etiologies include infarctions in the watershed between the middle and posterior cerebral arteries, tumor, trauma, multifocal degenerative processes such as Alzheimer's disease, and a wide variety of infectious conditions such as progressive multifocal leukoencephalopathy, HIV disease, and Creutzfeldt–Jakob disease.

1.3. Comment on Human–Monkey Homologies

Modern concepts on the organization of the monkey's striate and extrastriate cortex depend on experimental procedures such as tracer injections, ablations, and microelectrode recordings. These procedures are generally too invasive for application in the human, with few exceptions like the therapeutic ablation of brain tumors, and cortical penetrations with depth electrodes (e.g., Dobelle *et al.*, 1979) preceding epilepsy surgery. Consequently, it is important to question whether and to what extent a variety of biological schemes and information processing metaphors that have been used to model data obtained mostly in the monkey really apply to man. Examples of such models are concepts of serial or parallel processing streams, modularity, and multiplexing of information among several interconnected functional maps. Our assumption, given a variety of interspecies homologies (Hitchcock and Hickey, 1980; Horton and Hedley-White, 1984; Tolhurst and Ling, 1988; Burkhalter and Bernardo, 1989; Horton *et al.*, 1990; Hockfield *et al.*, 1990), is that they do apply to humans, to an appreciable extent. As indicated below, certain findings on the visual system in both species may be interpreted by making use of similar concepts.

1.4. Inputs to V1: Evidence for Parallel Subcortical Streams

In the subhuman primate, the segregation of visual inputs begins in the retina (Dowling, 1987; Livingstone and Hubel, 1987; Rodieck, 1988; Schiller and Logothetis, 1990). About 10% of the retinal ganglion cells, the "parasol," P alpha, A, or "phasic" cells, show large receptive fields, broadband sensitivities, transient responses, and faster-conducting axons. They form the "M" channel named for its cortical relays via magnocellular (M) layers 1 and 2 of the lateral geniculate nucleus (LGN). The "P" channel arises from color-opponent cells, the "midget," P beta, B, or "tonic" cells. They constitute about 80% of the ganglion cells, relay via parvocellular (P) layers 3–6 of the LGN, have smaller receptive fields, sustained responses, and slower-conducting axons (Livingstone and Hubel, 1987; Schiller and Logothetis, 1990). Retinal P gamma and epsilon fibers probably do not reach the LGN. Instead, they synapse in other subcortical destinations, and perhaps they help mediate "subcortical vision" (see Bullier *et al.*, this volume).

Histological studies show that the human visual system contains cells morphologically similar to those of the M and P channels of the monkey (Rodieck *et al.*, 1985; Rodieck, 1988). The functional properties of these cells have been inferred from the pattern of psychophysical performance in humans with deficits in particular cell populations (see Bassi and Lehmkuhle, 1990, for a recent review). Livingstone *et al.* (1991) and Galaburda *et al.* (1991) recently examined the brains of five subjects with developmental dyslexia. Their preliminary findings indicated a smaller size for putative magnocellular neurons in the LGN compared with normal control brains. They tied the apparent histological defect to psychophysical evidence, which indicates that dyslexics have M pathway-like dysfunction (Martin and Lovegrove, 1987). In a different example, psychophysical evidence suggests that certain forms of human amblyopia may selectively affect the P pathway. In anisometropic amblyopia, visual blurring of the eye at an early age causes a persistent impairment of vision, even after the initial conditions leading to the blur are corrected. The assumption is that the defect has become permanent because development of the visual system has progressed beyond a critical phase. The psychophysical findings of P pathway-like dysfunction in humans with anisometropic amblyopia mirror those in young monkeys after chronic atropine treatment. These monkeys show shrinkage of cells in the P cell layers in LGN, and reduced CO activity in layer 4C beta of V1 (Kiorpes *et al.*, 1987; Hendrickson *et al.*, 1987; Movshon *et al.*, 1987). The corresponding pathology in humans with anisometropic amblyopia has not been defined.

Human retina and optic nerve disorders, e.g., glaucoma (Atkin *et al.*, 1979; Tyler, 1981; Trick, 1985), and even Alzheimer's disease (Trick *et al.*, 1989), might preferentially affect large retinal cells and their axons (Sadun *et al.*, 1987), and have also been discussed in terms of M and P systems. In general, loss of high-frequency temporal resolution on tests of flicker or motion perception, with relative preservation of high-frequency spatial resolution and color discrimination, may be interpreted as evidence for damage to a human M pathway. This may correspond to the so-called "transient" channel hypothesized by Kulikowski and Tolhurst (1973). Loss of high spatial frequency vision or visual acuity and of color discrimination, but not of temporal resolution, have been taken to suggest disease in a "sustained" channel, perhaps a human P pathway. Recently, however, models of parallel processing in the visual system of primates have been challenged (Martin, 1992; Merigan and Maunsell, 1993).

In summary, the notion of parallel processing in the human visual system is plausible but remains a research issue. It appears as if human area V1 receives the afferent inputs from relatively parallel subcortical channels that, both structurally and functionally, resemble the M and P pathways described in the monkey. These similarities do not exclude the possibility that human striate cortex might also receive additional inputs from less well defined channels, which may arise from other, non-M, non-P, cells of the LGN.

2. The Identification of Human V1

The history of the identification of primary visual cortex in the human helps explain how modern concepts of the human visual system developed, and why it took so long to split off the functional contributions of extrastriate regions from those of area V1.

A century ago, the idea that the human brain contained any stable, point-to-point representation of the retinal impressions of the visual fields was not well formulated (Holmes, 1945, Polyak, 1957). There was "antilocalizationist" sentiment. According to Lepore (1983), Goltz had cited cortical blindness in dogs with extensive bihemispheric ablations as evidence against a specific cortical region for vision. Also, the localizationists were off the mark. Ferrier (1876) believed that a lesion in the angular gyrus was responsible for monocular blindness. Ewens (1893) thought the angular gyrus contained a representation of the central retina, while the striate cortex served the periphery. Von Monakow (1900) assumed a diffuse projection of the retina upon striate cortex and believed in a mobile retinal center. Only in the 1870s and 1880s did the physiologists Munk (1881) and Schafer (1888) correctly localize a representation of the visual fields in the occipital lobe. Then, in the 1890s, Henschen (1893) observed that patients with visual field defects during life showed damage of the striate cortex in postmortem examination.

However, a specific point-to-point projection of the human retina onto the striate cortex was not proposed until early in this century. Then, Inouye (1909), who worked during the Russo-Jàpanese war, followed by Holmes and Lister (1916), who worked during World War I, outlined a retinotopic projection in the human. Their research was based on careful analyses of the residual vision in soldiers with missile wounds of the occiput. Holmes (1945) explained that the "essential" observations had to be assembled with "time and labor" from the "irregular rubble" of clinical material that was "rarely so simple or clean cut." The findings went hand in hand with contemporaneous physiological and anatomical observations, such as the delineation of an area 17 in the monkey and human (Brodmann, 1909), which in contrast to his own work, Holmes saw as "built in ashlar or hewn stones which can be easily fitted together."

Once it was finally recognized, the idea of a retinotopic map in human striate cortex had powerful repercussions. The severity of the visual loss that followed lesions of the occipital lobe suggested that all basic visual "impressions" depended on striate cortex. This seemed to militate against the need to consider any localization of visual functions outside that region. This view hindered the identification of human extrastriate areas as cortical zones which receive an output from V1, and the understanding of the visual system in general.

Holmes imagined that the relative sparing of motion perception in an otherwise blind hemifield, the "statokinetic dissociation" described by Riddoch (1917), must reflect the recovery of function in partially damaged striate cortex or underlying optic radiations, but not elsewhere. He also doubted that color vision, another basic visual "impression," could be impaired by lesions outside striate cortex. Color defects, if they did occur, would be caused by early recovery of light and form sensibility in the penumbra of what Holmes thought was a striate scotoma, when the recovery of color lagged behind. This ran contrary to reports such as Verrey's (1888), of a patient with an acquired defect of color vision in the right hemifield, "hemachromatopsie droite absolue," following a cerebrovascular accident, in which the autopsy showed a lesion in the fusiform and lingual gyri in the left hemisphere. If Holmes was aware of this case, it could not have pleased him that Verrey assigned the role of primary visual cortex to the fusiform and lingual gyri.

In his brain-injured soldiers, Holmes (1918b) observed "that red and green objects could not be recognized in certain regions, often in homonymous halves of the visual fields." The problem was that "in every instance visual sensibility to white test objects of the same size was reduced." Thus, Holmes's specific objection to the interpretation of central achromatopsia was that "it has not been conclusively shown that colour perception may be completely lost in any part of the field when that of light or white is undisturbed." However, he is remembered for stating more generally that "isolated loss or dissociation of colour vision is not produced by cerebral lesions."

Holmes' conclusions may have reflected his experience with a skewed population (Holmes and Lister, 1916; Damasio, 1985a). The missile wounds of the lower occiput which might disturb color vision, would also tear the cerebral veins and dural venous sinuses, a circumstance incompatible with life. Survivors would have lesions in a more rostral location, affecting the superior visual association cortex. This would cause "disturbances of visual orientation" (Holmes, 1918a), or "disturbances of spatial orientation and visual attention, with loss of stereo-scopic vision" (Holmes and Horrax, 1919), in association with lower quadrant visual field defects, but not achromatopsia.

In a recent review of central achromatopsia, Zeki (1990, p. 1740) states that he is less "generous toward Holmes' mistake," implying an intellectual bias against existing evidence from other research. Yet, Holmes was not the only investigator hostile to the idea of color processing beyond striate cortex. Based on their analysis of the visual field defects in 46 of 203 soldiers with penetrating wounds of the brain, Teuber *et al.* (1960) stated that "we fully agree with Holmes' view." They found that "areas of diminished or absent color vision are areas of diminished acuity," and that they "never observed intact color discrimination combined with serious alteration or loss of form perception." Lashley (1948), Polyak (1957), Critchley (1965), and others expressed similar reservations.

3. Mapping the Retinotopic Organization of Human V1

To plot the visual field defects in soldiers with missile wounds of the occipital lobe, Holmes and Lister (1916) used reflective white or colored targets moved slowly, from a point beyond the field of view, inward toward fixation. The technique of the visual field examination had been in refinement since the 1850s (von Graefe, 1856). They correlated the shape and size of the visual field defects they measured with the loci of cortical damage, which could be verified if a soldier died. Otherwise the lesion loci were inferred by craniometric measurements, taking into account the entrance and exit wounds in the skull with the assumption of a straight path by a single missile in between. Ultimately, Holmes had to admit that this was a shaky assumption, given the propensity for secondary skull and bullet fragments to produce greater damage (Holmes and Horrax, 1919). Moreover, bleeding, swelling, and infection, encountered in the acute phase of the injury, played a role in the final extent of the lesion.

From this data, Holmes inferred a retinotopic projection in the occipital lobe (Holmes, 1918b). He reported that:

(1) The upper half of each retina is represented in the dorsal, and the lower in the ventral part of each visual area;
(2) The centre for macular or central vision lies in the posterior extremities of the visual areas, probably in the margins and lateral surfaces of the occipital poles;
(3) The centre for vision subserved by the periphery of the retinae is probably situated in the anterior end of the visual area, and the serial concentric zones of the retina from the macula to the periphery are probably represented in this order from behind forwards in the occipital cortex.

Holmes (1918b) indicated that the segment of the striate cortex located in the anteriormost depths of the calcarine fissure contained a monocular representation of the peripheralmost sector of the temporal fields of the contralateral eye. This was based on the finding that soldiers with occipital wounds which spared the anteriormost segments of the calcarine cortex would sometimes demonstrate a crescent of residual peripheral vision in the upper and lower temporal fields of the contralateral eye. This "monocular temporal crescent," which extends from 60° to 100° out from fixation, has no corresponding homologous region in the nasal fields of the opposite eye (Benton *et al.*, 1980), and it helps account for the greater area of the temporal visual fields compared with the nasal visual fields in normal individuals. The monocular temporal crescent depends on peripheral fibers of the nasal retina that project to the contralateral LGN, and then to the corresponding segment of V1, which was spared in some of Holmes's cases.

Holmes also pointed out the exaggerated cortical representation of the retinal fovea. Current estimates of this "cortical magnification" factor range from 8–11 to 20–25 mm/degree close to the fovea (Dobelle *et al.*, 1979; Fox *et al.*, 1986; Tolhurst and Ling, 1988; Horton and Hoyt, 1991a), not far from the 15 mm/degree estimated for the macaque. He concisely plotted his findings in his canonical scheme (see Holmes, 1918b; reproduced here as Fig. 2). This scheme has proven extremely useful, though to quote Holmes, it "does not claim to be in any respect accurate; it is merely a scheme." He quoted Inouye (1909) for support (Holmes and Lister, 1916; Holmes, 1918b), though later he wrote that there was "no definite evidence of the local representation of the retina" in "the striate area" "prior to 1914" (Holmes, 1945, p. 350), the year his own work commenced.

4. "Sparing" of the Foveal (Macular) Representation in V1

Repeated study of patients with unilateral occipital lesions has shown that contralateral field defects sometimes spare foveal or macular vision (Walsh and Hoyt, 1969). The fovea includes the central 1–3°, while the macula includes the central 10° or so of vision. The question is whether this finding is merely the

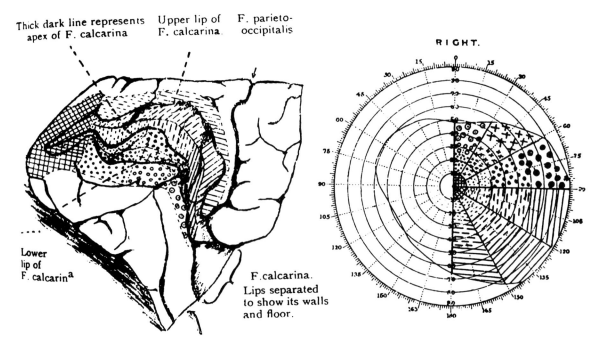

Figure 2. Holmes's retinotopic map. This early visual field plot shows the position of human V1 along the lips of the calcarine fissure. The foveal representation is located posteriorly toward the occipital pole. The peripheral fields are represented anteriorly in the depths of the calcarine fissure, in the mesial surface of the hemisphere (see text).

result of an incomplete cortical lesion, or whether there is some other biological factor at play. In some primate species it has been suggested that "foveal (macular) sparing" may be the consequence of a bilateral foveal (macular) representation. The standard assumption in the human model is that the nasal retina of each eye projects to the opposite hemisphere and that the temporal retinal projects ipsilaterally. Consequently, the striate cortex in each hemisphere has a representation of the contralateral visual field only.

Bunt and Minckler (1977) reported on HRP injections into the LGN of Old World monkeys. The pattern of retrograde labeling showed that ipsilaterally and contralaterally projecting retinal ganglion cells are mixed in a vertical strip that is about 1° wide across the vertical meridian, widening to 3° at the fovea. Leventhal *et al.* (1988) provided confirmation and further detail. They placed HRP injections into the LGN and superior colliculus in New and Old World monkeys to evaluate the projections of the different classes of retinal ganglion cells. Colliculus injections showed no ipsilateral projection from the retina. LGN injections showed a 0.5° ring of ipsilaterally projecting cells circling the nasal side, but not the temporal side of the foveal pit. Some ipsilaterally projecting cells were P beta cells with their dendritic arborizations close by. The results indicated that ipsilaterally projecting cells in and around the fovea can generate 2–3° of bilateral representation in geniculocortical pathways because they are intermingled with contralaterally projecting cells on the nasal side of the foveal pit.

The above results might help solve the "enigmatic" and "age-old problem of macular (foveal) sparing and splitting" (Leventhal *et al.*, 1988). Bunt and Minckler's results, if they apply in human, provide a basis for foveal sparing. However, they would also predict that foveal splitting does not occur with occipital lesions, when in fact it often does occur. Leventhal and colleagues' results would predict that striate lesions cause foveal splitting of the eye ipsilateral to the lesion, and foveal sparing in the eye contralateral to the lesion, at least for stimuli that trigger the P beta cells. If such a pattern occurs, it must be quite rare.

Alternative explanations for foveal (macular) sparing do not question that the human foveal (macular) representation is entirely crossed, but propose other explanations. First, there is the pattern of blood supply. The occipital lobe benefits from a dual blood supply and there is considerable interindividual variation (Smith and Richardson, 1966). The foveal (macular) representation can receive branches of the middle as well as posterior cerebral arteries (PCA) while the rest of lobe is restricted to a PCA supply. Consequently, strokes caused by PCA occlusions may spare the central few degrees of vision. Second, the foveal (macular) representation in the occipital lobe is relatively large. The central 10° of the visual fields subtends over 50% of occipital cortex (Horton and Hoyt, 1991b), and so is often incompletely damaged. Finally, it has been suggested that foveal (macular) sparing is an artifact of inaccurate monitoring of fixation during the visual field test or of light scatter into the "good" field.

5. Residual Vision in V1 Scotomata

Most investigators since Holmes have thought that patients with striate scotomata were absolutely blind. This is because patients with lesions of striate cortex report no conscious appreciation of vision in any portion of their scotoma. Consequently, it appears as if V1 scotomata do not permit the detection of any visual stimulus within their boundaries. However, Weiskrantz *et al.* (1974) argued against this view. They coined the term *blindsight* to convey the intriguing and controversial idea of residual vision in the fields of a human striate scotomata. As we shall see later, the issue is complex and even the very definition of blindsight is ambiguous and problematic (Weiskrantz, 1987; 1990).

Questions of residual vision in striate scotomata were spurred by experimental observations on animals (e.g., Denny-Brown and Chambers, 1955). After complete removal of the visual cortex, monkeys are not rendered blind. They no longer recognize other monkeys or discriminate patterns and colors, yet they can still reach for peripheral objects, negotiate obstacles, and track visual targets with eye movements. Zee *et al.* (1987) reported on occipital lobectomies in six monkeys that had been trained to fixate and follow small targets. At first, all of the monkeys acted blind. Spontaneous and end gaze nystagmus, absent postoperatively, also developed. But 6 months after surgery the monkeys made accurate saccades to visual targets. Smooth pursuit gain, a measure of pursuit accuracy (defined as eye velocity/target velocity), improved to normal levels (to 0.90 and 0.95), even in the two monkeys with complete occipital ablations. One way to explain the recovery is to invoke activity in visual structures and connections that had been assumed to play a vestigial role in primates. Such a system may include subcortical relays from the retina to the pulvinar and superior colliculus. Connections to prestriate cortex from pulvinar, colliculus, or even LGN might also be important, as long as prestriate cortex is not damaged (see Bullier *et al.*, this volume).

Do humans have alternative visual pathways that might permit vision in the absence of striate cortex, or is the finding restricted to the monkey, reflecting the greater importance of subcortical pathways in nonhuman primates? Weiskrantz *et al.* (1974) sought to address this issue in human subject D.B., later the subject of a monograph (Weiskrantz, 1987). D.B. had a portion of his visual cortex resected during surgery to remove a vascular malformation. "From the surgical notes it was estimated that the incision extended from the occipital pole forward by approximately 6 cm and was thought to include a major portion of the calcarine cortex—in which the striate cortex is situated—in the right hemisphere."

Immediately after surgery, D.B. could not detect light in the left half of both eyes, a homonymous hemianopia. Although this defect improved considerably, D.B. was left with a chronic defect in the left lower quadrant. It was in this field that Weiskrantz performed experiments which led him to hypothesize "blindsight." Subjects like D.B. deny the ability to see in a V1 scotoma, and are diagnosed as blind in that region according to classical practice. However, according to Weiskrantz, only awareness, but not detection of visual inputs, depends exclusively on V1. The key to demonstrating "blindsight" is an experimental design that incorporates forced-choice paradigms. It is not sufficient to ask "Do you see it?," because the answer is always "no." However, statistical analysis of forced-choice responses, especially on tasks of visuospatial localization, reveals a performance above threshold that might result from covert discrimination by a "second visual system" that bypasses V1.

One set of important criticisms of "blindsight" is often raised from concerns of psychophysical methodology. Campion *et al.* (1983) argued that light scatter from targets presented into a "blind" field can lead to detection, localization, or discrimination of stimuli. These functions are actually carried out by normal portions of the visual fields which have access to stray light caused by the effects of the scatter.

Other arguments against "blindsight" include that it relies on a yes–no procedure. Having patients report "yes" or "no" when asked if a target is present, might encourage a laxer response criterion than having them assert "I see it." Consequently, yes–no procedures would tend to inflate the hit rate in the bad visual field. Signal detection calculations provide an index of "true sensitivity" (d') that is independent of response bias (Green and Swets, 1966) and ought to be employed (Campion *et al.*, 1983). Another methodological issue concerns monitoring of fixation. It is important to verify true fixation as objectively as possible during "blindsight" experiments, e.g., by using electrooculography, to minimize the potential that the good field could "cheat" for the bad one. Campion *et al.* concluded that the unconscious aspect of "blindsight" is "essentially trivial" (p. 427), although Weiskrantz (1987) strongly objected.

Finally, Weiskrantz *et al.* raised the question of trainability of abnormal vision in the setting of rehabilitation. Zihl (1981) found that repeatedly orienting the eyes into the aberrant field measurably decreased the size of the visual field defect. It was hypothesized that training procedures such as forcing saccade movements to light targets presented in a perimetrically blind field, could lead to recovery. Balliet *et al.* (1985) found evidence to the contrary, suggesting that the improvement was in test-taking, not vision.

Our own experience is that patients with occipital lesions can orient their eyes successfully into the fields of a V1 scotoma (as long as the scotoma does not fall within a neglected hemifield [Rizzo and Hurtig, 1992]). The explanation for this behavior may have more to do with the circumstances of testing than with residual vision, as shown in Fig. 3. Figure 3 illustrates the eye movement behavior of a subject with a right occipital lesion and a complete left homonymous hemianopia who is looking for a target presented into the abnormal left visual field. Initially, the subject does not know the location of a target and has to search for it. The typical strategy is to make several small-amplitude saccade eye movements into the blind hemifield until the target is located. When the pattern of horizontal eye position versus time is graphically plotted, it resembles a "staircase." Once the target is located, its position is learned and the subject is able to return accurately to the target, even though he cannot see it, but this is only because the target is presented at regular intervals and to the same location. When the target is presented at unpredictable intervals and locations in the hemianopic field on successive trials, subjects fail to make accurate movements.

5.2. Anatomical Considerations

Anatomical documentation of visual system lesions is crucial to the argument of "blindsight." This would help to resolve whether residual vision measured in patients with occipital lesions is still mediated by conventional or geniculostriate pathways, or whether it depends on different structures. The best opportunity for in vivo verification of occipital lesions currently demands computerized tomography and magnetic resonance imaging with 3-D reconstructions. These modern techniques are more accurate than estimations obtained from the bloody field of surgery, but they have not been possible in subject D.B. because a surgical clip, presumably ferromagnetic, precludes them; earlier influential reports predate modern image analysis techniques (e.g., Poppel *et al.*, 1973).

Curiously, access to modern neuroimaging techniques in patients with "blindsight" has not definitively resolved the issue of whether the residual vision is mediated by a geniculostriate or an extrageniculostriate locus. The study of Blythe *et al.* (1987) is an example. They reported that only 5 of 25 patients with retinogeniculate lesions had residual vision in their aberrant fields. However, all 25 were aware of the stimuli they detected, and all had some portion of their visual cortex remaining. Case 23, who performed worse than the other patients in some regards, did not even have a demonstrable lesion near the occipital lobe. The authors found "no obvious correlation between the extent of neuronal damage as revealed by CT scans and the existence of residual vision." Fendrich *et al.* (1992) studied a man with left homonymous hemianopia following a right posterior cerebral artery stroke and found that a small, isolated, 1° island of residual vision could be mapped within the hemianopic field only by high-resolution computerized perimetry aided by retinal stabilization of test targets. Such islands of vision, which are not detected on standard perimetry, are presumably mediated by islands of functioning striate cortex. However, the MR images provided by the authors in the case they studied do not confirm the existence of such an island of cortex.

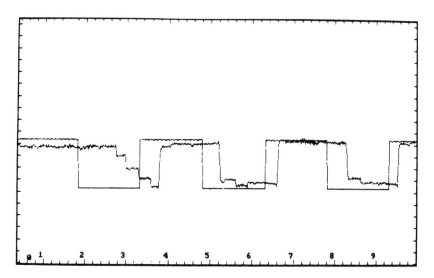

Figure 3. Attempt to track a moving light target in a patient with a right occipital lobe lesion. Electrooculographic recording. The ordinate tick scale is in 5° units. Target and eye position are superimposed. Leftward movement is down. The subject makes an exploratory staircase 20° into the blind left hemifield, quickly learns the location of the target, then follows the pattern with his eyes. This successful orienting into the field of a V1 defect probably can be attributed to the predictability of the target, not "blindsight."

Celesia *et al.* (1991) studied 12 patients with occipital lesions. Five underwent single photon emission tomography, and one had positron emission tomography with deoxyfluoroglucose. In each case, residual vision correlated with visually evoked activity in area 17. The authors concluded that residual vision following striate lesions, was probably mediated by striate cortex. In a similar vein, Hess and Pointer (1989) found no greater than chance performance on tests of spatial and temporal contrast sensitivity presented at several locations in the visual fields contralateral to striate cortex lesions.

It has been suggested that the early age of onset of an occipital lesion may favor the development of residual vision (Perenin and Jeannerod, 1978; Blythe *et al.*, 1987). One study in our laboratory concerning this issue concentrated on a 17-year-old subject with bilateral lesions of the visual cortex that included the foveal representation (Rizzo and Hurtig, 1989). The lesions occurred at birth. At first, he was thought to be blind, but by age 8 it became apparent to his family and to his doctors that the subject had rudimentary vision. When he was studied at the age of 17 he could tell light from dark, had visual acuity of 3/500, could detect motion, and could usually avoid obstacles while navigating new ground. Eye recordings showed he could make saccades, but these were inaccurate. The subject could not generate smooth pursuit to track moving targets, including his own hand. This suggested that the development of normal smooth pursuit depends on the influence of an intact foveal representation in striate cortex in human. Retinal photography showed that the subject had an unstable fixation point (see Fig. 4). It wandered eccentric and superior to the fovea, constrained to a region served by residual visual cortex, as verified by MRI. Figure 5 shows sagittal images of the brain of this subject. The calcarine fissure is encompassed by the lesion in both hemispheres and cannot be identified. The right occipital lobe is virtually absent; however, portions of the left superior occipital lobe remain. The rudimentary vision in the subject is probably mediated by the striate cortex in the residual portions of the occipital lobe. Thus, it was not necessary to hypothesize subcortical vision.

Hemispherectomy cases should provide some of the strongest evidence for rudimentary subcortical vision, but even these still have an element of ambiguity. Perenin and Jeannerod (1978) studied six subjects who had unilateral hemispherectomies in early life for intractable epilepsy. The surgery presumably included the entire neocortex. The subjects could point with a stick with fair accuracy to targets introduced into their blind hemifields. A potential source of error was poor fixation, observed in at least three subjects. Moreover, target presentation of up to 2 sec increased the chances for fixation lapses, compatible with the observation that increasing target durations correlated with improved performance. However, the accumulating evidence in hemispherectomy cases (e.g., Ptito *et al.*, 1991) appears to fall on the side of a weak residual visual capacity in the visual fields contralateral to the ablation.

5.3. Motion and Color Processing in "Blindsight"

In 1917, Riddoch reported early recovery of the ability to appreciate movement in the hemifield opposite to an occipital lesion, despite a patient's continuing inability to see color or static forms. This pattern has been recorded many times since (Walsh and Hoyt, 1969) and might reflect the selective recovery of

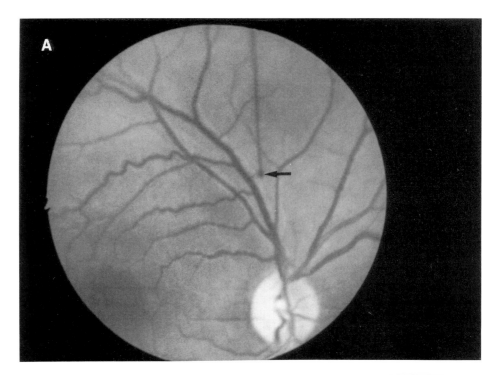

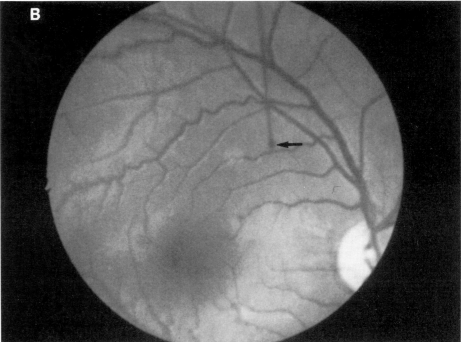

Figure 4. A 17-year-old subject had severe bilateral occipital damage that included the foveal representation of V1. The lesions were acquired at birth. He was thought to be blind, but at age 8 his residual vision included measurable acuity of 3/500. He could navigate unfamiliar terrain using vision. Eye movements were severely defective. Fundus photographs in the right eye show that fixation (indicated by the black arrows) was unstable (compare A and B) and constrained to portions of the superior retina eccentric to the fovea. This corresponded to MRI-verified residually intact striate cortex (see Fig. 5).

intrahemispheric pathways such as the one from V1 to MT, identified in the monkey. Alternatively, callosal connections might allow functionally effective access to the motion processing capabilities of the undamaged hemisphere. The residual motion detection might also reflect indirect access to intact extrastriate cortex via subcortical structures, or motion detection mediated by the colliculus. The latter possibility is important to consider since Ptito *et al.* (1991) now report movement perception in the blind hemifield of hemispherectomized patients. Residual, and presumably subcortical, motion processing in hemispherectomized patients would probably still depend on the processing of motion signals by P alpha retinal ganglion cells, although the contribution of other retinal ganglion cell types cannot be excluded. Finally, it is important to distinguish between preserved motion perception in the periphery of a homonymous hemianopia caused by sparing of the representation of the monocular temporal crescent in V1, and other mechanisms of residual motion perception.

The range of reported residual functions in "blindsight" now also includes "wavelength sensitivity," reported several times in the same three patients by Stoerig and Cowey (1989, 1991, 1992). CT/MR studies were obtained in these

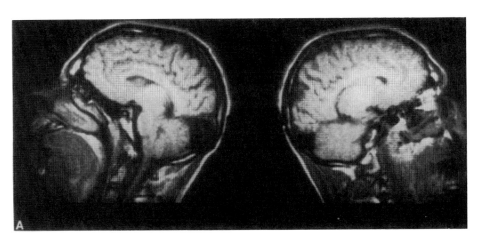

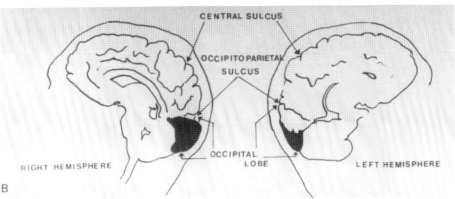

Figure 5. The brain of the same subject whose right fundus is depicted in Fig. 4 is shown. Sagittal images reveal the extend of the damage in both occipital lobes. The right occipital lobe is virtually absent. Portions of the left superior occipital lobe remain. (A) MRI spin echo T1 weighted sequences. (B) Schematic plot of the sagittal sections. Note that the calcarine fissure is surrounded by the lesion, and cannot be identified in either hemisphere. (Neuroanatomical study and illustration courtesy of Hanna Damasio.)

cases. However, the published data do not provide a detailed picture of the location and extent of the visual cortex lesions. Consequently, the anatomical basis of the findings remains ambiguous. The authors' discussion centered on which retinal projections, P alpha, P beta, or P gamma, could possibly be implicated in the residual functions they measured. The P beta cells possess color opponency, and would be the most likely candidate. The question is: how and where would the information from this channel be processed? There are several possibilities, similar to the ones invoked for the Riddoch phenomenon. One is selective recovery or preservation of a V1/V2 to V4 pathway in the damaged hemisphere. Alternatively, callosal connections might allow access to the processing capacity of the undamaged hemisphere. The residual ability might reflect indirect access to intact extrastriate cortex via subcortical structures, rudimentary color processing mediated by colliculus, or simply, residual vision in an extrastriate scotoma.

5.4. Issues of Definition

Another problem with blindsight is its very definition. Definitions of "blindsight" seem to differ between reports. Weiskrantz effectively expanded the concept of "blindsight" in 1987 (p. 166), defining it broadly as "visual capacity in a field defect in the absence of acknowledged awareness," and he later reiterated this position (Weiskrantz, 1990). This definition lacks anatomical specificity, and allows the entry of other forms of nonconscious visual processing that may have nothing to do with lesions of V1. Consequently, the concept can be taken to subsume knowledge without awareness in patients with agnosia for faces, although it is difficult to see what "blindsight" has to do with knowledge, and it has also been applied to visual perception in a neglected hemifield (Marshall and Halligan, 1988).

In summary, we should be wary of lumping together visual processing defects arising in different sectors of the nervous system, especially if they have different anatomical and behavioral underpinnings, all under the rubric of "blindsight." Nonconscious processing is likely to depend on several different mechanisms. Should the phenomenon by which normal individuals detect the position of a stimulus before they detect or recognize it be called "blindsight" (Meeres and Graves, 1990)?

6. Bilateral Visual Deficits with Unilateral V1 Lesions

Above, "blindsight" was addressed as a possible exception to the classical assumption that a unilateral lesion of the striate cortex results in a contralateral visual field scotoma that is absolutely blind. A different issue is whether patients with unilateral occipital lesions might have a subtle visual defect in the hemifield ipsilateral to the lesion, in the field that is presumably "normal." In our laboratory, pertinent data were collected in 12 patients with CT/MR-defined unilateral occipital lesions (Rizzo and Robin, 1992). They had V1-type visual field defects contralateral to the lesion as defined by Goldmann (Anderson, 1982) or threshold automated perimetry (Allergan Humphrey, 1987), and visual acuities of 20/30 or better. The subjects were tested for their ability to respond to transient signals, created by the onset or offset of a small light target, in both hemifields.

Older normal adults ($n = 20$) served as controls. The salient finding was that the performance in the ipsilateral fields, the supposed "normal" fields, was definitely reduced. In similar patients, Hess and Pointer (1989) found spatial and temporal contrast sensitivity deficits in the ipsilateral fields. The defects observed by Rizzo and Robin (1992) extended outside the central 1–3° of a hypothetical bilateral foveal representation in V1. Moreover, the type of transient stimuli used were not likely to have depended strongly on the P beta cells that might convey a bilateral foveal representation to V1 (Leventhal *et al.*, 1988). The findings are compatible with the existence of long-range perceptual interactions between the hemifields. Anatomical substrates could be any number of direct and indirect connections, such as callosal and feedback, impacting on V1. Several connectional systems could be involved.

The left and right areas 17 probably do not communicate directly with each other. The first direct transcallosal connections occur between the visual maps of areas 18 and 19 (Van Essen *et al.*, 1982; Cusick and Kaas, 1988). In the monkey there are direct transcallosal connections between portions of areas V2, V4, MT, and even heterotopic connections have been reported from area V2 to MT. Human occipital lesions would often affect such connections, as well as feedback connections to V1 from V2 (Rockland and Pandya, 1979). The overall result of damaging these and other systems could be to disturb within V1 of the synthesis of information from other vision areas including, perhaps, from both hemispheres, with a resultant mild disturbance in the ipsilateral field.

7. V1 versus V2/V3 Deficits

In reality, lesions such as Holmes reported do not affect V1 alone. Gunshot wounds exceed the cellular boundaries of V1, both in depth and in extent, as do other pathologies. Human occipital lesions often extend into extrastriate cortex (Brodmann's areas 18 and 19), into underlying white matter (the optic radiations, subcortical U-fibers), and into adjacent temporal and parietal regions. In fact, some of Holmes's patients had damage in the parietal lobe, as far anterior as the angular gyrus (Holmes, 1918a,b; Holmes and Horrax, 1919). Thus, the visual field defects plotted by Inouye, Holmes, and in most human patients with occipital lesions, are generally the result of combined damage in functional representations in visual cortex. So-called "striate scotomata" are frequently associated with damage in a human V2/V3 region, so designated because the borders between V2 and V3 are indistinct. Since V2/V3 shares a common border with V1 and concentrically surrounds it, dysfunction caused by V1 as opposed to V2/V3 lesions in the human can be very difficult to separate.

Relevant to this question are two patients with quadrantic field defects studied by Horton and Hoyt (1991b). Both patients had tumor resections, one in the region of the right cuneus, the lobule situated between the calcarine and parieto-occipital fissures on the mesial surface of the human brain, and the other in the visual association cortex of the right parieto-occipital lobe. Afterwards, both patients had lower left visual field defects. One patient could detect hand movements in the aberrant quadrant of vision, resembling the finding of Riddoch (1917). The second patient had macular sparing of the central 10° but could not detect movement. Histological analysis of the excised tissue in both cases showed sparing of striate cortex.

Horton and Hoyt (1991b) suggested their patients' quadrantic visual field defects might reflect V2/V3 damage rather than damage to optic radiations, the more conventional explanation that Holmes would have invoked. Because of the serial retinotopic projections from V1 to V2 and V3, a lesion of V2/V3 may be functionally equivalent to a lesion in V1. Horton and Hoyt (1991b) rejected the notion that the quadrantic defects they observed were caused by lesions of the optic radiations, mainly because the visual field defects they plotted closely followed the horizontal meridian. For ablations, or any other structural brain lesions, to produce such a pattern of visual loss, there would have to be an anatomical interval, or a spatial separation, between the upper and lower portions of the optic radiations. Otherwise a quadrantic defect would likely cross the horizontal meridian of the visual fields, however skilled the surgeon. But there is no such anatomical interval, at least not anterior to the calcarine fissure. The authors hypothesized that quadrantic visual field defects that respect the horizontal meridian are a signature of V2/V3 lesions.

Sometimes, however, humans with lesions of extrastriate cortex do not suffer visual field defects, at least not in the sense that Holmes described for "striate" scotomata. But they do have trouble seeing. Humans with lesions in dorsal portions of areas 18 and 19, where a human V2/V3 ought to lie, may report that objects disappear from view (Rizzo and Hurtig, 1987). Also, they may fail to detect stimuli presented at unpredictable temporal intervals and spatial locations, over an array, over time (Rizzo and Robin, 1990). Among other possibilities, the findings might be interpreted as a defect of visual attention. Lesions in human extrastriate cortex would damage attention-related neurons similar to those identified in the extrastriate regions of the monkey (Moran and Desimone, 1985; Spitzer *et al.*, 1988; Motter, 1992). The findings would also correspond with evidence that monkeys with ibotenic acid (IA) lesions of V2 (Nealy *et al.*, 1992) have abnormal vision, but are not blind, the caveat being that IA is primarily a neuronal toxin that spares white matter, does not cause retrograde degeneration, and basically causes less extensive damage than surgical lesions in the monkey or human.

Assuming the lesions Horton and Hoyt described did spare optic radiations, the presence of quadrantic field defects in their patients and its absence in others may relate to two factors. First, extrastriate cortex occupies a relatively large territory. The lesions may be more extensive in some patients than others, or they may occupy somewhat different portions of the extrastriate regions. Second, the locations of maps within extrastriate cortex might vary among individuals. Consider that area V3 in the monkey often has an incomplete representation of the visual fields, and that ventrally its foveal representation may be displaced by V4 (Felleman and Van Essen, 1991). Consequently, lesions that are similarly located by neuroimaging procedures, using gross cortical landmarks such as the calcarine and parieto-occipital fissures, might still affect different functional maps, or portions of those maps.

8. Extrastriate Lesions: What They Tell Us about Human V1

Felleman and Van Essen (1991) have catalogued over 30 functional representations of the visual fields in monkey. How this schema corresponds with the human arrangement is unclear, though homologies can be postulated, at least to the level of early prestriate cortex.

In the monkey, area V1 connects reciprocally with V2, V3, and surrounding regions in early visual association cortex (Felleman and Van Essen, 1991). It also connects directly and reciprocally with area V5 (MT) and at least part of V4 (see Bullier *et al.*, this volume). These two areas are in turn linked with more anterior visually related cortices. V4 receives a balance of M and P inputs, projects ventrally toward inferotemporal cortex (area IT), and may be biased to color and pattern processing (Desimone *et al.*, 1985). V4 also contains neurons modulated by attention, relevance, and perceptual context (Schiller and Lee, 1991; Motter, 1992). In contrast, V5 (MT) receives a predominance of M inputs, and projects dorsally toward the parieto-occipital regions. The neural complex that includes area MT and surrounding regions (MST, FST, and VIP) (Boussaoud *et al.*, 1990; Ungerleider and Desimone, 1986) has been regarded as a key unit for the processing of visuospatial percepts, notably motion. Experiments in binocular rivalry (Logothetis and Schall, 1989) suggest that area MT may also be concerned with aspects of attention.

The evidence from human brain lesion studies is consistent with the view that functional representations of the visual fields in human prestriate cortex segregate along a ventral–dorsal axis. Damage to the ventral visual system should include human homologues of the monkey's areas V4 and IT. Damage in the dorsolateral visual pathway should include a homologue of the monkey's MT complex and its projections to the parieto-occipital cortex.

8.1. A Ventral Pathway from V1

In humans the ventral visual cortices are located in the occipital lobe beneath the calcarine fissure and in adjacent temporal regions. Damage in these regions leads to defects of pattern processing and recognition such as visual object agnosia and prosopagnosia (Damasio, 1985a,b). This may be describable as a defect of "object attention" (Treisman and Gelade, 1980; Treisman, 1988), and can be contrasted with lesions both of area V1 and of the dorsal visual system, although these latter may impair recognition by causing low visibility or acuity (Frisen, 1980), or other perceptual defects. Such defects fall under the rubric of "apperceptive agnosia" (Lissauer, 1890). They are incompatible with Teuber's (1968) more modern definition of agnosia as a recognition impairment in which normal or near-normal percepts are stripped of their meanings.

As mentioned, lesions of the ventral visual system outside V1, may also lead to central achromatopsia. Figure 6 shows MR images including 3-D voxel reconstruction of the raw MR data (Damasio and Frank, 1992) of the brain of a patient with achromatopsia, prosopagnosia, and left homonymous hemianopia. The individual shown has bilateral lesions which involve the fusiform gyrus and undercut the most posterior segment of the lingual gyrus. The relationship of central achromatopsia to damage of these gyri (Meadows, 1974; Damasio *et al.*, 1980; Zeki, 1990; Plant, 1991) remains a research issue. The more extensive lesion in the patient's right hemisphere affected the white matter, in which optic radiations travel, above and below the calcarine fissure. It caused the functional equivalent of a V1 scotoma in the left visual hemifield. The lesion on the left hemisphere does not reach the surface of the hemispheres and can be seen only in

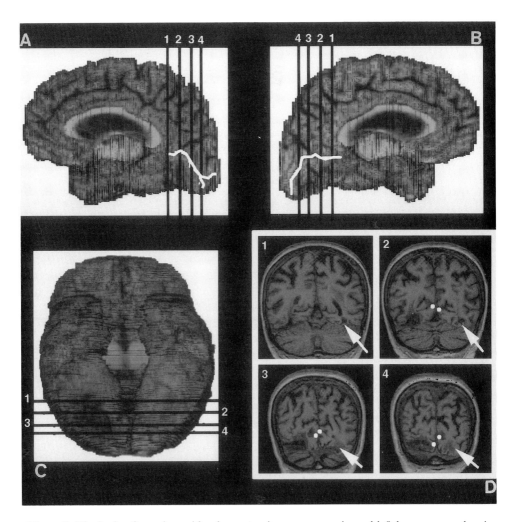

Figure 6. The brain of a patient with achromatopsia, prosopagnosia, and left homonymous hemianopia. The lesion analysis was performed by Hanna Damasio, and was based on the Brainvox technique (Damasio and Frank, 1992) which performs 3-D voxel reconstruction from raw MR data. The mesial surfaces of the right and left hemispheres, respectively, are reconstructed in panels A and B. The ventral surface is in panel C (the cerebellum and brain stem are not shown and were graphically removed). The white lines through the reconstructed brain show the relative position of the four coronal slices shown in panel D. The calcarine fissure is traced in white. This tracing automatically transfers to the MR slices to allow easy identification of the calcarine fissure. The individual shown has bilateral lesions which involve the fusiform gyrus and undercut the most posterior segment of the lingual gyrus. The more extensive lesion in the right hemisphere affected the white matter (in which optic radiations travel) above and below the calcarine fissure. It caused the functional equivalent of a V1 scotoma in the left visual hemifield. The lesion on the left hemisphere does not reach the surface of the hemispheres and can be seen only in the coronal sections (white arrows). It lies entirely beneath the calcarine fissure. We have hypothesized that such a lesion would cause damage in a possible human homologue of the monkey's area V4, or at least disrupt connections to and from such an area.

the coronal sections. It lies entirely beneath the calcarine fissure. We have hypothesized that such a lesion would cause damage in a possible human homologue of the monkey's area V4, or at least disrupt connections to and from such an area.

The nature of the color vision defect in central achromatopsia is another important research issue. Some patients describe the world in shades of gray. Others still see color, but it is highly degraded. In other words, central achromatopsia appears to comprise a range of impairments (Rizzo *et al.*, 1993a). It would be useful to characterize such defects along psychophysical color axes. On the red–green (R-G) axis the ratio of long-wavelength-sensitive ("red" or L-cone) to middle-wavelength-sensitive ("green" or M-cone) excitation varies at a fixed level of short-wavelength-sensitive ("blue" or S-cone) excitation. On the blue–yellow (B-Y) or tritan axis, the level of S-cone excitation varies with an unvarying ratio of L-cone to M-cone excitation (Boynton and Kambe, 1980; Krauskopf *et al.*, 1982; Derrington *et al.*, 1984). In our experience, patients with central achromatopsia may show combined impairments along both the R-G and B-Y tritan axes. Tritan mechanisms may be more affected because of the relative paucity of "blue" cones and, by inference, a relatively small and vulnerable central representation. But the profiles in central achromatopsia differ from the protanopic, deuteranopic, or tritanopic profiles of hereditary retinal cone disorders. In addition, patients may fail to see color, especially in smaller targets. This is an apparent exaggeration of the normal tendency to reduced color perception with reduced target size or increased spatial frequency (DeValois and DeValois, 1988).

The achromatopsic subject whose brain lesions are depicted in Fig. 6 performed like a "light meter": color brightness perception and color discrimination ability were severely impaired, yet achromatic discriminations were relatively normal, as in two other reports (Heywood *et al.*, 1987; Victor *et al.*, 1989). Properties of surface reflectance other than color, e.g., specularity, transparency, and iridescence, were relatively spared. We doubt that describing central achromatopsia as a defect of "color constancy" (Zeki, 1990) adequately characterizes such impairments. If central achromatopsia were really just a defect of "color constancy," then surface colors would appear to vary widely in different lighting conditions, something which generally does not occur in achromatopsic patients. However, even if color appearance did vary widely in different visual environments, those colors would still appear to have hue and they would not all appear either gray, as they do in patients with the most severe color deficits, or desaturated, as they do in the rest. Moreover, colors would still look relatively different from each other, that is, color differences might still be discriminable under many conditions, when color discrimination is generally quite poor and is sometimes completely absent in an achromatopsic field. Thus, a defect of color constancy is probably not necessary or sufficient to explain the findings in most cases of central achromatopsia.

8.2. A Dorsal Pathway from V1

The dorsolateral visual association cortices in the human are located in the occipital lobe above the calcarine fissure and in adjacent parietal and temporo-parieto-occipital regions. Damage in these regions may affect the control of eye and hand movements under visual guidance (Jeannerod, 1988), and a variety of other processes that fall under the rubric of spatial attention (Posner, 1980; Posner *et al.*, 1984). The defects include aspects of the symptom complex reported by Bálint (1909), Holmes (1918a), Hecaen and Ajuriaguerra (1954), Luria (1959), and others (Rizzo, 1993), as well as the perception of motion in its many respects.

In 1983, Zihl *et al.* reported a relatively "selective disturbance of movement vision after bilateral brain damage." The patient did not have prosopagnosia or achromatopsia. The patient, L.M., had a cortical venous sinus thrombosis that caused hemorrhagic strokes that widely surrounded the occipito-temporo-parietal junction in both hemispheres. A handful of cortical motion processing defects have since been reported in human subjects. These reports have been eagerly accepted, since they came after extrastriate maps had already been well defined in monkeys. Human visual motion processing disorders have been readily incorporated into the framework of experimental vision models, having fewer of the difficulties of central achromatopsia, although even here there are important issues which remain unresolved.

Because there are many roles of motion (Nakayama, 1985), so-called "central akinetopsia" [Zeki's (1991) term] may affect a variety of psychophysical motion mechanisms, among others, local or global, first or second order, long or short range. The disorder could impair the ability to generate smooth pursuit eye movements to track moving targets which follow sinusoidal, triangular, or step-ramp patterns, or the control of hand movements under visual guidance (Zihl *et al.*, 1983; Rizzo *et al.*, 1992). It might be associated with a reduction in the temporal resolution needed to perceive the flickering of a stationary light, or it might alter perception of apparent movement in a complex flickering display, such as a Ternus (1926) display, or even a Hollywood movie. The latter apparent motion effect is related to the "phi phenomenon" (Rock, 1983). Moreover, "central akinetopsia" may impair the abilities to perceive global motion among random background noise (Buffington *et al.*, 1987; Newsome and Paré, 1988), and 3-D structure from motion parallax, from motion and dynamic stereopsis (Nawrot and Blake, 1989, 1991), or from biological motion cues. Other issues are precise localization of an MT homologue in the human cortex, as well as the extent of that homologue and associated complex.

In the monkey, area MT lesions have been reported to cause bidirectional smooth pursuit defects for targets presented in the contralateral hemifield (Newsome *et al.*, 1985). Area MST lesions do the same and may also cause inability to track targets in the ipsilateral fields toward the side of the lesion. Thurston *et al.* (1988) measured an MT-like smooth pursuit defect in a single human subject. Anatomical analysis of the CT images suggested that the subject's lesions affected Brodmann's area 19 and adjacent 37. Otherwise, the lesions reported in association with defective motion perception have varied broadly around the junction of occipital, parietal, and temporal lobes, and have included lesions in a parieto-occipital location (e.g., Zihl *et al.*, 1983; Vaina, 1989; Vaina *et al.*, 1990). Thurston *et al.* (1988) hypothesized that the lesions they plotted might correspond to Flechsig's (1901) area 10, an area whose myelogenesis resembles that of the monkey's area MT (Allman, 1977).

If they exist in human, it seems likely that areas homologous to MT and MST and their connections are generally damaged in combination. Thus, the deficits caused by human brain lesions will probably not resemble the dysfunction caused by "pure" MT or MST lesions as described in the monkey. Furthermore, in humans, motion perception defects are likely to be worse with right than left hemisphere lesions. This correlates with the dominant role for spatiotemporal functions long attached to the right hemisphere.

We recently had an opportunity to perform new perceptual experiments in patient L.M. (Rizzo *et al.*, 1993b). Motion direction discrimination thresholds were elevated, as previously reported (Baker *et al.*, 1991). L.M. could derive structure-from-motion to describe objects such as a rotating transparent sphere, but that ability failed when moderate amounts of random noise were added. However, L.M. could not perceive 2-D shapes defined by motion cues, and had difficulty identifying 2-D shapes created by the transient onset or offset of random dots against a static background. Also, L.M. could not use dynamic stereo to perceive 2-D shapes or to determine disparity-defined direction of rotation of a 3-D figure, despite large disparities, even though static stereoacuity was 140 sec of arc. Compared to controls, L.M. also had trouble perceiving 2-D shapes defined by static density, stereo, and texture cues. The results imply that the dorsolateral visual system, which should include a human homologue of the monkey's area MT complex, plays a role not only in motion direction discrimination, but also in the perception of figure-from-ground based on complex motion and nonmotion cues. Apparently, the notion that the processing of form or "object vision" is the exclusive domain of the ventral visual system may be an oversimplification (Martin, 1992; Marigan and Maunsell, 1993).

8.3. Mixing of Visual Field Defects

Damage in the ventral visual association cortex that should contain a human homologue of V4 and IT is often associated with an upper quadrantanopsia. Damage in the dorsolateral visual association cortex that should contain a human homologue of the monkey's MT complex is often associated with a lower quadrantanopsia. When these quadrantic defects are present, they may reflect damage in V1, the optic radiations, or V2/V3. Sometimes, however, damage in a putative human V4 or MT homologue is not associated with a quadrantanopsia. When this occurs, it is most evident that the representations of the visual fields in extrastriate cortex have become skewed along the vertical axis, compared to the way they are in V1 (see Section 3). For example, a unilateral lesion in the ventral visual association cortex located entirely beneath the calcarine fissure (e.g., in regions of the fusiform and lingual gyri hypothesized to contain a human V4 homologue; see Fig. 6) may cause a color vision defect that encompasses the entire contralateral hemifield. A comparable lesion below the calcarine fissure, but restricted to area V1, would affect vision only in the contralateral superior quadrant.

8.4. Comment

The pattern of defects associated with human extrastriate lesions tells us that putative human M and P pathways are no longer segregated in the way that they were upon their arrival in V1. Judging by the pattern of deficits caused by human extrastriate lesions, the "transformation" of M and P pathways by a visual network that includes area V1 is inhomogeneous, and takes the form of different outputs to different extrastriate regions.

Cortical color perception and motion perception defects do not occur in isolation. This may be a reflection of the numerous interstream interactions between parallel channels in primate vision (Albright, 1991; Felleman and Van Essen, 1991). Schiller and Logothetis (1990) have outlined the combined defects encountered with MT and V4, as opposed to M and P, lesions in the monkey. Similar patterns may occur with human cortical lesions (Rizzo *et al.*, 1992).

9. Cerebral Activation Studies of Human Vision

Aside from human lesion studies, human cortical organization can now be investigated by a variety of techniques capable of mapping the patterns of activity in behaving humans. They include electroencephalographic techniques, such as evoked potential (EP) and brain electrical activity mapping (BEAM), and cerebral metabolism or blood flow techniques, such as single photon emission computed tomography (SPECT); dynamic MRI; and positron emission tomography (PET). Of these techniques, PET has attracted the most attention. However, like the lesion method, PET also has many technical and interpretative problems.

9.1. Procedures for PET

PET uses short-half-life positron-emitting radioisotopes such as O^{15} or deoxyfluoroglucose (DFG), generated in a cyclotron on site, as an injectable tracer substance. When positrons collide, they annihilate each other giving off gamma rays which can be detected. The subject must be positioned precisely in 3-D space, within the fields of the gamma detector array, and must remain nearly motionless for several hours. A few millimeters of head movement translates to a corresponding localization error in the subsequent data. Recalibration of head position may be necessary to avoid such errors.

Once a subject is lying in the scanner, a "transmission scan" is taken. This verifies the positioning of the individual and measures inherent baseline radioactivity. Next the subject engages in the desired behavioral task, and the radioisotope is injected intravenously. Each injection corresponds to a single PET scan. A single injection of O^{15} allows approximately 2 min of data collection, of which only the first 40 sec or so produces counts that are usable. Translating regional cerebral radioactive counts to blood flow measurements depends on pharmacokinetic models, which assume that the measured radial artery activity reflects the cerebral arterial activity.

A typical O^{15} PET study in a single individual consists of as many as 10 injections. Usually one of the injections is given in a "resting state," in which a subject may be lying quietly with eyes closed. Subsequent injections are associated with different activation tasks. There is at least a 10-min delay between successive scans (five half lives for O^{15}) to allow radioactivity to decay close to the baseline levels. The data set from each scan in each individual can be represented in several different planes of section. The most common published representations are axial views.

The strategies for analyzing PET data are complex. Raw PET data consist of radioactive counts in a 3-D coordinate system. However, to be of any interest, they must be represented on sections through the brain. The way this has been done is by correlating PET coordinates with appropriate landmarks. Stereotactic brain atlases, e.g., Talairach and Tournoux (1988), originally meant for surgical use, permit cerebral localization in a 3-D coordinate grid based on cranial and cerebral markings. In PET experiments, these landmarks have been derived from a priori lateral skull films, and even from PET data sets.

A different level of complication arises in the analysis of data between different injections in a given individual, say between the resting and activated state. A problem here is that global brain activity may vary considerably in an individual between injections, for a variety of uncontrollable reasons such as changes in arousal or in the arterial pCO_2. So far, it does not appear that changes in task demand associated with different cerebral activation procedures can themselves cause changes in overall blood flow. Consequently, this variability is addressed by statistical ("normalization") procedures that compensate, pixel by pixel, for differences in global blood flow to allow the comparison ("subtraction") of PET data sets collected in a subject under different experimental conditions. Overall brain activity also varies dramatically between subjects. Further statistical analyses are required for the pooling of data from several subjects performing in the same experiments.

9.3. PET Results in Vision

The results of PET studies on human vision have generally supported the findings of human brain lesion studies. Visual cortex is more activated with the eyes open and active, than with the eyes closed and resting. The differences between these two states have been reported to be on the order of about 20 to 30% with O^{15} (Zeki, 1991). It should be noted that the resting state activity in area V1 is relatively high. This may be a consequence of many feedback connections to area V1 from extrastriate cortex that are active in a thinking subject, or perhaps a high mitochondrial content in an area that stains heavily with cytochrome oxidase (see Wong-Riley, this volume).

Fox *et al.* (1986) derived a cortical magnification factor for striate cortex based on patterns of activation with different concentric targets. The results were in agreement with calculations based on the localization of phosphenes following cortical stimulation (Dobelle *et al.*, 1979) and with those based on correlations of MRI-verified lesions with perimetrically defined visual field defects (Horton and Hoyt, 1991b). A potential problem, however, is that activation of striate cortex does not proceed without concomitant activation of early prestriate cortex; PET does not distinguish between V1 and V2 (Zeki *et al.*, 1991).

Striate cortex is activated by motion, shape, color, and all other attributes of light, compatible with its key role in distributing information from the retinogeniculate system to the visual association cortices. However, the pattern of activation in extrastriate cortex is modulated by variations in task demand associated with changing attention to the different attributes of a stimulus (Corbetta *et al.*, 1991), and varies dramatically with the choice of stimulus. Stimulation with

Land color Mondrians, which are displays composed of overlapping color rectangles such as those created by the Dutch artist, Piet Mondrian (1947), and later used by Land and McCann in their research on "color constancy," reportedly activates human V1/V2 and V4, but not MT (Zeki *et al.*, 1991). Stimulation with moving stimuli activates V1/V2 and MT, but not V4. In addition, Haxby *et al.* (1991) and Grady *et al.* (1992) have found evidence to support different specialization for the upper and lower visual association cortices, as has also been noted in human lesion studies. For example, a task of dot localization activated the upper visual cortex, while a face discrimination task (Benton and Van Allen, 1985) activated the lower visual cortex.

PET studies of human subjects with lesions of the occipital lobe caused by stroke show improvement of blood flow in visual cortex over time, in concert with recovery of the visual field defects (Bosley *et al.*, 1987). Interestingly, the blood flow in the occipital lobe may be increased in patients with long-standing blindness (Veraart *et al.*, 1990). However, PET does not show the lateral geniculate or superior colliculus well enough to monitor the activity in those structures in experiments on blindsight.

9.4. Comment

PET offers a potentially exciting means of localizing patterns of visual function in human neocortex. However, the technique is labor intensive, expensive, and mildly invasive. The whole-body radiation expose on one study, a rad or so, is equivalent to the amount of radiation exposure involved in living in Denver for 6 months, or of flying a few round-trips from Los Angeles to New York. Although this is low, the dose still effectively limits the PET studies in a normal individual to about one per year, making it difficult to replicate the findings.

The temporal resolution of PET is poor. Many important cerebral processes evolve rapidly, within hundreds of milliseconds, as shown by the results of neuronal recordings, and by EEG techniques. As an example, the "P-100," an EEG potential recorded over the scalp in the occipital region in human subjects viewing stimuli such as a flickering light or a shifting checkerboard pattern, evolves over about 100 msec after the delivery of the stimulus. PET, which shows the average regional blood flow over tens of seconds, is blind to the development, and perhaps even to the existence of such rapid events.

It should be pointed out that potential errors are associated with PET studies that rely on the use of stereotactic atlases for anatomical localization. The problem is not necessarily one of low resolution of PET, which might be 5 mm for raw, and 8 mm for reconstructed data. The problem is one of placing the recorded activity in the 3-D gridwork of the brain. As an example, a recent report of activation in the anterior temporal pole in chemically induced anxiety was later found to be the result of blood flow changes in extracranial structures (Drevets *et al.*, 1992). Activity of the masseter and temporalis muscles associated with teeth clenching had been mislocalized to temporolimbic cortex.

There are several problems with transposing individuals' PET data into the framework of an idealized brain coordinate system. Small errors of angulation off the reference lines from which sulcal position is estimated can result in gross mislocalization. For example, a 5° angulation error off the anterior commissure (AC)/posterior commissure (PC) line can result in as much as 14 mm ante-

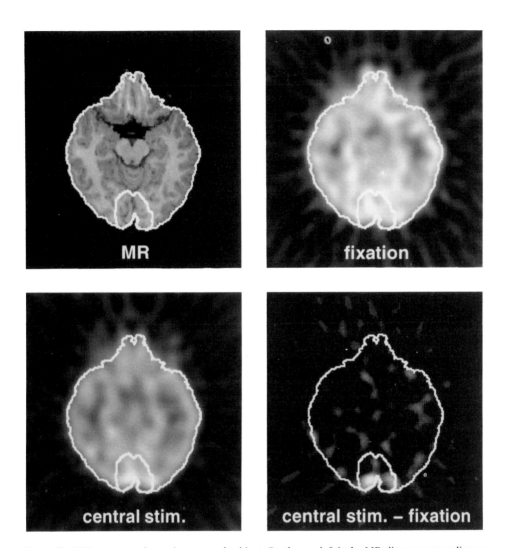

Figure 7. PET scan experiment in a normal subject. On the top left is the MR slice corresponding to the PET slice. The white lines of contour of the brain and of the region of interest containing the calcarine cortex in both hemispheres were traced on the MR and automatically transferred to the PET slices. "Fixation" shows the PET data while the subject was fixating on a tiny dot. "Central stim." shows the PET data while viewing concentric rings of red and black checkers which occupied a 10°-diameter circular aperture, 5° to either side of fixation, and underwent phase reversal at 10 Hz. "Central stim. − fixation" shows the subtraction of the "fixation" from the "central stim." PET data sets. In the subtraction image the main region with high flow values is the sector of V1 to which the macula projects. [Courtesy of H. Damasio, T. Grabowski, M. Rizzo, and M. Nawrot, Department of Neurology, and R. Hichwa, L. Ponto, and L. Watkins, PET facility, University of Iowa College of Medicine.]

roposterior variation in the determination of the superior temporal sulcus (Steinmetz *et al.*, 1989). More importantly, there are inherent errors relating to reliance on the coordinates of a single, "average" or "ideal" brain. The basic problem is that no two brains are alike. Variations of about 2 cm have been noted in the case of the parieto-occipital fissure. The position of the calcarine fissure varies among individuals, and as a consequence so does the position of the striate cortex which surrounds it (Rademacher *et al.*, 1992). Moreover, the corresponding markings of the left and right hemispheres may differ in an individual, which the Talairach and Tournoux (1988) atlas does not take into account. Interpretation of the published PET data on vision must be considered in light of these inherent difficulties.

Problems of anatomical localization may be mitigated by the use of 3-D reconstructions of the raw MR data of a subject's brain (Damasio and Frank, 1992). Cerebral regions of theoretical interest for a given function can be defined, a priori, with respect to the landmarks of an individual's own brain. Subsequent PET activation data can be translated into the MR matrix to test the hypotheses on visual function with respect to the regions of interest.

Figure 7 shows a vision experiment in a normal young adult volunteer by combined MR and PET methodologies. The upper left panel shows a transverse MR slice of the brain at the level of the visual cortex. The white lines of contour around the brain and around the region of interest containing the calcarine fissure and surrounding cortex in both hemispheres were traced in the MR and automatically transferred to the PET slices which occupy the same anatomical position as the MR slice. The panel labeled "fixation" shows the PET data as the subject fixated a small white dot on a videomonitor. The panel labeled "central" represents the PET data collected as the subject viewed a 10°-diameter disk composed of concentric rings of red and black checks which underwent phase reversal at 10 Hz. "Central-fixation" shows a "subtraction" image; the data collected in the "fixation" condition have been subtracted from the "central" data. The region of high flow that remains in the subtraction image is primarily the sector of V1 to which the retinal macula projects; the macular representation in V2, which lies directly adjacent to the macular representation in V1, may also be activated.

10. Conclusions

Converging evidence from lesion, PET, and comparative anatomical studies indicates that human area V1 serves as the main recipient of several parallel subcortical channels, and functions as a critical way station, regulator, and bottleneck for ascending inputs to functional maps in extrastriate cortex. These maps have multiple interconnections, which exert their influence among each other, and then play back upon V1. Some lesions of this network result in predictable clinical phenomena, although a variety of attention-related and other behavioral factors may complicate the interpretation of the data. The denial of perceptual impairments and covert processing of perceptual stimuli, for example, are important factors.

Evidence for subcortical human "M" and "P" inputs to striate cortex has been inferred from the pattern of histological and psychophysical deficits in conditions such as optic neuropathy and developmental dyslexia. The severity of visual loss following human V1 damage informs us about its critical position with respect to these inputs. Humans have no conscious awareness of seeing in V1 scotomata; there may be residual vision, described under the rubric of "blindsight," but it is rudimentary at best.

The visual loss that follows lesions of the human visual association cortex in Brodmann's areas 18 and 19 is generally less severe than that caused by a V1 lesion. Furthermore, the behavioral dissociations caused by these lesions no longer clearly resemble the dysfunction caused by lesions in putative human M or P pathways. Apparently these parallel subcortical M- and P-like inputs have been altered by a network of interactions in which V1 may be assumed to play a crucial role. As in the monkey, the outputs of V1 are distributed in different streams to prestriate regions that may contain human homologue of the monkey's areas V2, V3, V4 and V5 (MT).

Cerebral activation studies using PET do not distinguish between V1 and V2, which are generally coactivated any time the eyes are open. However, activity in extrastriate regions which contain the V4 or MT homologues, depends more on the physical nature of the stimulus and the attention paid to it. Results obtained through use of PET seem to support the localizations of visual function that were first determined by human lesion studies, but they also point to a highly interactive, synthetic organization in which multiple distributed regions act in concert to produce human vision.

ACKNOWLEDGMENTS. This work was supported by NINDS PONS 19632.

11. References

Albright, T. D., 1991, Color and the integration of motion signals, *Neurosci. Trends* **14**:266–269.

Allergan Humphrey, 1987, *The Visual Field Analyzer Primer,* 2nd ed., San Leandro.

Allman, J. M., 1977, Evolution of the visual system in early primates, in: *Progress in Psychology, Physiology, and Psychiatry* (J. Sprague and A. N. Epstein, eds.), Academic Press, New York, pp. 1–53.

Anderson, D. R., 1982, *Testing the Field of Vision,* Mosby, St. Louis, pp. 22–42, 180–196.

Anderson, S., and Rizzo, M., 1992, Visual hallucinations following occipital lobe damage, *International Neuropsychological Society Meeting,* San Diego, Calif., February 5–8, 1992.

Anton, F., 1899, Ueber die Selbstwharnehmungen der Herderkrankungen des Gehirns durch den Kranken bei Rinderblindheit und Rindentaubheit, *Arch. Psychiatr.* **32**:86–127.

Atkin, A., Bodis-Wollner, I., Wolstein, M., Moss, A., and Podos, S. M., 1979, Abnormalities of contrast sensitivity in glaucoma, *Am. J. Ophthalmol.* **88**:205–213.

Baker, C. L., Hess, R. F., and Zihl, J., 1991, Residual motion perception in a 'motion blind' patient assessed with limited-lifetime random dot stimuli. *J. Neurosci.* **11**:454–461.

Bálint, R., 1909, Seelenlahmung des "Schauens", optische Ataxie, raumliche Storung der Aufmerksamkeit, *Monatsschr. Psychiatr. Neurol.* **25**:51–181.

Balliet, R., Blood, K. M., and Bach-y-Rita, P., 1985, Visual field rehabilitation in the cortically blind? *J. Neurol. Neurosurg. Psychiatry* **48**:1113–1124.

Bassi, C. J., and Lehmkuhle, S., 1990, Clinical implications of parallel visual pathways, *J. Am. Optom. Assoc.* **61**:98–110.

Benton, A. L., and Van Allen, M. W., 1985, Visuoperceptual, visuospatial, and visuoconstructive disorders, in: *Clinical Neuropsychology,* 2nd ed. (K. M. Heilman and E. Valenstein, eds.), Oxford University Press, London, pp. 151–186.

Benton, S., Levy, I., and Swash, M., 1980, Vision in the temporal crescent in occipital infarction, *Brain* **103**:83–97.

Blythe, I. M., Kennard, C., and Ruddock, K. H., 1987, Residual vision in patients with retrogeniculate lesions of the visual pathways, *Brain* **110**:887–905.

Bosley, T. M., Dann, R., Silver, F. L., Alavi, A., Kushner, M., Chawluk, J. B., Savino, P. J., Sergott, R. C., Schatz, N. J., and Reivich, M., 1987, Recovery of vision after ischemic lesions: Positron emission tomography, *Ann. Neurol.* **21**(5):444–450.

Boussaoud, D., Ungerleider, L. G., and Desimone, R., 1990, Pathways for motion analysis: Connections of the medial superior temporal and fundus of the superior temporal visual areas in the macaque, *J. Comp. Neurol.* **296**:462–495.

Boynton, R. M., and Kambe, N., 1980, Chromatic difference steps of moderate size measured along theoretically critical axes, *Color Res. Appl.* **5**:13–23.

Brindley, G. S., 1970, Introduction to sensory experiments, in: *Physiology of the Retina and Visual Pathway*, 2nd ed., Williams and Wilkins, Baltimore, pp. 133–138.

Brodmann, K., 1909, *Vergleichende Lokalisationslehre der Grosshirnrinde in ihren Prinzipien dargestellt auf Grund des Zellenbaues*, J. A. Barth, Leipzig.

Buffington, J., Nawrot, M., and Sekuler, R., 1987, MacCinematogram 1.07d, software survey section, *Vision Res.* **27**:716–718.

Bunt, A. H., and Minckler, D. S., 1977, Foveal sparing, *Arch. Ophthalmol.* **95**:1445–1447.

Burkhalter, A., and Bernardo, K. L., 1989, Organization of corticocortical connections in human visual cortex, *Proc. Natl. Acad. Sci. USA* **86**:1071–1075.

Campion, J., Latto, R., and Smith, Y. M., 1983, Is blindsight an effect of scattered light, spared cortex, and near-threshold vision? *Behav. Brain Sci.* **6**:423–448.

Celesia, G. G., Bushnell, M. D., Toleikis, S. C., and Brigell, M. G., 1991, Cortical blindness and residual vision: Is the "second" visual system capable of more than rudimentary visual perception? *Neurology* **41**:862–869.

Clark, S., and Miklossy, J., 1990, Occipital cortex in man: Organization of callosal connections, related myelo- and cytoarchitecture, and putative boundaries of functional visual areas, *J. Comp. Neurol.* **298**:188–214.

Corbetta, M., Miezin, F. M., Dobmeyer, S., Shulman, G. L., and Petersen, S. E., 1991, Selective and divided attention during visual discriminations of shape, color, and speed: Functional anatomy by positron emission tomography, *J. Neurosci.* **11**:2382–2402.

Cowey, A., and Stoerig, P., 1991, The neurobiology of blindsight, *Trends Neurosci.* **14**:140–145.

Critchley, M., 1965, Acquired anomalies of colour perception of central origin, *Brain* **88**:711–724.

Cusick, C., and Kass, J., 1988, Cortical connections of area 18 and dorsolateral visual cortex in squirrel monkeys, *Visual Neurosci.* **1**:211–237.

Damasio, A. R., 1985a, Disorders of complex visual processing: Agnosias, achromatopsia, Balint's syndrome, and related difficulties of orientation and construction, in: *Principles of Behavioral Neurology* (M. -M. Mesulam, ed.), Davis, Philadelphia, pp. 259–288.

Damasio, A. R., 1985b, Prosopagnosia, *Trends Neurosci.* **8**:132–135.

Damasio, A., Yamada, T., Damasio, H., Corbett, J., and McKee, J., 1980, Central achromatopsia: Behavioral, anatomic and physiologic aspects, *Neurology* **30**:1064–1071.

Damasio, H., and Damasio, A., 1989, *Lesion Analysis in Neuropsychology*, Oxford University Press, London.

Damasio, H., and Frank, R., 1992, Three-dimensional in vivo mapping of brain lesions in humans, *Arch. Neurol.* **49**:137–143.

Denny-Brown, D., and Chambers, R. A., 1955, Visuo-motor function in the cerebral cortex, *J. Nerv. Ment. Dis.* **121**:288–289.

Derrington, A. M., Krauskopf, J., and Lennie, P., 1984, Chromatic mechanisms in lateral geniculate nucleus of macaque, *J. Physiol. (London)* **357**:241–265.

Desimone, R., Schein, S., Moran, J., and Ungerleider, L., 1985, Contour, color and shape analysis beyond the striate cortex, *Vision Res.* **25**:441–452.

DeValois, R. L., and DeValois, K. K., 1988, Sensitivity to spatial variations, in: *Spatial Vision*, Oxford University Press, London, pp. 212–238.

Dobelle, W. H., Turkel, J., Henderson, D. C., and Evans, J. R., 1979, Mapping the representation of the visual field by electrical stimulation of human visual cortex, *Am. J. Ophthalmol.* **88**:727–735.

Dowling, J. E., 1987, *The Retina: An Approachable Part of the Brain*, Belknap, Harvard.

Drevets, W. C., Videen, T. O., MacCleod, A. K., Haller, J. W., and Raichle, M. E., 1992, PET images of blood flow changes during anxiety: Correction, *Science* **256**:1696.

Ewens, G. F. W., 1893, A theory of cortical visual representation, *Brain* **16**:475–491.

Felleman, D. J., and Van Essen, D. C., 1991, Distributed hierarchical processing in the primate cerebral cortex, *Cereb. Cortex* **1**:1–47.

Fendrich, R., Wessinger, C. M., and Gazziniga, M. S., 1992, Residual vision in a scotoma: Implications for blindsight, *Science* **258**:1489–1491.

Ferrier, D., 1876, *The Functions of the Human Brain*, Smith, Elder, London, p. 164.

Flechsig, P., 1901, Developmental (myelogenetic) localisation of the cerebral cortex in the human subject, *Lancet* **2:**1027–1029.

Fox, P. T., Mintun, M. A., Raichle, M. E., Miezin, F. M., Allman, J. M., and Van Essen, D. C., 1986, Mapping human visual cortex with positron emission tomography, *Nature* **323:**806–809.

Frisen, L., 1980, The neurology of visual acuity, *Brain* **103:**639–670.

Galaburda, A. M., Rosen, G. D., Drislane, F. W., and Livingstone, M. S., 1991, Physiological and anatomical evidence for a magnocellular defect in developmental dyslexia, *Soc. Neurosci. Abstr.* **17:**20.

Grady, C. L., Haxby, J. V., Horwitz, B., Schapiro, M. B., Rapoport, S. I., Ungerleider, L. G., Mishkin, M., Carson, R. E., and Herscovitch, P., 1992, Dissociation of object and spatial vision in human extrastriate cortex: Age-related changes in activation of regional cerebral blood flow measured with [^{15}O] water and positron emission tomography, *J. Cogn. Neurosci.* **4:**23–34.

Green, D. M., and Swets, J. A., 1966, *Signal Detection Theory and Psychophysics*, Krieger, New York.

Haxby, J. V., Grady, C. L., Horwitz, B., Ungerleider, L. G., Mishkin, M., Carson, R. E. Herscovitch, P., Schapiro, M. B., and Rapoport, S. I., 1991, Dissociation of object and spatial visual processing pathways in human extrastriate cortex, *Proc. Natl. Acad. Sci. USA* **88:**1621–1625.

Hacaen, H., and Ajuriaguerra, J., 1954, Bálint's syndrome (psychic paralysis of visual fixation) and its minor forms, *Brain* **77:**373–400.

Hendrickson, A. E., Movshon, J. A., Eggers, H. M., Gizzi, M. S., Boothe, R. G., and Kiorpes, L., 1987, Effects of early unilateral blur on the macaque's visual system. II. Anatomical observations, *J. Neurosci.* **7:**1327–1339.

Henschen, S. E., 1893, On the visual path and center, *Brain* **16:**170–180.

Hess, R. F., and Pointer, J. S., 1989, Spatial and temporal processing sensitivity in hemianopia: A comparative study of the sighted and blind hemifields, *Brain* **112:**871–894.

Heywood, C. A., Wilson, B., and Cowey, A., 1987, A case of cortical colour "blindness" with relatively intact achromatic discrimination, *J. Neurol. Neurosurg. Psychiatry* **50:**22–29.

Hitchcock, P. F., and Hickey, T. L., 1980, Ocular dominance columns: Evidence for their presence in humans, *Brain Res.* **182:**176–179.

Hockfield, S., Tootell, R. B., and Zaremba, S., 1990, Molecular differences among neurons reveal an organization of human visual cortex, *Proc. Natl. Acad. Sci. USA* **87:**3027–3031.

Holmes, G., 1918a, Disturbances of visual orientation, *Br. J. Ophthalmol.* **2:**449–468, 506–516.

Holmes, G., 1918b, Disturbances of vision caused by cerebral lesions, *Br. J. Ophthalmol.* **2:**353–384.

Holmes, G., 1945, The organization of the visual cortex in man, *Proc. R. Soc. London Ser. B* **132:**348–361.

Holmes, G., and Horrax, G., 1919, Disturbances of spatial orientation and visual attention, with loss of stereoscopic vision, *Arch. Neurol. Psychiatry* **1:**385–407.

Holmes, G., and Lister, W. T., 1916, Disturbances of vision from cerebral lesions with special reference to the cortical representation of the macula, *Brain* **39:**34–73.

Horton, J. C., and Hedley-White, E. T., 1984, Mapping of cytochrome oxidase patches and ocular dominance columns in human visual cortex, *Philos. Trans. R. Soc. London Ser. B* **304:**255–272.

Horton, J. C., and Hoyt, W. F., 1991a, Quadratic visual field defects. A hallmark of lesions in extrastriate (V2/V3) cortex, *Brain* **114:**1703–1718.

Horton, J. C., and Hoyt, W. F., 1991b, The representation of the visual field in human striate cortex, *Arch. Ophthalmol.* **109:**816–824.

Horton, J. C., Dagi, L. R., McCrane, E. P., and de Monasterio, F. M., 1990, Arrangement of ocular dominance columns in human visual cortex, *Arch. Ophthalmol.* **108:**1025–1031.

Inouye, T., 1909, *Die Sehstorungen bei Schussverletzungen der kortikalen Sesphare*, Engelmann, Leipzig.

Jeannerod, M., 1988, *The Neural and Behavioral Organization of Goal-Directed Movements*, Oxford University Press, London.

Kiorpes, L., Boothe, R. G., Hendrickson, A. E., Movshon, J. A., Eggers, H. M., and Gizzi, M. S., 1987, Effects of early unilateral blur on the macaque's visual system. I. Behavioral observations, *J. Neurosci.* **7:**1318–1326.

Kolb, B., 1990, Recovery from occipital stroke: A self-report and an inquiry into visual processes, *Can. J. Psychol.* **44:**130–147.

Krauskopf, J., Williams, D., and Heely, D., 1982, Cardinal directions of color space, *Vision Res.* **22:**1123–1131.

Kulikowski, J. J., and Tolhurst, D. J., 1973, Psychophysical evidence for sustained and transient detectors in human vision, *J. Physiol. (London)* **232:**149–162.

Lashley, K. S., 1948, The mechanism of vision. XVIII. Effects of destroying the visual "associative areas" of the monkey, *Genet. Psychol. Monog.* **27:**107–166.

Lepore, F. E., 1983, Neuro-ophthalmology according to Gordon Holmes, in: *Neuro-Ophthalmology*, Volume 1 (C. H. Smith and R. W. Beck, eds.), Neurologic Clinics, pp. 789–805.

Leventhal, A. G., Ault, S. J., and Vitek, D. J., 1988, The nasotemporal division in primate retina: The neural bases of macular sparing and splitting, *Science* **240:**66–67.

Lissauer, H., 1890, Ein fall von Seelenblindheit nebst einem Bitrag zur Theorie derselben, *Arch. Psychiatr. Nervenkr.* **2:**22–70.

Livingstone, M. S., and Hubel, D. H., 1987, Psychophysical evidence for separate channels for the perception of form, color, movement and depth, *J. Neurosci.* **7:**3416–3468.

Livingstone, M. S., Rosen, G. D., Drislane, F. W., and Galaburda, A. M., 1991, Physiological and anatomical evidence for magnocellular defect in developmental dyslexia, *Proc. Natl. Acad. Sci. USA* **88:**7943–7947.

Logothetis, N. K., and Schall, J. D., 1989, Neuronal correlates of subjective visual perception, *Science* **245:**761–763.

Luria, A. R., 1959, Disorders of "simultaneous perception" in a case of bilateral occipito-parietal brain injury, *Brain* **82:**437–449.

Marinkovic, S. V., Milisavljevic, M. M., Lolic-Draganic, V., and Kovacevic, M. S., 1987, Distribution of the occipital branches of the posterior cerebral artery: Correlation with occipital lobe infarcts, *Stroke* **18:**728–732.

Marshall, J. C., and Halligan, P. W., 1988, Blindsight and insight in visuospatial neglect, *Nature* **336:**766–767.

Martin, F., and Lovegrove, W., 1987, Flicker contrast sensitivity in normal and specifically disabled readers, *Perception* **16:**215–221.

Martin, K. A. C., 1992, Parallel pathways converge, *Curr. Biol.* **2:**555–557.

Meadows, J. C., 1974, Disturbed perception of colors associated with localized cerebral lesions, *Brain* **97:**615–632.

Meeres, S. L., and Graves, R. E., 1990, Localization of unseen visual stimuli by humans with normal vision, *Neuropsyhologia* **28:**1231–1237.

Merigan, W. H., and Maunsell, J. H. R., 1993, How parallel are the primate visual pathways? *Ann. Rev. Neurosci.* **16:**369–402.

Mondrian, P., 1947, *Plastic Art and Pure Plastic Art. Documents of Modern Art*, Wittenborn, New York.

Moran, J., and Desimone, R., 1985, Selective attention gates visual processing in the extra-striate cortex, *Science* **229:**782–784.

Motter, B., 1992, Selective activation of V4 neurons in a color and orientation discrimination task, *Invest. Ophthalmol. Visual Sci.* **33:**1130.

Movshon, J. A., Eggers, H. M., Gizzi, M. S., Hendrickson, A. E., Kiorpes, L., and Boothe, R. G., 1987, Effects of early unilateral blur on the macaque's visual system. III. Physiological observations, *J. Neurosci.* **7:**1340–1351.

Munk, H., 1881, *Ueber die Funktionen der Grosshirnrinde: gessammelte Mitteilungen aus den Jarhen 1877–1880*, Hirschwald, Berlin.

Nakayama, K., 1985, Biological image motion processing: A review, *Vision Res.* **25:**625–660.

Nawrot, M., and Blake, R., 1989, Neural integration of information specifying structure from stereopsis and motion, *Science* **244:**716–718.

Nawrot, M., and Blake, R., 1991, The interplay between stereopsis and structure from motion, *Percept. Psychophys.* **49:**230–244.

Nealy, T., Maunsell, J. H. R., and Merigan, W. H., 1992, Visual dysfunction after a lesion of area V2 in the macaque. *Invest. Ophthalmol. Visual Sci.* **33:**1130a.

Newsome, W. T., and Paré, E. B., 1988, A selective impairment of motion perception following lesions of the middle temporal area (MT), *J. Neurosci.* **8:**2201–2211.

Newsome, W. T., Wurtz, R. H., Dursteler, M. R., and Mikami, A., 1985, Deficits in visual motion processing following ibotenic acid lesions of the middle temporal area of the macaque monkey, *J. Neurosci.* **5:**825–840.

Palca, J., 1990, Insights from broken brains, *Science* **248:**812–814.

Perenin, M. T., and Jeannerod, M., 1978, Visual function within the hemianopic field following early cerebral hemidecortication in man. I. Spatial localization, *Neuropsychologia* **10:**1–13.

Pessin, M. S., Lathi, E. S., Cohen, M. B., Kwan, E. S., and Hedges, T. R., 1987, Clinical features and mechanisms of occipital infarction, *Ann. Neurol.* **21:**290–299.

Plant, G. T., 1991, Disorders of colour vision in diseases of the nervous system, in: *Inherited and Acquired Colour Vision Deficiencies: Fundamental Aspects and Clinical Studies* (Cronly-Dillon, ed.), CRC Press, Boca Raton, Fla., pp. 173–198.

Polyak, S., 1957, *The Vertebrate Visual System*, University of Chicago, Chicago.

Poppel, E., Held, R., and Frost, D., 1973, Residual vision function after brain wounds involving the central visual pathways, *Nature* **243:**295–296.

Posner, M. I., 1980, Orienting of attention, *Q. J. Exp. Psychol.* **32**:3–25.

Posner, M. I., Walker, J. A., Friedrich, F. J., and Rafal, R. D., 1984, Effects of parietal lobe injury on covert orienting of visual attention, *J. Neurosci.* **4**:1863–1874.

Ptito, A., Lepore, F., Ptito, M., and Lassonde, M., 1991, Target detection and movement discrimination in the blind field of hemispherectomized patients, *Brain* **114**:497–512.

Rademacher, J., Galaburda, A. M., Kennedy, D. N., Filipek, P. A., and Caviness, V., 1992, Human cerebral cortex: Localization, parcellation, and morphometry with magnetic resonance imaging, *J. Cogn. Neurosci.* **4**:352–374.

Riddoch, G., 1917, Dissociation of visual perception due to occipital injuries with especial reference to appreciation of movement, *Brain* **40**:15–57.

Rizzo, M., 1993, "Balint's syndrome" and associated visuo-spatial disorders, in: *Visual Perceptual Defects: Bailliere's International Practice and Research* (C. Kennard, ed.), W. B. Saunders, London, pp. 415–437.

Rizzo, M., and Hurtig, R., 1987, Looking but not seeing: Attention, perception, and eye movements in simultanagnosia. *Neurology* **37**:1642–1648.

Rizzo, M., and Hurtig, R., 1989, The effect of bilateral visual cortex lesions on the development of eye movements and perception, *Neurology* **39**:406–413.

Rizzo, M., and Hurtig, R., 1992, Visual search in hemineglect: What stirs idle eyes? *Clin. Vision Sci.* **7**:39–52.

Rizzo, M., and Robin, D. A., 1990, Simultanagnosia: A defect of sustained attention yields insights on visual information processing, *Neurology* **40**:447–455.

Rizzo, M., and Robin, D., 1992, Visual dysfunction in the "good" fields in humans with unilateral occipital lesions, *Invest. Ophthalmol. Visual Sci.* **33**:1130.

Rizzo, M., Hurtig, R., and Damasio, A. R., 1987, The role of scanpaths in facial recognition and learning, *Ann. Neurol.* **22**:41–45.

Rizzo, M., Nawrot, M., Blake, R., and Damasio, A. R., 1992, A human disorder resembling area V4 dysfunction in the monkey, *Neurology* **42**:1175–1180.

Rizzo, M., Smith, V., Pokorny, J., and Damasio, A. R., 1993a, Color perception profiles in central achromatopsia, *Neurology* **43**:995–1001.

Rizzo, M., Nawrot, M., and Zihl, J., 1993b, Perception of structure-from-motion and non-motion cues in "central akinetopsia," *Association for Research in Vision and Ophthalmology* (abstract).

Rock, I., 1983, *The Logic of Perception*, MIT Press, Cambridge, Mass.

Rockland, K. S., and Pandya, D. N., 1979, Laminar origins and terminations of cortical connections of the occipital lobe in the rhesus monkey, *Brain Res.* **179**:3–20.

Rodieck, R. W., 1988, The primate retina, in: *Comparative Primate Biology,* Volume 4 (J. Horst, D. Steklis, and J. Erwin, eds.), Liss, New York, pp. 203–278.

Rodieck, R. W., Binmoeller, R. F., and Dineen, J., 1985, Parasol and midget cells of the human retina, *J. Comp. Neurol.* **233**:115–132.

Sadun, A. A., Borcher, M., De Vita, E., Hinton, D. R., and Bassi, C. J., 1987, Assessment of visual impairment in patients with Alzheimer's disease, *Am. J. Ophthalmol.* **104**:113–120.

Schafer, E. A., 1888, On the functions of the temporal and occipital lobes: A reply to Dr. Ferrier, *Brain* **11**:145–165.

Schiller, P. H., and Lee, K., 1991, The role of the primate extrastriate area V4 in vision, *Science* **251**:1251–1253.

Schiller, P. H., and Logothetis, N. K., 1990, The color-opponent and broad-based channels of the primate visual system, *Trends Neurosci.* **11**:392–398.

Smith, G., and Richardson, W., 1966, The course and distribution of the arteries supplying visual (striate) cortex, *Am. J. Ophthalmol.* **61**:1391–1396.

Spitzer, J., Desimone, R., and Moran, J., 1988, Increased attention enhances both behavioral and neuronal performance, *Science* **240**:338–340.

Steinmetz, H., Furst, G., and Freund, H. J., 1989, Cerebral cortical localization: Application and validation of the proportional grid system in MR in aging, *J. Comput. Assist. Tomogr.* **13**:10–19.

Stoerig, P., and Cowey, A., 1989, Wavelength sensitivity in blindsight, *Nature* **342**:916–918.

Stoerig, P., and Cowey, A., 1991, Increment-threshold spectral sensitivity in blindsight: Evidence for colour opponency, *Brain* **114**:1487–1512.

Stoerig, P., and Cowey, A., 1992, Wavelength discrimination in blindsight, *Brain* **115**:425–444.

Talairach, J., and Tournoux, P., 1988, *Co-planar Stereotaxic Atlas of the Human Brain. 3-Dimensional Proportional System: An Approach to Cerebral Imaging,* Thieme, Stuttgart.

Teller, D. Y., 1984, Linking propositions, *Vision Res.* **10**:1233–1246.

Ternus, J., 1926, Experimentelle Untersuchungen über phänomenale Identität, *Psychol. Forsch.* **7:**71–126. Translated and condensed in Ellis, W., (ed.), *Source Book of Gestalt Psychology*, Humanities Press, New York.

Teuber, H. L., 1968, Alteration of perception and memory in man, in: *Analysis of Behavioral Change* (L. Weiskrantz, eds.), Harper & Row, New York.

Teuber, H. L., Battersby, W. S., and Bender, M. B., 1960, *Visual Field Defects After Penetrating Missile Wounds of the Brain*, Harvard University Press, Cambridge, Mass.

Thurston, S. E., Leigh, R. J., Crawford, T., Thompson, A., and Kennard, C., 1988, Two distinct defects of visual tracking caused by unilateral lesions of cerebral cortex in humans, *Ann. Neurol.* **23:**266–273.

Tolhurst, D. J., and Ling, L., 1988, Magnification factors and the organization of the human striate cortex, *Hum. Neurobiol.* **6**(4):247–254.

Tranel, D., and Damasio, A. R., 1985, Knowledge without awareness: An autonomic index of facial recognition by prosopagnosics, *Science* **228:**1453–1454.

Treisman, A. M., 1988, Features and objects: The fourteenth Bartlett memorial lecture, *Q. J. Exp. Psychol.* **40:**201–237.

Treisman, A. M., and Gelade, G., 1980, A feature integration theory of attention, *Cogn. Psychol.* **12:**97–136.

Trick, G. L., 1985, Retinal potentials in patients with primary open-angle glaucoma: Physiological evidence for temporal frequency tuning defects, *Invest. Ophthalmol. Visual Sci.* **26:**1750–1759.

Trick, G., Barris, M. C., and Bickerl-Bluth, M., 1989, Abnormal pattern electroretinograms in patients with senile dementia of the Alzheimer's type, *Ann. Neurol.* **26:**226–231.

Tyler, C. W., 1981, Specific deficits in glaucoma and ocular hypertension, *Invest. Ophthalmol. Visual Sci.* **20:**204–212.

Ungerleider, L. G., and Desimone, R., 1986, Cortical connections of visual area MT in the macaque, *J. Comp. Neurol.* **248:**190–222.

Vaina, L. M., 1989, Selective impairment of visual motion interpretation following lesions of the right occipito-parietal area in humans, *Biol. Cybern*, **61:**347–359.

Vaina, L. M., Lemay, M., Bienfang, D. C., Choi, A. Y., and Nakayama, K., 1990, Intact "biological motion" and "structure from motion" perception in a patient with impaired motion mechanisms. A case study, *Visual Neurosci.* **5:**353–369.

Van Essen, D. C., Newsome, W. T., and Bixby, J. L., 1982, The pattern of interhemispheric connections and its relationship to extrastriate visual areas in the macaque monkey, *J. Neurosci.* **2:**265–283.

Veraart, C., DeVolder, A. G., Wanet-Defalque, M. C., Bol, A., Michel, C., and Goffinet, A. M., 1990, Glucose utilization in human visual cortex is abnormally elevated in blindness of early onset but decreased in blindness of late onset, *Brain Res.* **510:**115–121.

Verrey, D., 1888, Hemiachromatopsie droite absolue, *Arch. Ophthalmol. (Paris)* **8:**289–300.

Victor, J., Maiese, K., Shapley, R., Sidtis, J., and Gazziniga, M., 1989, Acquired central dyschromatopsia with presentation of color discrimination, *Clin. Visual Sci.* **3:**183–196.

von Graefe, A., 1856, Ueber die Untersuchung des Gesichtsfeldes bei amblyopischen Affectionen, *Arch. Ophthelmol.* **2**(2):258–298.

Von Monakow, C., 1900, Pathologische und anatomische Mittheilungen uber die optischen Centren des Menschen, *Neurol. Centralbl.* **19:**680–681.

Walsh, F. B., and Hoyt, W. F., 1969, *Clinical Neuro-Ophthalmology*, 3rd ed., Williams & Wilkins, Baltimore.

Weiskrantz, L., 1987, *Blindsight: A Case Study and Implications*, Oxford University Press, London.

Weiskrantz, L., 1990, The Ferrier Lecture, 1989. Outlooks for blindsight: Explicit methodologies for implicit processes, *Proc. R. Soc. London Ser. B* **239:**247–278.

Weiskrantz, L., Warrington, E. K., Sanders, M. D., and Marshall, J., 1974, Visual capacity in the hemianopic field following a restricted occipital ablation, *Brain* **97:**709–728.

Zee, D. S., Tusa, R. J., Herdman, S. J., Butler, P. H., and Gucer, G., 1987, Effects of occipital lobectomy upon eye movements in primate, *J. Neurophysiol.* **58:**883–907.

Zeki, S., 1990, A century of cerebral achromatopsia, *Brain* **113:**1727–1777.

Zeki, S., 1991, Cerebral akinetopsia (visual motion blindness): A review, *Brain* **114:**811–824.

Zeki, S., Watson, J. D., Lueck, C. J., Friston, K. J., Kennard, C., and Frackowiak, R. S., 1991, A direct demonstration of functional specialization in human visual cortex, *J. Neurosci.* **11:**641–649.

Zihl, J., 1981, Recovery of visual functions in patients with cerebral blindness: Effects of specific practice with saccadic localization, *Exp. Brain Res.* **44:**159–169.

Zihl, J., von Cramon, D., and Mai, N., 1983, Selective disturbance of movement vision after bilateral brain damage, *Brain* **106:**313–340.

Index

Page numbers followed by f and t indicate figures and tables respectively